APPLICATIONS OF LASER ABLATION - THIN FILM DEPOSITION, NANOMATERIAL SYNTHESIS AND SURFACE MODIFICATION

Edited by **Dongfang Yang**

Applications of Laser Ablation - Thin Film Deposition, Nanomaterial Synthesis and Surface Modification
http://dx.doi.org/10.5772/62615
Edited by Dongfang Yang

Contributors
Anastasiia Shabalina, Valery Svetlichnyi, Ivan Lapin, Daria Goncharova, Ralf Weiskirchen, Jagannathan Thirumalai, Sergey Sarkisov, Abdalla Darwish, Yonggang Tong, Xiubing Liang, Shuxin Bai, Qing H Qin, Ci-Ling Pan, Marius Enachescu, Calin Moise, Ion N. Mihailescu, Phuoc Huu Le, Chih Wei Luo, Armen Kuzanyan, Astghik Kuzanyan, Kenji Goya, Toshihiko Ooie, Yusuke Fuchiwaki, Masahiko Shiraishi, Kazuhiro Watanabe, Rozalina Zakaria, Seong Shan Yap, Thian Khok Yong, Chen Hon Nee, Teck Yong Tou, Sven Stauss, Hitoshi Muneoka, Keiichiro Urabe, Kazuo Terashima, Ikuko Akimoto, Sérgio De Lemos Campello, Walter mendes de Azevedo, Diego Leite Da Cunha, Leonardo Tadeu Boaes De Mendonça, Ohanna Maria Menezes Madeiro Da Costa, Abubaker Hassan Hamad, Andrei Popescu, Camelia Popescu, Gabriela Dorcioman

CBS, Edition **2017**

Published by InTech
Janeza Trdine 9, 51000 Rijeka, Croatia

Publishing Process Manager Maja Bozicevic

Technical Editor SPi Global

Cover Designer InTech Design Team

Additional hard copies can be obtained from orders@intechopen.com

Applications of Laser Ablation - Thin Film Deposition, Nanomaterial Synthesis and Surface Modification
Edited by Dongfang Yang
p. cm.
Print ISBN 978-953-51-2811-3
Online ISBN 978-953-51-2812-0

Contents

Preface

This book covers various aspects of using the laser ablation phenomenon for material processing including laser ablation applied for the deposition of thin films, for the synthesis of nanomaterials, for the chemical compositional analysis, and for the surface modification of materials. Through the 18 chapters written by experts from the international scientific community, the reader will have access to the most recent research and development findings on laser ablation through original research studies and literature reviews. Chapter 1 presents an overview on laser ablation for the synthesis of thin films including the description of three main types of laser thin film deposition techniques: pulsed laser deposition (PLD), combinatorial pulsed laser deposition (CPLD), and matrix assisted pulsed laser evaporation (MAPLE). The chapter uses a number of examples to illustrate the effect of processing parameters on the quality and property of thin films, as well as the advantages and disadvantages of thin film deposition by laser. Chapter 2 successfully exploits the use of PLD to grow excellent quality and high performance thin phosphor films of $Ca_0.5R_{1-}x(MoO_4)_2{:}xLn^{3+}$, M^+ (R^{3+} = La, Y), (Ln^{3+} = Eu, Tb, Dy) (M^+ = Li^+, K^+ and Na^+) for electroluminescence and display applications. Chapter 3 describes the use of PLD to grow and control the thermoelectric properties and novel topological surface states of bismuth chalcogenide (e.g., Bi_2Te_3, Bi_2Se_3, and Bi_3Se_2Te) thin films by varying ambient pressures and substrate temperatures. Chapter 4 presents an overview on the deposition of indium-tin oxide (ITO) films by PLD. The effect of laser wavelength, substrate temperature, and background gas on the resistivity and transmittance of ITO films, as well as ITO nanostructures, is systematically investigated. Chapter 5 presents the results of the characterization and in vitro evaluation of calcium phosphate films deposited by PLD or MAPLE. The results presented in Chapter 5 confirm that excellent quality and high-performance biomaterial thin films can be deposited by PLD and MAPLE for medical applications. Chapter 6 presents an exciting new method for making nanocomposite films based on multi-beam laser ablation/evaporation processes with the acronym MBMT-MAPLE/PLD. The feasibility of the new deposition technique is demonstrated for making polymer nanocomposite films with two inorganic additives: up-conversion fluorescent phosphor NaYF4:Yb3+, Er3+, and aluminum-doped ZnO. Chapter 7 describes methods to obtain uniform deposition of thin film over large areas on rotating substrates and moving ribbons by PLD technique. These methods, such as tilting targets and using shadow masks of different shapes, are simple, low cost, and can be easily adapted by existing PLD systems. Implementation of these methods not only can improve thickness uniformity and reduce micro/sub–micro-sized particles for PLD films but also can increase the utilization of targets. Chapter 8 gives an overview on using laser ablation to fabricate nanoparticles in different environments such as vacuum, ambient air, different liquid environments, and different background gases. The chapter analyzes in detail laser ablation mechanisms in each environment and suggests that each medium possesses some unique advantages and disadvantages for nanomaterial production or machining. Chapter 9 presents interesting works on using laser ablation in a liquid environment to synthesize several important compounds, such as hydrocerussite $Pb_3(CO_3)_2(OH)_2$; CdS quantum dots; magnetic materials such as Fe, FeO, $NaFeO_3$, Na_2FeO_4; and metal organic frameworks $Cu_2(BTC)_2(H_2O)_3$, $Eu(TMA)(H_2O)_4$. The chapter provides convincing mechanisms for each synthetic process and demonstrates that laser ablation in a liquid environment is a simple, low-cost, chemically clean, and one-pot synthetic route that enables fabrication of numerical practically important materials. Chapter 10 gives an overview on the use of pulsed laser ablation for the synthesis of nanomaterials under high-pressure gases, liquids, and supercritical fluids media. Chapter 11 provides detailed descriptions of the physical and chemical processes occurring during laser ablation in liquids for the synthesis of nanoparticles. The parameters that affect the formation, composition, and structure of the nanoparticles obtained are also discussed. Chapter 12 presents two very interesting applications of laser ablation in liquid phase: (i) organic nanoparticle formation by laser ablation of organic materials and (ii) hydrogen gas generation and nanoparticle generation by ablation of solid carbon in water. The chapter also reveals the mechanism of hydrogen generation by laser ablation. Chapter 13 presents the design of a pulsed laser vaporization reactor that can be used to synthesize single-wall carbon nanotubes, carbon nano-onions, and graphene by just tuning the experimental parameters. Chapter 14 presents the use of femtosecond for laser micromachining of fiber optics with a focus on investigation of the effects of laser parameters on surface qualities. The chapter describes different aspects of laser induced periodic surface structures and gives an example of using femtosecond laser for the fabrication of fiber optic sensors. Chapter 15 presents the use of dual-color laser that consists of the fundamental (ω) and its second harmonic (2ω) of a femtosecond Ti:sapphire for the ablation of PMMA. By changing the relative phase of the fundamental (ω) and second-harmonic (2ω) outputs, the chapter demonstrates that the laser ablated area can be modulated. Chapter 16 reports on the use of three lasers to modify the surface of the allyl-diglycol CR39 polymer in order to fabricate optical waveguides. Chapter 17 presents a new application of laser ablation for characterization of material ablation resistance by giving examples in ablation-resistance characterization of ultrahigh temperature ceramics and ceramic matrix composites and provides convincing interpretation of laser ablation behavior and mechanisms. Chapter 18 discusses the use of laser ablation for chemical compositional analysis of materials and exploits different methods for normalization of the laser ablation inductively coupled mass spectrometry in order to reduce measurement-related fluctuations.

This publication provides a snapshot of current efforts to extend the use of laser ablation for processing of materials and will be useful for the researchers and students who work in the area of laser and manufacturing. I gratefully acknowledge all chapter authors for their enthusiastic and collaborative contributions, and thanks to Ms. Maja Bozicevic, Publishing Process Manager, for her guidance and support in the preparation of this book.

Dongfang Yang, PhD
Senior Research Officer
Automotive and Surface Transportation
National Research Council Canada
London, Ontario, Canada

Section 1

Thin Film Deposition

Chapter 1

Laser Ablation Applied for Synthesis of Thin Films: Insights into Laser Deposition Methods

Camelia Popescu, Gabriela Dorcioman and
Andrei C. Popescu

Additional information is available at the end of the chapter

http://dx.doi.org/10.5772/65124

Abstract

This chapter will focus on laser ablation applied for thin film deposition. The first thin films deposition method based upon laser ablation was pulsed laser deposition (PLD), that could produce thin films out of metals, ceramics and even temperature resistant organics. The need of depositing increasingly complex and delicate materials, lead to radical modifications of PLD and allowed other laser ablation methods to develop. If complex libraries are to be synthesized two or more plasmas will be mixed and the thin films will have a variable composition over surface. This technique is called Combinatorial PLD (CPLD).

PLD/CPLD are however limited when it comes to organic materials transfer, because the high intensity laser beam can damage them. Matrix Assisted Pulsed Laser Evaporation (MAPLE) is a thin film deposition technique derived from PLD, able to transfer accurately fragile molecules from a target to a substrate, by using a frozen mix made of the material to be deposited and a protective buffer layer.

All these techniques will be discussed in detail with their advantages, drawbacks, influencing factors and applications, while relevant practical examples will be provided in order to make the information easily understandable for the new reader.

Keywords: laser ablation, pulsed laser deposition, matrix-assisted pulsed laser evaporation, thin films, nanoparticles

1. Introduction

A low wavelength of the laser beam and a very short pulse (ns-fs) duration induce instant local vaporization on the surface of a target material generating a plasma plume consisting of photons, electrons, ions, atoms, molecules, clusters, and liquid or solid particles. This phenomenon is known in the literature as '**laser ablation**' [1], a term derived from the Latin word '***ablatio***', meaning 'to carry away'.

Shortly, after the first laser functionality demonstration in 16 May 1960, numerous theoretical and experimental studies were performed concerning the interaction of the high intensity laser beam with solids [2–4], liquids [5], and gases [6].

Laser ablation is the base principle of most applications involving laser processing of materials: precise cutting, hole drilling, laser cleaning of surfaces, compositional analysis, and thin film deposition. The latter came as an obvious application, as a plate/slide/wafer can easily be positioned in front of the plasma plume, acting as a collector for the hot ablated material that condenses in the form of a thin film. This deposition method is known as pulsed laser deposition (PLD). The earliest attempt of thin film deposition was made in 1965 by Smith and Turner [7], but the true breakthrough was achieved by Dijjkamp et al. in 1987 [8], who succeeded the stoichiometric transfer of a compound with a complex molecular structure, very difficult to obtain using other deposition techniques. In this situation, it can be considered that a *congruent ablation* was attained. The decrease in the pulse duration meant laser beams with higher delivered energies that significantly increased the range of materials that could be ablated [9].

Historically, the method was known under several denominations [9]: pulsed laser evaporation, laser induced flash evaporation, laser molecular beam epitaxy, laser assisted deposition and annealing, and laser sputtering.

Some variations in PLD emerged out of necessity to deposit more complex materials or materials degradable at high temperatures. Instead of a single laser beam as in classical PLD, two laser beams can be used simultaneously to ablate two targets mounted on a carousel system, producing a mix of plasmas that will generate thin films with variable composition over the surface. This variation in PLD is known in the literature as combinatorial pulsed laser deposition (CPLD) [10].

Another variation in PLD developed out of necessity to protect compounds with long and fragile molecular chains is called matrix-assisted pulsed laser evaporation (MAPLE), and it uses as the target, a frozen mix consisting of the active material to be deposited and a buffer matrix that preponderantly absorbs the laser beam energy [11].

All these variations in PLD will be discussed in detail in the next chapters with relevant examples for their advantages and drawbacks.

2. Pulsed laser ablation techniques for deposition and compositional analysis

2.1. Pulsed laser deposition (PLD)

The material that is irradiated by the laser beam is called the 'target', while the collector is commonly referred to as the 'substrate'. They have to be placed plan-parallel in a deposition chamber, which is under vacuum conditions. A high intensity laser placed outside the deposition chamber is used as an energy source to ablate the target material and to deposit the thin film. The target vaporization is induced by photons, so no contamination/impurification occurs during the deposition process.

Contrary to this, simplicity of the experimental assembly, the laser-material interaction, which is the PLD base, is a very complex physical phenomenon that involves a succession of different processes [12]. They are:

a. Coupling of the optical energy to the target material.

b. Melting of the surface.

c. Vaporization in form of a plume of the thin upper layer of molten surface.

d. Photon absorption by the vaporized species, which eventually limits the laser fluence at the target surface.

e. Propagation of the plume in the direction normal to the target.

f. Return to the initial state after few nanoseconds from the end of the pulse, with a resolidified surface.

If deposition is made in reactive gas and the obtained film has a composition different from that of the target, the name of the synthesis process is reactive pulsed laser deposition (RPLD). The PLD/RPLD set-up is given in **Figure 1**.

2.1.1. Factors responsible for PLD deposition

Factors influencing the laser ablation process include the following: (i) **deposition conditions** (nature of ambient-ultra-high/high/vacuum, reactive gas, target-substrate separation distance, number of pulses); (ii) **laser beam parameters** (wavelength and pulse duration fluence); and (iii) **material properties of the target** (melting temperature, thermal diffusion rate, optical reflectivity).

Uniform ablation of the target is obtained through its rotation and translation with respect to the laser beam. The distance between the target and substrate is generally of a few centimetres. The film uniformity can be improved if the substrate is moved in the plasma direction, for example, by rotating the substrate holder. The substrate temperature is a very important parameter for the morphology, microstructure, and crystallinity of the deposited films.

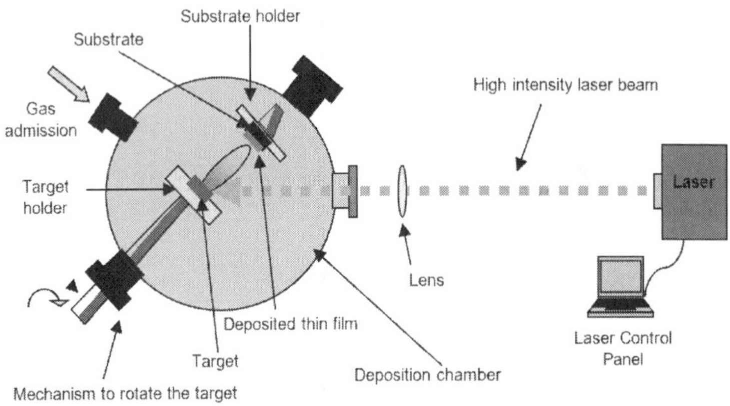

Figure 1. Experimental set-up PLD/RPLD.

2.1.1.1. Deposition conditions

2.1.1.1.1. Influence of the ambient gas inside the deposition chamber

Depending on the structure and composition of the thin films that one desire to achieve by PLD, in the deposition chamber a gas, which can be active or passive, can be introduced. In principle, the passive influence of the gas is necessary because it helps to compensate the eventual losses of the constituent elements. For example, the oxide thin films tend to be oxygen deficient.

We provide a relevant example of ZnO thin films synthesised in a vacuum (4×10^{-2} Pa) and in O_2 (13 Pa) ambient. The aspect of both films was radically different: the films deposited in a vacuum were opaque, dark-coloured (**Figure 2a**), while the films obtained in an oxygen flux were highly transparent (**Figure 2b**) [13].

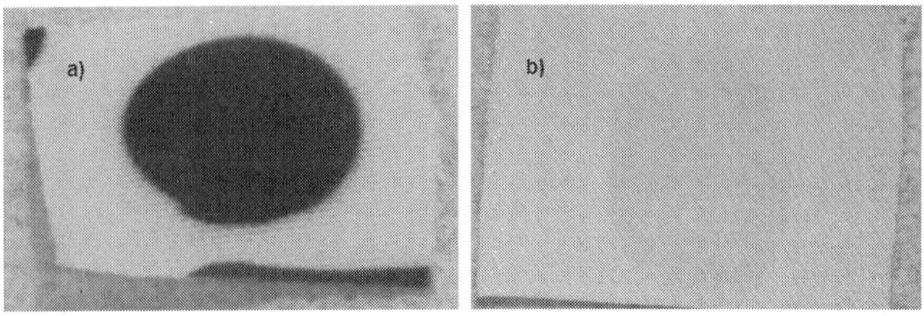

Figure 2. Textile material partially coated with ZnO films: (a) dark-coloured film deposited in vacuum and (b) transparent film deposited in a 13 Pa oxygen flux.

http://dx.doi.org/10.5772/65124

The explanation is that in oxygen ambient, due to the intense collisions with the environmental atoms, the ejected matter is confined to an elongated, 'cigar'- shaped plasma (**Figure 3a**). A thermal equilibrium is reached as a result of collisions during transfer and the substance condenses in large quantities forming compact thin films. In a vacuum, at much lower collision rates, the matter is ejected in all directions (**Figure 3b**), with high energies and speed. These high energetic species are bombarding the layers previously deposited and cause damage (by sputtering off atoms from the outer layers) or defects (dislocations, cracks, holes) on the deposited film. These bombardments occur for each pulse, resulting in a very disordered thin film that is full of defects. The defects are highly absorbent in the visible spectrum and hence the dark aspect.

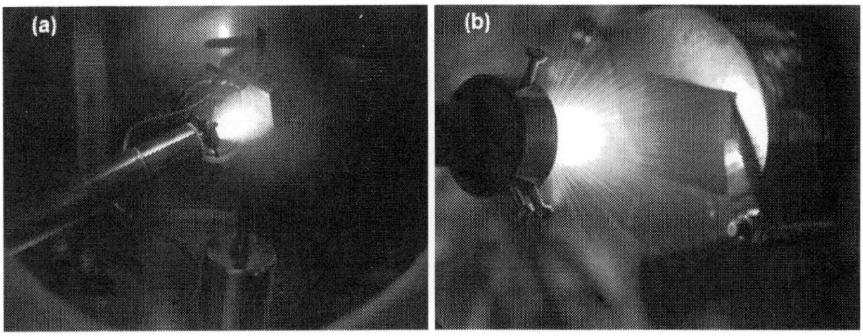

Figure 3. Plasma plume in PLD recorded in 13 Pa O_2 flux (a) and vacuum (b) (Reproduced with permission from Ref. [13]).

2.1.1.1.2. Influence of the target-substrate separation distance

The effect of the target-substrate distance is reflected by the angular scattering of the ejected flux. Different features can occur depending on the position of the substrate. The optimal position of the substrate in order to obtain stoichiometric structures is determined by the plasma evolution. The best depositions (in terms of stoichiometry, uniformity and homogeneity) are obtained when the plasma length is identical with the target-substrate separation distance [13]. To support this assertion we provide an example of ZnO deposition using a low number of pulses and three separation distances: 3, 4, and 5 cm. The plasma plume was 4 cm in length (**Figure 4**).

As shown in **Figure 4**, the largest number of ZnO nanoparticles was present on the surface of the sample placed at 4 cm from the target, while smaller amounts of ZnO nanoparticles were observed for the samples positioned at 3 and 5 cm. As it is known [14], the quantity of deposited substance in PLD is inversely proportional to the square of the target-substrate separation distance. However, this does not apparently apply in our case for the sample placed at 3 cm from the target. A possible explanation could be that the plasma plume deposited and removed ('washed') nanoparticles at the same time from the substrate because the separation distance, in this case, was too small (in any case, smaller than the plasma length).

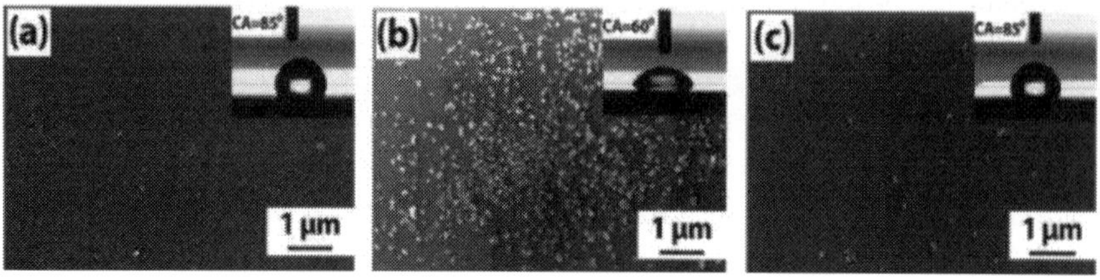

Figure 4. SEM micrographs of ZnO nanoparticles deposited in a 13 Pa O2 flux on a Si substrate. Target-substrate separation distance was of 3 cm (a), 4 cm (b) or 5 cm (c). Inset: water droplet in static mode and the measured CA.

For target-substrate separation distances longer than plasma length, the species in plasma lost their kinetic energy by collisions with other species and gas molecules from the ambient and therefore the ablation rate was significantly lower than for 4 cm.

2.1.1.1.3. Influence of number of pulses

A very low number of pulses (generally under 100) generate a deposition of nano/microparticles on the substrate surface. Slightly increasing the number of pulses produces islands of material. Upon increasing the number of pulses, the substrate is covered by a continuous thin film [15].

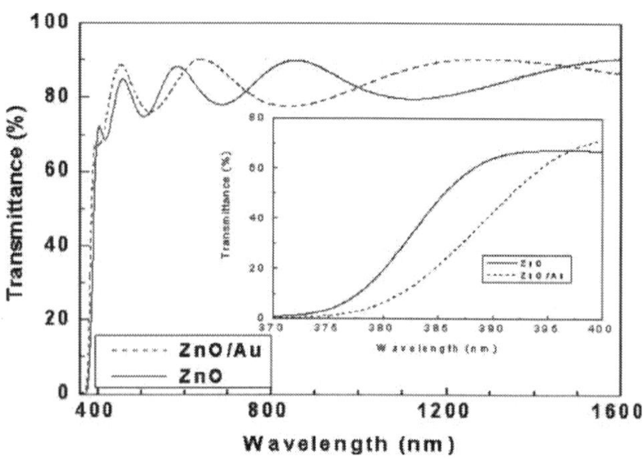

Figure 5. Typical transmission spectra recorded in the case of PLD simple ZnO films (solid curve), and films covered with Au nanoclusters after ablation by 100 pulses from a Au target (dashed curve). (Reproduced with permission from Ref. [15]).

We present a case when ZnO thin films were synthesised by PLD after applying 50,000 pulses to a ZnO target. The target was further switched on with a gold one and by irradiating it with 100 laser pulses Au nanoclusters were generated on the ZnO thin film surface. In **Figure 5**, it can be found that the transmittance spectrum was shifted toward longer wavelengths following the thin film with Au nanoclusters. The infrared band-gap energy was 3.26 eV for ZnO films covered with Au, which is slightly lower than that of simple ZnO films of 3.29 eV [15].

2.1.1.2. Laser beam parameters

2.1.1.2.1. Influence of wavelength and pulse duration

PLD can be applied to vaporize and deposit thin film from any kind of material if the absorbed power density is high enough. The amount of material that is ablated during laser irradiation can be estimated from the thermal diffusion depth, $l_\tau = \sqrt{D \cdot \tau}$, where D is the heat diffusivity in the solid target and τ is the pulse laser duration. The thermal diffusion depth decreases with the duration of the laser pulse.

The delivered laser energy is absorbed by the target material in a layer with thickness given by the formula, $l_s = \frac{1}{\alpha}$, where l_s is known as the optical penetration depth and α is the absorption coefficient for the respective laser wavelength.

The energy delivered by ultra-short laser pulses is absorbed in a thinner layer as compared to ns laser pulses, thus producing higher temperatures at surface level and faster vaporization of the target material [16].

The efficiency of laser beam absorption into the target is closely related to the wavelength that will be used. However, for numerous materials, absorption coefficient dependence on the wavelength can be more complex due to different absorption mechanisms, such as network vibration, free carrier absorption, impurities, and bandgap.

For exemplification, we present the case of Mg film deposition using laser sources with different wavelengths and pulse duration: 308 nm XeCl excimer laser (generating pulses of 30 ns) and 248 nm KrF excimer laser (with 5 ps and 500 fs). Electron microscopy analysis showed that the droplets spread and the density decreased, when using laser pulses with shorter duration [17].

The films deposited using an ns laser source had their surface covered by droplets (**Figure 6a**). These droplets were spherical and had an average diameter of 10 μm. Their presence and morphology are indicative of expulsion of molten material from the surface of the target [18]. In this case, the optical penetration depth $l_s = 2$ μm, (Mg ablated at 308 nm), was less than the thermal diffusion depth, $l_\tau = 17$ μm.

When using ps or fs pulses, the morphology of the Mg film surface changed from droplets covered to smooth surfaces, as shown in **Figure 6b** and **c**. When using ps laser pulses there were still particulates on film surface (not larger than 200 nm) but they completely disappeared when ablation was conducted with fs pulses. The smoothness of these film surfaces is a consequence of the removal of ablated material with reduced expulsion of melted particles [17].

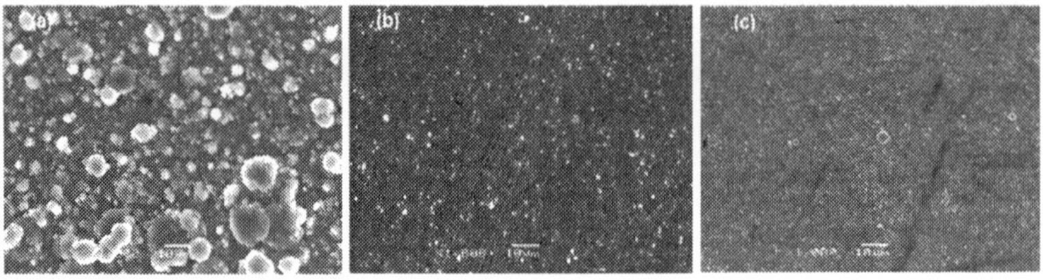

Figure 6. SEM images of Mg thin films deposited by PLD in different regimes: ns (a) ps (b) and fs (c) (Reproduced with permission from Ref. [17]).

2.1.1.2.2. *Influence of the laser fluence*

The laser pulse fluence can be defined as the optical energy that is delivered to a selected area on the target. Therefore, the fluence can be varied by changing the laser energy or the dimension of the spot area on target.

The coupling of the laser energy to the target surface is dependent on pulse parameters (duration and energy profile), and target characteristics (surface roughness, porosity and density). The fusion and vaporization processes occur only when the laser beam intensity is higher that a *threshold value* defined as the minimal energy of the laser pulse per surface unit that generates plasma ignition.

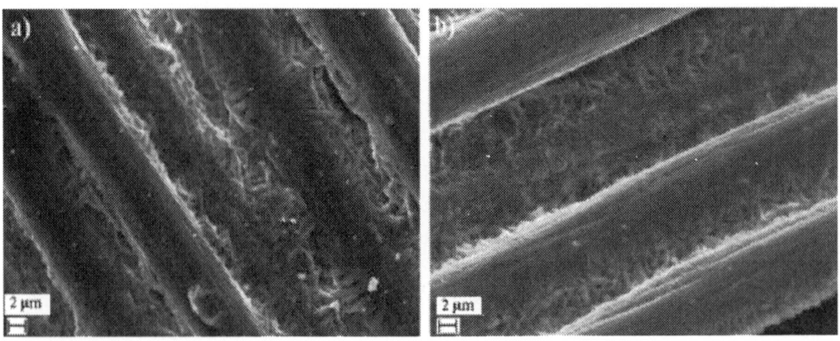

Figure 7. Typical SEM micrographs of TiO_2 nanoparticles deposited on carbon cloth substrate at a laser fluence of 5 (a) and 1 J/cm^2 (b) respectively (Reproduced with permission from Ref. [20]).

For a fixed wavelength and a chosen material, the fluence on the target will have a major effect on the particulate size and density [1]. We present an example where in order to obtain a porous gas diffusion layer, TiO_2 nanoparticles have been deposited at two different laser fluences on carbon cloth. In the case of 5 J/cm^2 laser fluence, a uniform spatial distribution of nanoparticles over the substrate surface with dimensions of tens to hundreds of nanometres (**Figure 7a** and **b**). Decreasing the laser fluence to 1 J/cm^2 (**Figure 7b**), the number of nanoparticles was considerably reduced and film protuberances were smoother [19].

2.1.2. Advantages of thin film deposition via PLD

1. A major advantage of PLD is related to its large versatility, that is, by control of the deposition parameters, one can obtain thin films with a completely different morphology, structure and/or functionality [20].

   We return again to our example of ZnO thin films synthesized by PLD in a vacuum or in oxygen ambient. Just by changing the ambient not only the aspect, but also the wettability behaviour of the films was completely different (**Figure 8**). The thin films were hydrophilic when deposited in an oxygen flux and superhydrophobic (157°) when synthesized in a vacuum. Different conditions changed the Zn and O arrangement in the crystal lattice that influenced the electrical behaviour of the surface [13, 21].

2. The target composition (stoichiometry) can be reproduced with relative ease in the thin films synthesized by PLD. Due to congruent vaporization, it is possible to deposit materials with complex chemical composition. A relevant example is that of hydroxyapatite [$Ca_{10}(PO_4)6(OH)_2$] which is the main constituent of the mineral part of the bone. Thin films of this material have been synthesized by PLD to cover metallic medical implants in view of increasing their bioactivity. EDX analyses revealed the Ca/P ratio of 1.6, very close to the nominal 1.64 value [22, 23].

3. The sequential nature of the PLD process allows for a control of the film thickness through the number of applied pulses.

   We provide an example with profiles of TiN films synthesized by PLD by applying to a TiN target 5,000, 10,000 or 20,000 laser pulses [24]. TiN is a hard material, quite difficult to ablate, so the thicknesses of films were quite low, even for a high number of applied pulses. A progressive increase of the TiN films thickness is evidenced in the profiles of **Figure 9**. Films synthesized with 5,000 pulses were of ~60 nm thickness, for 10,000 pulses the thickness was of ~86 nm, while for 20,000, it increased to ~133 nm.

4. Any type of material can be ablated, so the method is not limited to special classes of compounds.

   Ceramic, metallic and organic materials have been deposited by PLD. An exhaustive list can be found in the subchapter 2.1.4.

5. Using a carousel system with targets of different compositions, multi-layer films can be obtained. The combinations are endless and new structures with complementary properties can be obtained.

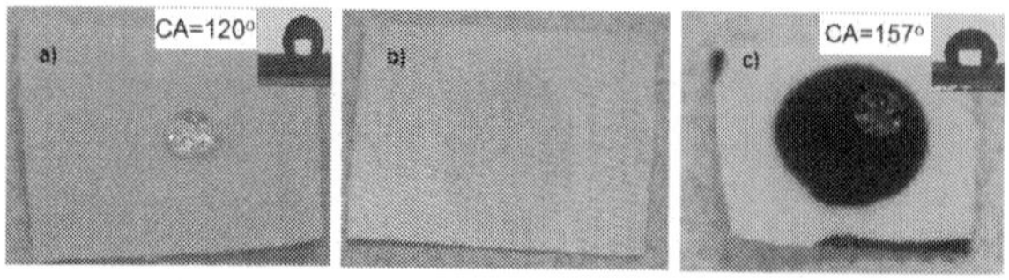

Figure 8. Textile material partially coated with ZnO nanostructures: (a) hydrophobic nanoparticle deposited in vacuum, (b) hydrophilic thin film deposited in 13 Pa oxygen flux, and (c) hydrophobic thin film deposited in vacuum. Inset (a) and (c): water droplet in static mode and the measured CA images were acquired with a EOS 50D digital camera (Canon).

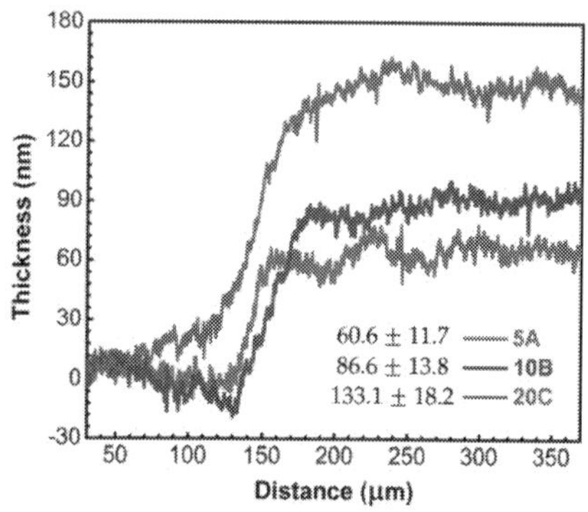

Figure 9. Thickness profiles of TiN layers recorded by profilometry (Reproduced with permission from Ref. [24]).

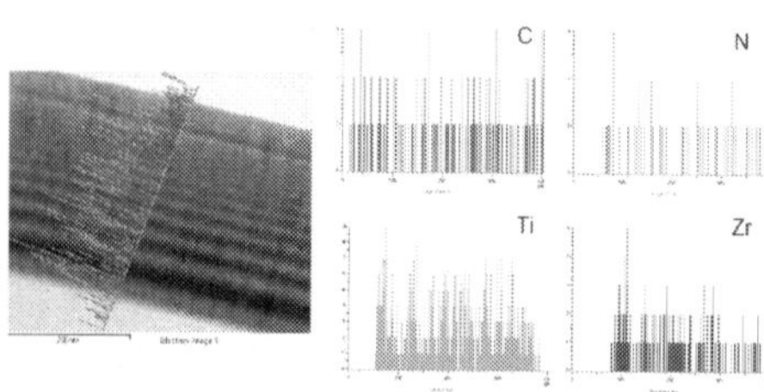

Figure 10. SEM/EDX images recorded for ZrC/TiN multi-layers deposited by PLD (Reproduced with permission from Ref. [26]).

A relevant case of multi-structures ZrC/ZrN and ZrC/TiN is given for exemplification (**Figure 10**). The purpose of this research was to increase the hardness and the elastic modulus of protective coatings. Out of ZrC, ZrN, and TiN single layers, the best results were obtained in case of ZrC with a hardness of 27.6 GPa and a reduced modulus of 228 GPa [25]. For multi-structures, the hardness and reduced modulus increased to similar values between 32.4 and 33.2 GPa and between 251 and 270 GPa, respectively [26].

2.1.3. *Drawbacks of PLD*

Because the plasma plume is expanding mainly in the *z* direction (see **Figure 11**), the deposition area is usually of a few square centimetres only. Special translation/rotation robotized substrate holders should be used for uniform coating of large substrate areas [9].

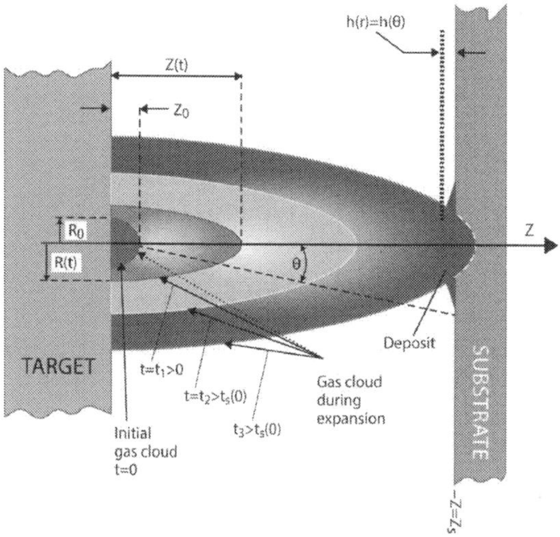

Figure 11. Schematic for the vapor cloud expansion after target irradiation by a laser pulse with energy over the ablation threshold.

Even though PLD deposition of organic materials was reported, generally, this deposition method is not suitable for such compounds—under the action of intense laser pulses, long organic chains can be broken, the deposited material being different from the original target material.

The most important disadvantage however, comes from the micronic and sub-micronic aggregates (known in the literature as particulates or droplets) that hinder applications in fields requiring high finesse (micro or nano-electronics). However, additional procedures in conjunction with PLD can drastically diminish the droplets in PLD films [27–29].

2.1.4. *Applications*

The idea to use PLD to deposit some of the most varied materials, simple or complex structures with the purpose of obtaining thin films has been rewarded with numerous results published in the literature.

Complex oxide thin films for superconductors ($YBa_2Cu_3O_7$ [30]; $Ba_2Co_2Fe_{12}O_{22}$ [31]), transparent conducting oxides TCO [32–34], active mediums (Er:YAG [35]) wide bandgap electronics (ZnO doped [36]) or in conjunction with other semiconductors [37]), thin films for gas sensors based on nanostructured tungsten oxide [38, 39], CNx/Si thin heterostructures [40]; complex $(As_2S_3)(100-x)(AgI)x$ chalcogenide glass [41]; vanadium oxide thin films with various crystal structures [42]; tin oxide for detecting NO_2 [43]; protective coatings and barriers (e.g., DLC [44], BN [45], TiN [46], ZrC [47], ZrN [48, 49], ZrC/TiN and ZrC/ZrN thin multi-layers [50]; TiN biocompatible coatings (prostheses coatings [51–53]); particles for drug delivery [54]; antimicrobial coatings [55]; tissue engineering [56]; organic thin films, i.e., polymethylmetacrylate (PMMA) [57–59]. For biosensor applications: CuO thin film for uric acid biosensor [60], gold-coating of silicon microcantilever for DNA biosensors [61].

2.2. Combinational pulsed laser deposition (CPLD)

One simple approach to study a binary or ternary system is to map all the possible compositions of the phase diagram. Of course, one can synthesize and test one composition at a time, but the disadvantages are the large number of experiments and lots of wasted time. By using a multi-target carousel holder and rhythmically changing the deposition targets during a PLD experiment or two laser beams that irradiate two targets at the same time, one can obtain alternating layers with different periodicities both vertically and horizontally, along the substrate surface (**Figure 12**) [62]. In a single CPLD experiment, thousands of different compositions can be synthesized on a substrate of a few square centimetres [10]. Different stoichiometries can give rise to a variety of different structures and properties. The feasibility and utility of this concept has been demonstrated in the discovery of a number of new materials with much improved physico-chemical properties than the precursors [63–65]. **Figure 12** shows a schematic of a CPLD deposition experiment. The experimental set-up can include two independent laser beams or a split laser beam that hits two targets alternatively or one target at a time, the targets being interchanged with the desired frequency via a mobile carousel.

2.2.1. *Advantages of CPLD*

1. Thin film libraries can be synthesized in relatively short time (minutes).
2. A large number of new binary or ternary compounds with different properties to study can be obtained.
3. By monitoring the number of laser pulses, one can control the deposition of materials at an atomic layer level. If targets of different nature are used, composite materials will be synthesized.

We give a practical example of IZO (indium-zinc-oxide) compositional libraries synthesized by CPLD [66–69]. Due to the reduced availability of indium, in order to minimize costs, Zn is used for partial replacement of this element. Normally, individual thin films should be synthesized by PLD for measuring their conductivity. By CPLD, we synthesise a library with hundreds of IZO compositions, identify the areas with high conductivity and then we assess the IZO composition using a punctual spectroscopic technique (laser induced breakdown spectroscopy-LIBS). Thus, a significant reduction of time devoted to film deposition and individual analyses could be achieved. LIBS allows fast optical spectra recording and line identification with an excellent spatial resolution (the laser beam is focused at a spot of ≈100 μm diameter), in addition to minimal damage to the film [70, 71]. The quantitative LIBS measurement method, based on the calculation of the spectral radiance of plasma in local thermal equilibrium, was used to measure the Zn/ (In+Zn) ratio and its variation over the length of samples synthesized by CPLD.

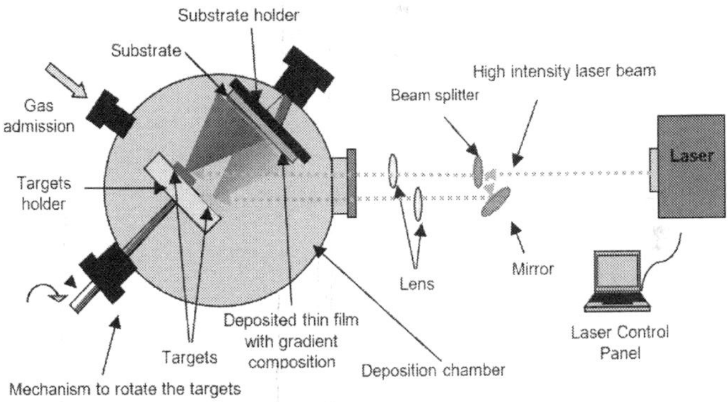

Figure 12. Schematic presentation of a CPLD experiment, along the line the plasmas overlap producing a compositional library.

From the study of the obtained compositional libraries, optimum values of the optical transmittance higher than 85%, resistivity around (5–7) × 10^{-4} Ohm cm and mobility in the (45–53) $cm^2/(V\ s)$ range, were inferred.

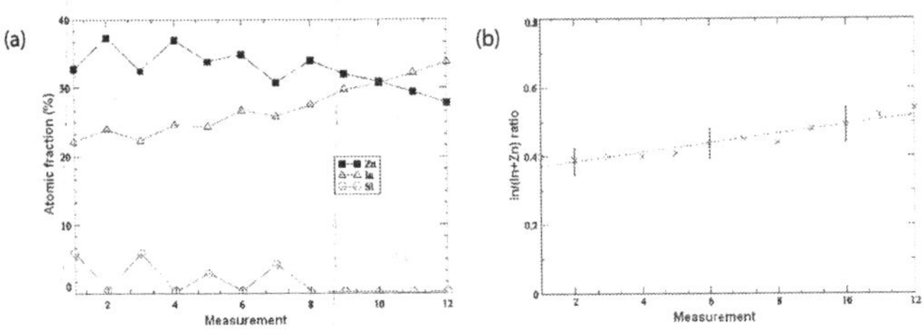

Figure 13. Evolution of (a) In and Zn concentrations and (b) Zn/In ratio on the transversal axis. Twelve measurements were performed on a distance of 6 cm from border towards the centre of the sample (Reproduced with permission from Ref. [72]).

The LIBS measurements started from the border of the glass plate where Zn has the largest concentration, towards the centre of the sample, along the transversal axis [72]. At the border A (close to the ZnO target) the In/(In+Zn) ratio was of 0.40. The trend for this ratio is to increase continuously. Up to measurement 6 (corresponding to 3 cm towards the centre of the sample), it increases by 8%, that is, to a value of 0.44. The Zn decrease and the In increase are almost linear (**Figure 13a**), the overall In/(In + Zn) ratio increases when shifted towards the centre of the sample. From 3 to 6 cm (in respect with the centre of the sample), the ratio continues its ascendant trend to reach a value of 0.50 at the centre of the film.

In **Figure 13b**, the evolution of the In/(In+Zn) ratio is presented. Starting from the glass border towards the centre, an ascendant trend from 0.40 to 0.54 is observed. Along a distance of 6.5 cm, the In/(In + Zn) ratio increased by 26%. The results are very similar to the values obtained by EDS investigations on the same CPLD samples [73].

2.2.2. *Drawback of CPLD*

The plasma spread is strongly influenced by the atomic weight of the species. For each type of the target material, new complete compositional tests and mapping should be performed on the resulting films.

2.2.3. *Applications*

The synthesis of new materials starting from precursors (i.e., BaF_2, SrF_2 and TiO_2 to form (Ba, Sr)TiO_3 [74] or BaF_2 and TiO_2 to form $BaTiO_3$ [75]); deposition of thin film libraries of doped oxide materials: $(Ba_{1-x}Sr_x)TiO_3$ doped with Ca, W, Cr, Mg, Mn, Y and La [74]; libraries of TiO_2 doped with Co [76]; artificial oxide lattices and heterojunctions controlled at an atomic scale [77] ($SrTiO_3$/$BaTiO_3$superlattices); composition spreads: $La_{1-x}Sr_xMnO_3$with $0 \leq x \leq 1$ [78], $Ba_{1-x}Sr_xTiO_3$ [79], $Mg_xZn_{1-x}O$ [80] and IZO [81].

2.3. Matrix-assisted pulsed laser evaporation (MAPLE)

For electronic, optical and biosensor device applications, the materials of choice cover polymeric materials for the fabrication and passivation of electronic coatings, organic dye molecules for non-linear and optical limiting applications, biocompatible and protein coatings for micro-array biosensor applications, and living cells for tissue engineering. The ability to deposit various classes of functional polymeric and organic materials using a single technique provides a significant advantage for their development and implementation.

When a laser interacts with an organic target under the usual conditions for PLD, the material, which is grown in a thin film form, is different from the starting material, the functional groups being often altered [11]. The organic chain can also be broken leaving the film to be made up of smaller polymeric pieces and with different functional groups terminating the ends. Even small changes in the number of functional groups or the degree of polymerization can preclude the use of these films for their desired application. Such modifications might be acceptable for some applications, but in general, the use of lasers for depositing thin films of polymeric and organic materials, requires more subtle approaches than those offered by PLD alone.

MAPLE, a laser based vapour deposition technique derived from PLD, has been developed, at the end of 1990s, to deposit thin organic and biologic films without decomposition or other major irreversible damage [82]. MAPLE is a physical vapour deposition technique capable of depositing uniform thin films over a larger area. Specific to MAPLE is the use of a cryogenic composite target made of a dilute mixture of the polymer/ biopolymer/ protein to be deposited and a light absorbent, and high vapour pressure solvent matrix.

The principal objectives that this technique has to perform are:

a. to avoid the damage of the organic molecules, that can be produced by photochemical and photothermal effects like in the case of PLD;

b. to ensure the deposition of a homogenous thin film, that cannot be obtained by other laser methods.

Changing the target to a frozen composite modifies the laser material interaction, so that the major part of the laser energy is absorbed by the solvent molecules and not by the fragile solute. The rapidly evaporating volatile solvent desorbs the fragile solute by soft collisions and deposits it as a uniform thin film whose properties, such as chemical structure and functionality, have been maintained (**Figure 14**). Since the receiving substrate is kept at room temperature and the sticking coefficient of the solvent is nearly zero, the evaporated solvent is efficiently pumped away by the vacuum system.

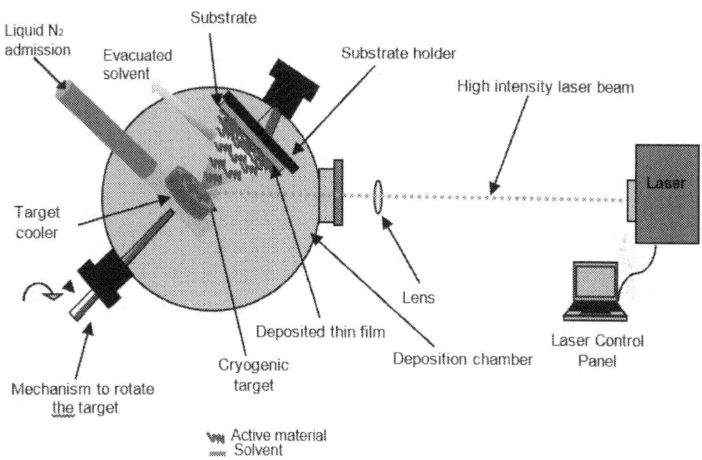

Figure 14. Schematic of material transfer by MAPLE technique.

The MAPLE target is composed of less than 10 wt% of the film material. Each film molecule is surrounded or shielded by a large amount of matrix. This structure prevents the direct thermal and photonic damage to the film material [83].

2.3.1. Factors influencing for MAPLE deposition

The nature of the organic compounds can influence the deposition of a thin film. The laser parameters (fluence, wavelength, and pulse duration) can affect the quality of the thin film.

Although it is derived from PLD, MAPLE differs in certain points, namely the method of preparation of the target and mechanism of laser-material interaction. Film roughness can be controlled by selecting correctly the appropriate laser fluence, solute concentration and substrate temperature.

2.3.1.1. Influence of the matrix

MAPLE efficiency is determined by the correct choice of the solvent matrix, which has to absorb the laser energy during the deposition, thus protecting the complex organic compound.

The solvent has to satisfy the following conditions:

i. to absorb the laser beam, even when frozen;

ii. to have a high melting point;

iii. to have a high vapour pressure and a high volatility at room temperature, in order to be evacuated very fast from the deposition chamber;

iv. not to be chemically active at laser beam exposure; and

v. to form a uniform solution with the complex organic material.

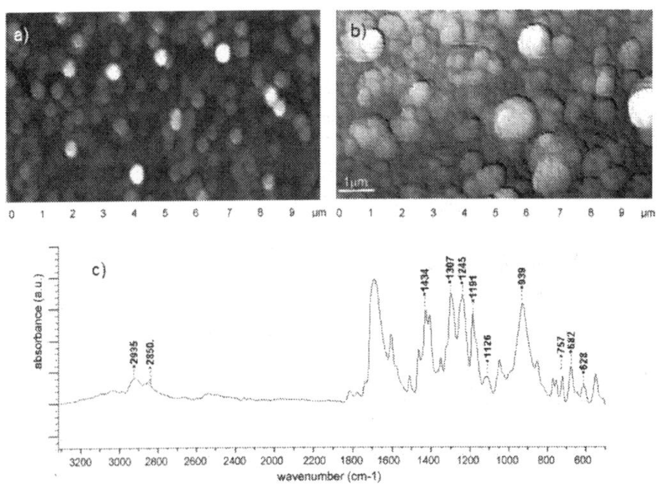

Figure 15. AFM micrographs of P(CPP:SA) 20:80 thin films deposited by MAPLE using ethyl acetate (a) and dimethyl chloride as matrix (b); typical FTIR spectra recorded for P(CPP:SA) 20:80 thin films obtained by MAPLE using ethyl acetate as matrix at a fluence of 0.3 J/cm^2(c). (Reproduced with permission from Ref. [84]).

We present an example of MAPLE thin film synthesis of poly(1,3-bis-(p-carboxyphenoxy, propane)-co-(sebacic anhydride)) (20:80) (P(CPP:SA)20:80) using two different solvents as a matrix to protect this hydrophobic anhydride copolymer [84]. One solvent was dimethyl chloride and the other, ethyl acetate. The copolymer was successfully transferred in both cases, but the morphology of the films was quite different. AFM images presented in **Figure 15a** for the case of ethyl acetate as the matrix reveal a granular morphology of thin films with individual grains with diameters of ~400 nm. In the case of dimethyl chloride as the matrix, the thin film seems to be compact with some circular (~500 nm in diameter) features splashed on the surface forming a layered plate-like structure (**Figure 15b**). Ethyl acetate was selected as the best solvent based on AFM data and FTIR spectra comparison (**Figure 15c**) that showed a high resemblance between the target material and the thin film.

2.3.1.2. *Influence of the laser fluence*

McGill and Chrisey [82] proposed in their patent that the laser energy is absorbed majorly by the solvent matrix and is converted in thermic energy producing solvent evaporation. The complex organic compound will reach a kinetic energy high enough that will ensure the transfer and the immobilization onto the substrate.

Georgiou and Kokkinaki [85] advanced the hypothesis that the process takes place due to a photomechanical process (material expulsion). The complex organic compound will be ejected into the gaseous phase only if the laser irradiation takes place at a fluence that surpasses the ablation threshold. When the laser fluence is under the ablation threshold a thermic vaporization process, that produces the solvent desorption, occurs.

Itina et al. [86] suggested that after the laser irradiation upon the organic compound, kinetic energy is a result of both thermic and mechanic phenomena. They proposed a theoretic study that simulates the initial steps of molecule ejection. They observed that when the laser fluence surpasses the ablation threshold, clusters would be ejected from the target. The most important observation was that during the ablation process the organic compound molecules are not fragmented.

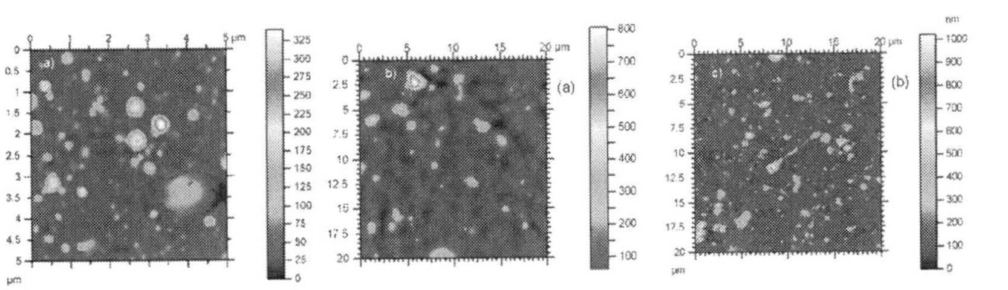

Figure 16. Top view AFM image of the RNase A thin films obtained from a frozen composite target containing 1% (w/v) biomaterial in buffer Hepes solvent, by irradiation with 15,000 subsequent laser pulses at 0.4 (a), 0.5 (b), 0.7 J/cm^2 (c) laser fluence (Reproduced with permission from Ref. [87]).

In their support, we present the example of RNaseA enzyme thin films obtained by MAPLE [87]. When irradiation of targets was conducted with a laser fluence of 0.4 J/cm^2 (**Figure 16a**) the films were constituted of tens of nanometre-sized particles probably generated after surface evaporation and cluster formation in transit towards the substrate. By increasing the laser fluence to 0.5 J/cm^2 (**Figure 16b**), both the mean diameter and the mean height of particles increase and double population could be identified: a majority population of 500 nm mean diameter and some large micronic particles. At 0.7 J/cm^2 (**Figure 16c**), the film is covered by micrometric particles, caused by the droplet expulsion from the target surface as a result of explosive evaporation or spallation mechanisms [88, 89] termed cold laser ablation [90, 91].

2.3.2. *Advantages of material deposition via MAPLE*

MAPLE was developed to surmount the difficulties in solvent-based coating technologies such as inhomogeneous films, inaccurate placement of material, and intricate or erroneous thickness control. The process utilizes a low fluence pulsed UV laser and a frozen target consisting of a dilute mixture of the material to be deposited and a high vapour-pressure solvent. The low fluence laser pulse interacts mainly with the volatile solvent, causing its evaporation. During the process, the solute desorbs intact, that is, without any significant decomposition, and is then uniformly deposited on the substrate.

- It enables the thin film deposition from a large amount of organic materials, such as polymers, proteins, enzymes and combination of organic-inorganic materials.

- It is a non-contact deposition technique free of pollution risks for the thin films. The molecular composition and structure of the material that is deposited by MAPLE are preserved during the transfer process.

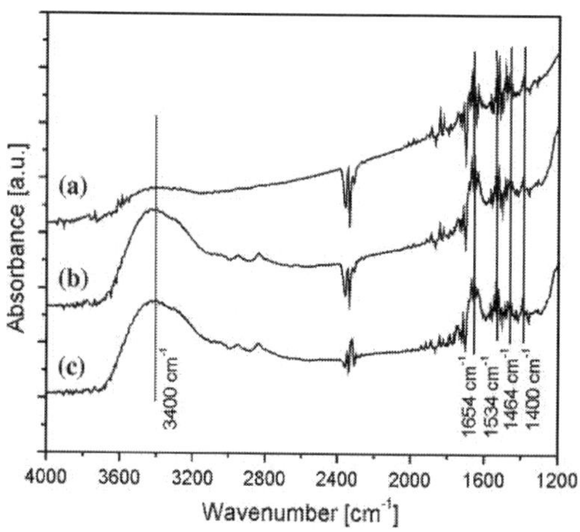

Figure 17. FTIR images of (a) laser immobilized RNase A obtained by the irradiation of 1 wt% frozen composite RNase A target, and the drop-cast samples of (b) initial and (c) final MAPLE target solutions of RNase A in buffer HEPES–NaOH, pH 7.5 (Reproduced with permission from Ref. [92]).

In support of this assertion, we give the FTIR spectra for a RNase A enzyme thin film (**Figure 17a**), for drop-cast of the initial RNase A solution used to prepare MAPLE targets (**Figure 17b**) and for the final RNase A solution, collected from the target holder after the laser irradiation experiments (**Figure 17c**) [92]. As can be seen in the **Figure 17** the spectrum of the transferred RNase A was comparable to that of the initial drop-cast and to that of the target after irradiation. The bands of RNase A target material were also present in the spectra of deposited films: at 3400 cm^{-1} there was a band characteristic to N–H stretching vibrations from amide I groups; the one at 1654 cm^{-1} corresponded to C=O and C–N stretching vibrations of amide I, while the band at 1464 cm^{-1} was characteristic to in-plane N–H bending as well as C–N stretching vibrations in the same functional group. At 1534 cm^{-1}, the band was associated with tyrosine amino acid residues, while the band at 1400 cm^{-1} could belong to the amide III region [93–95]. The lower band intensity, in case of the MAPLE thin films, was due to the five times lesser amount of RNase A enzyme in the film as compared to the amount present in the drop-cast samples.

2.3.3. *Drawbacks of MAPLE*

The solvent has to be UV absorbing, therefore the choice is limited and in many cases very difficult.

In order to obtain perfect transfer of the organic material from the target to the substrate, a process optimization procedure is mandatory. Laser beam fluence, repetition rate, and pressure inside the reaction chamber have to be tweaked for each new material to be deposited.

Toxic, volatile solvents are often the only choice for target preparation.

2.3.4. *Applications*

Biomaterials for drug delivery, antimicrobial systems and biosensors: polyethylene glycol [96–99], polyvinyl alcohol derivatives [100, 101], oligo(*p*-phenylene-ethynylene) [102], magnetite/salicylic acid/silica shell/antibiotics [103], porphyrin [104], laccase [105], urease [106], and lysozyme [107]. Polymers for electronic devices (electrically conductive polymers, light emitting thin films): polythiophene [108], polyfluorene [109], and hybrid composites [110]; CdS quantum dots onto TiO_2 nanotubes for solar cells [111]; Protective coatings: composite carbon/gold nanoparticle films [112].

3. Conclusions

Laser deposition methods have been used to deposit a countless simple or complex inorganic or organic materials. It is quite hard to find other alternative methods that are so easily tunable so that they could produce thin films of so many different materials while preserving in the same time their structure and properties. This is probably the main reason for the endurance of these techniques despite some big outcomes: (i) droplets on film surface, which reduce their use in high tech fields and (ii) small area depositions that prevent their real breakthrough in industry.

Major progress has been achieved to overcome these issues, numerous droplet filters having been proposed in the literature with good results and large scale depositions of up to square metres being reported.

This chapter aimed to show for the new reader just a small part of the myriad of parameters that can be tuned and how important small adjustments can become for the final aspect and the properties of thin films. The combination of various parameters is practically limitless and new structures with interesting properties that emerge from these combinations provide a long life for the research in this field.

The authors hope that after reading this chapter, one will get an idea about the countless applications where laser ablation can be used, for synthesis, tuning or characterization of thin films. It is expected that new methods based on laser ablation will continue to emerge and develop as science progresses and applications become more and more punctual. Laser-patterned deposition of thin films or living cells transfer in a matrix by laser-induced forward transfer, compositional analysis of organic compounds by matrix-assisted laser desorption/ionization are hot emerging fields that rotate around the laser ablation mechanisms.

Acknowledgements

The authors acknowledge financial support from the Romanian National Authority for Scientific Research through the PNII-RU-TE-2012-3-0379 (TE 16/2013) and the National Authority for Research and Innovation in the frame of Nucleus programme under contract 4N/2016.

Author details

Camelia Popescu, Gabriela Dorcioman and Andrei C. Popescu*

*Address all correspondence to: andrei.popescu@inflpr.ro

National Institute for Lasers, Plasma and Radiation Physics, Laser Department, "Laser-Surface-Plasma Interactions" Laboratory, Bucharest-Magurele, Romania

References

[1] Chrisey D.B., Hubler G.K., editors. Pulsed Laser Deposition of Thin Films. 1st ed. New York: Wiley; 1994. 648 p.

[2] Brech F., Cross L. Optical micromission stimulated by a ruby maser. Applied Spectroscopy. 1962;16(2):59

[3] Ready J.F. Development of plume of material vaporized by giant-pulse laser. Applied Physics Letter. 1963;3(1):11–13. DOI: 10.1063/1.1723555

[4] White R.M. Generation of elastic waves by transient surface heating. Journal of Applied Physics. 1963;34(12):3559–3567. DOI: 10.1063/1.1729258

[5] Askar'yan G.A., Prokhorov A.M., Chanturiya G.F., Shilnikov G.P. The effects of a laser beam in a liquid. Journal of Experimental and Theoretical Physics. 1963;17(6):1463–1465.

[6] Meyerand Jr. R.G., Haught, A.F. Gas breakdown at optical frequencies. Physical Review Letters. 1963;11(9):401–403. DOI: 10.1103/PhysRevLett.11.401

[7] Smith H.M., Turner A.F. Vacuum deposited thin films using a ruby laser. Applied Optics. 1965;4(1):147–148. DOI: 10.1364/AO.4.000147

[8] Dijkkamp D., Venkatesan T., Wu X.D., Shaheen S.A., Jisrawi N., Min-Lee Y.H. et al. Preparation of Y-Ba-Cu oxide superconductor thin films using pulsed laser evaporation from high Tc bulk material. Applied Physics Letters. 1987;51(8):619–621. DOI: 10.1063/1.98366

[9] Eason R., editor. Pulsed Laser Deposition of Thin Films: Applications-Lead Growth of Functional Materials. 1st ed. New Jersey: Wiley; 2006.

[10] Koinuma H., Takeuchi I. Combinatorial solid-state chemistry of inorganic materials. Nature Materials. 2004;3(7):429–438. DOI: 10.1038/nmat1157

[11] Chrisey D.B., Pique A., McGill R.A., Horwitz J.S., Ringeisen B.R., Bubb D.M. et al. Laser deposition of polymer and biomaterial films. Chemical Reviews. 2003;103(2):553–576. DOI: 10.1021/cr010428w

[12] Belouet C. Thin film growth by the pulsed laser assisted deposition technique. Applied Surface Science. 1996;96–98(2);630–642. DOI: 10.1016/0169-4332(95)00535-8

[13] Popescu A.C, Duta L., Dorcioman G., Mihailescu I.N., Stan G.E., Pasuk I. et al. Radical modification of the wetting behavior of textiles coated with ZnO thin films and nanoparticles when changing the ambient pressure in the pulsed laser deposition process. Journal of Applied Physics. 2011;110(6): 064321. DOI: 10.1063/1.3639297

[14] Mihailescu I.N., Gyorgy E. Pulsed laser deposition: an Overview. In: Asakura T., editor. International Trends in Optics and Photonics. 4th International Commission for Optics (ICO). Japan: Springer; 1999.

[15] Socol G., Axente E., Ristoscu C., Sima F., Popescu A., Stefan N. et al. Enhanced gas sensing of Au nanocluster-doped or-coated zinc oxide thin films. Journal of Applied Physics. 2007;102(8):083103. DOI: 10.1063/1.2798922

[16] Acquaviva S., Perrone A., Zocco A., Klini A., Fotakis C. Deposition of carbon nitride films by reactive sub-picosecond pulsed laser ablation. Thin Solid Films. 2000;373(1): 266–272. DOI: 10.1016/S0040-6090(00)01095-6

[17] Miglietta P., Cultrera L., Cojanu C., Papadopoulou E.L., Perrone A. Mg-based photocathodes prepared by ns, ps and fs PLD for the production of high brightness electron beams. Applied Surface Science. 2009;255(10):5228–5231. DOI: 10.1016/j.apsusc.2008.07.131

[18] Millon E., Albert O., Loulergue J.C., Etchepare J., Hulin D., Seiler W. et al. Growth of heteroepitaxial ZnO thin films by femtosecond pulsed-laser deposition. Journal of Applied Physics. 2000;88(11):6937–6939. DOI: 10.1063/1.1324679

[19] Dorcioman G., Ebrasu D., Enculescu I., Serban N., Axente E., Sima F. et al. Metal oxide nanoparticles synthesized by pulsed laser ablation for proton exchange membrane fuel cells. Journal of Power Sources. 2010;195(23):7776–7780. DOI: 10.1016/j.jpowsour.2009.09.060

[20] Papadopoulou E.L., Zorba V., Pagkozidis A., Barberoglou M., Stratakis E., Fotakis C. Reversible wettability of ZnO nanostructured thin films prepared by pulsed laser deposition. Thin Solid Films. 2009;518(4):1276–1270. DOI: 10.1016/j.tsf.2009.02.077.

[21] Duta L., Popescu A.C., Dorcioman G., Mihailescu I.N., Stan G., Zgura I. et al. ZnO thin films deposited on textile material substrates for biomedical applications. In: Vaseashta A., Braman E., Susmann P., editors. Technological Innovations in Sensing and Detection of Chemical, Biological, Radiological, Nuclear Threats and Ecological Terrorism. NATO Science for Peace and Security Series A: Chemistry and Biology ed. Netherlands: Springer; 2012

[22] Mihailescu N., Stan G.E., Duta L., Chifiriuc M.C., Bleotu C., Sopronyi M. et al. Structural, compositional, mechanical characterization and biological assessment of bovine-derived hydroxyapatite coatings reinforced with MgF_2 or MgO for implants functionalization. Materials Science and Engineering: C. 2016;59:863–874. DOI: 10.1016/j.msec.2015.10.078

[23] Eraković S., Janković A., Ristoscu C., Duta L., Serban N., Visan A. et al. Antifungal activity of Ag: hydroxyapatite thin films synthesized by pulsed laser deposition on Ti and Ti modified by TiO_2 nanotubes substrates. Applied Surface Science. 2014;293:37–45. DOI: 10.1016/j.apsusc.2013.12.029

[24] Duta L., Stan G.E., Popa A.C., Husanu M.A., Moga S., Socol M., et al. Thickness Influence on in vitro biocompatibility of titanium nitride thin films synthesized by pulsed laser deposition. Materials. 2016;9(1):38. DOI: 10.3390/ma9010038

[25] Craciun D., Socol G., Dorcioman G., Stefan N., Bourne G., Craciun V. High quality ZrC, ZrC/ZrN and ZrC/TiN thin films grown by pulsed laser deposition. Journal of Optoelectronics and Advanced Materials. 2010;12(3):461–465.

[26] Craciun D., Bourne G., Socol G., Stefan N., Dorcioman G., Lambers E. et al. Characteristics of ZrC/ZrN and ZrC/TiN multilayers grown by pulsed laser deposition. Applied Surface Science. 2011;257(12):5332–5336. DOI: 10.1016/j.apsusc.2010.11.106

[27] Yoshitake T., Nagayama K.. The velocity distribution of droplets ejected from Fe and Si targets by pulsed laser ablation in a vacuum and their elimination using a vane-type velocity filter. Vacuum. 2004;74(3–4):515–520. DOI: 10.1016/j.vacuum.2004.01.051

[28] Yoshitake T., Shiraishi G., Nagayama K.. Elimination of droplets using a vane velocity filter for pulsed laser ablation of $FeSi_2$. Applied Surface Science. 2002;197–198:379–383. DOI: 10.1016/S0169-4332(02)00344-6

[29] György E., Mihailescu I.N., Kompitsas M., Giannoudakos A. Deposition of particulate-free thin films by two synchronised laser sources: effects of ambient gas pressure and laser fluence. Thin Solid Films. 2004;446(2):178–183. DOI: 10.1016/j.tsf.2003.09.071

[30] Develos-Bagarinao K., Yamasaki H., Nakagawa Y., Endo, K. Relationship between composition and surface morphology in YBCO films deposited by large-area PLD. Physica C: Superconductivity. 2004;412–414(2):1286–1290. DOI: 10.1016/j.physc.2004.01.172

[31] Sudakar C., Subbanna G.N., Kutty T.R.N. Hexaferrite–FeCo nanocomposite particles and their electrical and magnetic properties at high frequencies. Journal of Applied Physics. 2003;94(9):6030–6033. DOI: 10.1063/1.1614848

[32] Gondoni P., Ghidelli M., Di Fonzo F., Bassi A.L., Casari C.S. Fabrication of nano-engineered transparent conducting oxides by pulsed laser deposition. Journal of Visualized Experiments. 2013;72:50297. DOI: 10.3791/50297

[33] Cho H., Douglas E.A., Gila B.P., Craciun V., Lambers E.S., Ren F. et al. Band offsets in $HfO_2/InGaZnO_4$ heterojunctions. Applied Physics Letters. 2012;100(1):012105. DOI: 10.1063/1.3673905

[34] Douglas E.A., Scheurmann A., Davies R.P., Gila B.P., Cho H., Craciun V. et al. Measurement of $SiO_2/InZnGaO_4$ heterojunction band offsets by X-ray photoelectron spectroscopy. Applied Physics Letters. 2011;98(24):242110. DOI: 10.1063/1.3600340

[35] Stanoi D., Popescu A., Ghica C., Socol G., Axente E., Ristoscu C. et al. Nanocrystalline Er: YAG thin films prepared by pulsed laser deposition: an electron microscopy study. Applied Surface Science. 2007;253(19):8268–8272. DOI: 10.1016/j.apsusc.2007.02.113

[36] Ohtomo A., Kawasaki M., Sakurai Y., Ohkubo I., Shiroki R., Yoshida Y. et al. Fabrication of alloys and superlattices based on ZnO towards ultraviolet laser. Materials Science and Engineering: B. 1998;56(2–3):263–266. DOI: 10.1016/S0921-5107(98)00218-9

[37] Ohta H., Hirano M., Nakahara K., Maruta H., Tanabe T., Kamiya M. et al. Fabrication and photoresponse of a pn-heterojunction diode composed of transparent oxide semiconductors, p-NiO and n-ZnO. Applied Physics Letters. 2003;83(5):1029–1031. DOI: 10.1063/1.1598624

[38] Palla-Papavlu A., Filipescu M., Schneider C.W., Antohe S., Ossi P.M., Radnoczi G. et al. Direct laser deposition of nanostructured tungsten oxide for sensing applications.

Journal of Physics D Applied Physics. 2016;49(20):205101. DOI: 10.1088/0022-3727/49/20/205101

[39] György E., Socol G., Mihailescu I.N., Ducu C., Ciuca, S. Structural and optical characterization of WO3 thin films for gas sensor applications. Journal of Applied Physics. 2005;97(9):093527. DOI: 10.1063/1.1889246

[40] Simeonov S., Szekeres A., Gyorgy E., Mihailescu I.N, A. Perrone A. CNx/Si thin heterostructures for miniaturized temperature sensors. Journal of Applied Physics. 2004;95(9):5111–5115. DOI: dx.doi.org/10.1063/1.1691482

[41] Kolev K., Popov C., Petkova T., Petkov P., Mihailescu I.N., Reithmaier J.P. Complex (As2S3)(100−x)(AgI)x chalcogenide glasses for gas sensors. Sensors and Actuators B: Chemical. 2009;143(1):395–399. DOI: 10.1016/j.snb.2009.08.016

[42] Huotari J., Lappalainen J., Puustinen J., Spetz A.L. Gas sensing properties of pulsed laser deposited vanadium oxide thin films with various crystal structures. Sensors and Actuators B: Chemical. 2013;187:386–394. DOI: 10.1016/j.snb.2012.12.067

[43] Starke T.K., Coles G.S., Ferkel H. High sensitivity NO2 sensors for environmental monitoring produced using laser ablated nanocrystalline metal oxides. Sensors and Actuators B: Chemical. 2002;85(3):239–245. DOI: 0.1016/S0925-4005(02)00114-4

[44] Mosaner P., Bonelli M., Miotello A. Pulsed laser deposition of diamond-like carbon films: reducing internal stress by thermal annealing. Applied Surface Science. 2003;208–209:561–565. DOI: 10.1016/S0169-4332(02)01383-1

[45] Marotta V., Orlando S., Parisi G.P., Santagata A. Boron nitride thin films deposited by RF plasma reactive pulsed laser ablation. Applied Surface Science. 2003;208:575–581. DOI: 10.1016/S0169-4332(02)01386-7

[46] Pelletier H., Carradò A., Faerber J., Mihailescu I.N. Microstructure and mechanical characteristics of hydroxyapatite coatings on Ti/TiN/Si substrates synthesized by pulsed laser deposition. Applied Physics A. 2011;102(3):629–640. DOI: 10.1007/s00339-010-6144-8

[47] Craciun D., Socol G., Stefan N., Mihailescu I.N., Bourne G., Craciun V. High-repetition rate pulsed laser deposition of ZrC thin films. Surface and Coatings Technology. 2009;203(8):1055–1058. DOI: 10.1016/j.surfcoat.2008.09.039

[48] Craciun D., Socol G., Dorcioman G., Simeone D., Gosset D., Behdad S. et al. Ar ions irradiation effects in ZrN thin films grown by pulsed laser deposition. Applied Surface Science. 2015;336:129–132. DOI: 10.1016/j.apsusc.2014.10.032

[49] Dorcioman G., Socol G., Craciun D., Argibay N., Lambers E., Hanna M. et al. Wear tests of ZrC and ZrN thin films grown by pulsed laser deposition. Applied Surface Science. 2014;306:33–36. DOI: 10.1016/j.apsusc.2013.12.048

[50] Craciun D., Socol G., Dorcioman G., Niculaie S., Bourne G., Zhang J. et al. Wear resistance of ZrC/TiN and ZrC/ZrN thin multilayers grown by pulsed laser deposition. Applied Physics A. 2013;110(3):717–722. DOI: 10.1007/s00339-012-7224-8

[51] Nelea V., Pelletier H., Iliescu M., Werckmann J., Craciun V., Mihailescu I.N. et al. Calcium phosphate thin film processing by pulsed laser deposition and in situ assisted ultraviolet pulsed laser deposition. Journal of Materials Science: Materials in Medicine. 2002;13(12):1167–1173. DOI: 10.1023/A:1021150207350

[52] Angelina J.T., Ganesan S., Panicker T.M., Narayani R., Paul Korath M., Jagadeesan K. Pulsed laser deposition of silver nanoparticles on prosthetic heart valve material to prevent bacterial infection. Materials Technology: Advanced Performance Materials. 2016. Forthcoming. DOI: 10.1080/10667857.2016.1160503

[53] Rajesh P., Mohan N., Yokogawa Y., Varma H. Pulsed laser deposition of hydroxyapatite on nanostructured titanium towards drug eluting implants. Materials Science and Engineering: C. 2013;33(5):2899–2904. DOI: 10.1016/j.msec.2013.03.013

[54] Singh R.K., Kim W.S., Ollinger M., Craciun V., Coowantwong I., Hochhaus G. et al. Laser based synthesis of nanofunctionalized particulates for pulmonary based controlled drug delivery applications. Applied Surface Science. 2002;197–198:610–614. DOI: 10.1016/S0169-4332(02)00408-7

[55] Gittard S.D., Narayan R.J., Jin C., Ovsianikov A., Chichkov B.N., Monteiro-Riviere N.A. et al. Pulsed laser deposition of antimicrobial silver coating on Ormocer® microneedles. Biofabrication. 2009;1(4):041001. DOI: 10.1088/1758-5082/1/4/041001

[56] Kuppuswamy H., Ganesan A. Structural, mechanical and in vitro studies on pulsed laser deposition of hydroxyapatite on additive manufactured polyamide substrate. International Journal of Bioprinting. 2016;2(2):85–94. DOI: 10.18063/IJB.2016.02.008

[57] Cristescu R., Socol G., Mihailescu I.N., Popescu M, Sava F., Ion E. et al. New results in pulsed laser deposition of poly-methyl-methacrylate thin films. Applied Surface Science. 2003;208–209:645–650. DOI: 10.1016/S0169-4332(02)01415-0

[58] Wang L.D., Kwok H.S. Pulsed laser deposition of organic thin films. Thin Solid Films. 2000;363(1–2):58–60. DOI: 10.1016/S0040-6090(99)00983-9

[59] Kecskemeti G., Smausz T., Kresz N., Toth Z., Hopp B., Chrisey D. et al. Pulsed laser deposition of polyhydroxybutyrate biodegradable polymer thin films using ArF excimer laser. Applied Surface Science. 2006;253(3):1185–1189. DOI: 10.1016/j.apsusc.2006.01.084

[60] Jindal K., Arora K., Tomar M., Gupta V.Uric acid biosensor based on pulsed laser deposited CuO thin film. Journal of Nanoscience Letters. 2012;2:28

[61] Rebollar E., Sanz M., Esteves C., Martínez N.F., Ahumada Ó., Castillejo M. Gold coating of micromechanical DNA biosensors by pulsed laser deposition. Journal of Applied Physics. 2012;112(8): 084330. DOI: 10.1063/1.4761986

[62] Craciun D., Socol G., Stefan N., Miroiu M., Mihailescu I.N., Galca A.C. et al. Structural investigations of ITO-ZnO films grown by the combinatorial pulsed laser deposition technique. Applied Surface Science. 2009;255(10):5288–5291. DOI: 10.1016/j.apsusc.2008.07.120

[63] Xiang X.D., Sun X., Briceno G., Lou Y. A combinatorial approach to materials discovery. Science. 1995;268(5218):1738–1740. DOI: 10.1126/science.268.5218.1738

[64] Briceno G., Chang H., Sun X., Schultz P.G., Xiang X.D. A class of cobalt oxide magnetoresistance materials discovered with combinatorial synthesis. Science. 1995;270(5234):273–275. DOI: 10.1126/science.270.5234.273

[65] Danielson E., Devenney M., Giaquinta D.M., Golden J.H., Haushalter R.C., McFarland E.W. et al. A rare-earth phosphor containing one-dimensional chains identified through combinatorial methods. Science. 1998;279(5352):837–839. DOI: 10.1126/science.279.5352.837

[66] Minami T. Substitution of transparent conducting oxide thin films for indium tin oxide transparent electrode applications. Thin Solid Films. 2008;516(7):1314–1321. DOI: 10.1016/j.tsf.2007.03.082

[67] Hanak J.J. A quantum leap in the development of new materials and devices. Applied Surface Science. 2004;223(1):1–8. DOI: 10.1016/S0169-4332(03)00902-4

[68] Bassim N.D., Schenck P.K., Otani M., Oguchi H. Model, prediction, and experimental verification of composition and thickness in continuous spread thin film combinatorial libraries grown by pulsed laser deposition. Review of Scientific Instruments. 2007;78(7):072203. DOI: 10.1063/1.2755783

[69] Widjonarko N.E., Perkins J.D., Leisch J.E., Parilla P.A., Curtis C.J., Ginley D.S. et al. Stoichiometric analysis of compositionally graded combinatorial amorphous thin film oxides using laser-induced breakdown spectroscopy. Review of Scientific Instruments. 2010;81(7):073103. DOI: 10.1063/1.3455218

[70] Cristoforetti G., De Giacomo A., Dell'Aglio M., Legnaioli S., Tognoni E., Palleschi V. et al. Local thermodynamic equilibrium in laser-induced breakdown spectroscopy: beyond the McWhirter criterion. Spectrochimica Acta Part B: Atomic Spectroscopy. 2010;65(1):86–95. DOI: 10.1016/j.sab.2009.11.005

[71] Aragón C., Aguilera J.A. Characterization of laser induced plasmas by optical emission spectroscopy: a review of experiments and methods. Spectrochimica Acta Part B: Atomic Spectroscopy. 2008;63(9):893–916. DOI: 10.1016/j.sab.2008.05.010

[72] Popescu A.C., Beldjilali S., Socol G., Craciun V., Mihailescu I.N., Hermann J. Analysis of indium zinc oxide thin films by laser-induced breakdown spectroscopy. Journal of Applied Physics. 2011;110(8):083116. DOI: 10.1063/1.3656448

[73] Socol G., Galca A.C., Luculescu C.R., Stanculescu A., Socol M., Stefan N. et al. Tailoring of optical, compositional and electrical properties of the $In_xZn_{1-x}O$ thin films obtained

by combinatorial pulsed laser deposition. Digest Journal of Nanomaterials and Biostructures. 2011;6(1):107–115.

[74] Chang H., Gao C., Takeuchi I., Yoo Y., Wang J., Schultz P.G. et al. Combinatorial synthesis and high throughput evaluation of ferroelectric/dielectric thin-film libraries for microwave applications. Applied Physics Letters. 1998;72(17):2185–2187. DOI: 10.1063/1.121316

[75] Takeuchi I., Chang K., Sharma R.P., Bendersky L.A., Chang H., Xiang X.D. et al. Microstructural properties of (Ba, Sr) TiO_3 films fabricated from BaF_2/SrF_2/TiO_2 amorphous multilayers using the combinatorial precursor method. Journal of Applied Physics. 2001;90(5):2474–2478. DOI: 10.1063/1.1388563

[76] Matsumoto Y., Takahashi R., Murakami M., Koida T., Fan X.J., Hasegawa T. et al. Ferromagnetism in Co-doped TiO_2 rutile thin films grown by laser molecular beam epitaxy. Japanese Journal of Applied Physics. 2001;40(11B):L1204–1206. DOI: 10.1143/JJAP.40.L1204

[77] Ohnishi T., Komiyama D., Koida T., Ohashi S., Stauter C., Koinuma H. et al. Parallel integration and characterization of nanoscaled epitaxial lattices by concurrent molecular layer epitaxy and diffractometry. Applied Physics Letters. 2001;79(4):536–538. DOI: 10.1063/1.1385587

[78] Fukumura T., Ohtani M., Kawasaki M., Okimoto Y., Kageyama T., Koida T. et al. Rapid construction of a phase diagram of doped Mott insulators with a composition-spread approach. Applied Physics Letters. 2000;77(21):3426–3428. DOI: 10.1063/1.1326847

[79] Chang K.S., Aronova M., Famodu O., Takeuchi I., Lofland S.E., Hattrick-Simpers J.R. et al. Multimode quantitative scanning microwave microscopy of in situ grown epitaxial $Ba_{1-x}Sr_xTiO_3$ Composition Spreads. Applied Physics Letters. 2001;79(26):4411–4413. DOI: 10.1063/1.1427438

[80] Takeuchi I., Yang W., Chang K.S., Aronova M.A., Venkatesan T., Vispute R.D. et al. Monolithic multichannel ultraviolet detector arrays and continuous phase evolution in $Mg_xZn_{1-x}O$ composition spreads. Journal of Applied Physics. 2003;94(11):7336–7340. DOI: 10.1063/1.1623923

[81] Taylor M.P., Readey D.W., Teplin C.W., van Hest M.F., Alleman J.L., Dabney M.S. et al. The electrical, optical and structural properties of $In_xZn_{1-x}O_y$ (0 ◎ x ◎ 1) thin films by combinatorial techniques. Measurement Science and Technology. 2004;16(1):90–94. DOI: 10.1088/0957-0233/16/1/012

[82] Mc Gill R.A., Chrisey D.B. Method of producing a film coating by matrix assisted pulsed laser deposition. US Patent 6025036. 2000.

[83] Piqué A., Auyeung R.C., Stepnowski J.L., Weir D.W., Arnold C.B., McGill R.A. et al. Laser processing of polymer thin films for chemical sensor applications. Surface and Coatings Technology. 2003;163–164:293–299. DOI: 10.1016/S0257-8972(02)00606-0

[84] Cristescu R., Cojanu C., Popescu A., Grigorescu S., Nastase C., Nastase F. et al. Processing of poly (1, 3-bis-(p-carboxyphenoxy propane)-co-(sebacic anhydride)) 20: 80 (P (CPP: SA) 20: 80) by matrix-assisted pulsed laser evaporation for drug delivery systems. Applied Surface Science. 2007;254(4):1160–1173. DOI: 10.1016/j.apsusc.2007.09.029

[85] Kokkinaki O., Georgiou S. Laser ablation of cryogenic films: implications to matrix-assisted pulsed laser deposition of biopolymers and dedicated applications in nanotechnology. Digest Journal of Nanomaterials and Biostructures. 2007;2(2):221–241.

[86] Itina T.E., Zhigilei L.V., Garrison B.J. Matrix-assisted pulsed laser evaporation of polymeric materials: a molecular dynamics study. Nuclear Instruments and Methods in Physics Research Section B: Beam Interactions with Materials and Atoms. 2001;180(1–4):238–244. DOI: 10.1016/S0168-583X(01)00423-2

[87] Popescu C., Roqueta J., Del Pino A.P., Moussaoui M., Nogués M.V., György E. Processing and immobilization of enzyme Ribonuclease A through laser irradiation. Journal of Materials Research. 2011;26(06):815–821. DOI: 10.1557/jmr.2010.55

[88] Dingus R.S., Scammon R.J. Grüneisen-stress-induced ablation of biological tissue. InOptics, Electro-Optics, and Laser Applications in Science and Engineering. In: Laser-Tissue Interaction II; January 20, 1991; Los Angeles. Proc. SPIE 1427; 1991. DOI: 10.1117/12.44088

[89] Cramer R., Haglund Jr. R.F., Hillenkamp F.. Matrix-assisted laser desorption and ionization in the O–H and C=O absorption bands of aliphatic and aromatic matrices: dependence on laser wavelength and temporal beam profile. International Journal of Mass Spectrometry and Ion Processes. 1997;169–170:51–67. DOI: 10.1016/S0168-1176(97)00223-1

[90] Leveugle E., Ivanov D.S., Zhigilei L.V. Photomechanical spallation of molecular and metal targets: molecular dynamics study. Applied Physics A. 2004;79(7):1643–1655. DOI: 10.1007/s00339-004-2682-2

[91] Mercado A.L., Allmond C.E., Hoekstra J.G., Fitz-Gerald J.M. Pulsed laser deposition vs. matrix assisted pulsed laser evaporation for growth of biodegradable polymer thin films. Applied Physics A. 2005;81(3):591–599. DOI: 10.1007/s00339-004-2994-2

[92] Popescu C., Popescu A.C., Iordache I., Motoc M., Pojoga D., Simon-Gruita A. et al . Structure and enzymatic activity of laser immobilized ribonuclease A. Journal of Materials Science. 2014;49(12):4371–4378. DOI: 10.1007/s10853-014-8136-0

[93] Neault J.F., Diamantoglou S., Beauregard M., Nafisi S., Tajmir-Riahi H.A. Protein unfolding in drug-RNase complexes. Journal of Biomolecular Structure and Dynamics. 2008;25(4):387–394. DOI: 10.1080/07391102.2008.10507187

[94] Georg H, Wharton CW, Siebert F. Temperature induced protein unfolding and folding of RNase A studied by time-resolved infrared spectroscopy. Laser Chemistry. 1999;19(1–4):233–235. DOI: 10.1155/1999/28202

[95] Kong J., Yu S. Fourier transform infrared spectroscopic analysis of protein secondary structures. Acta Biochimica et Biophysica Sinica. 2007;39(8):549–559. DOI: 10.1111/j.1745-7270.2007.00320.x

[96] Califano V., Bloisi F., Vicari L.R., Colombi P., Bontempi E., Depero L.E . MAPLE deposition of biomaterial multilayers. Applied Surface Science. 2008;254(22):7143–7148. DOI: 10.1016/j.apsusc.2008.05.295

[97] Cristescu R., Popescu C., Popescu A., Grigorescu S., Mihailescu I.N., Mihaiescu D. et al. Functional polyethylene glycol derivatives nanostructured thin films synthesized by matrix-assisted pulsed laser evaporation. Applied Surface Science. 2009;255(24):9873–9876. DOI: 10.1016/j.apsusc.2009.04.110

[98] Cristescu R., Popescu C., Popescu A.C., Grigorescu S., Duta L., Mihailescu I.N. et al. Laser processing of polyethylene glycol derivative and block copolymer thin films. Applied Surface Science. 2009;255(10):5605–5610. DOI: 10.1016/j.apsusc.2008.09.060

[99] Paun I.A., Moldovan A., Luculescu C.R., Dinescu M. Antibacterial polymeric coatings grown by matrix assisted pulsed laser evaporation. Applied Physics A. 2013;110(4):895–902. DOI: 10.1007/s00339-012-7193-y

[100] Grumezescu V., Holban A.M., Iordache F., Socol G., Mogoşanu G.D., Grumezescu A.M. et al. MAPLE fabricated magnetite@eugenol and (3-hidroxybutyric acid-co-3-hidroxyvaleric acid)–polyvinyl alcohol microspheres coated surfaces with anti-microbial properties. Applied Surface Science. 2014;306:16–22. DOI: 10.1016/j.apsusc.2014.01.126

[101] Cristescu R., Popescu C., Popescu A.C., Grigorescu S., Duta L., Mihailescu I.N. et al. Functionalized polyvinyl alcohol derivatives thin films for controlled drug release and targeting systems: MAPLE deposition and morphological, chemical and in vitro characterization. Applied Surface Science. 2009;255(10):5600–5604. DOI: 10.1016/j.apsusc.2014.01.126

[102] Ge W., Yu Q., López G.P., Stiff-Roberts A.D . Antimicrobial oligo (p-phenylene-ethynylene) film deposited by resonant infrared matrix-assisted pulsed laser evaporation. Colloids and Surfaces B: Biointerfaces. 2014;116:786–792. DOI: 10.1016/j.colsurfb.2014.01.033

[103] Mihaiescu D.E., Cristescu R., Dorcioman G., Popescu C.E., Nita C., Socol G. et al. Functionalized magnetite silica thin films fabricated by MAPLE with antibiofilm properties. Biofabrication. 2012;5(1):015007. DOI: 10.1088/1758-5082/5/1/015007

[104] Iordache S., Cristescu R., Popescu A.C., Popescu C.E., Dorcioman G., Mihailescu I.N. et al. Functionalized porphyrin conjugate thin films deposited by matrix assisted pulsed laser evaporation. Applied Surface Science. 2013;278:207–210. DOI: 10.1016/j.apsusc.2013.01.080

[105] Verrastro M., Cicco N., Crispo F., Morone A., Dinescu M., Dumitru M. et al. Amperometric biosensor based on Laccase immobilized onto a screen-printed electrode by

Matrix Assisted Pulsed Laser Evaporation. Talanta. 2016;154:438–445. DOI: 10.1016/j.talanta.2016.03.072

[106] György E., Sima F., Mihailescu I.N., Smausz T., Hopp B., Predoi D. et al. Biomolecular urease thin films grown by laser techniques for blood diagnostic applications. Materials Science and Engineering: C. 2010;30(4):537–541. DOI: 10.1016/j.msec.2010.02.003

[107] Purice A., Schou J., Kingshott P., Dinescu M. Production of active lysozyme films by matrix assisted pulsed laser evaporation at 355 nm. Chemical Physics Letters. 2007;435(4–6):350–353. DOI: 10.1016/j.cplett.2006.12.078

[108] Bloisi F., Cassinese A., Papa R., Vicari L., Califano V. Matrix-assisted pulsed laser evaporation of polythiophene films. Thin Solid Films. 2008;516(7):1594–1598. DOI: 10.1016/j.tsf.2007.03.159

[109] Tunno T., Caricato A.P., Caruso M.E., Luches A., Martino M., Romano F. et al. Matrix-assisted pulsed laser evaporation of polyfluorene thin films. Applied Surface Science. 2007;253(15):6461–6464. DOI: 10.1016/j.apsusc.2007.01.024

[110] Pate R., Lantz K.R., Stiff-Roberts A.D. Resonant infrared matrix-assisted pulsed laser evaporation of CdSe colloidal quantum dot/poly [2-methoxy-5-(2′-ethylhexyloxy)-1, 4-(1-cyano vinylene) phenylene] hybrid nanocomposite thin films. Thin Solid Films. 2009;517(24):6798–6802. DOI: 10.1016/j.tsf.2009.06.018

[111] Bjelajac A., Petrovic R., Socol G., Mihailescu I.N., Enculescu M., Grumezescu V. et al. CdS quantum dots sensitized TiO_2 nanotubes by matrix assisted pulsed laser evaporation method. Ceramics International. 2016;42(7):9011–9017. DOI: 10.1016/j.ceramint.2016.02.159

[112] Hunter C.N., Check M.H., Bultman J.E., Voevodin A.A. Development of matrix-assisted pulsed laser evaporation (MAPLE) for deposition of disperse films of carbon nanoparticles and gold/nanoparticle composite films. Surface and Coatings Technology. 2008;203(3):300–306. DOI: 10.1016/j.surfcoat.2008.09.003

Chapter 2

Influence of Alkali Metal Ions on Luminescence Behaviour of $Ca_{0.5}R_{1-x}(MoO_4)_2$:xLn^{3+} (R = Y, La), (Ln = Eu, Tb, Dy) Pulsed Laser Deposited Thin Phosphor Films

Jagannathan Thirumalai,
Venkatakrishnan Mahalingam and
Rajagopalan Krishnan

Additional information is available at the end of the chapter

http://dx.doi.org/10.5772/65006

Abstract

Thin phosphor films of $Ca_{0.5}R_{1-x}(MoO_4)_2$:xLn^{3+}, M^+ (R^{3+} = La, Y), (Ln^{3+} = Eu, Tb, Dy) (M^+ = Li^+, K^+ and Na^+) were deposited on quartz substrates by pulsed laser deposition (PLD) technique by ablation of a stoichiometric monocrystal target. The deposition was carried out using an Nd-YAG laser (λ = 1064 nm) in an ultra-high vacuum (UHV) with an oxygen back pressure of 300 mTorr at 600°C substrate temperatures. The laser-ablated films are optically active, as verified by the photoluminescence (PL) spectra, and the films exhibit smooth Stark levels. The photoluminescence of the $Ca_{0.5}R_{1-x}(MoO_4)_2$:xLn^{3+}, M^+ phosphors properties reveals characteristic visible emissions. Further, the co-doping of alkali metal chlorides MCl (M = Na, K, Li) into the $Ca_{0.5}R_{1-x}(MoO_4)_2$:xLn^{3+}, M^+ phosphor greatly improves the luminescence intensity, which can be explained by charge compensation effect. The fluorescence lifetime and photometric coordinates are discussed in detail.

Keywords: rare-earth and alkaline activated, thin phosphor films, pulsed laser deposition, surface morphology, luminescence

1. Introduction

The growth of good quality larger area thin film with homogeneous size distribution and morphology is still a demanding issue, and it is of significant attention towards the research fraternity. Remarkably, uniform micro/nano-structures have been paying global attention due to its potential application in high-performance luminescence and opto-electronic device based on community their novel optical and electronic properties. To synthesize novel thin film materials such as molybdates [1, 2], tungstates [3], vanadates [4], and fluorides [5], copious prominent techniques are extensively adopted, for example chemical bath deposition (CBD), successive immersion layer adsorption reaction (SILAR), polymerization, electrodeposition, sputtering, metal-organic chemical vapour deposition (MO-CVD), molecular beam epitaxy (MBE), atomic layer deposition (ALD), pulsed laser deposition (PLD). In the midst of all, PLD is a multitalented method to prepare multiconstituent thin film materials in which raster examining of high-energy pulsed laser ablates the target material and produces the plasma plume [6, 7]. In recent times, the PLD technique has created a widespread usage with an exceptionally astonishing result in materials preparation and fabrication of a device in the opto-electronics field. Albeit the fabrication of optical quality of the thin films and waveguides using PLD technique with various technical hitches and burning issues, till date, these issues have undeniably been lucratively conquered, and quite a few good-quality thin films were grown by PLD.

1.1. Recent research scenario

In contrast to conventional incandescent and fluorescent lamps, white light emitting diode (w-LEDs) is seemed to be an optimistic solid-state light source with a good-quality energy conversion luminescence device [8]. By coalesce into the GaN blue LED chip with yellow emitting phosphor YAG:Ce^{3+} which yield a white light emission using the conventional technique [9–11]. Nevertheless, the deficient of red emission cog ends with low colour rendering index (CRI) and luminous efficacy of radiation (LER) which restricts their pertinence towards a few ambits [9, 12]. To conquer this hindrance, red or orange-red emitting ion such as Pr^{3+}, Sm^{3+}, Cr^{3+} and Mn^{2+} is co-doped with YAG:Ce^{3+} [10]. The other one is combining YAG:Ce^{3+} with red or orange-red phosphors such as nitrides (M_2SiN_8:Eu^{2+}), sulphides (CaS:Eu^{2+}), oxynitrides ($MSi_2O_2N_2$:Eu^{2+}) (where M = Ca, Sr) [10]. Furthermore, the enhancement of intense emission in the host material can be engendered by co-doping of alkali metal-chloride results in strong emission, which may possibly be an opportune and a generally suitable approach to acquire the phosphors with sufficient intensity and excellent efficiency are a great essential deal for prospective solid-state lighting devices [2]. Therefore, it is necessary to discover a suitable phosphor material with a sufficient chemical permanence with enhanced efficiency. Rare earth-doped phosphor materials are paying attention towards the research problems based on its applications in all the prospects of science and technology. Molybdates and tungstates with metallic elements form an essential class of phosphor materials. They belong to the scheelite family having a space group $I_{41/a}$. In both molybdate and tungstate family, the alkaline earth-based rare-earth-activated double molybdates are very much highly significant efficient materials on the basis of its unique structural, optical properties have come across profound applications in technological aspects. Alkaline rare-earth-activated tungstates having a general formula ARE $(MoO_4)_2$ (RE = Y, La; A = Ba, Ca, Sr) are considered as better luminescent hosts investigated significantly for various purposes such as photocatalysts [11], displays [8] and acquire substantial hydrolytic and thermal permanence. Furthermore, the electroluminescent devices in the form of thin films from these micro/nano-architectures are to be built for the white light emitting diode applications. Among the aforesaid variety of fabrication of thin film techniques, pulsed laser deposition (PLD) is a viable method [2, 13, 14]. Nowadays, to fabricate homogeneous and large-scale thin films, laser rastering system attached into PLD technique has been used [13]. To consider the aspects of application, aforesaid reasons could make the PLD technique most unique and almost suitable for the growth and fabrication of good-quality micro/nano-thin films. It is interesting that the structural, optical, and photophysical properties of the micro/nano-architectures could be compared with the thin film phosphors and its bulk [2, 15] counter-parts.

In this viewpoint, we have prepared the single crystalline nano-thin phosphor films of $Ca_{0.5}R_{1-x}(MoO_4)_2$:xLn^{3+} (R^{3+} = Y, La), (Ln^{3+} = Eu, Tb, Dy) with co-doping of alkali metal chlorides (0.02 M of LiCl, KCl, NaCl) have been ablated on quartz substrates using the pulsed laser deposition method (PLD). For the first time, the luminescence properties of these alkali chloride-activated phosphors are studied. The as-prepared molybdate and tungstate powders were further deposited as thin films using the laser-ablation by forming a ceramic target under oxygen atmosphere. Followed by the as-prepared samples was analyzed using X-ray diffraction (XRD), atomic force microscopy (AFM), field emission scanning electron microscopy (FESEM), photoluminescence (PL) spectroscopy, photometric characteristics using commission internationale de i'eclairage (CIE) diagrams. The colour chromaticity coordinates and luminescence decay times have also been determined and discussed in reference to the effect of alkali metal ions.

2. Experimental details

By employing the PLD technique, for the first time, the nano-thin phosphor films of $Ca_{0.5}R_{1-x}(MoO_4)_2$:xLn^{3+} (x = 0.16 M) (R^{3+} = Y, La), (Ln^{3+} = Eu, Tb, Dy) with co-doping of alkali metal chlorides (0.02 M of LiCl, KCl, NaCl) were effectively coated on the quartz substrates by maintaining the substrate temperature of 600°C under oxygen atmospheric pressure (~300 mTorr). **Figure 1a–i** shows the images of different experimental procedures engaged for the growth of thin films.

2.1. Preparation of ceramic target, cleaning of substrates and growth of $Ca_{0.5}R_{1-x}(MoO_4)_2$:xLn^{3+} nano-thin phosphor films

To prepare a strong and extremely impenetrable ceramic (molybdate and tungstate) target for laser ablation, the starting precursors such as Na_2CO_3, La_2O_3, Y_2O_3, MoO_3 and Ln_2O_3 (Ln = Eu, Tb, Dy) were taken in stoichiometric ratios along with 0.02 M of alkali chlorides (LiCl, KCl and NaCl), followed by using the agate mortar pestle the powders were grounded for 2 h. The doping concentrations of Ln^{3+} were optimized in our previous work [15] and kept at constant (0.16 M) for all the Ln^{3+} ions. Without using any binders, the homogeneously mixed powders were pressed and pelletized in the form of disk (pellet) at a pressure of 6 tons. By eliminating the unstable contaminants, shun pores, crack, and endorse densification, to promote the diffusion in atomic level all through the preparation of target [16]. Furthermore, the as-prepared pellet was annealed at 900°C for 3 h to achieve a very strong, stable and thick pellet having a diameter of about 2.5 cm, and thickness of about 0.4 cm is attained. Then, the annealed target is used for laser ablation.

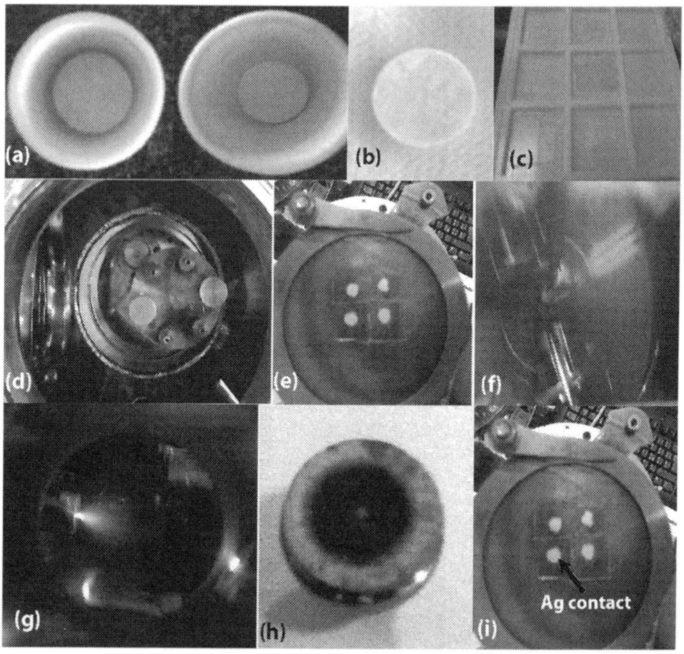

Figure 1. Photograph of different experimental methods demonstrating (a) homogeneously pelletized ceramic target, (b) high-temperature annealed target, (c) cleaning of substrates, (d) target loading, (e) loading of substrates, (f) impurities removal in UHV chamber (glow discharge), (g) laser ablation, (h) target after ablation, (i) thin films after deposition.

The procedure for predeposition cleaning of substrates and growth of nano-thin phosphor films were already discussed in detail on our previous work [2].

2.2. Characterization

The morphology of the product was analysed by field emission scanning electron microscope (FESEM-SUPRA 55). Using atomic force microscopy (NTEGRA PRIMA-NTMDT, USA), the surface topography of the thin phosphor films was studied. The crystal structure and phase purity of as-synthesized phosphor were recognized and confirmed by PANalytical's X'Pert PRO Materials Research X-ray Diffractometer (Almelo, USA) equipped with a CuKα radiation (λ = 0.154060 Å) at a scanning rate of 0.02°s^{-1} in a 2θ range of 15°–60°. Further, down-conversion PL excitation and emission studies and fluorescence decay time measurements were performed at room temperature using a Cary bench-top spectrophotometer (AGILENT Instruments, USA).

http://dx.doi.org/10.5772/65006

3. Results and discussion

3.1. Morphological and X-ray diffraction analysis

Figure 2a and **b** shows the scanning electron microscopy (SEM) images of $Ca_{0.5}R_{1-x}(MoO_4)_2{:}xLn^{3+}$ (R^{3+} = Y, La), (Ln^{3+} = Eu, Tb, Dy) thin phosphor film co-doped with Li^+, K^+ and Na^+ metal ions prepared at 600°C with 300 mTorr. The as-prepared phosphor film comprises homogeneous nearly circular grains with a typical grain size around 250 nm. Based on the AFM studies, the surface topography, line profile and roughness of the prepared thin film was estimated. **Figures 3a** and **4a** show the 3D AFM image of thin film of $Ca_{0.5}R(MoO_4)_2{:}Eu^{3+}$ (R = Y, La), and the scan was performed on 3 × 3 and 5 × 5 μm² area, respectively. From the 3D topographic image, the as-prepared thin phosphor film comprises of polished surface with uniform arrangement of the particles and with less agglomeration. The average roughness along with root-mean-square (rms) of the as-grown thin film was determined as 24.72 and 26.84 nm for $Ca_{0.5}R_{1-x}(MoO_4)_2{:}xEu^{3+}$ (R^{3+} = Y, La). The 3D surface topography, 2D surface scan image, line profile, and histogram analysis are shown in **Figures 3a–d** and **4a–d**, respectively. Based on the scaling process, the particle size would be reduced to nano-scale in thin film which could be efficiently used for display applications. The optimization of maintaining different substrate temperature on rare-earth doped phosphors has been well examined and reported in our earlier work [2].

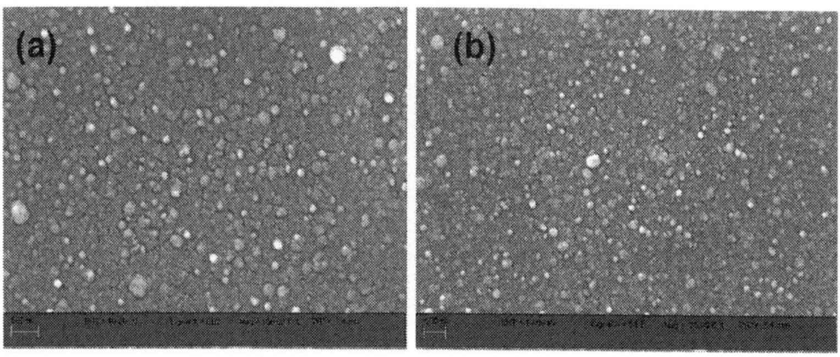

Figure 2. The SEM image (a and b) of $Ca_{0.5}R_{0.5}(MoO_4)_2{:}Eu^{3+}$ (R^{3+} = Y, La) thin film.

The crystallinity and phase purity of the prepared products were examined using indexed powder X-ray diffraction patterns **Figure 5(a** and **b)** for as grown thin film samples. The compound $Ca_{0.5}Y_{(1-x)}(MoO_4)_2{:}xRE^{3+}$ crystallizes in the scheelite tetragonal crystal structure with a space group of $I_{41/a}$. The unit cell of $Ca_{0.5}Y_{(1-x)}(MoO_4)^2$ consists of $[MoO_4]^{2-}$ anions and Ca^{2+} and Y^{3+}/La^{3+} cations. In this phase, Mo sites are occupied by the molybdenum atoms (Mo^{6+}) and located at the centres of tetrahedron and surrounded by four equivalent oxygen (O^{2-}) atoms. The divalent Ca^{2+} and trivalent Y^{3+}/La^{3+} occupies the dodecahedral sites associated with the tetrahedral symmetry. The degree of dodecahedral and tetrahedral distortions is discussed in the previous work [2, 15, 17].

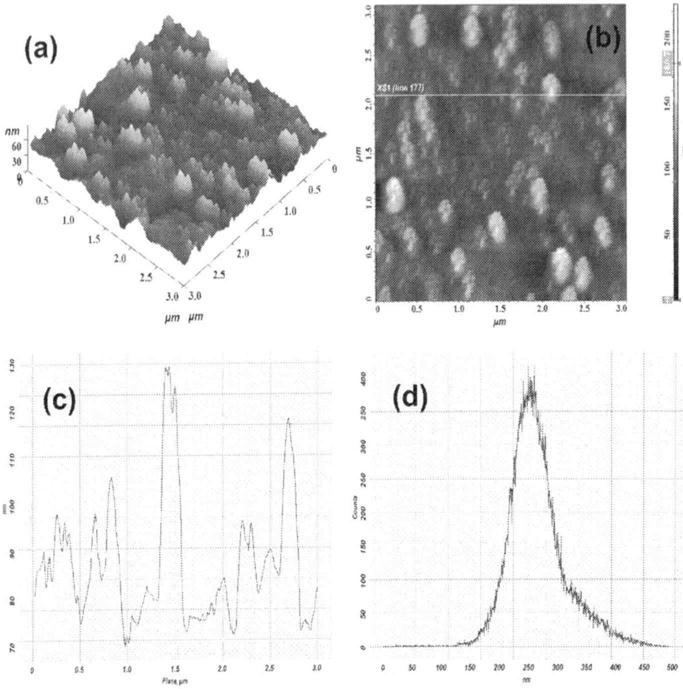

Figure 3. The AFM images of $Ca_{0.5}Y(MoO_4)_2:Eu^{3+}$ thin film (a) 3D surface topography, (b) 2D surface scan image, (c) line profile for vertical cross-section in 3 × 3 μm² scan area and (d) histogram analysis.

In the powder XRD pattern, all the peaks are indexed perfectly which indicates a pure tetragonal phase having scheelite crystal structure and the planes (1 0 1), (1 1 2), (0 0 4), (2 0 0), (2 0 4), (2 2 0), (1 1 6) and (1 3 2) are in well accordance with the JCPDS card no. 82-2369 of $NaY(MoO_4)_2$. No deleterious phases are found. The peak shift is not noticed with respect to doping. An intense peak with plane (1 1 2) is found at 28.95° [2, 15, 17].

3.2. Photoluminescence properties of laser-ablated thin phosphor films: $Ca_{0.5}R_{1-x}(MoO_4)_2$:$x$$Ln^{3+}$,$M^+$ (R = Y, La; Ln = Eu, Tb and Dy; M = Li, K and Na)

3.2.1. Enhancement of luminescence intensity by the persuade of alkali metal ions in $Ca_{0.5}R_{1-x}(MoO_4)_2$:$x$$Ln^{3+}$,$M^+$

In the phosphor material, by introducing the alkali metal chlorides, nitrates or fluorides, which substantially increase the luminescence intensity owing to the charge compensation effect between unequal ions [18]. On our previous work in $Ca_{0.5}R_{1-x}(MoO_4)_2$:$x$$Ln^{3+}$ (R = Y, La) phosphor, the doping of alkali ions appreciably improves the emission properties by charge compensation using solid-state reaction method [15, 19].

http://dx.doi.org/10.5772/65006

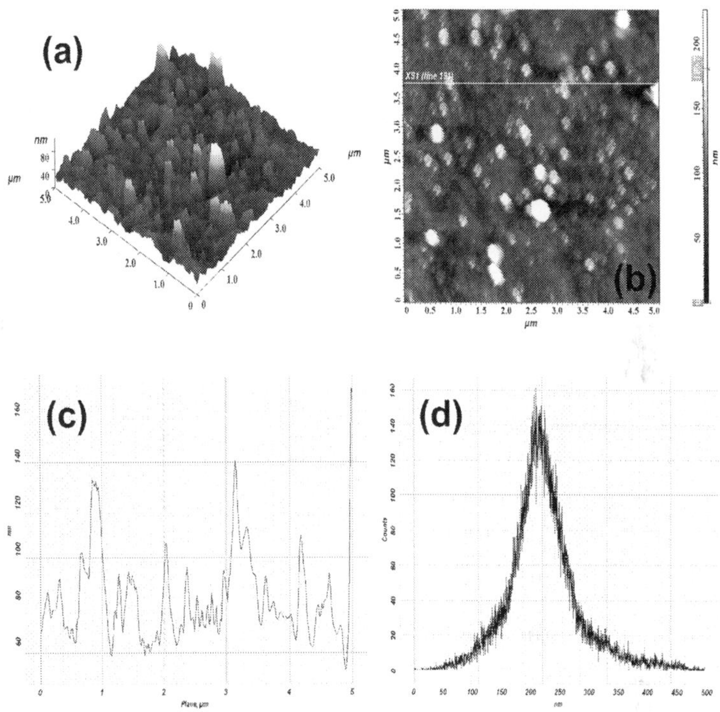

Figure 4. The AFM images of $Ca_{0.5}La(MoO_4)^2$:Eu^{3+} thin film (a) 3D surface topography, (b) 2D surface scan image, (c) line profile for vertical cross-section in 5 × 5 μm² scan area and (d) histogram analysis.

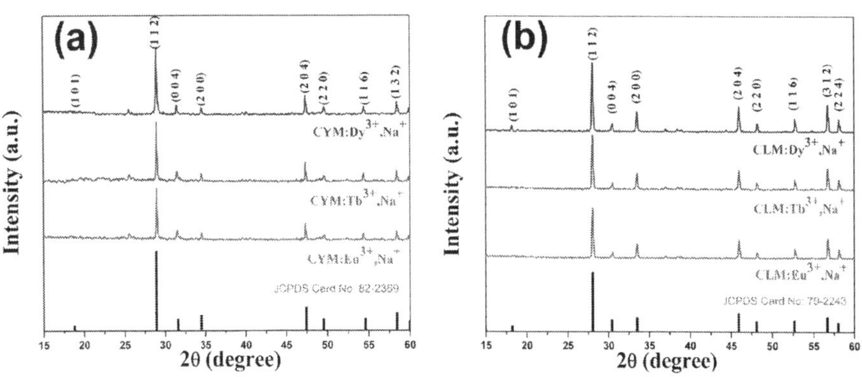

Figure 5. XRD patterns of (a) $Ca_{0.5}Y(MoO_4)_2$:Ln^{3+},Na^+, (b) $Ca_{0.5}La(MoO_4)_2$:Ln^{3+},Na^+ (Ln = Eu, Tb and Dy).

In our system, Eu^{3+}, Tb^{3+}, Dy^{3+} and M^+ co-doped in $Ca_{0.5}R(MoO_4)_2$ (R = Y, La) matrix would induce a distortion in lattice, and consequently, the lattice symmetry is desperately lowered [20]. The co-doped Eu^{3+}, Tb^{3+}, Dy^{3+} and M^+ at the Ca^{2+} sites in prepared thin film samples would play a role of dominance with enhanced luminescence intensity [21]. This is due to altering the symmetry and their surroundings in the locality of rare earth ions by adding the charge compensators of alkali metal ions [22]. **Figures 6** and **7** show the PL emission spectra of $Ca_{0.5}Y_{1-x}(MoO_4)_2$:xLn^{3+},M^+ (Ln = Eu, Tb and Dy; M = Li, K and Na) and $Ca_{0.5}La_{1-x}(MoO_4)_2$:$xLn^{3+}$, M^+ (Ln = Eu, Tb and Dy; M = Li, K and Na) thin film phosphors. The luminescence emission intensity is deliberately increased for Na^+ ion co-doped $Ca_{0.5}R_{1-x}(MoO_4)_2$:xLn^{3+} (R = Y, La; Ln = Eu, Tb and Dy). This could be owing to the charge compensation effect, and the proposed mechanism is $Ca^{2+} \rightarrow 2M^+$ ($M = Li^+, K^+, Na^+$) [21, 22]. Furthermore, the ionic radius of Na^+ (0.97 Å) is closer to the Ca^{2+} (1.12 Å) which is somewhat better than that of K^+ (1.33 Å) and Li^+ (0.59 Å) [3, 15]. Hence, there is an efficient replacement of Ca^{2+} ions by alkali metal ions. This forms the basis for the increased luminescence intensity, and obviously, Na^+ is having the best charge compensation effect. The electronic configurations and its transitions are explained in the subsequent sections.

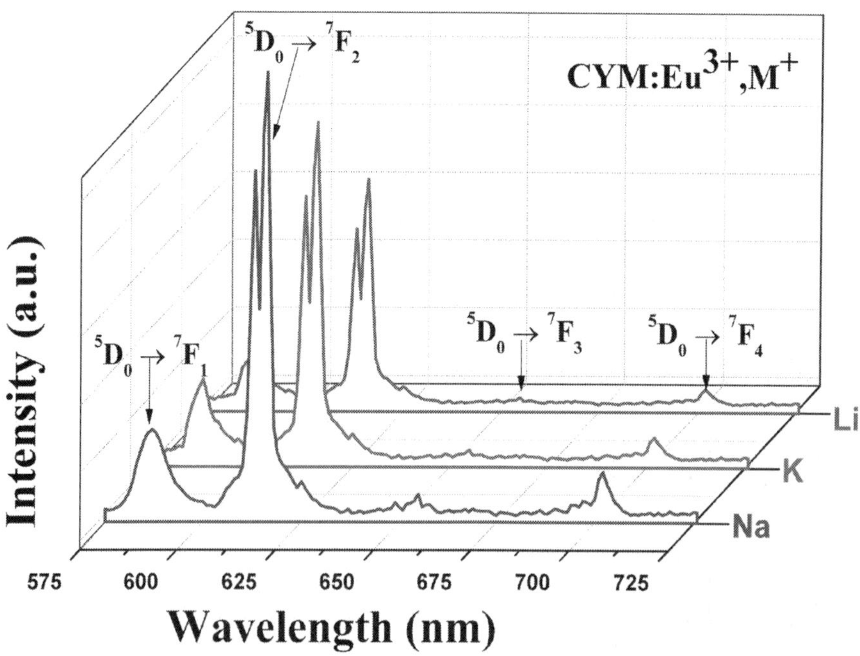

Figure 6. PL emission spectrum of the thin film phosphor $Ca_{0.5}Y(MoO_4)_2$:Eu^{3+} co-doped with various alkali metal ions [Li^+(Green), K^+(Red) and Na^+(Blue)].

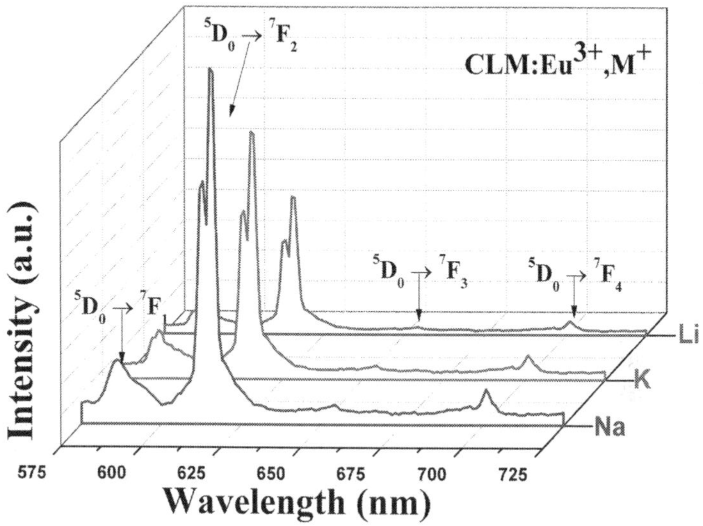

Figure 7. PL emission spectrum of the thin film phosphor $Ca_{0.5}La(MoO_4)_2$:Eu^{3+} co-doped with various alkali metal ions [Li^+(Green), K^+(Red) and Na^+(Blue)].

3.2.2. Photoluminescence excitation studies

3.2.2.1. $Ca_{0.5}Y_{1-x}(MoO_4)_2$:xLn^{3+},Na^+ (Ln = Eu, Tb and Dy)

Figure 8a shows the room temperature PL excitation spectra of $Ca_{0.5}Y_{1-x}(MoO_4)_2$:xLn^{3+} (Ln = Eu, Tb and Dy) thin film phosphors. The PL excitation spectrum of $Ca_{0.5}Y_{1-x}(MoO_4)_2$:xEu^{3+},Na^+ is having a wavelength range of 225–575 nm. It is showing two regions with a broad band and intense sharp peaks. The broad band is located from 225 to 350 nm with a centre at ~306 nm attributed to the O^{2-} to Eu^{3+} charge transfer band (CTB) and also designated as ligand-to-Eu^{3+} metal charge transfer transitions (LMCT) [23, 24]. Above 350 nm, intense sharp peaks are found at 362 nm ($^7F_0 \rightarrow {}^5D_4$), 382 nm ($^7F_0 \rightarrow {}^5L_7$), 395 nm ($^7F_0 \rightarrow {}^5L_6$), 416 nm ($^7F_0 \rightarrow {}^5D_3$), 465 nm ($^7F_0 \rightarrow {}^5D_2$) and 536 nm ($^7F_0 \rightarrow {}^5D_1$). Among which, the strong and intense peak is found at 395 nm. The characteristic configurations were attributed to the transition from the $^7F_0$ ground state of Eu^{3+} to the upper excited states ($^5D_{1,2,3,4}$ and $L_{6,7}$). This strongest peak in UV region is more suitable for exciting Eu^{3+} ions.

Figure 8b depicts the room temperature excitation spectrum of $Ca_{0.5}Y_{1-x}(MoO_4)_2$:xTb^{3+},Na^+ with a wavelength range of 270–390 nm in the UV region. It consists of two regions. One is from 270 to 360 nm, a highly intense and wide band designated as charge transfer band (CTB) having centred at ~295 nm is ascribed to the charge transfer of molybdate host lattice [17]. The other is due to f-f transitions of Tb^{3+} and its peak is at 376 nm ($^7F_6 \rightarrow {}^5G_6$) which is less intense than CTB. The energy transfer is being occurred from $4f^8$ to $4f^75d$ configuration of Tb^{3+} ions [17].

The excitation spectrum for the $Ca_{0.5}Y_{1-x}(MoO_4)_2$:$x$$Dy^{3+}$,$Na^+$ thin film phosphor is shown in **Figure 8c**. The wavelength of the excitation spectrum ranges from 240 to 480 nm. The strong broad band is ranging from 240 to 340 nm bears a centre at ~270 nm. Above 340 nm, the numerous intense f-f transitions of Dy^{3+} ions are found. The f-f transitions are located at 353 nm ($^6H_{15/2} \rightarrow {}^6P_{7/2}$), 367 nm ($^6H_{15/2} \rightarrow {}^6P_{5/2}$), 388 nm ($^6H_{15/2} \rightarrow {}^4I_{3/2}$), 425 nm ($^6H_{15/2} \rightarrow {}^4G_{11/2}$), 453 nm ($^6H_{15/2} \rightarrow {}^4I_{15/2}$) and 475 nm ($^6H_{15/2} \rightarrow {}^4F_{9/2}$) [25]. The highly intense peak is found at 353 nm which is the best candidate for exciting Dy^{3+} ions.

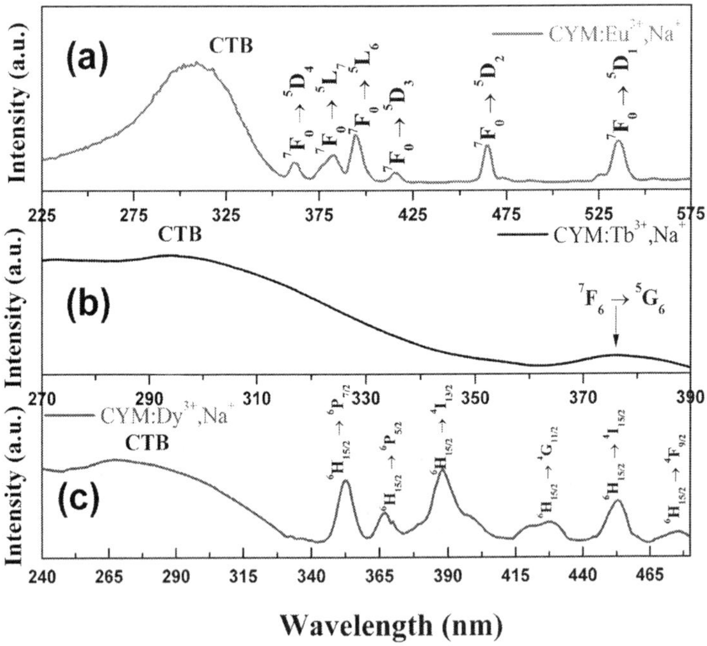

Figure 8. PL excitation spectrum (a, b, and c) for the thin film phosphors $Ca_{0.5}Y(MoO_4)_2$:Ln3+ (Ln = Eu, Tb and Dy).

3.2.2.2. *$Ca_{0.5}La_{1-x}(MoO_4)_2$:$xLn^{3+}$,$Na^+$ (Ln = Eu, Tb and Dy)*

The room temperature PL excitation spectra of $Ca_{0.5}La_{1-x}(MoO_4)_2$:$x$$Ln^{3+}$ (Ln = Eu, Tb and Dy) thin phosphor films are illustrated in **Figure 9a**. The PL excitation spectrum of $Ca_{0.5}Y_{1-x}(MoO_4)_2$:$x$$Eu^{3+}$,$Na^+$ comprises of wavelength range of 200–550 nm. It consists of two regions with a wide band and highly intense sharp peaks. The wide band is found from 220 to 350 nm with a centre at ~278 nm ascribed to the O^{2-} to Eu^{3+} ligand-to-Eu^{3+} metal charge transfer transitions (LMCT) [19]. Above 350 nm, intense sharp peaks are found at 360 nm ($^7F_0 \rightarrow {}^5D_4$), 382 nm ($^7F_0 \rightarrow {}^5L_7$), 394 nm ($^7F_0 \rightarrow {}^5L_6$), 415 nm ($^7F_0 \rightarrow {}^5D_3$), 464 nm ($^7F_0 \rightarrow {}^5D_2$) and 535 nm ($^7F_0 \rightarrow {}^5D_1$). In that, the strongest and highly intense peak is found at 394 nm. This strongest peak in UV region is good for exciting Eu^{3+} ions.

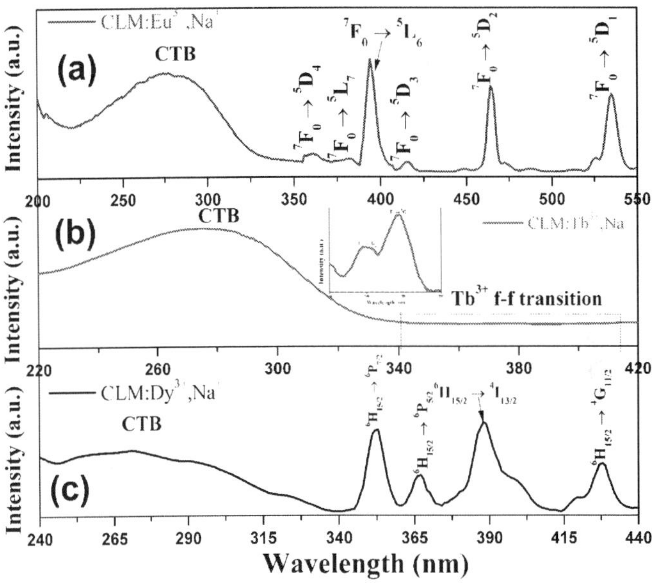

Figure 9. PL excitation spectrum (a, b, and c) for the thin film phosphors $Ca_{0.5}La(MoO_4)_2{:}Ln^{3+}$ (Ln = Eu, Tb and Dy).

Figure 9b shows that the photoluminescence excitation spectrum of $Ca_{0.5}La_{1-x}(MoO_4)_2{:}xTb^{3+},Na^+$ possess a wavelength range of 220–420 nm in the UV region. Among the two regions, the broad region is from 220 to 340 nm, ascribed to charge transfer band (CTB) which is centred at ~278 nm. The other sharp peaks are due to f-f transitions of Tb^{3+} with peaks at 369 nm ($^7F_6 \rightarrow {}^5G_5$) and 378 nm ($^7F_6 \rightarrow {}^5G_6$) is having lesser intensity than charge transfer band (CTB) [19].

The room temperature excitation spectrum for the $Ca_{0.5}La_{1-x}(MoO_4)_2{:}xDy^{3+},Na^+$ thin film phosphor is depicted in **Figure 9c**. The range of the excitation spectrum is from 240 to 440 nm. The broad band ranges from 240 to 340 nm bears a centre at ~271 nm. After 340 nm, a number of highly intense f-f transitions of Dy^{3+} ions are located. The f-f transitions are found at 352 nm ($^6H_{15/2} \rightarrow {}^6P_{7/2}$), 367 nm ($^6H_{15/2} \rightarrow {}^6P_{5/2}$), 388 nm ($^6H_{15/2} \rightarrow {}^4I_{3/2}$) and 428 nm ($^6H_{15/2} \rightarrow {}^4G_{11/2}$) [19]. The most intense peak that is situated at ~352 nm is fit for exciting Dy^{3+} ions.

3.2.3. Photoluminescence emission studies

3.2.3.1. $Ca_{0.5}Y_{1-x}(MoO_4)_2{:}xLn^{3+},Na^+$ (Ln = Eu, Tb and Dy)

The room temperature PL emission spectra for $Ca_{0.5}Y_{1-x}(MoO_4)_2{:}xLn^{3+},Na^+$ (Ln = Eu, Tb and Dy) thin phosphor films are shown in **Figure 10a**. The emission spectra monitored at 395 nm UV excitation for $Ca_{0.5}Y_{1-x}(MoO_4)_2{:}xEu^{3+},Na^+$ illustrates a number of intra-configurational 4f-4f transitions arising from Eu^{3+} $^5D_0$ excited state to the ^7F_J (J = 1, 2, 3 and 4) ground states [26–28]. Upon other excitations and also with LMCT, there is no significant change in emission spectra.

The strong and most intense emission peak is found at 616 nm upon 395 nm UV excitation is ascribed to the $^5D_0 \rightarrow {}^7F_2$ electric-dipole transition depicts hypersensitive red emission which is parity forbidden ($\Delta J = 2$) [26]. Also, it shows two sub-peaks arises due to Stark energy splitting, that is (2J + 1) Stark components of J-degeneracy splitting [27]. The predominant electric-dipole transition signifies that Eu^{3+} ions are found at sites without inversion symmetry. The other transitions found at 587 nm ($^5D_0 \rightarrow {}^7F_1$) show orange emission owing to magnetic-dipole transition. The other relatively weaker transitions are located at 655 nm ($^5D_0 \rightarrow {}^7F_3$) and 702 nm ($^5D_0 \rightarrow {}^7F_4$) [28]. The red emission to orange emission (R/O) ratio for the thin film phosphor is 5.2536. From these findings, it is evident that $Ca_{0.5}Y_{1-x}(MoO_4)_2{:}xEu^{3+},Na^+$ possess scheelite tetragonal structure having C_{3v} site symmetry could be used for display applications.

The PL emission spectrum for $Ca_{0.5}Y_{1-x}(MoO_4)_2{:}xTb^{3+},Na^+$ upon ~295 nm UV excitation is shown in **Figure 10b**, comprises four PL emission bands having peaks at 489 nm ($^5D_4 \rightarrow {}^7F_6$), 545 nm ($^5D_4 \rightarrow {}^7F_5$), 587 nm ($^5D_4 \rightarrow {}^7F_4$) and 621 nm ($^5D_4 \rightarrow {}^7F_3$). In these emission peaks, the highly remarkable green colour is located at 545 nm related to the predominant transition $^5D_4 \rightarrow {}^7F_5$ [27]. It is due to the energy transfer from the host which is populating only $^5D_4$ level. The dominant emission peak which is having two sub peaks is due to the Stark energy splitting and forms the suitable candidate for display applications.

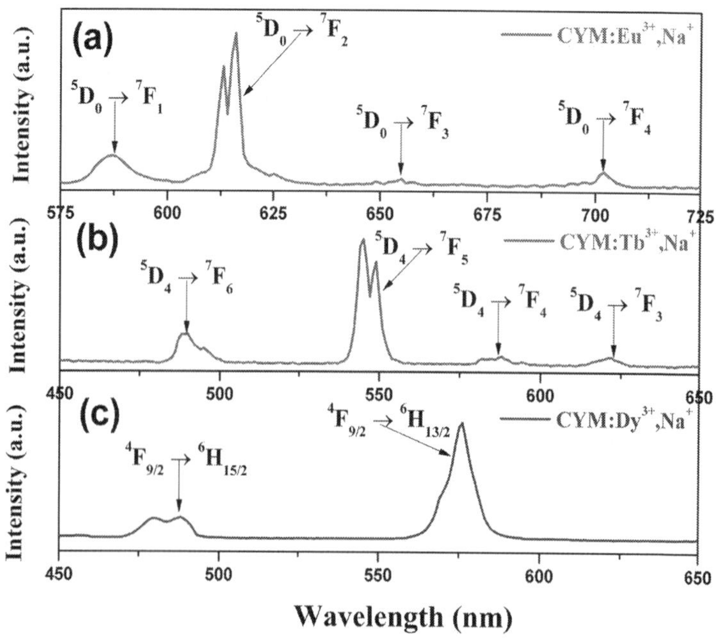

Figure 10. PL emission spectra (a, b, and c) of the thin film phosphors $Ca_{0.5}Y(MoO_4)_2{:}Ln^{3+},M^+$ (Ln = Eu, Tb and Dy; M = Na).

http://dx.doi.org/10.5772/65006

The room temperature PL emission spectrum (**Figure 10c**) for the thin film phosphor $Ca_{0.5}Y_{1-x}(MoO_4)_2$:xDy^{3+},Na^+ excited upon 353 nm excitation wavelength. The emission spectrum consists of two major peaks with respective peaks at 485 nm ascribed to magnetic dipole transition of ($^4F_{9/2} \rightarrow {}^6H_{15/2}$) and 576 nm related to forced electric dipole transition of $^4F_{9/2} \rightarrow {}^6H_{13/2}$. The magnetic dipole transition is lesser sensitive to the coordination environment [27, 28]. The forced electric dipole transition is appeared only in the case of Dy^{3+} ions which are found at the local sites without inversion centre symmetry [28]. The respective blue emission having a centre at 485 nm is relatively lower intense than predominant yellow emission. The yellow-to-blue line ratio is 6.8026 which signifies that forced electric dipole transition is in dominance thereby indicating that the Dy^{3+} ions would found at the local sites with non-inversion centre symmetry in the $Ca_{0.5}Y_{1-x}(MoO_4)_2$ host.

3.2.3.2. *$Ca_{0.5}La_{1-x}(MoO_4)_2$:$xLn^{3+}$,$Na^+$ (Ln = Eu, Tb and Dy)*

The room temperature PL excitation spectrum of $Ca_{0.5}La_{1-x}(MoO_4)_2$:$xEu^{3+}$ thin phosphor films monitored at 394 nm excitation wavelength is shown in **Figure 11a**, consists of numerous intra-configurational 4f-4f transitions. As the Eu^{3+} concentration in the $Ca_{0.5}La(MoO_4)_2$ host lattices increases, the photoluminescence emission of the host is suppressed due to the overcoming of Eu^{3+} ions. The intensities of the different $^5D_0 \rightarrow {}^7F_J$ transitions might depend on the local symmetry of crystal field of Eu^{3+} ions [19]. The highly intense and strong emission peak is located at 615 nm upon 394 nm UV excitation corresponds to the $^5D_0 \rightarrow {}^7F_2$ electric-dipole transition showing hypersensitive red emission with parity forbidden ($\Delta J = 2$). The split-up in the peaks is due to the Stark energy splitting, which is having (2J + 1) Stark components of J-degeneracy splitting [28]. The predominant electric-dipole transition implies that Eu^{3+} ions would be situated at sites with non-inversion symmetry [15]. The transitions located at 586 nm show emission in the orange region which is associated with magnetic-dipole transition ($^5D_0 \rightarrow {}^7F_1$) and is not affected by the chemical surroundings of Eu^{3+}. The remaining transitions at 654 nm ($^5D_0 \rightarrow {}^7F_3$) and 701 nm ($^5D_0 \rightarrow {}^7F_4$) are the weakest ones. The red-to-orange emission (R/O) ratio of the phosphor is 5.5311 which suggests the sites symmetry of the respective Eu^{3+} ions. Based on these observations, it is suggested that $Ca_{0.5}La_{1-x}(MoO_4)_2$:$xEu^{3+}$,$Na^+$ might be a suitable phosphor candidate for solid-state lighting applications.

The PL emission spectrum for $Ca_{0.5}La_{1-x}(MoO_4)_2$:$xTb^{3+}$,$Na^+$ recorded at ~278 nm UV excitation is shown in **Figure 11b** which consists of four PL emission peaks at 489 nm ($^5D_4 \rightarrow {}^7F_6$), 545 nm ($^5D_4 \rightarrow {}^7F_5$), 585 nm ($^5D_4 \rightarrow {}^7F_4$) and 621 nm ($^5D_4 \rightarrow {}^7F_3$). Among these emission peaks, the sensitive green colour is located at 545 nm associated to the predominant transition $^5D_4 \rightarrow {}^7F_5$ [19]. It is based on the energy transfer from the host populates only $^5D_4$ level. The dominant emission peak possesses two sub peaks which correspond to the Stark energy splitting.

The room temperature PL emission spectrum (**Figure 11c**) for the thin phosphor film $Ca_{0.5}La_{1-x}(MoO_4)_2$:$xDy^{3+}$,$Na^+$ excited with 352 nm excitation wavelength. The emission spectrum comprises of two major peaks with peak positions at 477 nm attributed to magnetic dipole transition of ($^4F_{9/2} \rightarrow {}^6H_{15/2}$) and 570 nm corresponds to forced electric dipole transition of $^4F_{9/2} \rightarrow {}^6H_{13/2}$ [25]. The magnetic dipole transition is least sensitive to the coordination environment. The forced electric dipole transition would be found only in the case of Dy^{3+} ions situated at the local sites without inversion symmetry [19]. The emission in the blue region having a centre at 477 nm is having relatively least intense than predominant yellow emission. The yellow-to-blue line ratio is 6.0076 which implies that the forced electric dipole transition is most dominant hence indicating that the Dy^{3+} ions are situated at the local sites with non-inversion symmetry in the $Ca_{0.5}La_{1-x}(MoO_4)_2$ host.

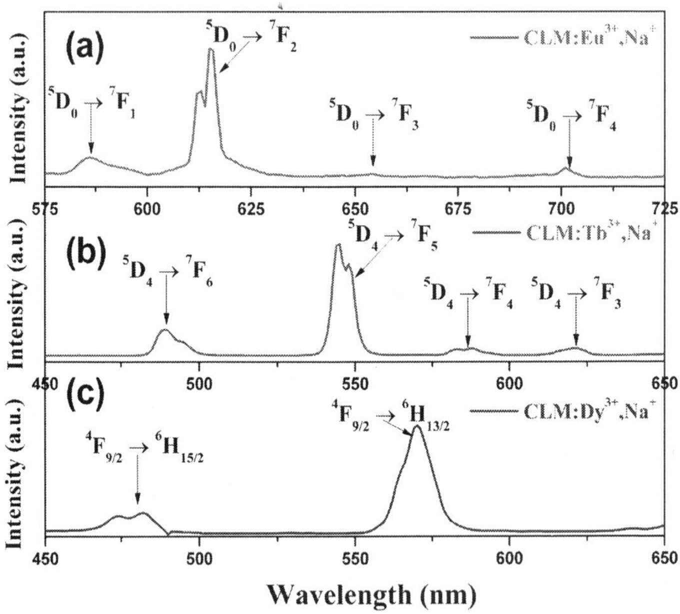

Figure 11. PL emission spectra (a, b, and c) of the thin film phosphors $Ca_{0.5}La(MoO_4)_2:Ln^{3+},M^+$ (Ln = Eu, Tb and Dy; M = Na).

3.3. Photometric characterization and decay-time analysis

Figures 12a, b and **13a, b** show the decay time profile and Commission Internationale de I'Eclairage (CIE) colour chromaticity coordinates of the $Ca_{0.5}Y_{1-x}(MoO_4)_2:xLn^{3+},Na^+$ (Ln = Eu, Tb and Dy) and $Ca_{0.5}La_{1-x}(MoO_4)_2:xLn^{3+},Na^+$ (Ln = Eu, Tb and Dy) phosphors. The CIE colour chromaticity coordinate of these phosphors was estimated and is given in **Table 1**. Furthermore, to know about the characteristic emission of these phosphors, the value of colour purity was derived by the equation [24]

$$\text{Colourpurity} = \frac{\sqrt{(x-x_i)^2+(y-y_i)^2}}{\sqrt{(x_d-x_i)^2+(y_d-y_i)^2}}100 \quad (1)$$

where (x, y) is denoted as CIE chromaticity coordinate of the synthesized sample, (x_i, y_i) is the CIE white illumination, and (x_d, y_d) is the CIE chromaticity coordinate of the dominant wavelength. Thus, the colour purities of the $Ca_{0.5}Y_{1-x}(MoO_4)_2$:xEu^{3+},Na^+, $Ca_{0.5}Y_{1-x}(MoO_4)_2$:xTb^{3+},Na^+ and $Ca_{0.5}Y_{1-x}(MoO_4)_2$:xDy^{3+},Na^+ phosphors are 90.0, 87.5 and 81.3%, respectively. Similarly, the colour purities of the $Ca_{0.5}La_{1-x}(MoO_4)_2$:$xEu^{3+}$,$Na^+$ $Ca_{0.5}La_{1-x}(MoO_4)_2$:$xTb^{3+}$,$Na^+$ and $Ca_{0.5}La_{1-x}(MoO_4)_2$:$xDy^{3+}$,$Na^+$ phosphors are 95.0, 91.9 and 81.7%. From the results, it is suggested that these phosphors with remarkable CIE chromaticity coordinate with high colour purities might be suitable for applications in display devices as the best red, green and yellow emitting phosphors.

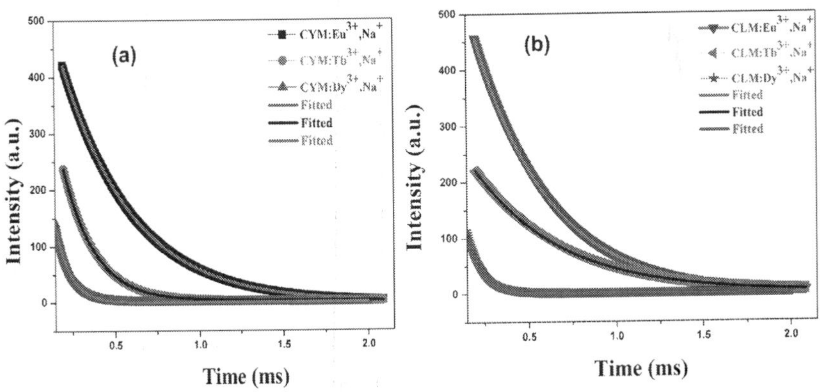

Figure 12. Luminescence decay profiles (a and b) for the thin film phosphors $Ca_{0.5}R_{1-x}(MoO_4)_2$:xLn^{3+} (R^{3+} = Y, La),Na^+ (Ln = Eu, Tb and Dy).

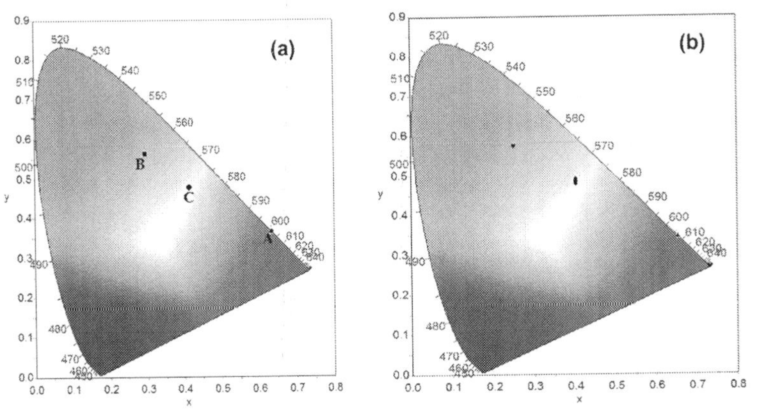

Figure 13. CIE diagram (a, b) for the thin film phosphors $Ca_{0.5}Y(MoO_4)_2$:Ln^{3+},Na^+ (Ln = (A) Eu, (B) Tb and (C) Dy).

The representative PL decay curves for luminescence emission for the phosphors $Ca_{0.5}Y_{1-x}(MoO_4)_2$:$x$$Ln^{3+}$,$Na^+$ (Ln = Eu, Tb and Dy) and $Ca_{0.5}La_{1-x}(MoO_4)_2$:$x$$Ln^{3+}$,$Na^+$ (Ln = Eu, Tb and Dy) are shown in **Figure 12a, b**. This can be fitted well into a single exponential function [15, 27] as

$$I = I_0 \exp\left(\frac{-t}{\tau}\right) \tag{2}$$

where I_0 is the luminescence intensity at times $t = 0$ and τ is its associated luminescence lifetime. The photometric quantities and luminescence decay time values are given in **Table 1**.

Phosphor	CCT (K)	CRI	Colour coordinates		LER (lm W^{-1})	Colour purity (%)	τ (ms)
			x	y			
$Ca_{0.5}Y(MoO_4)_2$:Eu^{3+},Na^+	1149	33	0.635	0.365	162	90.0	0.462
$Ca_{0.5}Y(MoO_4)_2$:Tb^{3+},Na^+	N/A	26	0.296	0.564	491	87.5	0.455
$Ca_{0.5}Y(MoO_4)_2$:Dy^{3+},Na^+	3658	18	0.414	0.478	525	81.3	0.172
$Ca_{0.5}La(MoO_4)_2$:Eu^{3+},Na^+	1196	42	0.656	0.343	321	95.0	0.481
$Ca_{0.5}La(MoO_4)_2$:Tb^{3+},Na^+	6850	12	0.251	0.570	530	91.9	0.485
$Ca_{0.5}La(MoO_4)_2$:Dy^{3+},Na^+	4195	16	0.404	0.485	462	81.7	0.187

Table 1. Photometric parameters, color purity and luminescence decay time for the phosphors $Ca_{0.5}R_{1-x}(MoO_4)_2$:$x$$Ln^{3+}$, Na^+ (R = Y, La; Ln = Eu, Tb and Dy).

3.4. Photoluminescence emission studies from nano-architectures

The thin phosphor films grown from nano-powder are being synthesized by the hydrothermal method, and the synthesis procedure is described previously by our group [17]. The luminescence emission intensity is being enhanced by co-doping of alkali metal ions. Furthermore, for the co-doping of alkali precursors, instead of using alkali chloride, alkali carbonates were taken and converted them into alkali nitrates. These alkali nitrates were co-doped with the existing precursors following the hydrothermal method nano-powders were synthesized and thin films were deposited from these powders [17]. The room temperature PL emission spectrum for $Ca_{0.5}R_{1-x}(MoO_4)_2$:$x$$Eu^{3+}$,$Na^+$ (R = Y, La) as the representative thin phosphor films are shown in **Figure 14**. The emission spectra monitored at 395 nm UV excitation for both the phosphors shows a number of intra-configurational f-f transitions. The strong and most intense emission peak is found at 616 nm for $Ca_{0.5}Y_{1-x}(MoO_4)_2$:$x$$Eu^{3+}$,$Na^+$ and 615 nm for $Ca_{0.5}La_{1-x}(MoO_4)_2$:$x$$Eu^{3+}$,$Na^+$ is attributed to the $^5D_0 \rightarrow {}^7F_2$ electric-dipole transition possess hypersensitive red emission [15, 28, 29]. The Stark energy splitting is mildly shown for both the phosphors. It is noticed that the splitting of the electric dipole transition is uniform and homogeneous between the two thin phosphor films. The intensity of the spectral peaks for the nano-thin phosphor film is nearly close to that of those from the bulk thin phosphor film, as peak intensity is related to reduced particle size and improved homogeneity [17, 29]. The dominant electric-dipole transition suggests that Eu^{3+} ions are found at sites with non-inversion symmetry [30]. The other transitions found at 587 nm ($^5D_0 \rightarrow {}^7F_1$) for $Ca_{0.5}Y_{1-x}(MoO_4)_2$:xEu^{3+},Na^+ and 586 nm ($^5D_0 \rightarrow {}^7F_1$) for $Ca_{0.5}La_{1-x}(MoO_4)_2$:$xEu^{3+}$,$Na^+$ show orange emission due to magnetic-dipole transition. The other relatively weaker transitions are found at 654 nm ($^5D_0 \rightarrow {}^7F_3$) and 702 nm ($^5D_0 \rightarrow {}^7F_4$) for both the nano-thin phosphor films. The photometric parameters for both the phosphors are under further investigation. From these results, it is indicated that $Ca_{0.5}R_{1-x}(MoO_4)_2$:xEu^{3+},Na^+ (R = Y, La) phosphors are best candidates for display applications.

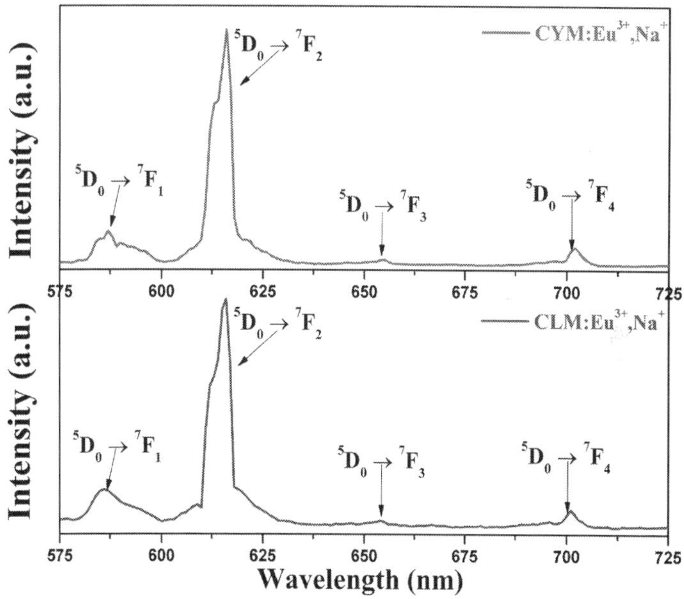

Figure 14. PL emission spectrum of the thin film phosphors $Ca_{0.5}Y(MoO_4)_2$:Eu^{3+},Na^+ and $Ca_{0.5}La(MoO_4)_2$:Eu^{3+},Na^+ prepared from nano-phosphors.

4. Conclusion

In conclusion, the nano-sized single crystalline $Ca_{0.5}La_{1-x}(MoO_4)_2$:$xEu^{3+}$,$M^+$ ceramic thin phosphor films deposited on quartz substrates by pulsed laser deposition technique using Nd-YAG laser source in an ultra-high vacuum (UHV). The FESEM images exhibited the spherical-shaped phosphor particles. XRD patterns revealed the scheelite-type crystal structure without any impurity phases. By using AFM, the surface topographies and distributions of grains were investigated. Upon optical excitation, Eu-, Tb, and Dy-doped $Ca_{0.5}La_{1-x}(MoO_4)_2$:$x$$Eu^{3+}$,$M^+$ thin phosphor films showed characteristic emissions in the bright-red, green and yellow regions, respectively. The obtained results suggested that the deposited thin film phosphors could serve as efficient materials for electroluminescence and display applications.

Conflict of interest

The authors declare that there is no conflict of interests regarding the publication of this chapter.

Acknowledgements

Work incorporated in this chapter was supported by Science and Engineering Research Board (SERB), (SR/FTP/PS-135/2011) Govt. of India. The authors apologize for inadvertent omission of any pertinent references.

Author details

Jagannathan Thirumalai[1*], Venkatakrishnan Mahalingam[1] and Rajagopalan Krishnan[2]

*Address all correspondence to: thirumalaijg@gmail.com; jthirumalai@bsauniv.ac.in

1 Department of Physics, School of Physical and Chemical Sciences, B. S. Abdur Rahman University, Vandalur, Chennai, Tamil Nadu, India

2 Department of Physics, Rajalakshmi Institute of Technology, Kuthambakkam, Chennai, Tamil Nadu, India

References

[1] Park S W, Moon B K, Choi B C, Jeong J H, Bae J S, Kim K H: Red photoluminescence of pulsed laser deposited Eu:$NaY(MoO_4)_2$ thin film phosphors on sapphire substrates. Curr. Appl. Phys. 2012;12:S150–S155. doi:10.1016/j.cap.2012.02.048

[2] Krishnan R, Thirumalai J: Synthesis and up/down conversion luminescence properties of $Na_{0.5}R_{0.5}MoO_4$:Ln^{3+} (R^{3+} = La, Gd), (Ln^{3+} = Eu, Tb, Dy, Yb/Er) thin phosphor films grown by pulsed laser deposition technique. RSC Adv. 2014;4:64258–64266. doi:10.1039/c4ra11274a

[3] Mahalingam V, Thirumalai J, Krishnan R, Mantha S: Up/down conversion luminescence and charge compensation investigation of $Ca_{0.5}Y_{1-x}(WO_4)_2$:xLn^{3+} (Ln = Pr, Sm, Eu, Tb, Dy, Yb/Er) phosphors. Spectrochim. Acta A. 2016;152:172–180. doi:10.1016/j.saa.2015.06.129

[4] Barreca D, Depero L E, Noto V D, Rizzi G A, Sangaletti L, Tondello E: Thin films of bismuth vanadates with modifiable conduction properties. Chem. Mater. 1999;11:255–261. doi:10.1021/cm980725q

[5] Yanagihara M, Yusop M Z, Tanemura M, Ono S, Nagami T, Fukuda K, Suyama T, Yokota Y, Yanagida T, Yoshikawa A: Vacuum ultraviolet field emission lamp utilizing $KMgF_3$ thin film phosphor. APL Mater. 2014;2:046110(1–6). doi: 10.1063/1.4871915

[6] Shaikh P A, Thakare V P, Late D J, Ogale S: A back-to-back MOS-Schottky (Pt-SiO_2-Si-C-Pt) nano-heterojunction device as an efficient self-powered photodetector: one step fabrication by pulsed laser deposition. Nanoscale. 2014;6:3550–3556. doi:10.1039/C3NR06525A

[7] Patel P K, Yadav K L, Kaur G: Reduced dielectric loss in $Ba_{0.95}Sr_{0.05}(Fe_{0.5}Nb_{0.5})O_3$ thin film grown by pulsed laser deposition. RSC Adv. 2014;4:28056–28061. doi:10.1039/C4RA03502J

[8] Lin C C, Liu R S: Advances in phosphors for light-emitting diodes. J. Phys. Chem. Lett. 2011;11:1268–1277. doi:10.1021/jz2002452

[9] Li Y Q, van Steen J E J, van Krevel J W H, Botty G, Delsing A C A, Disalvo F J, de With G, Hintzen H T: Luminescence properties of red-emitting $M_2Si_5N_8$:Eu^{2+}(M = Ca, Sr, Ba) LED conversion phosphors. J. Alloys Compd. 2006;417:273–279. doi:10.1016/j.jallcom.2005.09.041

[10] Aboulaich A, Michalska M, Schneider R, Potdevin A, Deschamps J, Deloncle R, Chadeyron G, Mahiou R: Ce-doped YAG nanophosphor and red emitting $CuInS_2$/ZnS core/shell quantum dots for warm white light-emitting diode with high color rendering index. Appl. Mater. Interfaces. 2014;6:252–258. doi:10.1021/am404108n

[11] Zhang L, Lu Z, Han P, Wang L, Zhang Q: Synthesis and photoluminescence of Eu^{3+}-activated double perovskite NaGdMg(W, Mo)O_6—a potential red phosphor for solid state lighting. J. Mater. Chem. C. 2013;1:54–57. doi:10.1039/C2TC00189F

[12] Li K, Geng D, Shang M, Zhang Y, Lian H, Lin J: Color-tunable luminescence and energy transfer properties of $Ca_9Mg(PO_4)_6F_2$:Eu^{2+},Mn^{2+} phosphors for UV-LEDs. J. Phys. Chem. C. 2014;118:11026–11034. doi:10.1021/jp501949m

[13] Yang H, Wang W, Liu Z, Li G: Homogeneous epitaxial growth of AlN single-crystalline films on 2 inch-diameter Si (1 1 1) substrates by pulsed laser deposition. Cryst. Eng. Comm. 2013;15:7171–7176. doi:10.1039/C3CE40886H

[14] Atanasov P A, Tomov R I, Perriere J, Eason R W, Vainos N, Klini A, Zherikhin A, Millon E: Growth of Nd: potassium gadolinium tungstate thin-film waveguides by pulsed laser deposition. Appl. Phys. Lett. 2000;76:2490–2492. doi:10.1063/1.126385

[15] Mahalingam V, Thirumalai J, Krishnan R, Chandramohan R: Potential visible light emitting rare-earth activated $Ca_{0.5}Y_{1-x}(MoO_4)_2$:$x$$RE^{3+}$ (RE = Pr, Sm, Eu, Tb, Dy) phosphors for solid state lighting applications. J. Mater. Sci. Mater. Electron. 2015;26:842–852. doi: 10.1007/s10854-014-2473-2

[16] Eason R, editor. Pulsed laser deposition of thin films: Applications–led growth of functional materials. John Wiley & Sons, Inc., Hoboken, New Jersey : Wiley; 2007 p. 1–682. DOI: 10.1002/0470052120

[17] Xia Z, Chen D: Synthesis and luminescence properties of $BaMoO_4$:Sm^{3+} phosphors. J. Am. Ceram. Soc. 2010;93:1397–1401. doi:10.1111/j.1551-2916.2009.03574.x

[18] Krishnan R, Thirumalai J, Mahalingam V, Mantha S, Lavanya M: Synthesis, luminescence and photometric characteristics of $Ca_{0.5}La(MoO_4)_2$:Ln^{3+} (Ln = Eu, Tb, Dy) phosphors. Mater. Chem. Phys. 2015;162:41–49. doi:10.1016/j.matchemphys.2015.04.050

[19] Du P, Yu J S: Dual-enhancement of photoluminescence and cathodoluminescence in Eu^{3+}-activated $SrMoO_4$ phosphors by Na^+ doping. RSC Adv. 2015;5:60121–60127. doi: 10.1039/c5ra06053b

[20] Liu X, Li L, Noh H M, Jeong J H, Jang K, Shin D S: Controllable synthesis of uniform $CaMoO_4$:Eu^{3+},M^+ (M = Li, Na, K) microspheres and optimum luminescence properties. RSC Adv. 2015;5:9441–9454. doi:10.1039/c4ra12183j

[21] Luitel H N, Chand R, Watari T: $CaMoO_4$:RE^{3+},Yb^{3+},M^+ phosphor with controlled morphology and color tunable upconversion. Displays. 2016;42:1–8. doi:10.1016/j.displa.2015.12.004

[22] Blasse G, Grabmaier B C, editors. Luminescent materials. Berlin: Springer; 1994.

[23] Som S, Das S, Dutta S, Visser H G, Pandey M K, Kumar P, Dubey R K, Sharma S K: Synthesis of strong red emitting Y_2O_3:Eu^{3+} phosphor by potential chemical routes: comparative investigations on the structural evolutions, photometric properties and Judd-Ofelt analysis. RSC Adv. 2015;5:70887–70898. doi:10.1039/c5ra13247a

[24] Mahalingam V, Thirumalai J, Krishnan R, Chandramohan R: Controlled synthesis and luminescence properties of $Ca_{0.5}Y_{1-x}(MoO_4)$:$x$$RE^{3+}$ (RE = Eu, Pr, Sm, Tb, Dy, Yb/Er, Yb/Tm and Yb/Ho) phosphors by hydrothermal method versus pulsed laser deposition. Electron. Mater. Lett. 2016;12:32–47. doi:10.1007/s13391-015-5248-x

[25] Dutta S, Som S, Sharma S K: Luminescence and photometric characterization of K^+ compensated $CaMoO_4$:Dy^{3+} nanophosphors. Dalton Trans. 2013;42:9654–9661. doi: 10.1039/c3dt50780g

[26] Saraf R, Shivakumara C, Dhananjaya N, Behera S, Nagabhushana H: Photoluminescence properties of Eu^{3+} activated $CaMoO_4$ phosphors for WLEDs applications and its Judd-Ofelt analysis. J. Mater. Sci. 2015;50:287–298. doi:10.1007/s10853-014-8587-3

[27] Krishnan R, Thirumalai J, Thomas S, Gowri M: Luminescence and magnetic behaviour of almond like $(Na_{0.5}La_{0.5})MoO_4$:RE^{3+} (RE = Eu, Tb, Dy) nanostructures. J. Alloys Compd. 2014;604:20–30. doi:10.1016/j.jallcom.2014.03.065

[28] Krishnan R, Thirumalai J: Up/down conversion luminescence properties of $(Na_{0.5}Gd_{0.5})MoO$:Ln^{3+} (Ln = Eu, Tb, Dy, Yb/Er, Yb/Tm and Yb/Ho) microstructures: synthesis, morphology, structural and magnetic investigation. New J. Chem. 2014;38:3480–3491. doi:10.1039/c4nj00165f

[29] Gupta S K, Sahu M, Ghosh P S, Tyagi D, Saxena M K, Kadam R M: Energy transfer dynamics and luminescence properties of Eu^{3+} in $CaMoO_4$ and $SrMoO_4$. Dalton Trans. 2015;44:18957–18969. doi:10.1039/c5dt03280f

[30] Singh B P, Parchur A K, Ningthoujam R S, Ansari A A, Singh P, Rai S B: Enhanced photoluminescence in $CaMoO_4$ by Gd^{3+} co-doping. Dalton Trans. 2014;43:4779–4789. doi:10.1039/c3dt53408a

Chapter 3

Thermoelectric and Topological Insulator Bismuth Chalcogenide Thin Films Grown Using Pulsed Laser Deposition

Phuoc Huu Le and Chih Wei Luo

Additional information is available at the end of the chapter

http://dx.doi.org/10.5772/65898

Abstract

Bismuth chalcogenides have been intensively studied for their high-performance thermoelectric properties and their novel topological surface states, which could significantly benefit novel applications in fields such as TE devices, spintronics, and quantum computing. This chapter reports recent advances in pulsed laser deposition (PLD) for the growth of bismuth chalcogenide (e.g., Bi_2Te_3, Bi_2Se_3, and Bi_3Se_2Te) thin films and their novel properties. It covers a wide range of fields such as thin film growth using PLD for fabricating polycrystalline and epitaxial films with different thermoelectric, nanomechanical, and magnetotransport properties as a function of the PLD processing conditions. Moreover, the proximity-induced superconductivities in Bi inclusions/bismuth chalcogenide thin films are also reported and discussed in detail.

Keywords: pulsed laser deposition, thermoelectrics, topological insulators, bismuth chalcogenides, superconductivity, magnetoresistance

1. Introduction

Bismuth chalcogenide thin films are of great interest because of the exciting properties of topological insulators (TIs) and their applications to thermoelectrics (TEs). These materials have been applied in integrated TE cooling devices working at near room temperature [1, 2]. TIs are exotic materials with an insulating bulk and topologically protected states on the surface which could be used in different applications, such as spintronics and quantum computing [3–6]. The topological surface states (TSSs) exhibit Dirac linear energy dispersion inside the bulk gap, spin-polarization by spin-momentum locking nature, and weak anti-localization (WAL) due to the strong spin-orbit coupling [3–6]. Thus, the WAL through magnetotransport studies has been widely used as a signature of TI materials [7–9].

For application purposes, thin film growth techniques for TE and TI materials are required. Among physical vapor deposition techniques, pulsed laser deposition (PLD) offers great versatility in growing polycrystalline and epitaxial thin films with high growth rates, multiple elements, and diverse structural morphologies for both fundamental studies and applications. The purpose of this chapter is to outline recent advances in the PLD growths of bismuth chalcogenide thin films with desired properties for TE/TI applications and fundamental studies.

2. Thin film growth using pulsed laser deposition (PLD)

Thin film growth consists of nucleation, growth, and coalescence (**Figure 1a**). In a physical vapor deposition, an extremely nonequilibrium process takes place at high supersaturations and at comparatively high concentrations of impure atoms [10]. Nucleation takes place at high supersaturations S (defined as $S = p/p_e$, where p is the vapor pressure of the deposited material evaporated from the source at temperature T and p_e is the equilibrium vapor pressure of the substrate material at temperature T_S). The incident vapor arrives at the surface of substrates and then forms small but highly mobile clusters or islands with uniform distribution. In this stage, the impinging atoms and subcritical clusters are incorporated and consequently increase their sizes, while the island density rapidly saturates. In the following stage, the islands are emerged via a coalescence phenomenon which is liquid-like for some cases, especially at high substrate temperatures. Coalescence leads to a decrease in island density and forms local denuding positions on the surface of substrates where further nucleation can then occur (**Figure 1a**) [11]. The sequence of film nucleation and growth events can be well appreciated in the transmission electron microscopy (TEM) images in **Figure 1b–d** [11].

2.1. Basic growth modes

For all phase transitions, the formation of thin films is characterized by the formation of nuclei and their growth. Depending on the interaction energies of substrate atoms and film atoms, any of three growth modes in **Figure 2a–c** can occur:

- 2D Frank-van der Merwe mode: layer-by-layer growth, in which the interaction between substrate and atoms of film is greater than that between adjacent atoms of film.

- 3D Volmer-Weber mode: separated islands form on the surface of substrates, in which the interaction between atoms of film is greater than that between a substrate and the adjacent atoms of film.

- Stranski-Krastanov mode: layer plus island, in which one or two monolayers form first and then grow individually.

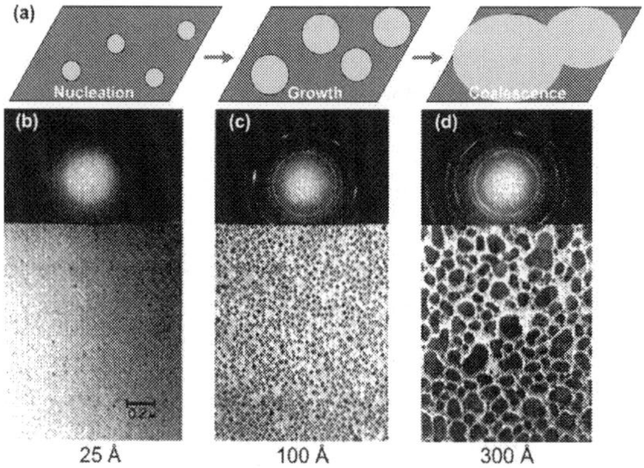

Figure 1. (a) Schematics of thin film growth processes: nucleation, growth, coalescence. Transmission electron microscope images of (b) nucleation, (c) growth, and (d) coalescence of Ag films on (1 1 1) NaCl substrates. Corresponding diffraction patterns are shown.

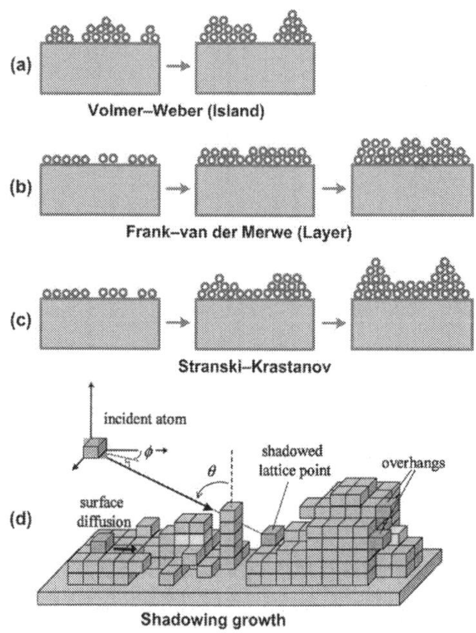

Figure 2. Basic modes of thin film growth: (a) island in the Volmer-Weber mode, (b) layer by layer in the two-dimensional Frank-van der Merwe mode, (c) layer plus island in the Stranski-Krastanov mode. (d) Shadowing growth: a schematic of three-dimensional Monte Carlo simulations for oblique angle deposition [13].

After an initially random nucleation of islands on the surface of the substrates, the deposition on the top of the islands is faster than that in the valleys due to the oblique incident flux (the so-called **shadowing effect**) [12, 13]. Isolated columns are therefore formed on these islands during subsequent growths (**Figure 2d**) [13].

2.2. Epitaxy growth

Epitaxy refers to the growth of a single crystal film on top of a single crystal substrate. The deposited film is denoted as an epitaxial film or epitaxial layer. The growth is called homoepitaxy if the film and the substrate are the same material, and it is called heteroepitaxy if they are different materials. Epitaxial relationship is determined as: (*HKL*) | | (*hkl*); [*UVW*] | | [*uvw*], where (*hkl*) and (*HKL*) are the Miller indices of the overgrowth plane and substrate at the common interface. The corresponding parallel directions in the overgrowth and substrate planes, denoted by [*wuw*] and [*UVW*], respectively, must also be specified.

2.3. Factors governing the epitaxy growth

The key factors governing epitaxy growths are structural compatibility, chemical compatibility, and growth temperatures.

- Structural compatibility: The structures of a film and a substrate should have good lattice matching in terms of crystal structures ($a_{0,\ sub}$) and lattice constants ($a_{0,\ film}$), that is, small lattice misfit.

Lattice misfit *f*:

$$f = \frac{a_{0,sub} - a_{0,film}}{a_{0,film}} \approx \frac{a_{0,sub} - a_{0,film}}{a_{0,sub}}\sqrt{2}. \quad (1)$$

- Chemical compatibility: This includes chemical bonding and chemical diffusion.

- Growth temperatures: Good epitaxy growth is obtained at above or around the well-defined elevated substrate temperature (T_e). T_e depends on the deposition rate, particle energy, and surface contamination. Generally, a higher temperature is recommended to reduce surface contamination by desorption (or enhance surface mobility) of atoms to reach the favorable sites and also enhance the diffusivity in deposition for favoring re-crystallization and defect annihilation.

2.4. Preparation of bismuth chalcogenide films by PLD

PLD is one of the most convenient thin film growth techniques that uses a high-intensity pulsed laser beam as an external energy source to ablate a target, form a plume, and then deposit thin films onto a substrate. **Figure 3** shows a typical PLD system for preparing TE and TI thin films. A substrate is heated and maintained at a desired substrate temperature (T_S) using a thermocouple and a proportional-integral-derivative temperature controller. The thermocouple was buried inside a stainless steel substrate holder, which is heated by a tungsten lamp just behind the holder. The pressure of ambient gas (He/Ar) can be fine-tuned by the needle valve. A KrF excimer laser beam (λ = 248 nm, pulsed duration of 15–20 ns, repetition rate in the range of 1–15 Hz, and fluence of 1–10 J/cm^2) is guided by several UV mirrors and focused on a stoichiometric polycrystalline or a single crystal target (e.g., Bi_2Te_3, Bi_2Se_3, Bi_2Se_2Te) within the vacuum chamber by a UV lens. The target-to-substrate distance was 40 mm. During the deposition of Bi_2Se_3 films, pure (6N) He/Ar gas was introduced into the vacuum chamber, which was evacuated to a base pressure of 4 × 10^{-4} Pa (or 3 × 10^{-6} Torr) and maintained at a certain constant pressure (P), using a differential evacuation system.

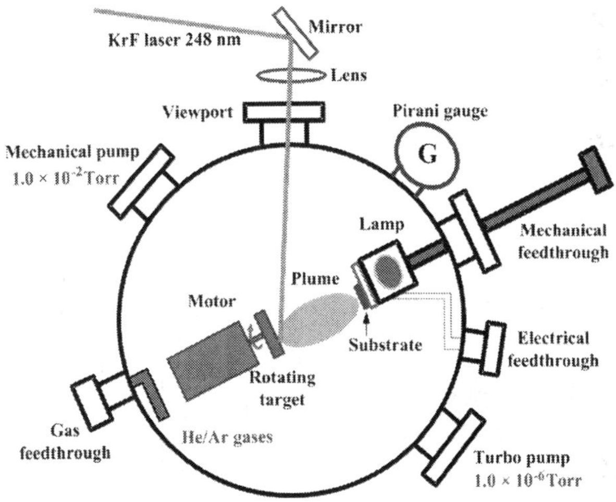

Figure 3. Schematic illustration of a pulsed laser deposition (PLD) system. G: gauge.

The surface of substrates should be atomically clean and free from impurities because the contaminants can interact with the thin films being deposited and substantially degrade its quality and adhesion to the substrates. The presence of unwanted surface contaminants can also influence the growth and orientation of the films in an undesired manner. In the depositions for TE thin films, an approximately 300-nm-thick SiO_2 layer was thermally grown on the Si wafers (thickness 525 μm) for electrical isolation purpose. The wafers were cut into 1.5 cm × 1.5 cm substrates. The substrates were cleaned with acetone to dissolve any contaminants adhering to the surface of substrates such as grease and oils. This was followed by rinsing with methanol to remove any residues left behind after cleaning with acetone. Afterward, the substrates were rinsed in distilled water and dried with nitrogen flow. The substrates were then used for the deposition of TE thin films.

Here are some examples of PLD growth of TE films. For Bi_2Se_3 thin films, the depositions were at T_S of 200–350°C and helium ambient pressure (P) of 0.7–173 Pa. The number of laser pulses was 9000 and the deposition took 30 min. The average growth rate was approximately 0.46 Å/

pulse [14]. For the growth of Bi_2Te_3 thin films, T_S was varied from room temperature (30°C) to 380°C and the Ar ambient pressure (P_{Ar}) was at 80 Pa. The number of laser pulses was 12,000 and the deposition took 40 min. The average growth rate was approximately 0.52 Å/pulse [15]. For the growth of Bi-Se-Te thin films, the depositions were at T_S of 200–350°C and a helium ambient pressure (P_{He}) of 0.027–86.7 Pa. The number of laser pulses was 9000 and the deposition took 15 min. The average growth rate was approximately 0.6 Å/pulse [16].

3. Thermoelectric bismuth chalcogenide thin films

3.1. Crystal structures of bismuth chalcogenides

The crystal structures of bismuth chalcogenides (e.g., Bi_2Te_3, Bi_2Se_3, and Bi_3Se_2Te) are usually described by a hexagonal cell consisting of 15 layers of atoms stacking along the *c*-axis with a sequence shown below [17], as shown in **Figure 4a** and **b**.

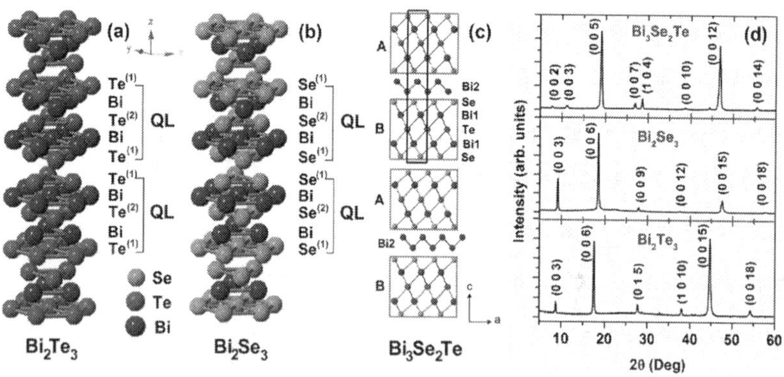

Figure 4. Crystal structures of (a) Bi_2Te_3, (b) Bi_2Se_3, and (c) Bi_3Se_2Te in the form of $(Bi_2)_m(Bi_2Se_2Te)_n$ (m = 1, n = 2) homologous series. The unit cells are marked with black thick boxes. (d) X-ray diffraction patterns of the typical TE bismuth chalcogenide thin films grown at 300°C on SiO_2/Si substrates.

A 5-atomic-layer-thick lamellae of -($Te^{(1)}$-Bi-$Te^{(2)}$-Bi-$Te^{(1)}$)- or -($Se^{(1)}$-Bi-$Se^{(2)}$-Bi-$Se^{(1)}$)- is called a quintuple layer, QL, in which the $Te^{(1)}$-Bi and Bi-$Te^{(2)}$ or $Se^{(1)}$-Bi and Bi-$Se^{(2)}$ are ionic-covalent bonds. Because of the weak binding (i.e., Van der Waals force) between Te or Se layers, bismuth chalcogenides could be cleaved easily along the plane perpendicular to the *c*-axis. It also induces the anisotropic thermal/electrical transport properties. For example, the thermal conductivity along the plane perpendicular to the *c*-axis (~1.5 W m^{-1} K^{-1}) is nearly two times higher than that along the *c*-axis direction (~0.7 W m^{-1} K^{-1}) [17, 18].

The crystal structures of Bi_3Se_2Te can be formed by ordered stacking of Bi_2Se_2Te and Bi_2 building blocks, that is, $(Bi_2)_m(Bi_2Se_2Te)_n$ (m = 1, n = 2) [19, 20], in which the covalently connected double layers of bismuth (Bi–Bi) lie between two QL (Se-Bi-Te-Bi-Se) blocks (**Figure 4c**); the (Bi–Bi) strictly alternates with two QL (Se–Bi–Te–Bi–Se) blocks [9, 20]. **Figure 4d** shows the typical XRD patterns of Bi_2Te_3, Bi_2Se_3, and Bi_3Se_2Te thin films grown on SiO_2/Si substrates at $T_S = 300°C$. They exhibit the dominance of (0 0 1) family planes of the rhombohedral phases of Bi_2Te_3 (PDF#82-0358), Bi_2Se_3 (PDF#33-0214), and Bi_3Se_2Te (JCPDS 00-053-1190), indicating that the films are highly c-axis oriented (i.e., textured films).

3.2. Introduction to thermoelectrics and applications

Thermoelectric materials are solid-state energy converters in which the combination of thermal, electrical, and semiconducting properties allows them to be used to convert waste heat into electricity or electrical power directly into cooling and heating [21].

3.2.1. The thermoelectric figure of merit (ZT)

The performance of the thermoelectric materials is often denoted as figure of merit Z whose unit is K^{-1} or ZT with the dimensionless unit [17, 22].

$$ZT = \frac{\alpha^2\sigma}{\kappa}T = \frac{\alpha^2\sigma}{\kappa_E + \kappa_L}T \tag{2}$$

where σ, α, κ, and T are the electrical conductivity, Seebeck coefficient, thermal conductivity, and absolute temperature, respectively. The total thermal conductivity can be split into a lattice contribution (κ_L) and an electronic contribution (κ_E). The quantity $\alpha^2\sigma$ is commonly used to represent the thermoelectric power factor (*PF*).

The efficiency of a thermoelectric device is directly related to ZT. For power generation, the maximum efficiency (η) is expressed by [23]

$$\eta = \frac{T_h - T_c}{T_h} \cdot \frac{\sqrt{1+Z\bar{T}} - 1}{\sqrt{1+Z\bar{T}} + \frac{T_c}{T_h}} \tag{3}$$

and for air-conditioning or refrigeration, the coefficient of performance is [23]

$$COP = \frac{T_c}{T_h - T_c} \cdot \frac{\sqrt{1+Z\bar{T}} - \frac{T_h}{T_c}}{\sqrt{1+Z\bar{T}} + 1} \tag{4}$$

where T_h and T_c are the hot-end and cold-end temperatures of the thermoelectric materials, respectively, and $\bar{T}$ is the average temperature of T_h and T_c. Therefore, the enhanced ZT value of TE materials is important to increase the COP for practical applications.

3.2.2. *Conflicting properties in thermoelectric materials*

Maximizing ZT is challenging due to the interdependence of the TE parameters. An increased power factor $\alpha^2\sigma$ by optimizing the carrier concentration n and/or a reduced lattice thermal conductivity κ_L by introducing the scattering centers are necessary to enhance ZT value. The dependences of these parameters with scattering factor r, carrier effective mass m^*, carrier mobility μ, and their interconnectivity limits the ZT to about 1 in large bulk materials [24].

The electrical conductivity (σ) and electrical resistivity (ρ) are related to n through the carrier mobility μ:

$$1/\rho = \sigma = ne\mu \tag{5}$$

The Wiedemann-Franz Law [2] states that the electronic contribution to the thermal conductivity is proportional to the electrical conductivity (σ) of the materials, with the relationship being

$$\kappa_e = L\sigma T = ne\mu LT \tag{6}$$

where e is electron charge, and L is the Lorenz factor of 2.48×10^{-8} J^2/K^2 C^2 for free electrons and this can vary particularly with carrier concentration [2].

The kinetic definition of α is the energy difference between the average energy of mobile carriers and the Fermi energy [25]. An increase in n leads to the increase in both the Fermi energy and the average energy, but the former increases more rapidly than the latter and thus results in a decrease in α value and a reduction factor of $\alpha^2 n$. Thus, in attempting to increase ZT for most of the homogeneous materials, the carrier concentration (n) increases electrical conductivity (σ) but reduces the Seebeck coefficient (α). For this reason, in metals and degenerate semiconductors (energy-independent scattering approximation), the Seebeck coefficient can be expressed as [2]:

$$\alpha = \frac{8\pi^2 k_B^2}{3eh^2} m^* T \left(\frac{\pi}{3n}\right)^{2/3} \tag{7}$$

The high m^* causes the Seebeck coefficient to rise according to Eq. (7). High m^* materials generally possess low μ which limits the enhancement of power factor by $(m^*)^{3/2}\mu$. Noticeably, the defect scatters are not only the phonons but also the electrons which lead to reduce κ_L as well as μ. Therefore, the ratio of μ/κ_L determines the improvement in ZT [17, 24]. Although the increase in the ratio is usually experimentally achieved through a greater reduction in κ_L rather than that in μ, some fundamental issues in this mechanism are not understood well [24].

Figure 5 shows the compromise between large α and high σ in thermoelectric materials that must be struck to maximize the figure of merit ZT. Meanwhile, the low carrier concentration will result in lower electrical conductivity with decreasing ZT. The ZT and PF peaks typically occur at carrier concentrations between 10^{19} and 10^{21} carriers per cm^3 (depending on the material system), which fall in between common metals and semiconductors, that is, the concentrations found in heavily doped semiconductors [2].

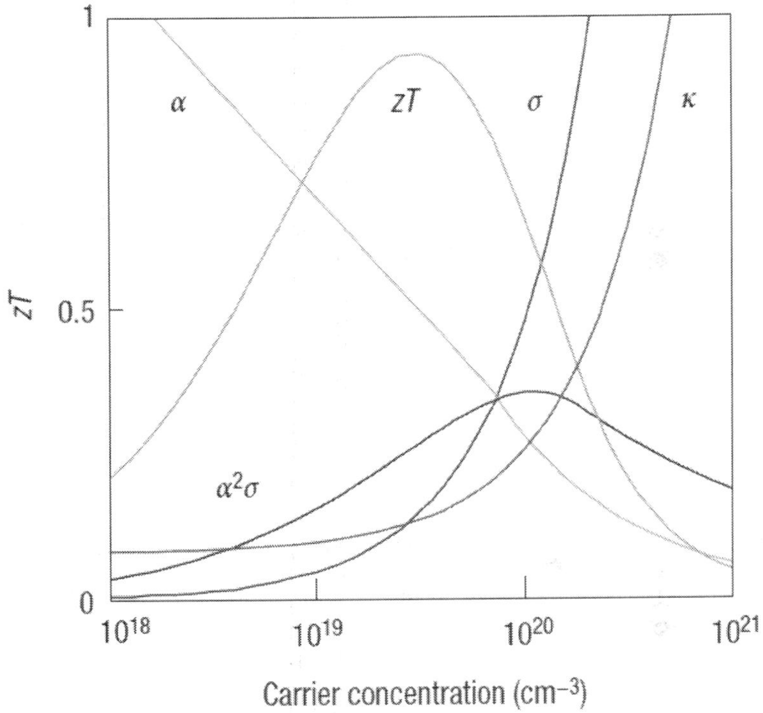

Figure 5. Maximizing the efficiency (ZT) of a thermoelectric device involves a compromise of thermal conductivity (κ, plotted on the y-axis from 0 to a top value of 10 W m^{-1} K^{-1}) and Seebeck coefficient (α, 0–500 μV K^{-1}) with electrical conductivity (σ, 0–5000 Ω^{-1} cm^{-1}) [2].

3.2.3. *Overview of thermoelectric applications*

TE devices have unique features: no moving parts, substantially less maintenance, quiet operation, high power density, low environmental impact, and high reliability [26]. Commercial use has been made mostly from Peltier thermoelectric cooling (TEC) effect, such as in small refrigerator devices used for camping and outdoor activities, automotive climate control seats, and localized cooling at the hot spots of chips. **Figure 6** gives an overview of the present and potential applications of thermoelectric generators (TEGs) [27]. Indeed, TEGs have been used for the power in miniaturized autarkic sensor systems, automotive waste heat recovery systems, ventilated wood stove, heating systems, water boilers, and heat recovery in industry.

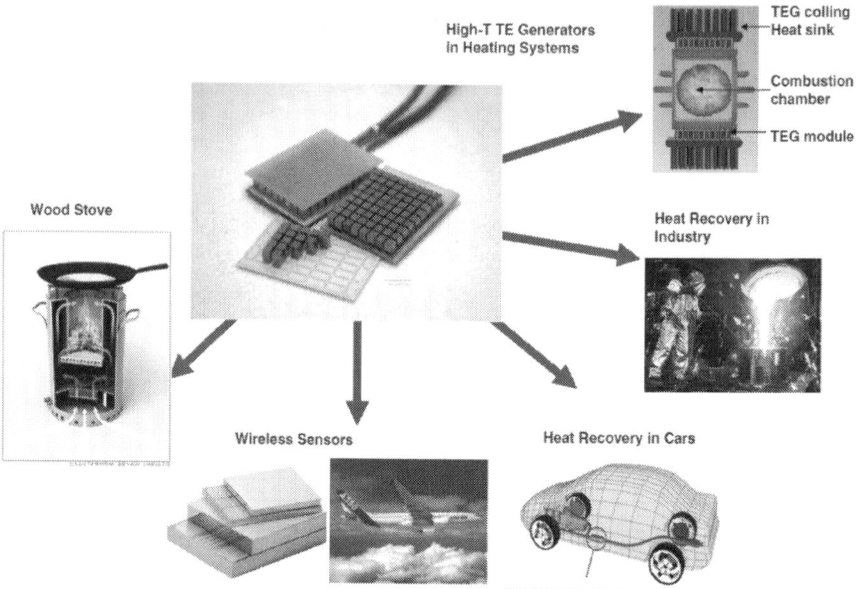

Figure 6. Overview of the potential applications of thermoelectric generators [27].

3.3. Thermoelectric properties of polycrystalline Bi_2Te_3, Bi_2Se_3, and Bi_3Se_2Te thin films with controlled structure morphology

Some typical HRTEM images of Bi_2Te_3, Bi_2Se_3, and Bi_3Se_2Te grown using PLD are shown in **Figure 7** [14–16]. HRTEM images performed on a high μ Bi_2Te_3 film with nanodisk-like morphology grown at 220°C are shown in **Figure 7a**. Clearly, the lower inset in **Figure 7a** shows the film with uniform thickness of approximately 295 nm and a SiO_2 layer with a thickness of 300 nm. It shows that projected period of 0.508 nm along the *c*-axis corresponds to the lattice spacing of the (0 0 6) planes. The highly (0 0 1)-orientated and crystallized structures of the film should facilitate the transport of charge carriers. The *c*-axis lattice constant of the Bi_2Te_3 film is 30.48 Å, which agrees closely with the value (30.44 Å) presented in JCPDS 82-0358. The other Bi_2Te_3 films grown at $T_S \geq 220$°C also display similar HRTEM results.

For a Bi_2Se_3 film deposited at 300°C and 40 Pa, an HRTEM image taken at the boundary of three platelets (P1, P2, and P3) revealed the granular-polycrystalline structure of the films (**Figure 7b**). Moreover, P1 and P2 partly overlapped and the corresponding fast Fourier transform (FFT) of this overlapping region indexed by {0 0 3} patterns of [0 1 0] zone axis was performed from the dashed-square area (**Figure 7b**, inset). The projected period along the *c*-axes of both P1 and P2 was 9.60 Å, corresponding to (0 0 3) planes, which was close to the reported value of 9.55 Å in Ref. [28].

HRTEM images of a Bi_3Se_2Te film deposited at 250°C and 40 Pa are shown in **Figure 7c** and **d**. Nanocrystallites with sizes of 10–20 nm are clearly observed in **Figure 7c**, confirming the nanocrystalline type of the Bi_3Se_2Te films. The interplanar spacing of the Bi_3Se_2Te (0 0 5) planes in the nanocrystallites is approximately 0.464 nm. Therefore, the *c*-axis lattice constant is determined to be 23.2 Å, closely agreeing with the value of 23.25 Å for Bi_3Se_2Te bulk (JCPDS 00-053-1190). In addition, the white lines in **Figure 7c** indicate the orientations of the (0 0 5) planes. It is seen that the overall orientation of the crystallites is disorganized. Intriguingly, near the interface of the film and substrate, the film has some nanoinclusions with sizes of 12–17 nm, as shown in **Figure 7d**. The EDS analysis shows that these are Bi semimetal nanoprecipitates (the inset in **Figure 7d**). The lattice spacing of Bi nanoinclusions is ~0.32 nm, corresponding to the Bi (0 1 2) planes. It has been found that Bi nanoinclusions can lead to the enhanced Seebeck coefficient and reduced lattice thermal conductivity owing to the low-energy electron filtering and phonon scattering at the nanoinclusions, respectively [29–31].

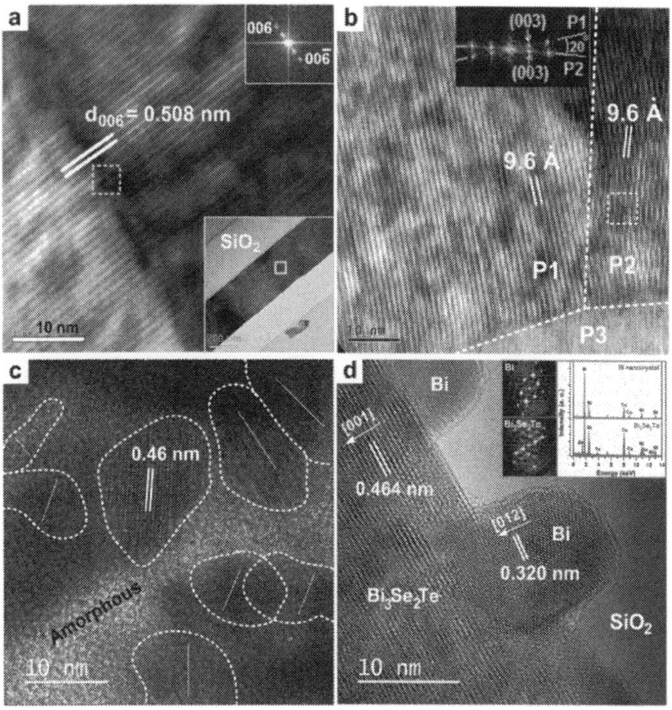

Figure 7. (a) HRTEM images of a high carrier mobility (μ) Bi_2Te_3 film with nanodisk-like morphology grown at 220°C and P_{Ar} of 80 Pa. (b) HRTEM image of an optimized Bi_2Se_3 film deposited at 300°C and P_{He} of 40 Pa. The inset shows the FFT patterns of the dashed-square area in the HRTEM image. (c and d) HRTEM images of the Bi_3Se_2Te film grown at 250°C and P_{He} of 40 Pa. The white lines in (c) indicate the (0 0 5)-orientation of nanograins. Inset in (d): FFT patterns and EDS spectra performed at film and Bi nanoinclusion positions.

Figure 8a shows the T_S-dependent α, σ, and *PF* (= $\alpha^2\sigma$) of some nanostructured Bi_2Te_3 films [15]. The σ value gradually increased from 34.5 ± 0.1 to 814.3 ± 1.5 S/cm when T_S was increased from 30 to 300°C, and then sharply decreased to 647.3 ± 0.4 S/cm at 340°C and 414.0 ± 1.2 S/cm at 380°C. The enhanced σ (= 647.3 – 814.3 S/cm) of the films grown at 220–340°C originated from the substantially enhanced μ because the n exhibited a slight decrease [15]. Although the coupled relationship between σ (= $ne\mu$) and $|\alpha|$ ($\sim n^{-2/3}$) generally constrains the concurrent enhancement of σ and $|\alpha|$, a reduction in n and a substantial increase in μ in the same optimal range of T_S (= 220–340°C) could lead to high values of both σ and $|\alpha|$. Consequently, the *PF* of the stoichiometric Bi_2Te_3 films grown in the range of 220–340°C reached remarkably high values, ranging between 18.2 ± 0.25 and 24.3 ± 0.44 μW cm^{-1} K^{-2}, whereas the *PF* was low (≤0.44 μW cm^{-1} K^{-2}) in the case of nonstoichiometric films deposited at $T_S \leq 120$ or 380°C (**Figure 8a**).

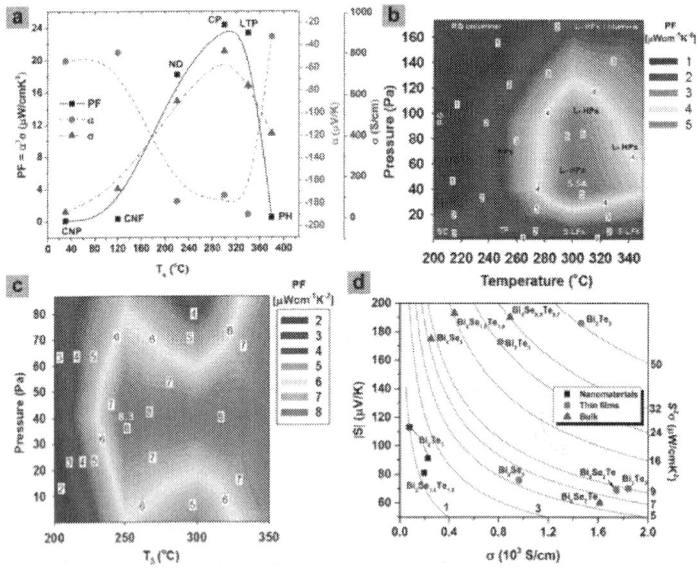

Figure 8. (a) Substrate temperature (T_S) dependence of room temperature Seebeck coefficient α (red circles), electrical conductivity σ (blue triangulars), and power factor (*PF* = $\alpha^2\sigma$, black squares) of the Bi_2Te_3 and Bi_4Te_3 (for "PH" point) films. The morphology abbreviations: CNP, columnar nanoparticle; CNF, columnar nanoflower; ND, nanodisk; CP, compact polycrystalline; LTP, layered triangular platelet; PH, polyhedral. (b) Contour plot of the Bi_2Se_3 film's *PF* as a function of P_{He} and T_S. The morphology abbreviations: SC, smooth and compact; RG, rice grain; TP, triangular polygonal; S-LFs, super-layered flakes; L-HPs, layered hexagonal platelets. (c) Contour plot of the film's *PF* as a function of T_S from 200 to 350°C and P_{He} from 0.027 to 86.7 Pa. (d) $|S|$ vs. σ of the films in this study and the relevant novel TE materials in the literature, listed in **Table 1**. Solid curves denote different *PF*s from 1 to 50 μW cm^{-1} K^{-2}.

In order to check the evolution of the *PF* (= $\alpha^2\sigma$) as a function of P_{He} and T_S, the contour plot is illustrated **Figure 8b**. The *PF* of Bi_2Se_3 films increased with increasing T_S from 200 to 300°C because σ became considerably larger but the Seebeck coefficient diminished only slightly. However, for films deposited at 350°C, *PF* was lowered primarily because of the reduction in

S and not the increase in *σ*. At intermediate pressures (40–93 Pa), the Bi_2Se_3 films remained stoichiometric or slightly Se-rich compositions, which in turn led to the reduced carrier concentrations and significantly enhanced the *α* values [14]. Thus, the *PF* of Bi_2Se_3 films grown at intermediate pressure was typically higher than the *PF* of films grown at a low or high pressure. The optimal value of *PF* was 5.54 ± 0.34 μW cm^{-1} K^{-2} for the layered hexagonal platelet Bi_2Se_3 films deposited at 300°C and 40 Pa [14].

System	Type	Method	*n* (10^{19} cm^{-3})	*μ* (cm^2/Vs)	*σ* (S/cm)	*α* (μV/K)	*PF* (μW cm^{-1} K^{-2})	References
Bi_2Te_3	Layered smooth film	PLD	10.1	90.6	1464	−186	50.6	[39]
Bi_2Te_3	Layered compact polycrystalline	PLD	5.0	102	814.3	−172.8	24.3	[15]
Bi_2Te_3	Nanoparticle film	PLD	9.7	14.8	230	−91	1.9	[34]
Bi_2Te_3	Super-assembled film	PLD	4.0	12.4	79	−113	1.0	[33]
Bi_2Se_3	Layered hexagonal platelets	PLD	7.4	81.4	963.8	−75.8	5.5	[14]
Bi_3Se_2Te	Nanocrystalline film	PLD	35.5	34.4	1747.5	−68.8	8.3	[16]
Bi_2Te_3	Layered structure	Sputtering	95	12.1	1840	−70	8.8	[35]
Bi_2Se_3	Bulk	Melting and hot-pressing	–	–	251.9	−175	7.7	[32]
$Bi_2Se_{0.3}Te_{2.7}$	Bulk	Ball milling hot pressing	–	–	892	−190	32.2	[36]
$Bi_2Se_{1.5}Te_{1.5}$	Bulk	Zone melting	1.2	230	441.6	−193	16.5	[37]
$Bi_2Se_{1.8}Te_{1.2}$	Nanoplatelet bulk	Polyol method	–	–	199.6	−80.9	1.3	[38]
Bi_2Se_2Te	Bulk	Ball milling hot pressing	–	–	1613	−60	5.8	[36]

Table 1. Material, type, method, carrier concentration (*n*), mobility (*μ*), electrical conductivity (*σ*), Seebeck coefficient (*S*), power factor ($PF = S^2\sigma$) of the optimal bismuth chalcogenide films in this study as compared to properties of Bi_2Te_3, Bi_2Se_3, $Bi_2Se_xTe_{1-x}$ bulk and films reported in the literature. All the selected values were recorded at room temperature.

The T_S- and P_{He}-dependent *PF* of nanocrystalline Bi_3Se_2Te films is further shown in **Figure 8c**. The films grown at 200°C only have *PF*s of 1.0–2.8 μW cm^{-1} K^{-2}. The *PF*s of the films grown at higher T_S are significantly enhanced because of their high *σ* values. Around T_S = 250–350°C and P_{He} = 40 Pa, a window for high *PF* is clearly observed. An optimal *PF* of 8.3 μW cm^{-1} K^{-2} is achieved for a Bi_3Se_2Te film deposited at 250°C and 40 Pa.

Table 1 summarizes the transport and room-temperature TE properties of bismuth chalcogenides in the literature [14, 15, 32–39]. For PLD growths, the highly (0 0 1)-oriented layered Bi_2Te_3 films achieved a *PF* of 50.6 μW cm^{-1} K^{-2} [39], and the layered compact polycrystalline film possessed a *PF* value of 24.3 μW cm^{-1} K^{-2} [15]. The Bi_2Se_3 films generally have lower TE properties than those of Bi_2Te_3 films. For example, the optimal *PF* of the Bi_2Se_3 films grown by PLD was 5.5 μW cm^{-1} K^{-2} [14], which was slightly lower than the *PF* of Bi_2Se_3 bulk (*PF* ≈ 7.7 μW cm^{-1} K^{-2}) [32]. The nanocrystalline Bi_3Se_2Te films had an optimal *PF* of 8.3 μW cm^{-1} K^{-2} [16]. Further, PLD growth allows fabrication of nanostructured TE films with different morphologies of nanoparticle Bi_2Te_3 film (*PF* = 1.9 μW cm^{-1} K^{-2}) [34] and super-assembled Bi_2Te_3 film (*PF* = 1.0 μW cm^{-1} K^{-2}) [33]. The Bi_2Te_3 film deposited by the sputtering technique had *PF* of 8.8 μW cm^{-1} K^{-2} [35]. There are some reports of TE properties for bulk materials of bismuth chalcogenides, such as $Bi_2Se_{1.8}Te_{1.2}$ nanoplatelet (*PF* ≈ 1.3 μW cm^{-1} K^{-2}) [38], Bi_2Se_2Te (*PF* ≈ 5.8 μW cm^{-1} K^{-2}), $Bi_2Se_{1.5}Te_{1.5}$ (*PF* ≈ 16.5 μW cm^{-1} K^{-2}) [37], and $Bi_2Se_{0.3}Te_{2.7}$ (*PF* ≈ 32.2 μW cm^{-1} K^{-2}) [36]. Unfortunately, the thermal conductivity κ of the films is missed in the reports to fully evaluate the TE performance of the films. Nevertheless, the κ of polycrystalline films with small grain sizes should be reduced thanks to the extensive phonon scattering at interfaces and grain boundaries.

Finally, **Figure 8d** shows the $|S|$ vs. σ plot for the list in **Table 1**. The solid curves denote different values of *PF*s (= $S^2\sigma$). It can be found that TE nanomaterials usually possess low σ values due to the separating or voided structure morphology, but bulk and thin films have superior σ. Note, the significant reduction in thermal conductivity κ is the key factor for employing nanostructured materials in the TE field.

4. Nanomechanical properties of Bi_2Te_3 and Bi_3Se_2Te thin films

Effects of helium ambient pressure (in PLD) on the nanomechanical properties of Bi_2Te_3 and Bi_3Se_2Te thin films have been investigated [9, 40]. The Bi_2Te_3 thin films were grown at T_S of 250°C on c-plane sapphire substrates using excimer laser PLD with a power density of 5 J/cm^2, at a repetition rate of 2 Hz. The helium pressures in PLD growth varied from 2×10^{-5} to 6.5×10^{-3} Torr. Similarly, Bi_3Se_2Te thin films were deposited on Al_2O_3 (0 0 0 1) substrates at a fixed substrate temperature of 250°C and P_{He} ranging from 2×10^{-5} to 6.5×10^{-1} Torr through PLD. The light source of the PLD system was a KrF excimer laser with λ = 248 nm, pulse duration of 20 ns, fluence of 7.0 J/cm^2, and repetition rate of 2 Hz. The target-to-substrate distance, the number of laser pulses and deposition time were 40 mm, 3000, and 25 min, respectively. The grown films had the average thickness of 200 nm (the average growth rate was ~0.67 Å/pulse). Nanoindentation experiments were performed on a MTS Nano Indenter® XP system with a three-sided pyramidal Berkovich indenter tip using the continuous stiffness measurement (CSM) technique [41]. The hardness and Young's modulus of the Bi_2Te_3 and Bi_3Se_2Te thin films were determined from the load-displacement results through the analytical method proposed by Oliver and Pharr [42].

As shown in **Figure 9a**, the hardness monotonically increased from 2.92 ± 0.12 to 4.02 ± 0.14 GPa for Bi_2Te_3 films, and from 2.5 ± 0.2 to 3.0 ± 0.1 GPa for Bi_3Se_2Te films when P_{He} was increased from 2.0×10^{-5} to 2.0×10^{-3} Torr. Similarly, the Young's modulus of Bi_2Te_3 thin films was 106.31 ± 0.63, 115.51 ± 1.92, and 127.46 ± 9.21 GPa for P_{He} at 2.0×10^{-5}, 2.0×10^{-4}, and 2.0×10^{-3} Torr, respectively. For $P_{He} > 2.0 \times 10^{-3}$ Torr, the hardness (Young's modulus) of Bi_3Se_2Te films continues to increase with increasing P_{He}, namely 3.2 ± 0.1 GPa (105.2 ± 10.2 GPa) at 2.0×10^{-1} Torr and 5.8 ± 0.2 GPa (188.5 ± 4.3 GPa) at 6.5×10^{-1} Torr (**Figure 9b**).

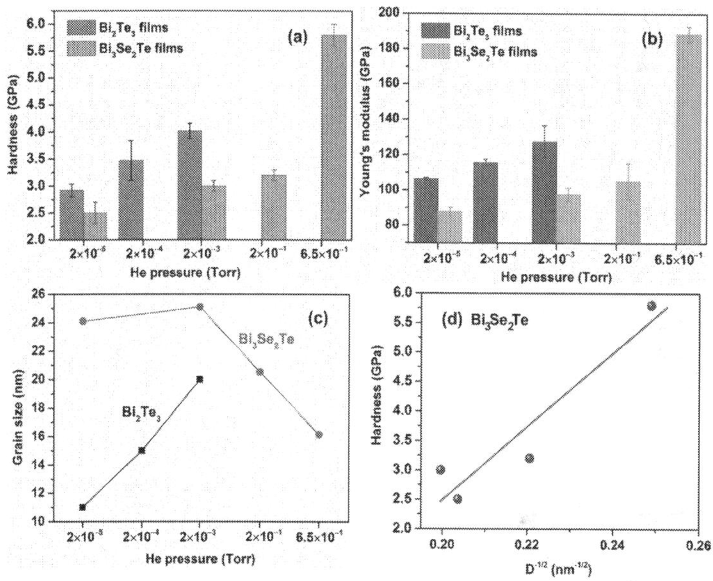

Figure 9. Material and nanomechanical properties of Bi_2Te_3 and Bi_3Se_2Te thin films grown on Al_2O_3 (0 0 0 1) substrates at T_S of 250°C and various helium pressures (P_{He}) between 2.0×10^{-5} and 6.5×10^{-1} Torr [9, 40]: (a and b) the hardness and Young's modulus, (c) grain size (D), (d) the Hall-Petch behavior observed on the Bi_3Se_2Te thin films, in which the hardness is observed to increase approximately with $D^{-1/2}$ (D is grain size).

The crystallite sizes (D) of the films were estimated using the Scherrer equation $D = 0.9\lambda/B\cos\theta$, where λ, B, and θ are the X-ray wavelength, full width at the half maximum of the Bi_2Te_3 (0 0 15) peak or Bi_3Se_2Te (0 0 5) peak, and Bragg diffraction angle, respectively. The P_{He}-dependent D of the Bi_2Te_3 and Bi_3Se_2Te films is shown in **Figure 9c**. The grain size increases monotonically from 11.0 to 20.0 nm for Bi_2Te_3 and from 16.1 to 20.5 nm for Bi_3Se_2Te with increasing P_{He} from 2.0×10^{-5} to 2.0×10^{-3} Torr. In the nanoscale, grain size can affect significantly the mechanical properties of materials. The dislocation activities can be drastically suppressed in a polycrystalline material when the grain size is decreased, and thus the grain boundary sliding and/or grain rotations become the dominant deformation behavior, which in turn would lead to the manifestations of the inverse Hall-Petch effect [43]. Softening caused by grain boundary sliding is mainly attributed to large amount of defects in grain boundaries, which allow rapid diffusion of atoms and vacancies under stress [44]. Consequently, the plastic deformation of Bi_2Te_3 films should be dominated by the grain boundary sliding and/or grain rotation rather than the dislocation activity because of $D \leq 20$ nm [40], which is consistent with the results in Refs. [45–48].

In contrast, the 200-nm-thick Bi_3Se_2Te films with D of 16.1–25.1 nm grown at a larger P_{He} range of 2.0×10^{-5} to 6.5×10^{-1} Torr exhibited the nanomechanical followed Hall-Petch relationship [44, 49]. The hardness and Young's modulus of the Bi_3Se_2Te thin films monotonically increased with increasing P_{He} because of a corresponding decrease in grain sizes (**Figure 9a–c**). **Figure 9d** shows that hardness (H) increased linearly with $D^{-1/2}$ (where D is the grain size of the Bi_3Se_2Te films in the nanoscale regime) which is the typical Hall-Petch relationship [44, 49]. This is because the multiplication and mobility of dislocations are hindered by reducing the grain size [44]. It is reasonable for the observed phenomenon when the present grain sizes ranged between 25.1 and 16.1 nm which is larger than the typical critical D_c of 10 nm [44, 49]. It is demonstrated that the hardness and Young's modulus of the Bi_2Te_3 and Bi_3Se_2Te thin films can be enhanced by proper selection of the ambient pressure in PLD growths.

5. Topological insulator bismuth chalcogenide thin films and their novel properties

5.1. The epitaxial growths of bismuth chalcogenide thin films

Topological insulator (TI), a new class of quantum matter, possesses insulating in bulk and robust gapless topological surface states (TSSs) in which the spin of the electron is locked perpendicular to its momentum by strong spin-orbit interaction [6, 50, 51]. TIs have been identified as promising materials for exploiting exciting physics and developing potential applications in optoelectronics, spintronics, and quantum computations [50–54]. Dirac fermions in TIs have also been studied by angle-resolved photoemission spectroscopy [55–57] or scanning tunneling microscopy [58, 59]. In magnetotransport studies, TSS can be studied by weak antilocalization (WAL) [4, 9, 60, 61] and Shubnikov-de Haas oscillations [3, 62].

Topological insulator bismuth chalcogenide thin films have been grown epitaxially on various substrates using PLD. Onose et al. reported the epitaxial growth of Bi_2Se_3 thin films on InP (1 1 1) substrates (the lattice mismatch of 0.2%) [7]. A designed Se-rich target with an atomic ratio of Bi:Se = 2:8 was used to compensate for the issue of high doping carriers and to avoid unwanted Se-deficient phases. The pulsed laser power and repetition were 140 mJ and 10–20 Hz, respectively. The Bi_2Se_3 films obtained a small full-width at half-maximum (FWHM) for the XRD rocking curve of (0 0 0 6) peak. The surfaces of the films are composed of triangular pyramids with step-and-terrace structures, reflecting the hexagonal symmetry of Bi_2Se_3. The epitaxial relationship is Bi_2Se_3 (0 0 1) || InP (1 1 1) and Bi_2Se_3 [1 1 $\bar{2}$ 0]||InP [0 0 $\bar{1}$]. Le et al. reported the epitaxial growth of Bi_2Se_3 films on $SrTiO_3$ (STO) (1 1 1) substrates using PLD at T_S of 300 and 350°C [63]. The PLD conditions were the pulse fluence of 3.7 J/cm^2, helium pressure of 40 Pa (300 mTorr), and repetition rate of 2 Hz. The laser source was KrF excimer laser (λ = 248 nm, duration 20 ns). By comparing the Bi_2Se_3 {0 1 5} and STO {2 0 0} diffraction peaks, the epitaxial relationship between the film and substrate was determined to be Bi_2Se_3 (0 0 1) || STO (1 1 1) and Bi_2Se_3 [1 1 0] || STO [$\bar{1}$ 1 0] [63].

Figure 10 presents the PLD epitaxial growths of Bi_2Te_3, Bi_2Se_3, and Bi_3Se_2Te thin films on large-misfit substrates [9, 53, 64]. The PLD conditions for growing Bi_2Te_3 films on STO (1 0 0) were as follows: substrate temperature of 300°C; helium ambient pressure of 40 Pa; repetition rate of 2 Hz; pulsed fluence of approximately 3.4 J/cm². As shown in **Figure 10a**, a ϕ-scan was conducted on the (0 1 5) plane of a 200-nm-thick Bi_2Te_3 film and the (1 1 1) plane of an STO (1 0 0) substrate in skew symmetric geometry by tilting the samples. The in-plane orientation of a hexagonal h-Bi_2Te_3/STO (1 0 0) film displayed a 12-fold symmetry instead of the expected six-

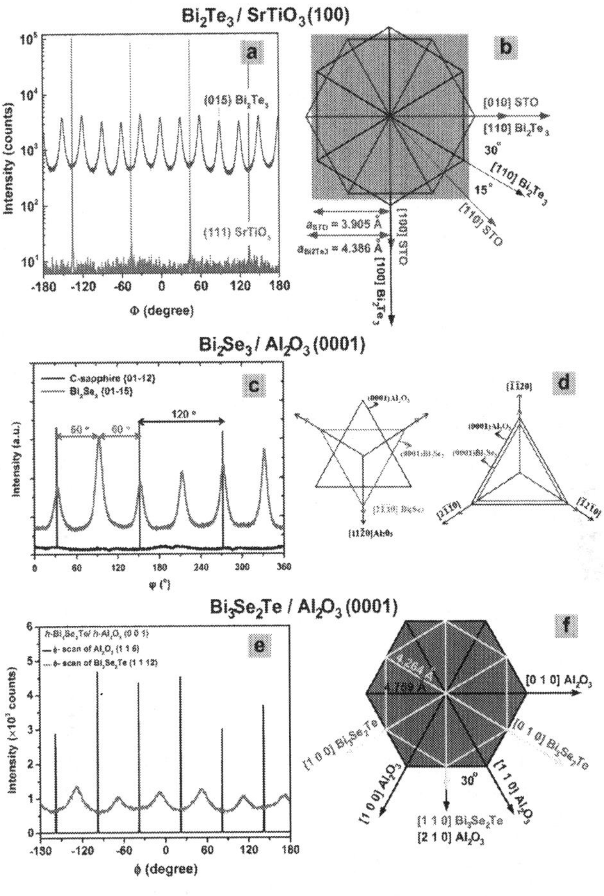

Figure 10. Bi_2Te_3 films grown on $SrTiO_3$ (1 0 0) substrates [53]: (a) XRD ϕ-scan patterns of the 200-nm-thick Bi_2Te_3 (0 1 5) plane and the $SrTiO_3$ (1 1 1) plane; (b) schematics of the in-plane arrangement of h-Bi_2Te_3 / $SrTiO_3$ (1 0 0). Bi_2Se_3 films grown on c-plane sapphire substrates [64]; (c) in-plane ϕ-scan $\{0\,1\,\bar{1}\,5\}$ planes of the film and $\{0\,1\,\bar{1}\,2\}$ planes of the substrate; (d) the schematic depicting twin/domain growth: in one case, the basal plane of Bi_2Se_3 is aligned with that of Al_2O_3, and in the other, there is a rotation of 60°/180°. Bi_3Se_2Te films grown on c-plane sapphire substrates [9]; (e) ϕ-scan patterns of hexagonal h-Bi_3Se_2Te thin films grown on Al_2O_3 (0 0 0 1) substrates; and (f) schematics of the in-plane arrangement of hexagonal h-Bi_3Se_2Te/h-Al_2O_3 (0 0 0 1).

fold symmetry of the (0 1 5) plane in Bi_2Te_3. **Figure 10b** shows a schematic drawing of the in-plane atomic arrangement between an h-Bi_2Te_3 film and an STO (1 0 0) substrate. Since the principal crystallographic orientations of h-Bi_2Te_3 films grown on STO (1 0 0) substrates can be aligned along either the STO [1 0 0] or STO [0 1 0] directions, the in-plane arrangements result in an observed 12-fold symmetry. The angle differences between STO [0 1 0] and the two orientations of h-Bi_2Te_3 [1 1 0] were 30° and 0°, respectively, as shown in **Figure 10b**. In other words, the in-plane relationships were Bi_2Te_3 [1 1 0] || STO [0 1 0] and Bi_2Te_3 [1 0 0] || STO [1 0 0]. Lee et al. reported epitaxial growth via domain matching epitaxy of Bi_2Se_3 thin films on Al_2O_3 (0 0 0 1) substrates with over 13% lattice misfit and a critical thickness of less than one monolayer [64]. A relatively low repetition rate of 0.2 Hz and low T_S of 250°C are key parameters of the PLD growth to achieving high-quality Bi_2Se_3 epitaxial films. **Figure 10c** shows ϕ-scan XRD results performed on $\{0\ 1\ \bar{1}\ 5\}$ planes of the film and $\{0\ 1\ \bar{1}\ 2\}$ planes of the substrate. Clearly, the presence of six peaks corresponding to the Bi_2Se_3 $\{0\ 1\ \bar{1}\ 5\}$ planes confirmed the epitaxy. A schematic depicting twin/domain growth and direction in the study is illustrated in **Figure 10d** [64]. In one case, the Bi_2Se_3 basal plane is aligned with that of Al_2O_3, and in the other, there is a rotation of 60°/180°. The epitaxial relationships are written as (0 0 0 1) Bi_2Se_3 || (0 0 0 1) Al_2O_3 (out-of-plane) and $[2\ \bar{1}\ \bar{1}\ 0]$ Bi_2Se_3 || $[2\ \bar{1}\ \bar{1}\ 0]$ Al_2O_3 or $[2\ \bar{1}\ \bar{1}\ 0]$ Bi_2Se_3 || $[1\ 1\ \bar{2}\ 0]$ Al_2O_3 (in-plane). **Figure 10e** shows the typical ϕ-scan patterns of hexagonal h-Bi_3Se_2Te/h-Al_2O_3 (0 0 0 1) [9]. With the skew symmetric geometry, the ϕ-scan measurements were performed on the (1 1 6) plane of Al_2O_3 substrates and the (1 1 12) plane of Bi_3Se_2Te films. As shown in **Figure 10e**, the in-plane orientations of both Al_2O_3 (1 1 6) and Bi_3Se_2Te (1 1 12) exhibited six-fold symmetries with a 30° difference. **Figure 10f** illustrates the in-plane atomic arrangement between h-Bi_3Se_2Te and h-Al_2O_3 (0 0 0 1). The epitaxial relationship between the films and substrates is (0 0 0 1) Bi_3Se_2Te || (0 0 0 1) Al_2O_3 and [1 1 0] Bi_3Se_2Te || [2 1 0] Al_2O_3 [9]. This in-plane orientation was established to obtain the optimal lattice matching.

5.2. Magnetotransport properties of bismuth chalcogenide thin films

The weak antilocalization (WAL) which is a negative quantum correction to classical MR caused by the wave nature of electrons is used as a signature of TSS. In TIs, WAL is induced by both the helicity of the surface state and the spin-orbit coupling of bulk [4, 61, 65, 66]. In a low B field, the 2D WAL MR of a system with strong spin-orbit interaction can be described using the Hikami-Larkin-Nagaoka model [65], which is [4, 60, 61]

$$\frac{\Delta R_W(B)}{[R_W(0)]^2} = -\alpha\frac{e^2}{2\pi^2\hbar}\left[\Psi\left(\frac{1}{2}+\frac{B_\phi}{B}\right)-\ln\left(\frac{B_\phi}{B}\right)\right] \tag{8}$$

where R_W is the sheet resistance, $\Delta R_W = R_W(B) - R_W(0)$, $\Psi(x)$ is the digamma function, $B_\phi = \hbar/(4eL_\phi^2)$ is a magnetic field varying with the coherence length L_ϕ, α is a parameter and reflects the number of conduction channels. In a 3D TI, $\alpha = -1/2$ for a single coherent transport channel in the 2D surface states, and $\alpha = -1$ for two independent coherent transport channels with similar L_ϕ in the 2D surface states [60, 65].

The typical magnetoresistance (MR) results of some bismuth chalcogenides (i.e., Bi_2Te_3, Bi_2Se_3, and Bi_3Se_2Te) thin films grown by PLD are presented in **Figure 11** [9, 64, 67]. The Bi_2Te_3, Bi_2Se_3,

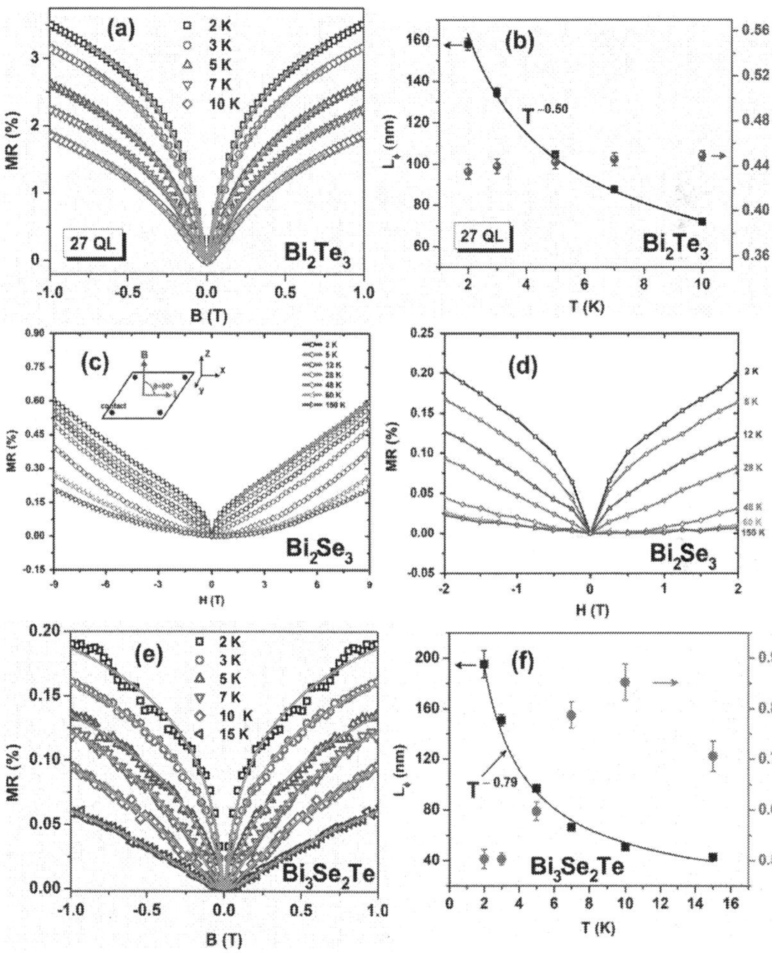

Figure 11. Magnetoresistance (MR) results of the Bi_2Te_3, Bi_2Se_3, and Bi_3Se_2Te thin films grown by PLD [9, 64, 67]. (a and e) The MR ($B = \pm 1$ T) of a 27-nm-thick Bi_2Te_3 and a 200-nm-thick Bi_3Se_2Te films grown on Al_2O_3 (0 0 0 1) substrates. (b and f) Variation in the extracted electron dephasing length L_ϕ and parameter $-\alpha$ of the films as a function of temperatures. (c and d) MR of the epitaxial Bi_2Se_3 films grown on Al_2O_3 (0 0 0 1) substrates at T_S = 250°C as a function of temperatures [64]. The solid green lines in (a) and (e) in low B are the theoretical predictions of 2D weak antilocalization (WAL) using Eq. (8).

and Bi_3Se_2Te thin films were grown on Al_2O_3 (0 0 0 1) substrates using PLD at T_S of 225, 250, and 250°C, respectively [9, 64, 67]. The WAL effect which presents as a sharp increase in resistance when B increases in the low magnetic field B regime is clearly observed on the films. **Figure 11a, b, e** and **f** show MR curves at several temperatures (T) and the extracted $\alpha(T)$ and $L_\phi(T)$ values using Eq. (8) for the 27-QL (~27 nm)-thick Bi_2Te_3 and 200-nm-thick Bi_3Se_2Te thin films [9, 67]. At 2 K, L_ϕ are 158.1 and 195.2 nm, and L_ϕ decreases monotonically with increasing T, obeying the power laws, as $L_\phi \sim T^{-0.50}$ and $L_\phi \sim T^{-0.79}$ for the Bi_2Te_3 and Bi_3Se_2Te films, respectively (**Figure 11b** and **f**). Theoretically, the result of $L_\phi \sim T^{-0.50}$ observed in the 27-QL-thick Bi_2Te_3 thin film indicates the predominant electron-electron scattering in 2D weakly disordered systems. The $L_\phi \sim T^{-0.79}$ is closed to the $L_\phi \sim T^{-0.75}$ for 3D systems, if e-e scattering is the dominant dephasing source [9]. The electron screening effect in 3D system is more effective than that in 2D systems. However, the e-e scattering is strongly weakened with high carrier densities ($n = 10^{20}$ cm^{-3} for the Bi_3Se_2Te films [9]). In 3D disordered conductors (namely, in the bulk state), dephasing by electron-phonon (e-ph) scattering would be significant and dominant. The e-ph scattering also causes a faster decay rate on L_ϕ, that is, $L_\phi \sim T^{-1.0}$ [68]. Additionally, the combination of 2D e-e scattering in the TSS thin layer and e-ph scattering in the bulk would further result in the $L_\phi \sim T^{-0.79}$. Consequently, e-ph scattering could be the force of channel separation as temperatures increase. It is worthy of mentioning that the MR result is a signature (not conclusive evidence) for the presence of TSS on Bi_3Se_2Te films. Further theoretical calculations and experiments are needed for reaching a final conclusion.

Figure 11b and **f** also present the $-\alpha(T)$ results of the Bi_2Te_3 and Bi_3Se_2Te films. The $-\alpha$ values of the Bi_2Te_3 film increase with increasing T from 0.43 at 2 K to 0.45 at 10 K, indicating the existence of a single coherent transport channel (i.e., likely a single surface state) [61]. Meanwhile, the $-\alpha$ values of the 200-nm-thick Bi_3Se_2Te film increase with increasing T from 0.5 at 2 K to 0.85 at 10 K and to 0.71 at 15 K, suggesting increased channel separation with T [69]. In WAL, the independent phase-coherent channels occur when the carriers in one channel lose phase coherence before being scattered into the other channel.

In **Figure 11c**, the MR (B,T) of a 50-nm-thick Bi_2Se_3 film was measured with B perpendicular to the film plane and ranging from −9 T to +9 T. At low B and T, the distinctive dips of WAL are clearly observed. The WAL effect results from strong spin-orbit coupling, showing the absence of backscattering giving rise to the destructive interference between the two time reversal symmetry loops when there is no magnetic field [64]. Resistance increases sharply with increasing B because the quantum interference is destroyed and backscattering increases (**Figure 11c**). **Figure 11d** is the enlarged MR at low B (−2 T to +2 T), and it reveals the existence of WAL as a function of T in a discernible way [64]. WAL gradually weakens as temperature increases, eventually disappearing entirely at $T = 48$ K; thus, the dependence on B is quadratic-like at low field. The MR cusp feature at low B is broadened and finally disappears with increasing temperature owing to the decrease in the phase coherence length. In addition to the WAL effect, **Figure 11c** shows a 2D, non-saturating linear MR at high B, which usually occurs with several TI materials of Bi_2Se_3 [70], Bi_2Te_3 [71], Bi_2Te_2Se [72], and Bi_3Se_2Te [9]. Theoretical models propose that the linear MR can appear in the gapless linear-dispersive energy spectrum when only the first landau level is filled [73, 74], or in the presence of both the gapless linear spectrum and Landau level overlaps [75]. Noticeably, the WAL and linear MR simultaneously reflect the 3D contribution of spin-orbit coupling in bulk and the Dirac nature of the 2D surface states. Because the magneto transport is a bulk sensitive measurement, it remains a major challenge to directly probe the topological nature [64].

5.3. Proximity-induced superconductivities in Bi inclusions/bismuth chalcogenide thin films

Recent studies have shown a two-dimensional interface state between TIs and superconductors resulting from the superconducting proximity effect that supports Majorana fermions [76, 77]. Majorana fermions, novel particles which are their own antiparticles, can potentially be applied to topological quantum computing, which has motivated intense interest in TIs [53].

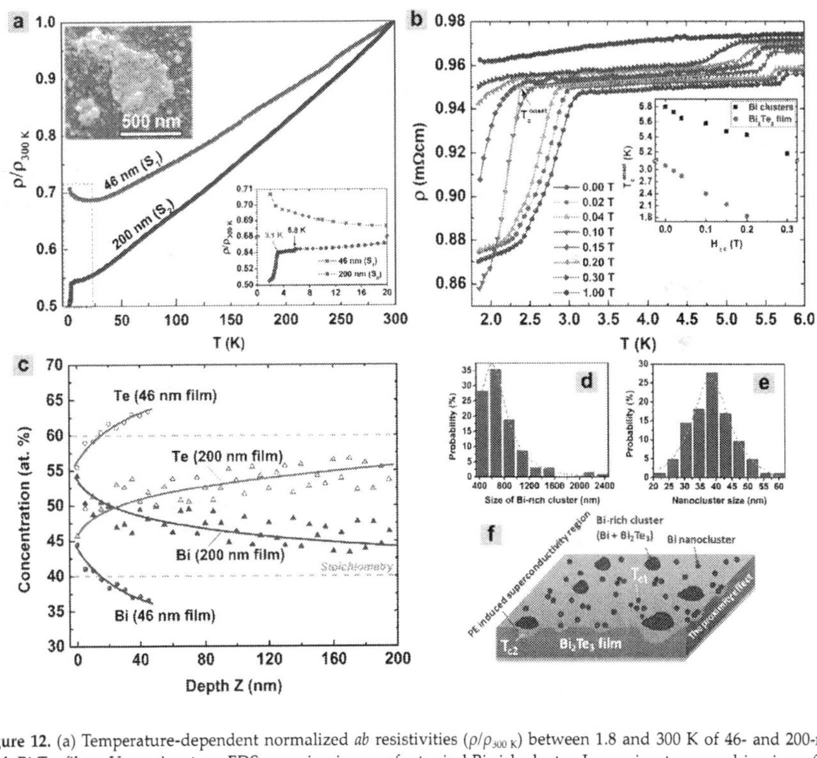

Figure 12. (a) Temperature-dependent normalized *ab* resistivities ($\rho/\rho_{300\ K}$) between 1.8 and 300 K of 46- and 200-nm-thick Bi_2Te_3 films. Upper inset: an EDS mapping image of a typical Bi-rich cluster. Lower inset: zoomed-in view of the $\rho/\rho_{300\ K}$ in the low temperature range. (b) $\rho(T)$ in 1.75–6.0 K of the 200 nm film at various $H_{||c}$ from 0 to 1 T. Inset: the onset T_c of the two superconducting transitions as a function of magnetic field. (c) Auger electron spectroscopy (AES) elemental depth profiling of a non-superconducting (46-nm-thick) and a superconducting (200-nm-thick) Bi_2Te_3 films. (d and e) The size distribution of Bi-rich clusters and Bi nanoclusters inside the clusters. (f) Schematics of the surface characteristics and a suggested superconducting mechanism in the Bi_2Te_3 films [53].

Koren et al. observed the local superconductivity in Bi_2Te_2Se and Bi_2Se_3 films below 2–3 K, which was naturally induced by small amounts of superconducting Bi inclusions or precipitations on the surface [78]. Moreover, Le et al. reported superconductivity at an onset critical temperature of approximately 3.1 K in a topological insulator 200-nm-thick Bi_2Te_3 thin film grown by pulsed laser deposition [53]. Indeed, **Figure 12a** shows the normalized resistivity $\rho/\rho_{300\ K}$ of a 46-nm-thick Bi_2Te_3 film (S_1) and a 200-nm-thick Bi_2Te_3 film (S_2) as functions of temperatures (T) from 1.8 to 300 K. Both films show a decrease in resistivity (ρ) with decreasing T in the range of 20–300 K, implying that the films exhibit weak metallic properties commonly seen in narrow band-gap semiconductors with high carrier concentrations [53]. Below 20 K (**Figure 12a**, the lower inset), the $\rho/\rho_{300\ K}$ of S_1 shows a gentle upturn because of the weak localization of electrons [7], whereas the $\rho/\rho_{300\ K}$ of S_2 reaches a plateau before dropping slightly at $T_{c1} \approx 5.8$ K and then sharply at $T_{c2} \approx 3.1$ K. **Figure 12b** further shows the $H_{\parallel c}$-dependent $\rho(T)$ of S_2 in the low T regime, where $H_{\parallel c}$ is the applied magnetic field along the c-axis of the film. At $H_{\parallel c} = 0$, ρ drops abruptly by 8% below T_{c2}^{onset} but does not go down to zero, even at $T = 1.8$ K. This nonzero ρ at low T indicates that the superconducting volume ratio is not 100%. The inset in **Figure 12b** shows that the $T_{c2}^{onset}(T_{c1}^{onset})$ decreases from 3.1 (5.8) to 1.8 (5.4) K with increasing $H_{\parallel c}$ from 0 to 0.2 T, strongly indicating that both transitions are superconducting in nature.

The detailed investigations of S_2 strongly suggest the existence of superconducting Bi nanoclusters on the surface that induce the $T_{c1} \sim 5.8$ K. EDS lateral elemental mapping revealed that the distributions of Te and Bi were not uniform, and many Bi-rich (47–54 at.%) clusters were visible (green color), as shown in the upper inset in **Figure 12a**, differing substantially from the uniform distribution and cluster-free surface observed in film S_1. The size distribution and the most probable size of Bi-rich clusters are in the range of 400–2400 nm and 560–772 nm, respectively. Intriguingly, a closer inspection reveals that Bi-rich clusters are composed by some Bi nanoclusters (or nanograins) with a size of 20–62 nm.

The Bi-rich environment on the film surface is confirmed by AES analysis (**Figure 12c**) [53]. This is because the vapor pressure of Te (at $T_S = 300°C$) is approximately 10^5 times higher than that of Bi, and therefore, more Te atoms are re-evaporated from the 300°C substrates [79]. The loss of Te is more severe in film S_2 than in film S_1 (14.3% in S_2 and 4.5% in S_1 at depth Z = 0, **Figure 12c**) because of the six times longer deposition time of S_2 (60 min) compared to S_1 (10 min). In addition, the nonstoichiometric effect is strongly depth-dependent (**Figure 12c**). The Te/Bi ratio gradually increases toward the stoichiometric ratio of 3/2 in 200-nm-thick films or slightly exceeds it (Te-rich) in 46-nm-thick films when the depth (Z) of the films increases. Under such a sufficiently high surface concentration of Bi atoms, the Bi clusters precipitate and segregate readily on the S_2 surface to minimize overall free energy, as long as the substrate temperature of 300°C is higher than the 271°C melting point of Bi, as demonstrated in **Figure 12f**. Notably, the Bi clusters can only be observed in highly Bi-rich (14.3% at Z = 0) films (S_2) and not in low Bi-rich (4.5% at Z = 0) films (S_1), suggesting a critical Bi-rich concentration for Bi precipitation (separating a Bi phase) in a Bi_2Te_3 film [53]. The T_{c1} at ~5.8 K found in our samples should be induced by the superconducting transition of the Bi nanoclusters, which is closely consistent with the T_c of 6.3 K for the surface Bi islands observed in Bi_2Te_2Se films [78]. The tiny resistivity drop at T_{c1} = 5.8 K (by approximately 0.5%, **Figure 12b**) indicates that the amount of superconducting Bi nanoclusters in S_2 is likely small and, therefore, the Josephson coupling between these islands is extremely weak. Since the superconductivity of Bi nanoclusters survived until $H_{||c}$ = 1.0 T (**Figure 12b**), the critical field of Bi nanoclusters is in between 0.3 and 1.0 T. This section demonstrates that natural defects generated during PLD growths, namely superconducting Bi nanoclusters or Bi inclusions, can substantially induce non-superconducting TI thin films (i.e., Bi_2Te_3, Bi_2Se_3, and Bi_2Te_2Se) into superconducting states at low temperatures.

6. Conclusion

This chapter provides the effects of ambient pressures and substrate temperatures in PLD growths on the structural-morphology, thermoelectric, nanomechanical, and magnetoresistance properties of bismuth chalcogenide thin films. The thermoelectric power factor of the stoichiometric Bi_2Te_3 films grown in the range of 220–340°C and P_{Ar} of 80 Pa reached remarkably high values, ranging between 18.2 ± 0.25 and 24.3 ± 0.44 μW cm^{-1} K^{-2}. The optimal *PF* values were 5.54 ± 0.34 μW cm^{-1} K^{-2} for the layered hexagonal platelet Bi_2Se_3 films deposited at 300°C and P_{He} of 40 Pa and 8.3 μW cm^{-1} K^{-2} for the nanocrystalline Bi_3Se_2Te films deposited at 250°C and P_{He} of 40 Pa. We also reported the effects of P_{He} in PLD on nanomechanical properties (i.e., hardness and Young's modulus) of Bi_2Te_3 and Bi_3Se_2Te thin films. It was observed that the hardness and Young's modulus increased with increasing P_{He}, depending on the grain sizes following the inverse Hall-Petch effect for Bi_2Te_3 films grown at $P_{He} \leq 2.0 \times 10^{-3}$ Torr and following the Hall-Petch relationship for Bi_3Se_2Te grown at P_{He} of 2.0×10^{-5} to 6.5×10^{-1} Torr. PLD has been successfully employed to grow epitaxially bismuth chalcogenide thin films on large-misfit substrates, for example, $Bi_2Te_3/SrTiO_3$ (1 0 0), Bi_2Se_3/Al_2O_3 (0 0 0 1), and Bi_3Se_2Te/Al_2O_3 (0 0 0 1). The magnetotransport studies show that the bismuth chalcogenide thin films such as Bi_2Te_3, Bi_2Se_3, and Bi_3Se_2Te films present a two-dimensional weak antilocalization effect in a low magnetic field (*B*) regime and linear magnetoresistance in a high *B* regime, which could be attributed to the topological insulator surface states. Furthermore, proximity-induced superconductivities in Bi_2Te_3 thin films have an onset T_c of approximately 3.1 K, evidently induced by Bi inclusions (nanoclusters with onset T_c at 5.8 K) segregated on the surface of films.

Acknowledgements

Financial support from Vietnam National Foundation for Science and Technology Development (NAFOSTED) under Grant number 103.99–2015.17, the Ministry of Science and Technology, Taiwan under Contract Nos.: 103-2923-M-009-001-MY3, 103-2628-M-009-002-MY3, 103-2119-M-009-004-MY3, and the Ministry of Education (MOE-ATU plan at National Chiao Tung University) are gratefully acknowledged.

Author details

Phuoc Huu Le[1*] and Chih Wei Luo[2,3*]

*Address all correspondence to: lhuuphuoc@ctump.edu.vn and cwluo@mail.nctu.edu.tw

1 Faculty of Basic Sciences, Can Tho University of Medicine and Pharmacy, Can Tho, Vietnam

2 Department of Electrophysics, National Chiao Tung University, Hsinchu, Taiwan

3 Taiwan Consortium of Emergent Crystalline Materials, Ministry of Science and Technology, Taipei, Taiwan

References

[1] Venkatasubramanian R, Siivola E, Colpitts T, O'Quinn B. Thin-Film thermoelectric devices with high room-temperature figures of merit. Nature. 2001; 413: 597–602. doi: 10.1038/35098012

[2] Snyder GJ, Toberer ES. Complex thermoelectric materials. Nat. Mater. 2008; 7: 105–114. doi:10.1038/nmat2090

[3] Qu D-X, Hor YS, Xiong J, Cava RJ, Ong NP. Quantum oscillations and hall anomaly of surface states in the topological insulator Bi_2Te_3. Science. 2010; 329: 821–824. doi: 10.1126/science.1189792

[4] Bao L, He L, Meyer N, Kou X, Zhang P, Chen Z, Fedorov AV, Zou J, Riedemann TM, Lograsso TA, Wang KL, Tuttle G, Xiu F. Weak anti-localization and quantum oscillations of surface states in topological insulator Bi_2Se_2Te. Sci. Rep. 2012; 2: 726 (1–7). doi:10.1038/srep00726

[5] Hsieh D, Qian D, Wray L, Xia Y, Hor YS, Cava RJ, Hasan MZ. A topological Dirac insulator in a quantum spin Hall phase. Nature. 2008; 452: 970–974. doi:10.1038/nature06843

[6] Moore JE. The birth of topological insulators. Nature. 2010; 464: 194–198. doi:10.1038/nature08916

[7] Onose Y, Yoshimi R, Tsukazaki A, Yuan H, Hidaka T, Iwasa Y, Kawasaki M, Tokura Y. Pulsed laser deposition and ionic liquid gate control of epitaxial Bi_2Se_3 thin films. Appl. Phys. Express. 2011; 4: 083001. doi:10.1143/APEX.4.083001

[8] Zhang HB, Yu HL, Bao DH, Li SW, Wang CX, Yang GW. Weak localization bulk state in a topological insulator Bi_2Te_3 film. Phys. Rev. B. 2012; 86: 075102. doi:10.1103/PhysRevB.86.075102

[9] Le PH, Chiu S-P, Jian S-R, Luo CW, Lin J-Y, Lin J-J, Wu KH, Gospodinov M. Nanomechanical, structural, and transport properties of Bi_3Se_2Te thin films. J. Alloys Compd. 2016; 679: 350–357. doi:10.1016/j.jallcom.2016.03.226

[10] Kaiser N. Review of the fundamentals of thin-film growth. Appl. Opt. 2002; 41: 3053–3060. doi:10.1364/AO.41.003053

[11] Rowe DM. Thermoelectrics handbook: macro to nano. In: Rowe DM (Ed) Boca Raton: CRC Press; 2006. 1014 p. doi:10.1201/9781420038903

[12] Thornton JA. Influence of apparatus geometry and deposition conditions on the structure and topography of thick sputtered coatings. J. Vac. Sci. Technol. 1974; 11: 666–670. doi:10.1116/1.1312732

[13] Karabacak T, Singh JP, Zhao Y-P, Wang G-C, Lu T-M. Scaling during shadowing growth of isolated nanocolumns. Phys. Rev. B. 2003; 68: 125408. doi:10.1103/PhysRevB.68.125408

[14] Le PH, Liao C-N, Luo CW, Lin J-Y, Leu J. Thermoelectric properties of bismuth-selenide films with controlled morphology and texture grown using pulsed laser deposition. Appl. Surf. Sci. 2013; 285: 657–663. doi:10.1016/j.apsusc.2013.08.107

[15] Le PH, Liao C-N, Luo CW, Leu J. Thermoelectric properties of nanostructured bismuth–telluride thin films grown using pulsed laser deposition. J. Alloy. Compd. 2014; 615: 546–552. doi:10.1016/j.jallcom.2014.07.018

[16] Tuyen LTC, Le PH, Luo CW, Leu J. Thermoelectric properties of nanocrystalline Bi_3Se_2Te thin films grown using pulsed laser deposition. J. Alloy. Compd. 2016; 673: 107–114. doi:10.1016/j.jallcom.2016.03.006

[17] Ohring M. The materials science of thin films. 2nd ed. San Diego: Academic Press; 2001. 794 p. ISBN: 978-0-12-524975-1

[18] Tritt TM. Thermoelectric phenomena, materials, and applications. Annu. Rev. Mater. Res. 2011; 41: 433–448. doi:10.1146/annurev-matsci-062910-100453

[19] Lind H, Lidin S, Häussermann U. Structure and bonding properties of $(Bi_2Se_3)_m(Bi_2)_n$ stacks by first-principles density functional theory. Phys. Rev. B. 2005; 72: 184101. doi: 10.1103/PhysRevB.72.184101

[20] Bos JWG, Zandbergen HW, Lee M-H, Ong NP, Cava RJ. Structures and thermoelectric properties of the infinitely adaptive series $(Bi_2)_m(Bi_2Te_3)_n$. Phys. Rev. B. 2007; 75: 195203. doi:10.1103/PhysRevB.75.195203

[21] Bell LE. Cooling, heating, generating power, and recovering waste heat with thermoelectric systems. Science. 2008; 321: 1457–1461. doi:10.1126/science.1158899

[22] Nolas GS, Sharp J, Goldsmid HJ. Thermoelectrics basic principles and new materials developments. 1st ed. Berlin: Springer; 2001. 293 p. doi:10.1007/978-3-662-04569-5

[23] Lan Y, Minnich AJ, Chen G, Ren Z. Enhancement of thermoelectric figure-of-merit by a bulk nanostructuring approach. Adv. Funct. Mater. 2010; 20: 357–376. doi:10.1002/adfm.200901512

[24] Alam H and Ramakrishna S. A review on the enhancement of figure of merit from bulk to nano-thermoelectric materials. Nano Energy. 2013; 2: 190–212. doi:10.1016/j.nanoen.2012.10.005

[25] Pichanusakorn P, Bandaru P. Nanostructured thermoelectrics. Mater. Sci. Eng. R. 2010; 67: 19–63. doi:10.1016/j.mser.2009.10.001

[26] Heremans JP, Jovovic V, Toberer ES, Saramat A, Kurosaki K, Charoenphakdee A, Yamanaka S, Snyder GJ. Enhancement of thermoelectric efficiency in PbTe by distortion of the electronic density of states. Science. 2008; 321: 554–557. doi:10.1126/science.1159725

[27] Nielsch K, Bachmann J, Kimling J, Böttner H. Thermoelectric nanostructures: from physical model systems towards nanograined composites. Adv. Energy Mater. 2011; 1: 713–731. doi:10.1002/aenm.201100207

[28] He L, Xiu F, Wang Y, Fedorov AV, Huang G, Kou X, Lang M, Beyermann WP, Zou J, Wang KL. Epitaxial growth of Bi_2Se_3 topological insulator thin films on Si (1 1 1). J. Appl. Phys. 2011; 109: 103702. doi:10.1063/1.3585673

[29] Faleev SV, Léonard F. Theory of enhancement of thermoelectric properties of materials with nanoinclusions. Phys. Rev. B. 2008; 77: 214304. doi:10.1103/PhysRevB.77.214304

[30] Zebarjadi M, Esfarjani K, Shakouri A, Bian Z, Bahk JH, Zeng G, Bowers J, Lu H, Zide J, Gossard A. Effect of nanoparticles on electron and thermoelectric transport. J. Electron. Mater. 2009; 38: 954–959. doi:10.1007/s11664-008-0656-4

[31] Sumithra S, Takas NJ, Misra DK, Nolting WM, Poudeu PFP, Stokes KL. Enhancement in thermoelectric figure of merit in nanostructured Bi_2Te_3 with semimetal nanoinclusions. Adv. Energy Mater. 2011; 1: 1141–1147. doi:10.1002/aenm.201100338

[32] Sun G, Qin X, Li D, Zhang J, Ren B, Zou T, Xin H, Paschen SB, Yan X. Enhanced thermoelectric performance of n-type Bi_2Se_3 doped with Cu. J. Alloys Compd. 2015; 639: 9–14. doi:10.1016/j.jallcom.2015.03.124

[33] Chang H-C, Chen T-H, Whang W-T, Chen C-H. Superassembling of Bi_2Te_3 hierarchical nanostructures for enhanced thermoelectric performance. J. Mater. Chem. A 2015; 3: 10459–10465. doi:10.1039/c5ta00911a

[34] Chang H-C, Chen C-H. Self-assembled bismuth telluride films with well-aligned zero- to three-dimensional nanoblocks for thermoelectric applications. CrystEngComm. 2011; 13: 5956. doi:10.1039/c1ce05350g

[35] Deng Y, Liang H-M, Wang Y, Zhang Z-W, Tan M, Cui J-L. Growth and transport properties of oriented bismuth telluride films. J. Alloys Compd. 2011; 509: 5683–4587. doi:10.1016/j.jallcom.2011.02.123

[36] Liu W, Lukas KC, McEnaney K, Lee S, Zhang Q, Opeil CP, Chen G, Ren Z. Studies on the Bi_2Te_3–Bi_2Se_3–Bi_2S_3 system for mid-temperature thermoelectric energy conversion. Energy Environ. Sci. 2013; 6: 552–560. doi:10.1039/c2ee23549h

[37] Wang S, Tan G, Xie W, Zheng G, Li H, Yang J, Tang X. Enhanced thermoelectric properties of $Bi_2(Te_{1-x}Se_x)_3$-based compounds as n-type legs for low-temperature power generation. J. Mater. Chem. 2012; 22: 20943. doi:10.1039/c2jm34608g

[38] Soni A, Yanyuan Z, Ligen Y, Aik MKK, Dresselhaus MS, Xiong Q. Enhanced thermoelectric properties of solution grown $Bi_2Te_{3-x}Se_x$ nanoplatelet composites. Nano Lett. 2012; 12: 1203–1209. doi:10.1021/nl2034859

[39] Li Bassi A, Bailini A, Casari CS, Donati F, Mantegazza A, Passoni M, Russo V, Bottani CE. Thermoelectric properties of Bi–Te films with controlled structure and morphology. J. Appl. Phys. 2009; 105: 124307. doi:10.1063/1.3147870

[40] Tasi C-H, Tseng Y-C, Jian S-R, Liao Y-Y, Lin C-M, Yang P-F, Chen D-L, Chen H-J, Luo C-W, Juang J-Y. Nanomechanical properties of Bi_2Te_3 thin films by nanoindentation. J. Alloys Compd. 2015; 619: 834–838. doi:10.1016/j.jallcom.2014.09.028

[41] Li X, Bhushan B. A review of nanoindentation continuous stiffness measurement technique and its applications. Mater. Charact. 2002; 48: 11–36. doi:10.1016/S1044-5803(02)00192-4

[42] Oliver WC, Pharr GM. An improved technique for determining hardness and elastic modulus using load and displacement sensing indentation experiments. J. Mater. Res. 1992; 7: 1564–1583. doi:10.1557/JMR.1992.1564

[43] Schiøtz J, Vegge T, Di Tolla FD, Jacobsen KW. Atomic-scale simulations of the mechanical deformation of nanocrystalline metals. Phys. Rev. B. 1999; 60: 11971–11983. doi: 10.1103/PhysRevB.60.11971

[44] Zhang S, Sun D, Fu Y, Du H. Recent advances of superhard nanocomposite coatings: a review. Surf. Coatings Technol. 2003; 167: 113–119. doi:10.1016/S0257-8972(02)00903-9

[45] Masumura RA, Hazzledine PM, Pande CS. Yield stress of fine grained materials. Acta Mater. 1998; 46: 4527–4534. doi:10.1016/S1359-6454(98)00150-5

[46] Greer JR, De Hosson JTM. Plasticity in small-sized metallic systems: intrinsic versus extrinsic size effect. Prog. Mater. Sci. 2011; 56: 654–724. doi:10.1016/j.pmatsci.2011.01.005

[47] Chang SY, Chang T-K. Grain size effect on nanomechanical properties and deformation behavior of copper under nanoindentation test. J. Appl. Phys. 2007; 101: 033507. doi: 10.1063/1.2432873

[48] Kumar KS, Suresh S, Chisholm MF, Horton JA, Wang P. Deformation of electrodeposited nanocrystalline nickel. Acta Mater. 2003; 51: 387–405. doi:10.1016/S1359-6454(02)00421-4

[49] Conrad H, Narayan J. On the grain size softening in nanocrystalline materials. Scr. Mater. 2000; 42: 1025–1030. doi:10.1016/S1359-6462(00)00320-1

[50] Hasan MZ, Kane CL. Colloquium: topological insulators. Rev. Mod. Phys. 2010; 82: 3045–3067. doi:10.1103/RevModPhys.82.3045

[51] Hasan MZ, Moore JE. Three-dimensional topological insulators. Annu. Rev. Condens. Matter Phys. 2011; 2: 55–78. doi:10.1146/annurev-conmatphys-062910-140432

[52] Luo CW, Wang HJ, Ku SA, Chen H-J, Yeh TT, Lin J-Y, Wu KH, Juang JY, Young BL, Kobayashi T, Cheng C-M, Chen CH, Tsuei KD, Sankar R, Chou FC, Kokh KA, Tereshchenko OE, Chulkov EV, Andreev YM, Gu GD. Snapshots of Dirac fermions near the Dirac point in topological insulators. Nano Lett. 2013; 13: 5797–802. doi:10.1021/nl4021842

[53] Le PH, Tzeng W-Y, Chen H-J, Luo CW, Lin J-Y, Leu J. Superconductivity in textured Bi clusters/Bi_2Te_3 films. APL Mater. 2014; 2: 096105. doi:10.1063/1.4894779

[54] Tu C-M, Yeh T-T, Tzeng W-Y, Chen Y-R, Chen H-J, Ku S-A, Luo C-W, Lin J-Y, Wu K-H, Juang J-Y, Kobayashi T, Cheng C-M, Tsuei K-D, Berger H, Sankar R, Chou F-C. Manifestation of a second Dirac surface state and bulk bands in THz radiation from topological insulators. Sci. Rep. 2015; 5: 14128. doi:10.1038/srep14128

[55] Chen YL, Analytis JG, Chu J-H, Liu ZK, Mo S-K, Qi XL, Zhang HJ, Lu DH, Dai X, Fang Z, Zhang SC, Fisher IR, Hussain Z, Shen Z-X. Experimental realization of a three-dimensional topological insulator, Bi_2Te_3. Science. 2009; 325: 178–181. doi:10.1126/science.1173034

[56] Hsieh D, Xia Y, Qian D, Wray L, Meier F, Dil JH, Osterwalder J, Patthey L, Fedorov AV, Lin H, Bansil A, Grauer D, Hor YS, Cava RJ, Hasan MZ. Observation of time-reversal-protected single-Dirac-cone topological-insulator states in Bi_2Te_3 and Sb_2Te_3. Phys. Rev. Lett. 2009; 103: 146401. doi:10.1103/PhysRevLett.103.146401

[57] Wang G, Zhu X-G, Sun Y-Y, Li Y-Y, Zhang T, Wen J, Chen X, He K, Wang L-L, Ma X-C, Jia J-F, Zhang SB, Xue Q-K. Topological insulator thin films of Bi_2Te_3 with controlled electronic structure. Adv. Mater. 2011; 23: 2929–2932. doi:10.1002/adma.201100678

[58] Zhang T, Cheng P, Chen X, Jia J-F, Ma X, He K, Wang L, Zhang H, Dai X, Fang Z, Xie X, Xue Q-K. Experimental demonstration of topological surface states protected by time-reversal symmetry. Phys. Rev. Lett. 2009; 103: 266803. doi:10.1103/PhysRevLett.103.266803

[59] Alpichshev Z, Analytis JG, Chu J-H, Fisher IR, Chen YL, Shen ZX, Fang A, Kapitulnik A. STM imaging of electronic waves on the surface of Bi_2Te_3: topologically protected

surface states and hexagonal warping effects. Phys. Rev. Lett. 2010; 104: 016401. doi: 10.1103/PhysRevLett.104.016401

[60] Chiu SP, Lin JJ. Weak antilocalization in topological insulator Bi_2Te_3 microflakes. Phys. Rev. B. 2013; 87: 035122. doi:10.1103/PhysRevB.87.035122

[61] Cha JJ, Kong D, Hong S-S, Analytis JG, Lai K, Cui Y. Weak antilocalization in $Bi_2(Se_xTe_{1-x})_3$ nanoribbons and nanoplates. Nano Lett. 2012; 12: 1107–1111. doi:10.1021/nl300018j

[62] Yu X, He L, Lang M, Jiang W, Xiu F, Liao Z, Wang Y, Kou X, Zhang P, Tang J, Huang G, Zou J, Wang KL. Separation of top and bottom surface conduction in Bi_2Te_3 thin films. Nanotechnology. 2013; 24: 015705. doi:10.1088/0957-4484/24/1/015705

[63] Le PH, Wu KH, Luo CW, Leu J. Growth and characterization of topological insulator Bi_2Se_3 thin films on $SrTiO_3$ using pulsed laser deposition. Thin Solid Films. 2013; 534: 659–665. doi:10.1016/j.tsf.2013.01.104

[64] Lee YF, Punugupati S, Wu F, Jin Z, Narayan J, Schwartz J. Evidence for topological surface states in epitaxial Bi_2Se_3 thin film grown by pulsed laser deposition through magneto-transport measurements. Curr. Opin. Solid State Mater. Sci. 2014; 18: 279–285. doi:10.1016/j.cossms.2014.07.001

[65] Hikami S, Larkin AI, Nagaoka Y. Spin-orbit interaction and magnetoresistance in the two dimensional random system. Prog. Theor. Phys. 1980; 63: 707–710. doi:10.1143/PTP.63.707

[66] Suzuura H, Ando T. Crossover from symplectic to orthogonal class in a two-dimensional honeycomb lattice. Phys. Rev. Lett. 2002; 89: 266603. doi:10.1103/PhysRevLett.89.266603

[67] Le PH, Liu P-T, Luo CW, Lin J-Y, Wu KH. Thickness-dependent magnetotransport properties and terahertz response of topological insulator Bi_2Te_3 thin films. J. Alloys Compd. doi:10.1016/j.jallcom.2016.09.109

[68] Lin JJ, Bird JP. Recent experimental studies of electron dephasing in metal and semiconductor mesoscopic structures. J. Phys. Condens. Matter. 2002; 14: R501–596. doi: 10.1088/0953-8984/14/18/201

[69] Steinberg H, Laloë J-B, Fatemi V, Moodera JS, Jarillo-Herrero P. Electrically tunable surface-to-bulk coherent coupling in topological insulator thin films. Phys. Rev. B. 2011; 84: 233101. doi:10.1103/PhysRevB.84.233101

[70] Tang H, Liang D, Qiu RLJ, Gao XPA. Two-dimensional transport-induced linear magnetoresistance in topological insulator Bi_2Se_3 nanoribbons. ACS Nano. 2011; 5: 7510–7516. doi:10.1021/nn2024607

[71] Wang X, Du Y, Dou S, Zhang C. Room temperature giant and linear magnetoresistance in topological insulator Bi_2Te_3 nanosheets. Phys. Rev. Lett. 2012; 108: 266806. doi: 10.1103/PhysRevLett.108.266806

[72] Assaf BA, Cardinal T, Wei P, Katmis F, Moodera JS, Heiman D. Linear magnetoresistance in topological insulator thin films: quantum phase coherence effects at high temperatures. Appl. Phys. Lett. 2013; 102: 012102. doi:10.1063/1.4773207

[73] Abrikosov AA. Quantum magnetoresistance. Phys. Rev. B. 1998; 58: 2788–2794. doi:10.1103/PhysRevB.58.2788

[74] Abrikosov AA. Quantum linear magnetoresistance; solution of an old mystery. J. Phys. A Math. Gen. 2003; 36: 9119–9131. doi:10.1088/0305-4470/36/35/301

[75] Wang CM, Lei XL. Linear magnetoresistance on the topological surface. Phys. Rev. B. 2012; 86: 035442. doi:10.1103/PhysRevB.86.035442

[76] Qi X-L, Hughes TL, Raghu S, Zhang S-C. Time-reversal-invariant topological superconductors and superfluids in two and three dimensions. Phys. Rev. Lett. 2009; 102: 187001. doi:10.1103/PhysRevLett.102.187001

[77] Wilczek F. Majorana returns. Nat. Phys. 2009; 5: 614–618. doi:10.1038/nphys1380

[78] Koren G, Kirzhner T, Lahoud E, Chashka KB, Kanigel A. Proximity-induced superconductivity in topological Bi_2Te_2Se and Bi_2Se_3 films: robust zero-energy bound state possibly due to Majorana fermions. Phys. Rev. B. 2011; 84: 224521. doi:10.1103/PhysRevB.84.224521

[79] Noro H, Sato K, Kagechika H. The thermoelectric properties and crystallography of Bi-Sb-Te-Se thin films grown by ion beam sputtering. J. Appl. Phys. 1993; 73: 1252–1260. doi:10.1063/1.353266

Chapter 4

Pulsed Laser Deposition of ITO: From Films to Nanostructures

Seong Shan Yap, Thian Khok Yong,
Chen Hon Nee and Teck Yong Tou

Additional information is available at the end of the chapter

http://dx.doi.org/10.5772/65897

Abstract

Indium-tin oxide (ITO) films have been deposited by pulsed laser deposition (PLD) to achieve low resistivity and high transmittance in visible region. Important parameters governing the growth of ITO films, which include laser wavelength, substrate temperature, and the background gas pressure, are discussed. By utilizing the energetic plasma in laser ablation of an ITO target, relatively low substrate temperature growth has been demonstrated. Room temperature deposition enables ITO films to be deposited on the polymer substrate. In addition, deposition in different background gases promotes the catalyst-free growth of nanostructured ITO films. In particular, deposition in Ar or He at optimized pressures enables the growth of highly crystalline ITO nanostructures, which include nanorods and nanowires due to the self-catalyzed growth from the plasma plume. The conditions which allow the pulsed laser deposition of ITO thin films and the growth of nanostructured ITO are reviewed and discussed.

Keywords: pulsed laser deposition, ITO, nanostructures, nanowires, nanorods, TCO

1. Introduction

Sn-doped indium oxide or indium tin oxide (ITO) plays important roles in optoelectronics applications by providing contact layer with high transparency in the visible. Currently, ITO remains as the transparent conducting oxide (TCO) with the lowest resistivity on a commercial scale ($1–2 \times 10^{-4}\ \Omega$ cm) [1]. In order to obtain both conductivity and transparency, the carrier density should not be too high to maintain its transparency with sufficiently low resistivity/sheet resistance. In addition, the carrier mobility can be maximized, which will then provide active carriers for conduction. **Figure 1** shows the band structure of the undoped In_2O_3 and

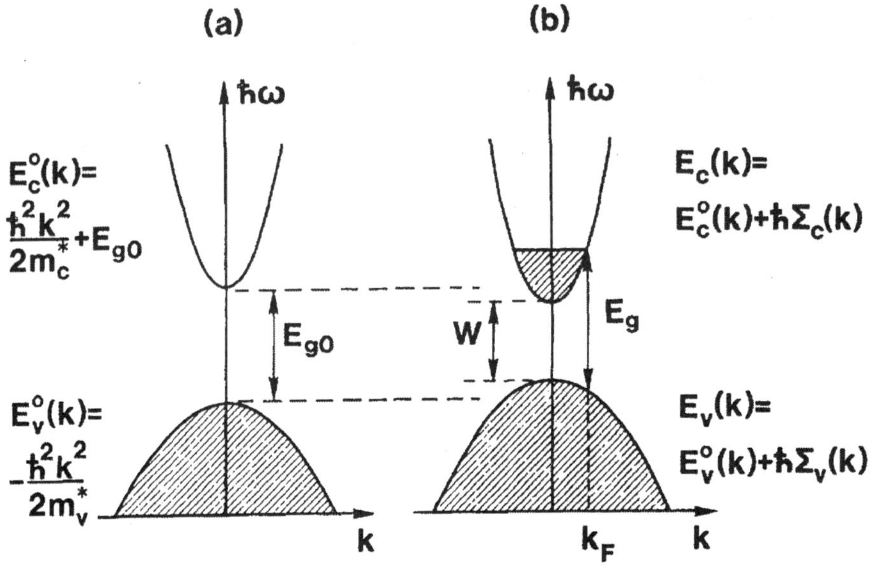

Figure 1. Schematic illustration of the assumed parabolic band structure of (a) undoped and (b) heavily Sn-doped In_2O_3 in the vicinity of the top of the valence band and the bottom of the conduction band with a band gap E_{go}. A vertical shift in the conduction and valence bands is apparent after heavy Sn doping in (b) assumed to have the sole effect of widening the band gap by a Burstein-Moss shift *W* due to filling up of the lowest states in the conduction band. Shaded areas denote occupied states. Band gaps, Fermi wave number k_F, and dispersion relations are indicated [2].

heavily Sn-doped In_2O_3 that was proposed by Hamberg et al. [2]. The band structure for In_2O_3 has a parabolic band that is characterized by effective mass m_c^* for the conduction band and m_v^* for the valence band. The direct band gap (E_{go}) is around 3.75 eV. After optimized Sn doping, the band gap (E_g), the carrier density, and mobility are increased, giving rise to the increase in carriers for conduction while maintaining the transparency. The change in carrier density, defect states, and crystallinity thus affects the final quality of the films.

ITO films can be grown by various methods. **Table 1** gives an overview of the requirements and performance of the different deposition techniques assessed against various different fundamental criteria. The deposition conditions of ITO crucially affect the dopant density, crystallinity, nanostructures formation, and defects in the films. Some of the common thin film deposition techniques are thermal evaporation, sputtering, pulsed laser deposition (PLD), chemical vapor deposition (CVD), and atomic layer deposition (ALD) which are becoming popular in recent years. These techniques can be broadly compared in terms of the energy of depositing specie, substrate temperature and possibly the annealing process after the film is deposited, and some specific features of the techniques.

The energy of depositing species, or the kinetic energy, is determined by the method for vapor creation. For TE, CVD, and ALD, the vapor is largely created via heating effect, which involves thermalization among the vapor species. Generally, a thermalized vapor is

	Thermal evaporation (TE)	Chemical vapor deposition (CVD)	Atomic layer deposition (ALD)	Sputtering (S)	Pulsed laser deposition (PLD)
Depositing species	Atoms/ions	Precursors molecules	Precursors molecules	Atoms/ions	Atoms/ion/clusters
Energy of depositing species	Low (0.1–0.5 eV)	Low	Low	Normally low, 1–100 eV (with biasing)	Low to very high 1–1000 eV
Substrate temperature	None	Typically high	Varies with precursor	Low to moderate	Low to moderate
Features	High growth rate, simple	Continuous growth	Step growth-layer by layer, scalable	Nonreactive and reactive deposition DC or RF sputtering	Nonreactive and reactive deposition, suitable for stoichiometric films, large control range for different dopants

Table 1. Different deposition techniques.

ideal for better control of the thin film morphology and microstructures; hence, a uniform thin film or coating is always achieved. The TE method remains one of the popular choices for metallic thin film deposition, such as the metal contact or electrodes. The disadvantage of TE method is always the contaminations due to out-diffusion from the walls of deposition chamber as a result of high temperature required to vaporize the targets. To reduce these contaminations, thin film deposition by TE always necessitates high vacuum condition, typically 10^{-4} Pa or better. On the other hand, CVD is operated at relatively lower temperatures, owing to dissociation energy of the precursors, but requires high substrate heating. As the precursors are constantly supplied from external sources, the film growth is a continuous process, easily multi-element control, good stoichiometry, and epitaxial. The precursors are usually volatile and/or toxic in gaseous state, and thus, the requirement for safety and environmental contamination is important. To obtain high quality and purity of the epitaxial thin films, the high vacuum requirement of CVD is usually more stringent than that for TE.

For deposition of metal oxide–based thin films, however, TE is generally not used due to refractory high temperature for vaporization from solid targets. However, the CVD methods are usually applicable for metal oxide thin film deposition from the metal organic precursors at relatively low temperature of about 200°C. In this respect, the ALD method has gained high popularity for ultrathin oxide deposition for encapsulation of devices against moisture and oxygen contaminations in recent years. With advent of nanoscale devices, which demand higher degrees of accuracy and control in terms of conformity and film thickness at Angstrom level, ALD has superseded the capability of CVD. In addition, ALD has established to be a better method for film composition for a wide range of materials, which include metals, semiconductor, and insulators of both crystalline and amorphous phase for different electronic applications in industries and manufacturing. However, for car industry and energy-efficiency buildings, smart windows with photocatalytic functions are proposed, which require not just cost-effective but also high throughput of coating method. Thus, the atmospheric pressure CVD or APCVD method is applied for transition metal oxide thin film deposition.

PLD [3] and sputtering are considered more "versatile" methods for normal thin film deposition mainly because they employ direct vaporization to produce the required deposition flux. In particular, PLD has attracted much interest mainly due to its ability to deposit of high T_c superconducting films with necessary stoichiometry. While PLD can be operated between high vacuum and low pressure, the sputtering method is usually operated at low pressures with a continuous gas flow. For both methods, the production of a wide range of kinetic energy for the depositing specie ensures thin film depositions from different types of solid targets including doped target such as ITO. The film stoichiometry or multicomponent control is often obtained with the use of background gases for moderation of kinetic energy. While a suitable energy range can facilitate the growth of more uniform and crystalline thin films, the presence of background gas often prevents the film damage by energetic ion bombardment. For PLD, the background gases at suitable pressures may also introduce gas-phase reactions which may generate new species and/or formation of nanoclusters prior to the deposition on the substrate.

2. Pulsed laser deposition of ITO films

A typical experimental setup of PLD is shown in **Figure 2** [4] where ablation can be obtained by laser with different wavelengths or pulse duration, in either background gas or vacuum. Deposition of ITO is normally performed in O_2 environment to preserve the stoichiometry, similar to the deposition of superconducting oxide. In addition, the substrates can be heated

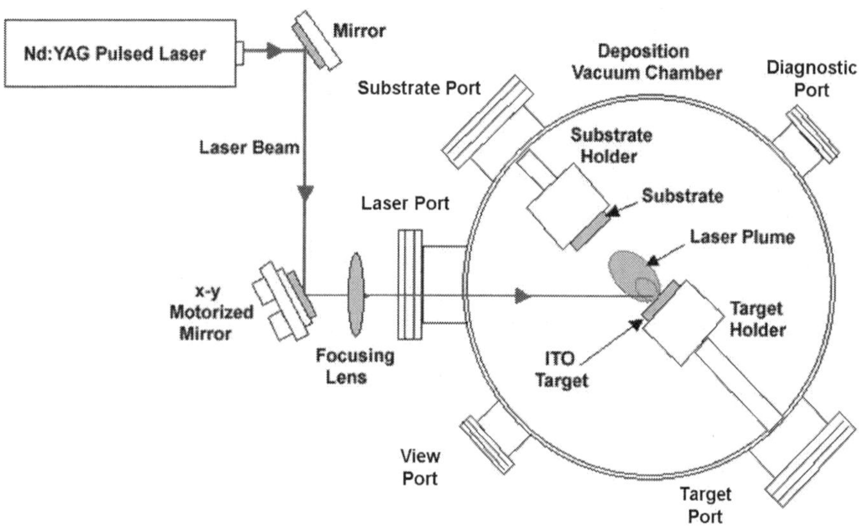

Figure 2. PLD setup for ITO growth. Deposition is often carried out in O_2 background.

http://dx.doi.org/10.5772/65897

References	Substrate	ITO (wt.% of SnO_2)	Lowest resistivity ($\times 10^{-4}$ Ω cm)	Optical transmittance	P_{O2} (mTorr)	d_{TS} (cm)	t (nm)	Laser parameters
[12, 13]	Glass	5	RT: 4 300°C: 2	RT: ~85% 300°C: ~92%	10	4.7	RT: 150 300°C: 170	Pulsed duration = 30 ns; Repetition rate = 10 Hz; Fluence ~2 J/cm^2
[14]	PET	5	RT: 7 100°C: 4.1	RT: ~87% 100°C: ~90%	40	5.8	200	
[15]	Glass	10	RT: 5.35 200°C: 1.75	>85%	10	7	100	
[16]	Fused-Quartz	5	RT: 4.76	>90%	10	7	100 ± 20	
[16]	Fused-Quartz	10	RT: 5.56	>90%	10	7	100 ± 20	
[17]	PC	5	RT: 2.5	NA	9.75	7	100	
[18]	Glass	NA	RT: 5	>90%	10	10.5	NA	Repetition rate = 5 Hz; Fluence ~2 J/cm^2 *UV assisted
[19]	Glass	10	RT: 4	~88%	10	3.45	400	Repetition rate = 10 Hz; Fluence ~2 J/cm^2 *N_2
[20]	Quartz	10	RT: 5 300°C: 2.5	88%	10	NA	180	Fluence ~2 J/cm^2
[21]	Glass	5	RT: 5.5 400°C: 2.3	>85%	10	NA	NA	Repetition rate = 10 Hz; Fluence ~4 J/cm^2

P_{O2}: oxygen gas pressure; d_{TS}: target-to-substrate distance; t: film thickness; RT: room temperature; NA: the data are not available.

Table 2. Summary of experimental conditions and properties of ITO films deposited with PLD using a 248 nm KrF excimer laser.

for films growth, and targets with different compositions can be sintered and used. In 1993, Zheng and Kwok [5] first reported the deposition of ITO by the pulsed laser ablation technique using an excimer laser at wavelength of 193 nm. The ITO target used was a 90 wt% In_2O_3 + 10 wt% SnO_2 sintered ceramic. The best electrical resistivity of the ITO films grown on glass substrate was 5.6×10^{-4} Ω cm and 1.4×10^{-4} Ω cm at substrate temperature of 20 and 310°C, respectively. The optical transmittance was greater than 90% for the optimized film between the wavelength ranges of 600 and 800 nm.

After the first successful deposition of ITO thin films on glass using the PLD technique was reported by Zheng and Kwok [5, 6], highly conductive ITO was deposited on InP substrates by PLD. The sample was then fabricated the ITO/p-InP device for the photovoltaic solar cell applications [7]. Thereafter, numerous studies on the ITO film initial growth and electrical conduction mechanism on different substrates using the same laser were reported [8–10]. When 248 nm KrF laser was used, ITO films with 85% optical transmittance in the visible and sheet resistance less than 1 Ω/sq were obtained at a substrate temperature of 200°C [11]. The properties of the ITO films were similar to the work done by using a 193 nm ArF laser. The properties of ITO films deposited at different conditions by using a KrF laser are summarized in **Table 2**.

An obvious disadvantage with excimer lasers is that its operation requires the handling of reactive gases. As an alternative, Nd:YAG solid-state laser has also been used for ITO deposition, especially after frequency doubled at 532 nm and frequency tripled at 355 nm. However, when the wavelength and pulse length are varied, it affects the laser penetration depth and the degree of absorption in laser-material interaction [3]. **Table 3** summarizes the experimental conditions and ITO thin films properties deposited by using an Nd:YAG pulsed laser at 355 and 532 nm in our previous reports as compared to others.

As a comparison, the effects due to the difference in laser wavelengths between 193 and 248 nm on the properties of the deposited ITO films are fairly small. A larger difference is observed for 193/248 nm deposition of ITO as compared to 355 nm deposition at room temperature. The resistivity that can be obtained by 193/248 nm laser is in the range of 10^{-4} Ω cm, while optimization in terms of ITO doping concentration [24] or substrate temperature [4, 26, 29] is needed in order to achieve the resistivity in the same range for 355 nm laser deposition.

Besides the need for sufficient Sn dopant in the ITO film, the deposition parameters: substrate temperature and background pressure are crucial to achieve the desired films properties. In the early reports, the resistivity of the films grown at lower substrate temperature was affected greatly by the background pressure as compared to those grown at higher temperature [5, 6]. The main reason for such an observation was ascribed to the film growth mechanisms in PLD. Upon ablation, the ablated plasma plume, consisting of Sn, In, and O species, undergoes collisions with the background gas that resultant in the final velocity distribution for each species. At higher substrate temperature, the low kinetic energy species will still be able to rearrange to form uniform films because of thermally induced migration from the heated substrate. Thus, deposition of high-quality films is possible covering a wider pressure range. However, at low substrate temperature, uniform films can only be obtained at an optimal velocity distribution which is affected greatly by the degree of collisions with the background gas. Thus, film

http://dx.doi.org/10.5772/65897

References	Substrate	ITO (wt.% of SnO_2)	Lowest resistivity	Optical transmittance	P_{O2} (mTorr)	d_{TS} (cm)	t nm)	Laser wavelength (nm)	Laser parameters
[22]	Glass	10	RT: 5×10^{-2} Ω cm 200°C: 1×10^{-3} Ω cm	RT: ~80% 200°C: ~90%	23	6	110	355	Pulsed duration = 6 ns; Repetition rate = 10 Hz; Fluence ~0.6 J/cm ^{2} *Deposition in various gas
[23]	Glass	10	RT: 1.1×10^{-3} Ω cm 200°C: 3.9×10^{-4} Ω cm	RT: >85% 200°C: >82%	23	6	110	355	Pulsed duration = 6 ns; Repetition rate = 10 Hz; Fluence ~2 J/cm^2
[24]	Glass	5	RT: 4.3×10^{-4} Ω cm200°C: 1.3×10^{-4} Ω cm	>80%	20	5	NA	355	Fluence ~2 J/cm^2
[25]	Silicon	In:Sn 9:1 alloy	RT: 500 Ω200°C: 80 Ω	NA	3750	3	NA	532	Fluence = 2 J/cm ^{2} Pulsed duration = 7 ns Repetition rate = 10 Hz
Our work [4, 26–28]	Glass PC	10	RT: 1×10^{-3} Ω cm 250°C: 3×10^{-4} Ω cm RT: 1.5×10^{-3} Ω cm	92% >85%	30	5	200	355	Fluence = 2.5 J/cm ^{2} Pulsed duration = 4.7 ns Repetition rate = 10 Hz
Our work [4, 26–28]	PC	10	RT: 3×10^{-3} Ω cm	>70%	30	5	200	532	Fluence = 2.5 J/cm ^{2} Pulsed duration = 4.7 ns Repetition rate = 10 Hz

Table 3. Summary of experimental conditions and properties of ITO films deposited with PLD using an Nd:YAG pulsed laser at 355 and 532 nm.

growth is rather sensitive to the background gas pressure. Similar trend has been observed in the deposition using excimer lasers [7, 13, 15] and the depositions using 355 nm laser [22, 30].

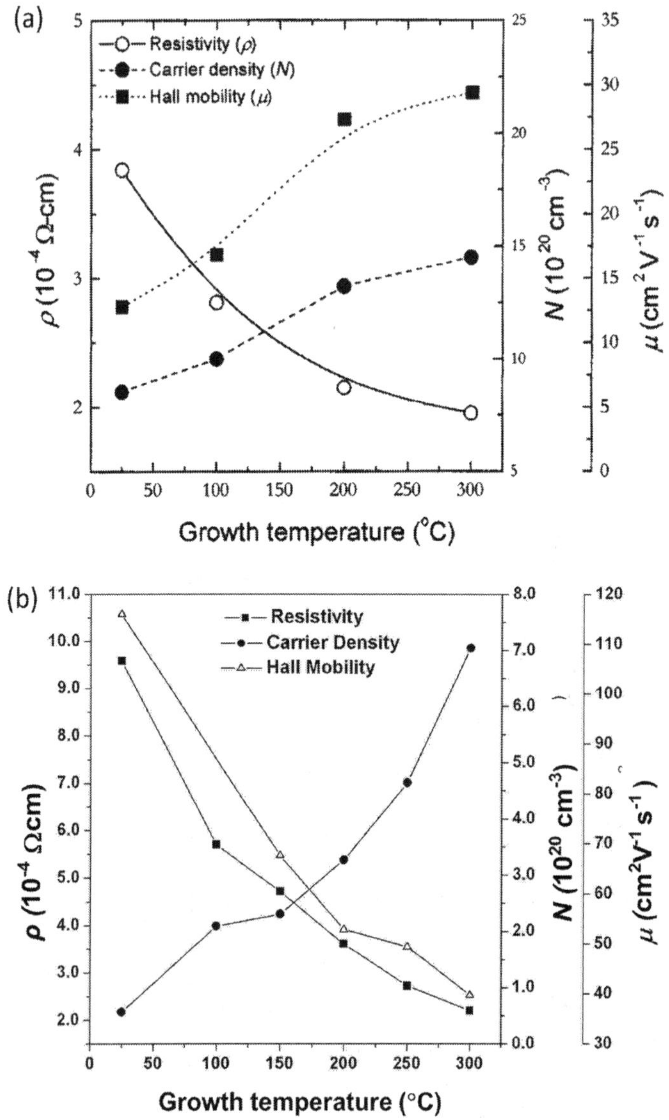

Figure 3. Effects of substrate temperature on film resistivity, carrier density, and Hall mobility deposited by (a) 248 nm laser at 10 mTorr [12] and (b) 355 nm laser at 30 mTorr [26].

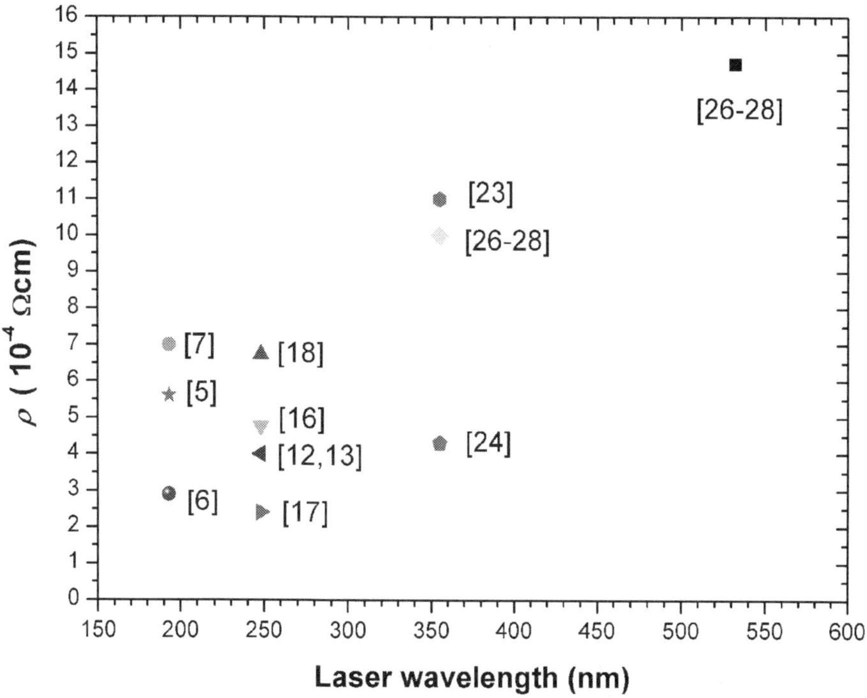

Figure 4. Resistivity of ITO films deposited in O_2 at RT as a function of laser wavelength.

In addition to the formation of a uniform film, substrate temperature affects the carrier density of the films and thus the resistivity of the ITO films. **Figure 3** shows the resistivity and the Hall effects measurement results of ITO films deposited by 248 nm [12] and a 355 nm laser [26]. As the substrate temperature was changed from room temperature (RT) to 300°C, the carrier density was doubled for the ITO films deposited by using 248 nm laser, while it was increased by 14 times for ITO deposited by using the 355 nm laser. At the same time, the Hall mobility was also increased by 2 times for ITO deposited by 248 nm laser, but the value was decreased by ~3 times for ITO deposited by 355 nm laser. Thus, effectively, the resistivity of ITO films in both cases was reduced at higher substrate temperature to $\sim 2 \times 10^{-4}$ Ω cm at 300°C. The results indicate that 248 nm laser was superior to 355 nm laser in the pristine deposition condition where the substrate was not heated, possibly due to the higher photon energy of 248 nm laser and its more congruent ablation with a smaller penetration depth [3]. Thus, ITO film properties that were inferior to those obtained by the shorter laser wavelengths were also obtained in 532 nm deposition [25, 28]. The resistivity of ITO films deposited at different laser wavelengths at RT is shown in **Figure 4**. However, substrate heating successfully improved the thin film property, and lower resistivity can be achieved for the case of 355 nm laser. The crystallinity of ITO films is also affected by substrate heating during the deposition process. The ITO films

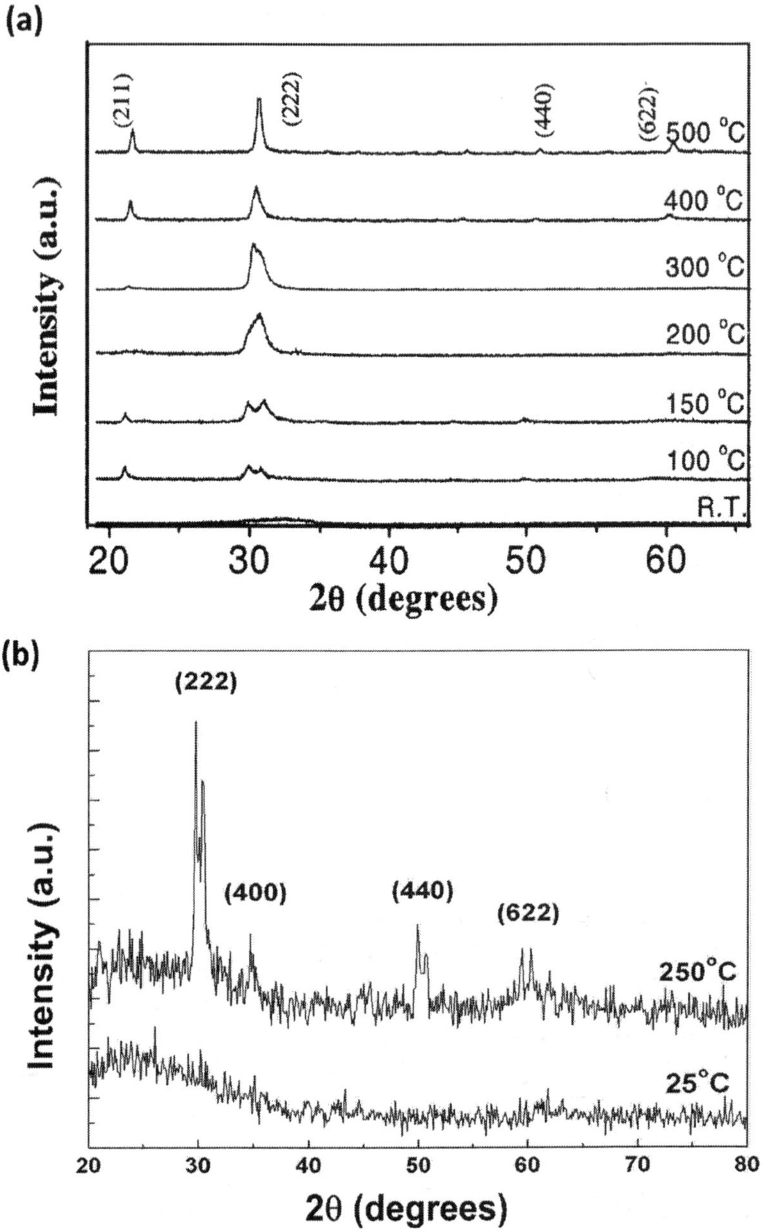

Figure 5. Effects of substrate temperature on the crystallinity of the ITO films deposited by (a) 248 nm laser [12] and (b) 355 nm laser [26].

are amorphous when deposited at RT by both 248 and 355 nm lasers. Crystalline films are obtained when deposited at above 100°C as shown in the XRD spectra in **Figure 5**.

ITO films deposited by laser ablation of ITO target have been applied in various applications such as solar cell [7, 31, 32], organic light-emitting devices [4, 12, 26], and more recently as low loss alternative plasmonic materials in the near-infrared region [33]. Our results show that ITO films with a range of properties can be obtained, controlled by the deposition parameters. PLD is capable of growing low resistivity, high transmittance ITO films at relatively low temperature or even room temperature that are beneficial for device application onto delicate polymer substrates or active materials.

3. Pulsed laser deposition of ITO nanostructures.

The growth of ITO nanostructures (nanowires, nanorods, nanowhiskers, and nanocrystals) is desirable because of the distinct nanoscale effects in addition to the advantages in large contact areas offered by nanomaterials. The nanoeffects of ITO nanostructures have been demonstrated in various aspects based on the UV light-emitting properties [34], terahertz and far-infrared transmitting characteristics [35], and field emission properties [36]. PLD, as a flexible and versatile tool for materials deposition, was first shown to be capable for the growth of nanowires by using excimer laser ablation in 2006 [37]. ITO nanowires were grown on catalyst-free oxidized silicon substrates at 500°C in nitrogen atmosphere. The growth of nanostructures was not observed by others although the growth involving non-oxygen gases such as N_2 [19], rare gas Ne, Ar, and Xe [22] (**Tables 2** and **3**) has been reported. In this report, ITO films with lower resistivity and higher transmittance were obtained when deposited in O_2. However, in the report by Savu and Joanni [37], when higher substrate temperate and Si substrate were used, nanostructural growth of ITO was observed in N_2. The morphology of the nanostructures was found to be highly dependent on the N_2 gas pressure as shown in

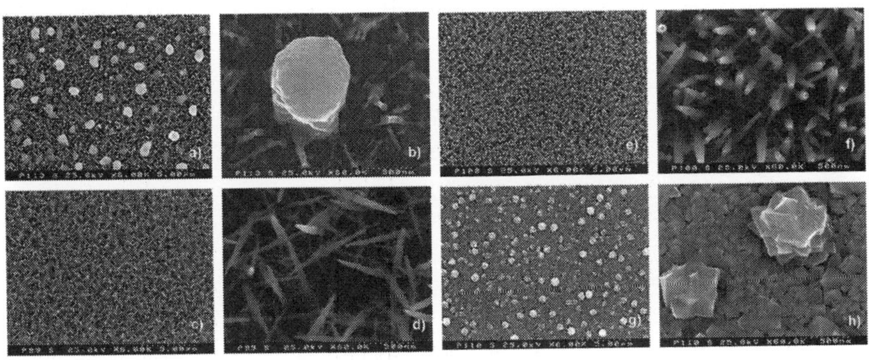

Figure 6. Surface images of nanostructured films deposited at 0.1 mbar (a, b), 0.5 mbar (c, d), 1 mbar (e, f), and 2 mbar (g, h) [37].

Figure 6. The growth mechanisms were proposed to be related to vapor-liquid-solid growth as a large amount of liquid was formed, especially at low pressure, thus promoting the growth of thin, branched nanowires. As the deposition pressure increases, the amount of liquid phase decreases, resulting in formation of nanowires having fewer branches. At a pressure of 1 mbar, the wires were almost perpendicular to the substrate and free of branches. At 2 mbar, columnar dense film is formed with large pyramidal and triangular structures.

Based on a standard PLD setup as shown in **Figure 2**, we studied the effects of different background gases Ar, He, N_2, and O_2 in details [38–40]. Glass substrates were used, and the growth was performed at a lower substrate temperature of 250°C. ITO nanostructures were formed, and the morphology of the nanostructures formed in different gases is shown in **Figure 7** [40]. The ITO film was uniform in size when deposited in O_2, while ITO deposited in Ar consisted of ultrafine nano-grains with a size of < 50 nm. For ITO deposited in N_2, the nanostructure consisted of porous network of nanorods of about 30 nm in diameter and 300 nm in length. Larger structures were obtained when deposited in He. ITO nanostructures formed in Ar, He, and O_2 were highly crystalline and possess higher transmittance than those obtained in N_2. The resistivity for ITO nanostructures deposited in N_2 was also higher than those deposited in Ar, He, and O_2. The results show that nanostructures can be obtained under specific conditions, which is not limited to higher temperature range and Si substrates in the first report by Savu and Joanni [37].

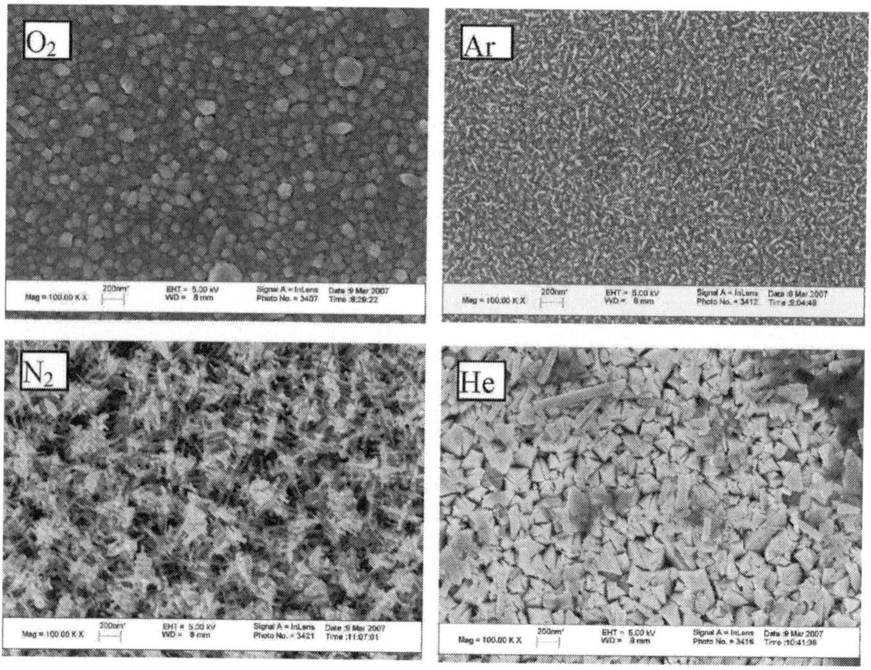

Figure 7. Effects of the background gas on the ITO microstructures. (a) O_2, (b) Ar, (c) N_2, and (d) He [40].

http://dx.doi.org/10.5772/65897

Further studies were performed for the growth in He and Ar at different background pressures based on the same setup [39]. The pressure range was chosen by considering the molecular

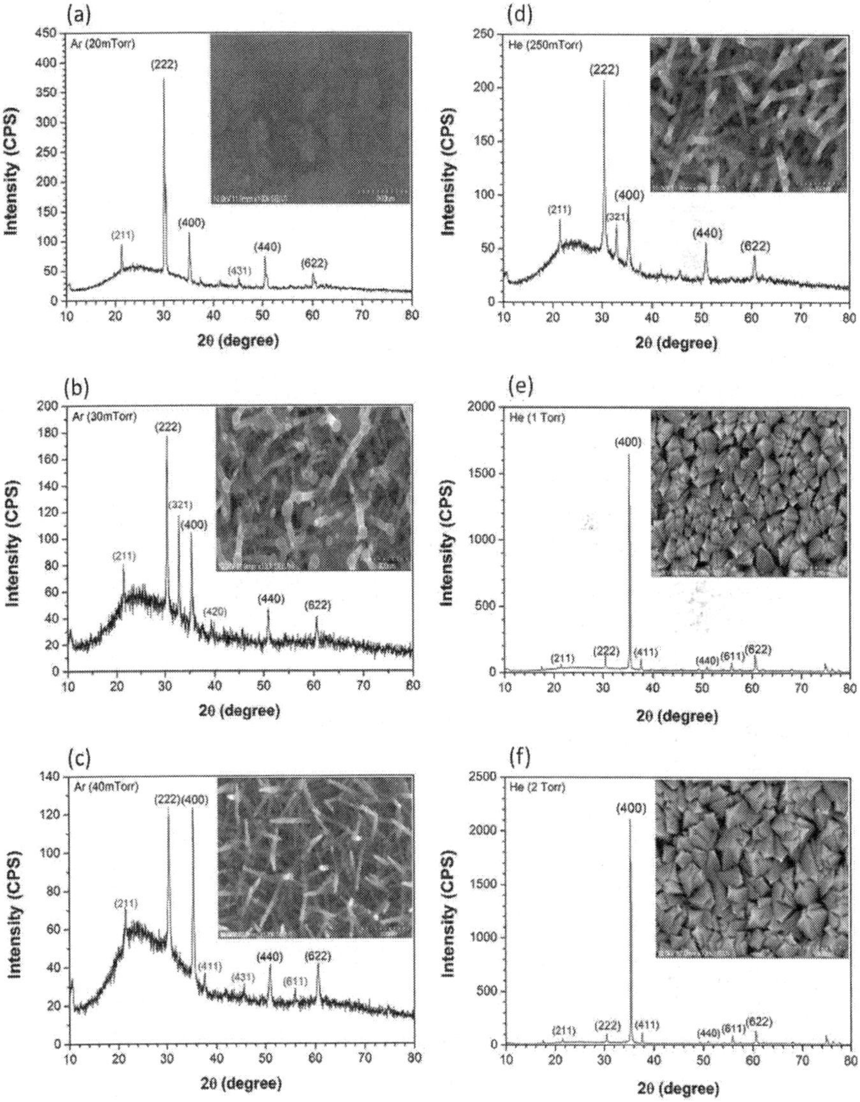

Figure 8. XRD and SEM of ITO samples grown in Ar ambient at (a) 20 mTorr, (b) 30 mTorr, and (c) 40 mTorr and grown in He at (d) 250 mTorr, (e) 1 Torr, and (f) 2 Torr [39].

ITO samples	Resistivity ($\times 10^{-4} \Omega$ cm)	Carrier density ($\times 10^{20}$ cm^{-3})	Hall mobility (cm^2/Vs)
Commercial ITO	1.97	18.7	16.6
Ar (30 mTorr, 250°C)	1.61	10.5	36.7
Ar (40 mTorr, 250°C)	2.09	6.65	44.9
He (250 mTorr, 250°C)	9.37	13.2	5.06
He (1000 mTorr, 250°C)	1.91	6.91	4.75
He (2000 mTorr, 250°C)	7.75	9.50	8.78

Table 4. Resistivity, carrier density, and Hall mobility of commercial ITO and ITO samples grown in Ar and He ambient.

weight of the gas that will affect the collision with ablated species. The growth was performed by PLD using a 355 nm laser at a substrate temperature of 250°C. The results show that nanostructures growth was dependent critically on the background pressure (**Figure 8**), similar to those reported in N_2 [37]. ITO nanowires were obtained in both gases. For ITO deposited in Ar, XRD diffraction patterns corresponding to (222), (400), (440), and (622) orientations of cubic bixbyite structure of ITO were detected. As the pressure of Ar increased, the (400) diffraction peak became relatively stronger, indicating the increase in crystalline orientation.

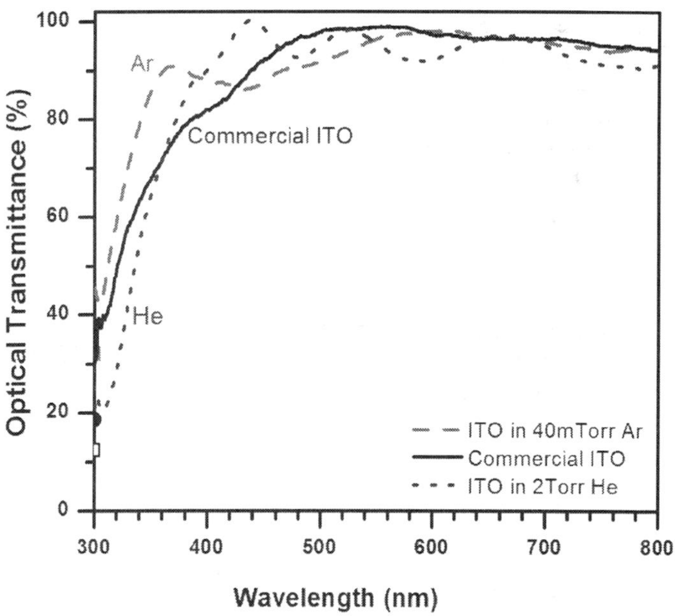

Figure 9. Optical transmittance of commercial ITO and ITO samples grown in Ar and He background gases [39].

http://dx.doi.org/10.5772/65897

At 30 mTorr, ITO nanowires were formed, and some spherical particles were observed on their tips, which suggest that they are formed by the vapor-liquid-solid (VLS) mechanism. At higher pressure of 40 mTorr, nanowires with smaller diameters were observed. No spherical particles were observed, and the tips were sharp, unlike those obtained at lower pressure. When deposited in He, nanowires were obtained at 250 mTorr, and spherical tips were formed on the tips. As the pressure increased, larger pyramid shape crystals were obtained, and the crystals orientation are aligned to (400). Both nanostructures grown in Ar and He exhibited good resistivity in the range of 10^{-4} Ω cm, and nanowires grown in Ar exhibited even higher carrier mobility than those measured from a commercial ITO sample. The values are shown in **Table 4**. **Figure 9** shows the optical transmittance of ITO nanostructures grown in Ar and He as compared to a commercial ITO sample. In addition, ITO nanowires grown by PLD are tested as TCO layer for a standard OLED device. ITO nanowires with larger contact areas and higher charge injection have led to higher emission current density for the devices [39].

4. Conclusion

Laser ablation of ITO target was studied in various aspects. Pulsed laser ablations by 193, 248, or 355 nm lasers in low vacuum conditions afforded the deposition of high-quality ITO films at relatively low temperatures. This was largely ascribed to the production of homogenous, energetic, and atomized plasma plume. The effects of substrate temperature, background gas pressure, in particular the O_2 on the electrical and optical properties of ITO films have been presented and discussed. In addition to the normal consensus that ITO depositions are to be performed in O_2 to preserve the stoichiometry of the final films, pulsed laser ablation of ITO target, when performed in various gases such as Ar, He, or N_2, could promote the growth of ITO nanostructures. At optimized pressure, the gas-phase condensation successfully induced nucleation and growth of the nanostructures such as nanorods and nanowires on glass substrate at relatively low substrate temperature.

Acknowledgements

We acknowledge the financial support from Penang State Government, Telekom Malaysia, the Ministry of Science, Technology and Innovation, and the Ministry of Higher Education of Malaysia in the form of research grants.

Author details

Seong Shan Yap[1,*], Thian Khok Yong[2], Chen Hon Nee[1] and Teck Yong Tou[1]

*Address all correspondence to: seongshan@gmail.com

1 Faculty of Engineering, Multimedia University, Cyberjaya, Selangor, Malaysia

2 Lee Kong Chian Faculty of Engineering and Science, Universiti Tunku Abdul Rahman, Bandar Sungai Long, Kajang, Selangor, Malaysia

References

[1] Ellmer K. 2012. Past achievements and future challenges in the development of optically transparent electrodes. Nat. Photon. 6:809–17

[2] Hamberg I, Granqvist C G, Berggren K-F, Sernelius B E and Engström L. 1984. Band-gap widening in heavily Sn-doped In_2O_3. Phys. Rev. B 30:3240–9

[3] Chrisey D B and Hubler G K. editors. Pulsed Laser Deposition of Thin Films. Wiley; 1994. 648 p. ISBN 0471592188/9780471592181.

[4] Yong T K, Yap S S, Sáfrán G and Tou T Y. 2007. Pulsed Nd:YAG laser depositions of ITO and DLC films for OLED applications. Appl. Surf. Sci. 253:4955–9

[5] Zheng J P and Kwok H S. 1993. Low resistivity indium tin oxide films by pulsed laser deposition. Appl. Phys. Lett. 63:1–3

[6] Zheng J P and Kwok H S. 1993. Preparation of indium tin oxide films at room temperature by pulsed laser deposition. Thin Solid Films. 232:99–104

[7] Jia Q X, Zheng J P, Kwok H S and Anderson W A. 1995. Indium tin oxide on InP by pulsed-laser deposition. Thin Solid Films. 258:260–3

[8] Sun X W, Huang H C and Kwok H S. 1996. In-situ resistance measurements during pulsed laser deposition of ultrathin films. Appl. Surf. Sci. 106:51–4

[9] Sun X W, Kim D H and Kwok H S. 1998. Ultra thin indium tin oxide films on various substrates by pulsed laser deposition Mater. Res. Soc. Symp. Proc. 485:267–72

[10] Sun X W, Huang H C and Kwok H S. 1996. On the initial growth of indium tin oxide on glass. Appl. Phys. Lett. 68:2663–5

[11] Hanus F, Jadin A and Laude L D. 1996. Pulsed laser deposition of high quality ITO thin films. Appl. Surf. Sci. 96–98:807–10

[12] Kim H, Gilmore C M, Piqué A, Horwitz J S, Mattoussi H, Murata H, Kafafi Z H and Chrisey D B. 1999. Electrical, optical, and structural properties of indium-tin-oxide thin films for organic light-emitting devices. J. Appl. Phys. 86:6451–61

[13] Kim H, Horwitz J S, Piqué A, Gilmore C M and Chrisey D B. 1999. Electrical and optical properties of indium tin oxide thin films grown by pulsed laser deposition. Appl. Phys. A Mater. Sci. Process. 69:S447–50

[14] Kim H, Horwitz J S, Kushto G P, Kafafi Z H and Chrisey D B. 2001. Indium tin oxide thin films grown on flexible plastic substrates by pulsed-laser deposition for organic light-emitting diodes. Appl. Phys. Lett. 79:284–6

[15] Adurodija F O, Izumi H, Ishihara T, Yoshioka H, Yamada K, Matsui H and Motoyama M. 1999. Highly conducting indium tin oxide (ITO) thin films deposited by pulsed laser ablation. Thin Solid Films. 350:79–84

[16] Adurodija F O, Izumi H, Ishihara T, Yoshioka H and Motoyama M. 2002. The electro-optical properties of amorphous indium tin oxide films prepared at room temperature by pulsed laser deposition. Sol. Energy Mater. Sol. Cells 71:1–8

[17] Izumi H, Ishihara T, Yoshioka H and Motoyama M. 2002. Electrical properties of crystalline ITO films prepared at room temperature by pulsed laser deposition on plastic substrates. Thin Solid Films. 411:32–5

[18] Craciun V, Craciun D, Chen Z, Hwang J and Singh R K. 2000. Room temperature growth of indium tin oxide thin films by ultraviolet-assisted pulsed laser deposition. Appl. Surf. Sci. 168:118–22

[19] Morales-Paliza M, Huang M and Feldman L. 2003. Nitrogen as background gas in pulsed-laser deposition growth of indium tin oxide films at room temperature. Thin Solid Films. 429:220–4

[20] Kim S H, Park N M, Kim T and Sung G. 2005. Electrical and optical characteristics of ITO films by pulsed laser deposition using a 10 wt.% SnO_2-doped In_2O_3 ceramic target. Thin Solid Films. 475:262–6

[21] Petukhov I A, Shatokhin A N, Putilin F N, Rumyantseva M N, Kozlovskii V F, Gaskov A M, Zuev D A, Lotin A A, Novodvorsky O A and Khramova A. D. 2012. Pulsed laser deposition of conductive indium tin oxide thin films. Inorg. Mater. 48:1020–5

[22] Thestrup B, Schou J, Nordskov A and Larsen N B. 1999. Electrical and optical properties of thin indium tin oxide films produced by pulsed laser ablation in oxygen or rare gas atmospheres. Appl. Surf. Sci. 142:248–52

[23] Thestrup B and Schou J. 1999. Transparent conducting AZO and ITO films produced by pulsed laser ablation at 355 nm. Appl. Phys. A 69:807–10

[24] Choi J B, Kim J H, Jeon K A and Lee S Y. 2003. Properties of ITO films on glass fabricated by pulsed laser deposition. Mater. Sci. Eng. B 102:376–9

[25] Teghil R, Marotta V, Guidoni A G, Di Palma T M and Flamini C. 1999. Reactive pulsed laser ablation and deposition of thin indium tin oxide films for solid state compact sensors. Appl. Surf. Sci. 138 522–6

[26] Yong T-K, Yap S-S and Tou T-Y. 2008. Parametric studies of pulsed Nd:YAG laser deposition of ITO for OLED application. Rev. Laser Eng. 36:1242–5

[27] Yong T-K, Kee Y-Y, Tan S-S, Yap S-S, Siew W-O and Tou T-Y. 2010. The effects of sodium in ITO by pulsed laser deposition on organic light-emitting diodes. Appl. Phys. A 101:621–6

[28] Yong T K, Tou T Y and Teo B S. 2005. Pulsed laser deposition of tin-doped indium oxide (ITO) on polycarbonate. Appl. Surf. Sci. 248:388–91

[29] Holmelund E, Thestrup B, Schou J, Larsen N B, Nielsen M M, Johnson E and Tougaard S. 2002. Deposition and characterization of ITO films produced by laser ablation at 355 nm. Appl. Phys. A: Mater. Sci. Process. 74:147–52

[30] Tou T-Y, Yong T-K, Yap S-S, Yang R-B, Siew W-O and Yow H-K. 2009. Parametric studies of pulsed laser deposition of indium tin oxide and ultra-thin diamond-like carbon for organic light-emitting devices. J. Opt. Soc. Korea 13:65–74

[31] Trehame R E, Seymour-Pierce A, Durose K, Hutchings K, Roncallo S and Lane D. 2011. Optical design and fabrication of fully sputtered CdTe/CdS solar cells. J. Phys. Conf. Ser. 286:12038

[32] Wu C, Zhang Z, Wu Y, Lv P, Nie B, Luo L, Wang L, Hu J and Jie J. 2013. Flexible CuS nanotubes-ITO film Schottky junction solar cells with enhanced light harvesting by using an Ag mirror. Nanotechnology 24:45402

[33] Fang X, Mak C L, Zhang S, Wang Z, Yuan W and Ye H. 2016. Pulsed laser deposited indium tin oxides as alternatives to noble metals in the near-infrared region. J. Phys. Condens. Matter 28:224009

[34] Gao J, Chen R, Li D H, Jiang L, Ye J C, Ma X C, Chen X D, Xiong Q H, Sun H D and Wu T. 2011. UV light emitting transparent conducting tin-doped indium oxide (ITO) nanowires. Nanotechnology 22:195706

[35] Yang C, Chang C, Chen P, Yu P and Pan C-L. 2013. Broadband terahertz conductivity and optical transmission of indium-tin-oxide (ITO) nanomaterials. Opt. Express 21:16670–82

[36] Wan Q, Feng P and Wang T H. 2006. Vertically aligned tin-doped indium oxide nanowire arrays: epitaxial growth and electron field emission properties. Appl. Phys. Lett. 89:3–5

[37] Savu R and Joanni E. 2006. Low-temperature, self-nucleated growth of indium-tin oxide nanostructures by pulsed laser deposition on amorphous substrates. Scr. Mater. 55:979–81

[38] Yong T-K, Tan S-S, Nee C-H, Yap S-S, Kee Y-Y, Sáfrán G, Horváth Z E, Moscatello J, Yap Y-K and Tou T-Y. 2012. Pulsed laser deposition of indium tin oxide nanowires in argon and helium. Mater. Lett. 66:280–1

[39] Kee Y Y, Tan S S, Yong T K, Nee C H, Yap S S, Tou T Y, Sáfrán G, Horváth Z E, Moscatello J P and Yap Y K. 2012. Low-temperature synthesis of indium tin oxide nanowires as the transparent electrodes for organic light emitting devices. Nanotechnology 23:25706

[40] Yong T K, Nee C H, Yap S S, Siew W O, Sáfran G, Yap Y K and Tou T Y. 2010. Pulsed laser deposition of nanostructured indium-tin-oxide film. In R J Martin-Palma, Y-J Jen and A Lakhtakia (eds) Proceedings of the SPIE, vol. 7766, pp 776615–6

Chapter 5

Laser Ablation of Biomaterials

Gianina-Florentina Popescu-Pelin,

Carmen Ristoscu and Ion N. Mihailescu

Additional information is available at the end of the chapter

http://dx.doi.org/10.5772/65197

Abstract

Biomaterials, defined by high biocompatibility and biodegradability, are intensively used in medical applications, mainly to replace partial or total, damaged or destroyed hard or soft tissues. Most of them are used not only as coatings for implant coverage but also as parts for some medical devices. In the last decades, researchers sought to find the optimum processing methods and parameters to modify or deposit the biomaterial of interest. An important family of techniques, used to process a biomaterial, is represented by laser techniques, based upon laser ablation phenomenon. Laser ablation of biomaterials ensures the transference or modification with good precision and without or with minimal disruptions generated. To obtain thin coatings from biomaterials, one can use deposition techniques: pulsed laser deposition (PLD) or matrix-assisted pulsed laser evaporation (MAPLE). These techniques are chosen according to the selected biomaterial and desired performances of the obtained coating. Therefore, some sensitive biomaterials can be transferred only by MAPLE. Some results in the field of calcium phosphates deposited by PLD or MAPLE are presented, proving the usefulness of these biomaterials for medical applications.

Keywords: laser ablation, pulsed laser methods, spallation, biomaterials and medical applications

1. Introduction

In the last few years, the research on microfabricated devices or implants for biomedical applications has quickly advanced [1, 2]. The aim of biological implants is to reinforce or replace the damaged or diseased tissue [3]. The presence of a biomaterial in the body after implantation procedure always causes a biological reaction [4]. This is a host response to the new living conditions and suggests that the body is trying to adapt at the given situation [5].

Every year, for a large number of patients, biomaterials save lives, relieve suffering, and improve the quality of life [6]. To achieve the needs of the biomedical society, materials comprise everything from metals and ceramics to glasses and polymers have been investigated [7].

Biomaterial performances can be improved by selectively modifying the surface properties. One simple solution could be the deposition of thin films useful for changing the chemical and physical properties of biomaterials and important in achieving the optimal surface [8]. There are many studies which demonstrated that the cell response is dependent on surface topography [4].

Laser processing of natural or synthetic biomaterials has been rapidly developed to produce biomimetic artificial organs, tissue engineering scaffolds, and other biomedical constructs [9]. The laser processing techniques are commonly grouped into three categories: polymerization (use of laser to induce cross-linking between biomaterial polymer chains), ablation (use of laser to selectively remove part of the biomaterial by thermal or chemical effects), and activation (use of laser to activate certain parts on the polymer chains for specific application) [9]. Relative to other methods, laser processing of biomaterials presents some advantages: reduced surface contamination and mechanical damage and the capability to produce three-dimensional components with complicated geometries by controlling the surface structuring [4].

The physics behind the interaction of laser pulse with the solid surfaces can help us to better understand the laser ablation mechanism [10].

This chapter will provide a brief overview in the field of laser ablation of biomaterials, starting with the explanation of ablation mechanism and then presenting some of its applications.

2. Biomaterials

Biomaterials science is a multidisciplinary domain, which involves various features of materials science, clinical science, chemistry, physics, biology, and medicine [11, 12]. It is also an exciting and quickly growing field. It is to be mentioned that a biomaterial is different from a biological material, the last being produced by a biological system [13]. A biomaterial is a natural or synthetic material used in a medical device, projected to interact with biological systems for direct medical treatment [6, 7, 14]. Their usage within a physiologic medium needs some specific characteristics such as biocompatibility (to perform the function with an appropriate host response), efficiency, and reliability [7, 14]. These representative features have provided with an appropriate combination of chemical, mechanical, physical, and biological properties [14]. Biomaterials used in medical applications are rarely used as simple material, being mostly integrated into devices [1].

The biocompatibility of a biomaterial implies its acceptance by the surrounding tissues and the whole body. Consequently, implanted biomaterials should not irritate the injured area,

http://dx.doi.org/10.5772/65197

provoke an abnormal inflammatory response, stimulate allergic or immunologic reactions, and cause other diseases [15]. Besides these characteristics, the selection of biomaterials for the manufacturing of an implant device and/or as thin films for the functionality improvement of implant should also depend on appropriate mechanical properties (strength, stiffness, and fatigue), proper optical properties, proper density, sterility, and long-standing storage [15].

The most commonly used metallic implants are based on titanium and its alloys due to their excellent mechanical properties and corrosion stability [16]. Unfortunately, studies in the field showed that bare titanium or other metallic implants can determine ion release into body [17]. To prevent, control or diminish the ion release, such implants are covered with thin films.

Nowadays, there are a lot of experimental studies focused on the growing of thin, uniform, and adherent films from polymers, organic materials, and biomaterials [18].

2.1. Hydroxyapatite

Hydroxyapatite (HA) named by Berndt et al. as "hydrated calcium phosphate mineral" is the main inorganic component of the hard tissues (bone and teeth) and is the most extensively studied materials for bone healing [1, 19, 20]. HA belongs to the "apatite" group of compounds, having the chemical formula $Ca_{10}(PO_4)_6(OH)_2$ and a Ca/P ratio of 1.67 [19, 21]. It can be termed as hydroxylapatite, calcium hydroxyapatite, or apatite and has a calculated density of 3.22 g/cm^3 [19]. The vacancies or substitutions, which could occur within the structural lattice, have therefore the non-stoichiometry (deficiency of Ca^{2+} and OH^- ions) of biological apatite. Research in the field demonstrated the poor crystallinity of the biological apatite [22]. On the contrary, synthetic HA is considered to be a stoichiometric material [19]. It is to be mentioned that the presence of some crystallographic sites in the structure of HA allows the atomic exchanges of specific elements with different ionic charges (Na^+, Mg^{2+}, HPO_4, K^+, CO_3^{2-}, F^-, Cl^-, or trace elements such as Sr^{2+}, Ba^{2+}, and Pb^{2+}) [19, 22].

The maximum sintering temperature of HA, up to which its structure remains unperturbed, is 1300°C. If one exceeds this threshold, the HA decomposition cannot be avoided [23].

Due to the similarity with the human bone, hydroxyapatite was largely used in medical applications. The influence of some HA parameters with respect to the structural, mechanical, and biological properties has been investigated [19].

One important feature of HA, used as thin film for medical implants, was the stability in the physiological media, presenting a dissolution rate of 0.1 mg/year [19, 24].

Hydroxyapatite can be considered a "smart" ceramic due to its functionalities such as fixation improvement and stabilization of implant [19].

2.2. Octacalcium phosphate

Octacalcium phosphate (OCP), with the chemical formula $Ca_8(HPO_4)_2(PO_4)_4{\cdot}5H_2O$ and a structure similar to calcium OH-apatite, is more soluble, being present in the first stage of tissue growth [25, 26]. OCP has a Ca/P ratio of about 1.33 and can induce the ectopic development of a new bone tissue [26, 27]. A disadvantage of OCP is represented by the thermal instability of this material. The decomposition takes place at temperatures lower than those needed for ceramic sintering. It was observed that:

– at ~90°C, OCP lose water;

– at 180°C, the OCP structure undergoes changes by interlayer bond breaking;

– at 300°C, the pyrophosphate appears from OCP [28].

The immersion of OCP in physiological fluids induces the conversion into bone-like apatite spontaneously and irreversibly [29].

Related studies revealed the potential use of synthetic OCP as bone substitute material in different forms, such as coatings and granules. Furthermore, the stoichiometry of synthetic OCP controls the osteoconductivity and biodegradability of this material and recommends it to use in bone regeneration [30].

Boanini et al. evaluate the possibility to deposit thin films of magnesium substituted OCP (Mg:OCP) and strontium substituted OCP (Sr:OCP) on Ti substrates [31].

2.3. Carbonated hydroxyapatite doped with manganese

Manganese is essential for a normal growth and function of bones and muscles [32]. Furthermore, Mn^{2+} ions increase the capacity of integrins (transmembrane receptors that mediate cellular interactions) to bind and facilitate cell adhesion [33]. The doping of hydroxyapatites with Mn^{2+} ensures a better connection of bone-implant and makes easier the regeneration of bone tissue. It was demonstrated that Mn-doped HA has a higher thermal stability than pure HA [32]. The results obtained by György et al. demonstrated that carbonated HA doped with manganese (Mn-CHA) films should determine a faster osteointegration of the implanted device [34].

To modify the biomaterial surface properties, several methods were applied [3]. One can mention chemical treatments, ion beam implantation, liquid immersion, thermal and plasma spray, electrophoretic processes or laser processing methods based on laser ablation [3].

3. Laser ablation

Laser, a versatile source of energy, is for sure an attractive tool with prospective applications in communication and military technology, industry, scientific research, and medicine [35, 36]. The continuous development of lasers and the impossibility of other techniques to deposit or modify some materials leaded to use lasers for material processing [3, 36]. The interaction between laser beam and material is a complex action, which needs the understanding of capabilities and limitations of this process. It involves the incidence of electromagnetic radiation on the material surface, followed by reflection, refraction, absorption, scattering, and/or transmission phenomena (**Figure 1**) [37]. Absorption of radiation in materials is an important and desired phenomenon and is governed by the excitation of free or bound electrons, inside the bulk material [35, 37].

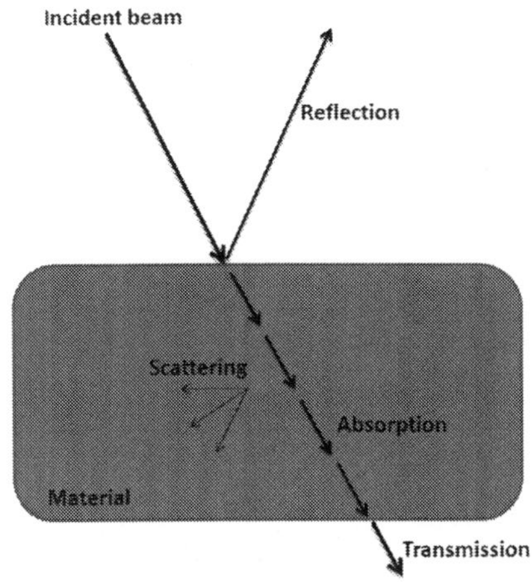

Figure 1. Possible phenomena generated by the interaction of laser beam with material.

Being the base of laser material processing, laser ablation consists in the removal of material from a substrate/target by absorption of laser energy [18, 38]. The ablation process always depends on the laser parameters like wavelength, energy, fluence, pulse duration, and the material properties, such as chemical structure, melting temperature, thermal diffusion rate, optical reflectivity, and the ambient medium [39, 40]. Laser ablation has significant applications in surface modification and deposition of thin films. Ablation is commonly determined by a phase transition from the liquid state to the vapor state. In function of the physical processes that take place, a classification of laser ablation in the following:

– thermal ablation: the predominant phenomenon is heat induced by laser and vaporization;

– photophysical ablation: the ablation rate is directly influenced by non-thermic excitation;

– photochemical ablation: the chemical bonds are interrupted by direct dissociation or by indirect transfer of energy via defects and impurities [41].

Laser ablation, also known as ablative photodecomposition, is of key importance in the field of material processing [38, 42]. Depending on the surface (from flat to rough), the ablation condition would gradually change [43]. The ablation process is also influenced by the laser beam parameters, thermo-optical properties, and ambient conditions [43]. The selection of proper parameters helps us to achieve the desired material modification [3]. In laser ablation, the stoichiometric transfer and controlled delivery of target composition to the substrate are possible, this process being a non-equilibrium process [44].

Laser ablation of biomaterials is physically defined by four main parameters: $d(F)$ (ablation rate per pulse), α_{eff} (effective absorption coefficient), F (irradiation fluence), and F_{th} (ablation threshold fluence), being frequently described by the equation [45, 46]:

$$d(F) = \frac{1}{\alpha_{eff}} ln\left(\frac{F}{F_{th}}\right). \tag{1}$$

The ablation threshold fluence initiates the plasma ignition and is described as the minimal energy of the laser pulse per surface unit [41].

Depending on the selected biomaterial and of laser processing parameters (wavelength, energy, and target-substrate distance), material removal during laser ablation is accompanied by a variety of mechanisms [3]. At high-laser intensities, a significant volume of the bulk material is directly excited producing plasma [3]. This can be considered as the fourth state of matter and consists of extremely excited atomic, molecular, ionic, and radical species [8]. The propagation and expansion processes of plasma are dominated by mechanical interactions [43].

It is to be mentioned that a complex cascade of strongly related processes happens in the vicinity of the target surface during and after laser-material interaction [47]. Laser ablation of biomaterials is also based on the strong absorption of laser photons by the investigated material. To achieve a maximum absorption, the wavelength has to be carefully chosen [48].

In the nanosecond regime, laser ablation is induced by rapid heating of the surface layer [49]. In our case, all experimental results were obtained on coatings deposited using a KrF* excimer laser (λ = 248 nm and pulse duration τ = 25 ns).

3.1. Pulsed laser deposition

The common method used for thin film synthesis and associated to laser ablation is pulsed laser deposition (PLD). In this process, the ablated material from target to substrate is pushed by the PLD plasma, which acts as a piston (**Figure 2**). Furthermore, the absorbed photons by irradiating a solid target with a laser beam of high intensity can generate the fusion and local vaporization of target. In PLD, ablation can be defined as the rapid boiling of material in a localized interaction volume on and in the vicinity of target surface (**Figure 3**) [50].

Figure 2. HA plasma expansion.

http://dx.doi.org/10.5772/65197

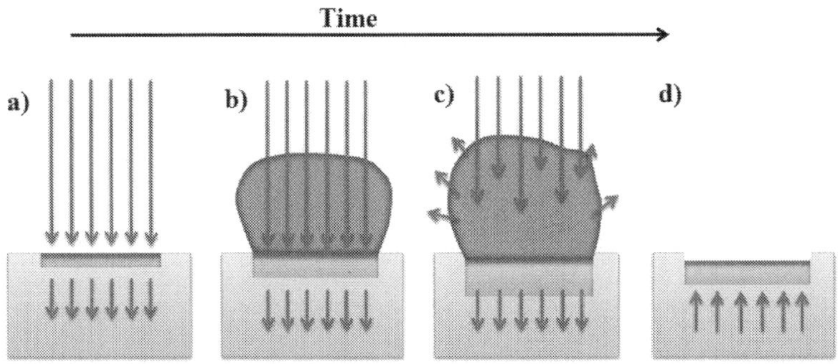

Figure 3. Schematically representation of laser ablation process. (a) The initial absorption of laser radiation (long arrows), the beginning of melting and vaporization. (b) The melted material propagates into solid, vaporization continues, and the interaction laser-plume becomes visible. (c) Absorption of incident laser radiation and plasma formation. (d) The melted material is ejected and re-solidification takes place [50].

The product of this interaction, known as laser ablation phenomenon, produces plasma of complex composition: photons, electrons, ions, atoms, molecules, clusters, and even liquid or solid particles [51]. The ejected material from the target is in a thermodynamic equilibrium and can be associated to thermic and non-thermic mechanisms. A substrate parallel with the irradiated target was used to collect the vaporized material. The deposition of a thin layer on the substrate is the result of repeated laser pulses [52]. The target ablation and deposition rate are dependent on the material or combination of materials present in the target. The ablation threshold is dependent of absorption coefficient and wavelength [53]. As compared to films deposited by other methods, PLD films also exhibit superior performances.

An important feature of PLD is the stoichiometry preservation of the target in the deposited thin films by choosing proper ablation and deposition parameters [54, 55]. Furthermore, in PLD, the deposited material exhibits an excellent adherence to substrate, with a controlled growth rate and without contaminations. This technique also offers the possibility to deposit multi-layers and doped films with great versatility [54]. The variation in the experimental parameters offers the possibility to deposit crystalline structures from various complex materials deposited at room temperature [55].

Nevertheless, PLD presents some limitations as low deposition rate, use of compounds that are not sensitive to thermal decomposition and degradation, and restricted deposition area [52].

3.2. Matrix-assisted pulsed laser evaporation

A specific tool appropriate for ablation of organic and inorganic materials is matrix-assisted pulsed laser evaporation (MAPLE; **Figure 4**). The material ejection and film formation in MAPLE process are also generated by photophysical interaction between laser and target material [56]. MAPLE technique proved to be an appropriate method to obtain high quality thin films by gentle laser transfer onto any substrate of interest [57]. MAPLE emerged as an alternative to PLD in order to preserve the chemical bonding or conformation of delicate materials. On the other words, MAPLE, a less damaging technique, is used to transfer, from the condensed phase to the vapor phase, organic and polymeric compounds, including small and large molecular weight species [58].

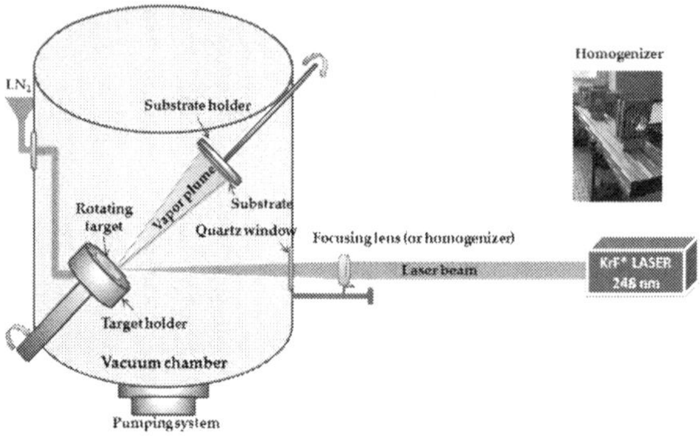

Figure 4. Typical MAPLE experimental set-up.

MAPLE, a non-contact technique, preserves the PLD advantages, allowing a better control of film thickness and morphology and enhancing the adherence of film to substrate [59].

To avoid significant material decomposition, one can optimize the MAPLE deposition conditions: laser wavelength, fluence, solvent type, concentration, target-collector separation distance, temperature, repetition rate, background gas, and pressure. The selected solvent should absorb most of the laser energy and should not react with the studied material [60].

In case of MAPLE, a cryogenic complex target of a dilute/suspension mixture of a material to be deposited and an appropriate solvent is irradiated using a laser beam of low energy. The target is maintained frozen during the laser irradiation and deposition process using liquid nitrogen [59]. This technique is governed by two photothermal processes, which takes place in the matrix, the evaporation of frozen complex target and material release into the chamber. The conversion of photon energy (absorbed by solvent) in thermal energy generates material heating and solvent vaporization [61].

MAPLE offers the possibility to produce uniform, ultrastable, and nanostructured coatings [59].

3.3. Laser spallation

In an extension of laser ablation, laser spallation is a progressive process in which the laser pulse is applied on the rear surface [62]. As known, spallation is defined as a damage occurred at the interface between inner film and substrate. Some examples are interface delamination, film spallation, and film expulsion. As illustrated in **Figure 5**, spallation can be described as a dynamic tensile fracture [63].

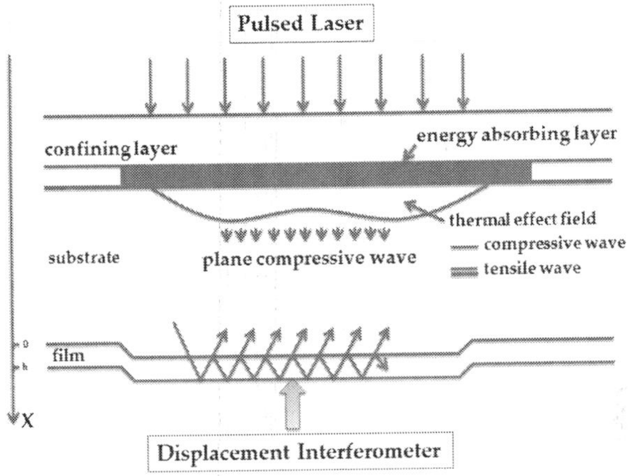

Figure 5. Laser-induced film spallation process: schematic diagram [63].

In 1992, Gupta et al. [64] used laser spallation to measure the strength of planar interfaces. To achieve this, a laser pulse of a high enough energy and a pre-determined duration was converted into a pressure pulse of a critical amplitude and width that was sent through the substrate toward the free surface with the coating [64].

Extreme superheating/deformation conditions generated by laser processing are related to mechanisms such melting and/or photomechanical damage/spallation [65]. By increasing the fluence over the spallation threshold, the composition of the ejected plume rapidly changes. The modifications consist in conversion from liquid deposits and large droplets to a blend of individual atoms, small droplets, and atomic clusters [66].

Nevertheless, it was found that certain laser processing conditions can generate undesired and very detrimental rear surface spallation effect at the textured front surface, even if there are used laser conditions well below the spallation threshold for planar surfaces [67].

4. Study of thin film growth mechanisms by laser ablation

One should take into account that coatings are obtained by successive ejected material from the target by each laser pulse. We will start the discussion by presenting some representative results from our studies related to the processing of hydroxyapatite by laser ablation.

In **Figure 6**, the surface of a HA target before and after laser irradiation is presented. The morphology is characteristic to a material melted and then resolidified. The details presented in **Figure 6c**, at a higher magnification, are specific to HA solidification, in fractal form. All circular bumps indicated the presence of the bubbling phenomenon. Moreover, the cracks appeared on target surface are due to expansion-contraction cycles as a result of repeated heating/cooling processes.

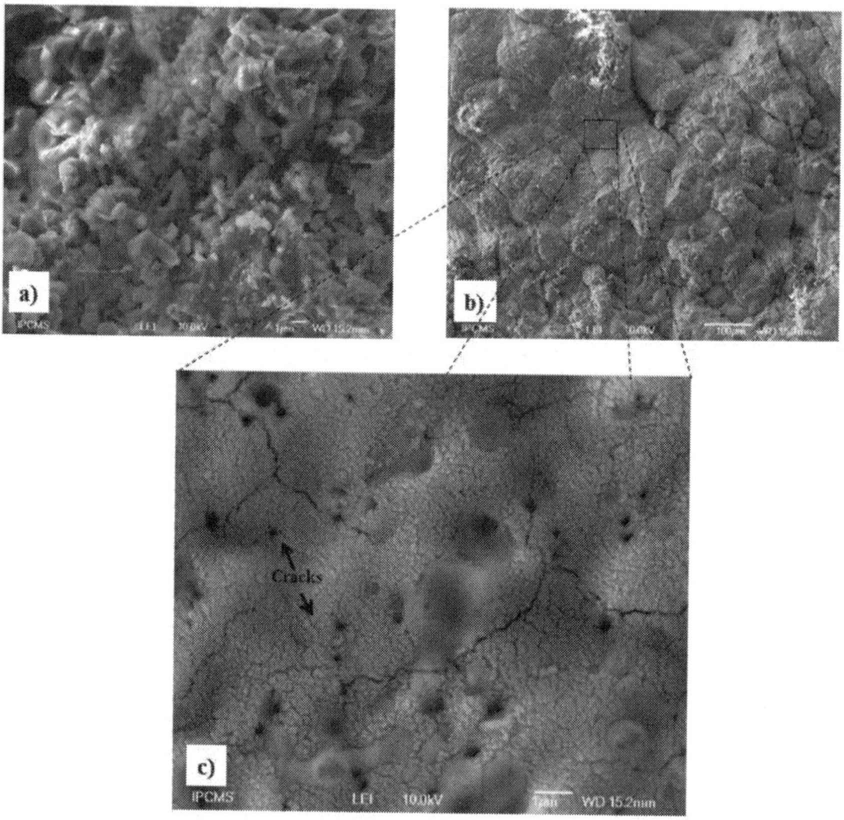

Figure 6. SEM micrographs of HA target before (a) and after (b and c) laser irradiation.

Typical HA structures deposited by laser ablation at different laser fluences and studied by Scanning Electron Microscopy (SEM) are presented in **Figure 7**. It is obvious that the aspect of deposited coatings varies from acicular at low fluence to cauliflower aspect at high fluence, respectively.

In case of samples HA1 and HA2, the coatings seem to be the product of material condensation originated from plasma. The droplets, even if they are present, are of small dimensions.

These coatings were also investigated By Energy Dispersive X-ray Spectroscopy (EDS) and X-ray Photoelectron Spectroscopy (XPS). EDS gives us information about the composition, on the entire thickness of the layer, while XPS provided information exclusively from surface.

Only three of the samples (HA1, HA3, and HA5) have been biologically investigated *in vitro* by cell growth. The cells used for these analyses were osteoprogenitor human cells (HOp) from bone marrow. HOp were cultured in ISCOVE (Sigma I 3390) medium, supplemented with 10%

(v/v) fetal bovine serum, 1% (v/v) glutamine, and 1% (v/v) penicillin and streptomycin. After 5 days, the samples were prepared for SEM investigation (**Figure 8**).

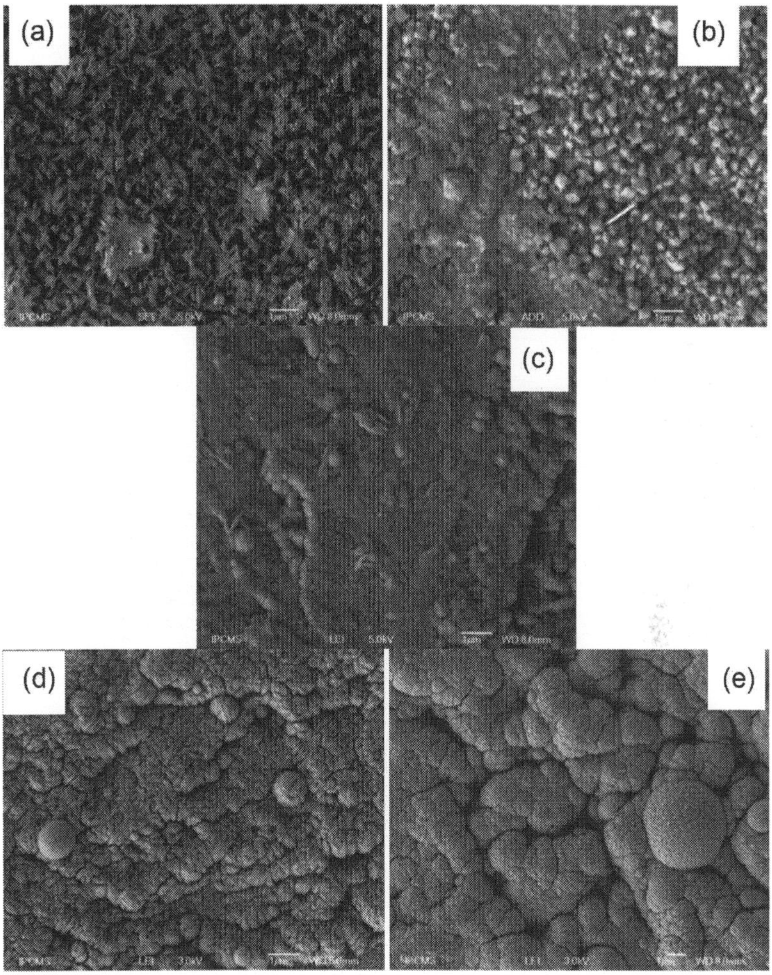

Figure 7. SEM micrographs for HA coatings deposited at (a) 1.2 J/cm^2 (HA1), (b) 1.8 J/cm^2 (HA2), (c) 2.7 J/cm^2 (HA3), (d) 5 J/cm^2 (HA4), and (e) 7.5 J/cm^2 (HA5).

From **Figure 8a**, one can observe that HA1 sample was destroyed, the coating being completed and dissolved in the culture medium. We could not identify cells on the sample surface. **Figure 8b** demonstrated a partial dissolution of HA3 coating. No cell was present on the sample surface. In case of HA5, one can observe an important coverage rate of the coatings deposited at 7.5 J/cm^2 in the presence of cells. The morphology of the coating was not modified, while the osteoblasts present a polygonal flattened form.

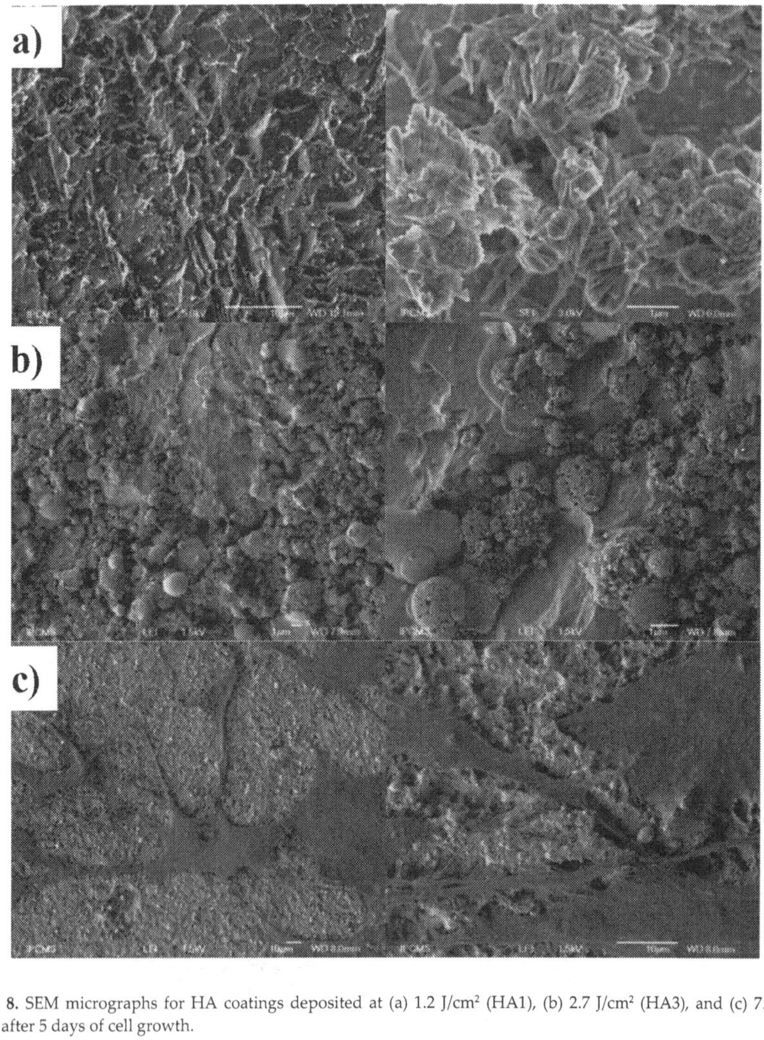

Figure 8. SEM micrographs for HA coatings deposited at (a) 1.2 J/cm^2 (HA1), (b) 2.7 J/cm^2 (HA3), and (c) 7.5 J/cm^2 (HA5) after 5 days of cell growth.

A similar investigations on HA coatings were conducted by Zhu et al. They analyzed the behavior of MC3T3-E1 cells cultured on the specimens after 7 days [68].

In another study, some of the HA coatings grown by PLD using a laser fluence of 3 J/cm^2 were thermally treated (at 400°C for 6h) in order to improve the crystallinity.

http://dx.doi.org/10.5772/65197

After a first examination, the SEM images of this type of structures did not revealed significant differences (**Figure 9**).

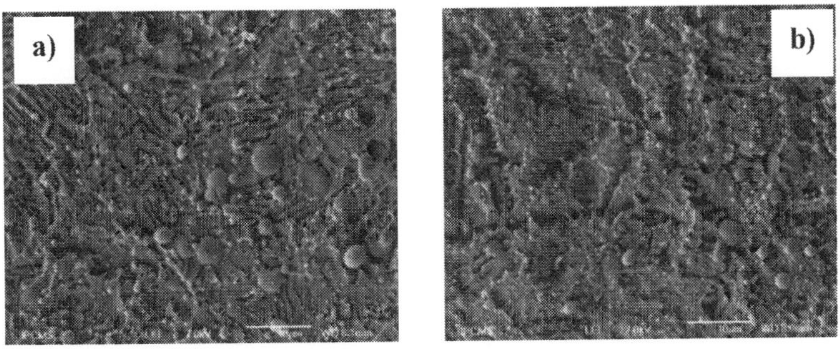

Figure 9. SEM micrographs of HA coatings deposited by PLD (a) with (HAt) and (b) without thermal treatment (HA).

SEM micrographs, at higher magnification, showed the differences induced by the thermal treatment (**Figure 10**).

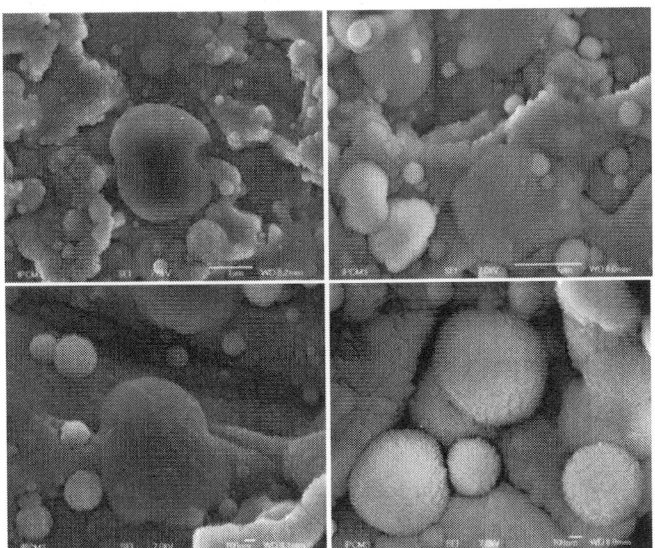

HAt (treated) **HA (untreated)**

Figure 10. SEM micrographs of HA coatings synthesized by PLD with (left) and without (right) thermal treatment.

It can be seen that HA coatings present a more rough morphology at nano level. The droplets' shape is similar to that of snowballs, and the surface is made up of parallelepipedic structure.

The thermal treatment induced the surface structure reorganization. There are no longer irregularities, and the appearance of the droplets is smooth.

The Atomic Force Microscopy (AFM) analysis revealed differences between the two types of structures related to their roughness (**Figure 11**). The decrease in the rough value from 1.01 to 0.8 nm, at nanometric scale, proves the smoothing of the target.

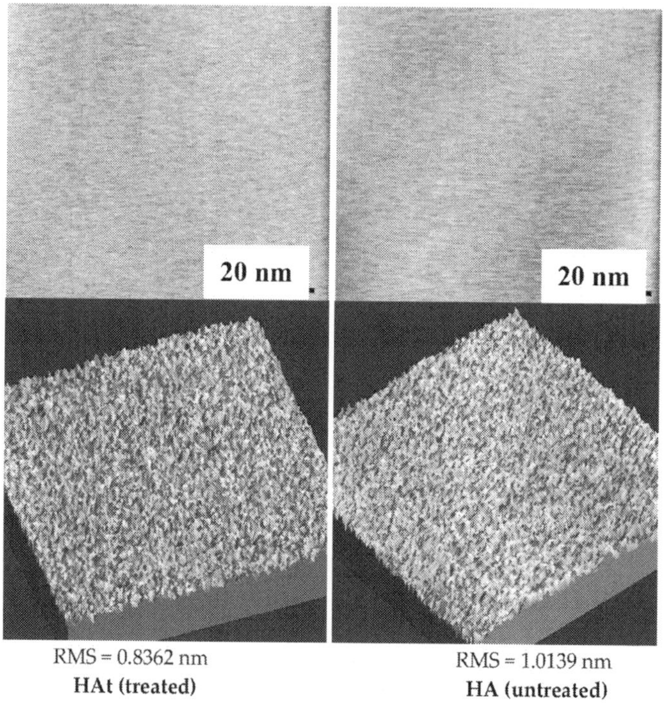

Figure 11. AFM imaged for HA coatings deposited by PLD with (left) and without (right) thermal treatment.

EDS results showed that the value of Ca/P ratio diminishes from 2.04 (untreated sample - HA) to 1.63 (treated sample - HAt). The corroborated results demonstrated the importance of thermal treatment in obtaining crystalline hydroxyapatite, biocompatible, having a structure similar to stoichiometric HA.

HA coatings were also grown by other techniques, such as thermal spray, high velocity oxy-fuel (HVOF) techniques, and plasma spraying trying to find, as in PLD depositions, the optimal conditions for good film with applications in medicine [19, 69].

A thermal treatment was also applied to Mn-CHA and OCP films obtained by PLD. In case of Mn-CHA coating, the Ca/P atomic ratio obtained by XPS and EDS investigations was 1.64–1.66, close to the stoichiometric values.

The morphologies of the two structures, OCP and Mn-CHA, are quite different. OCP has a porous and arborescent-like structure (**Figure 12a** and **b**), and Mn-CHA has a granular and more compact (**Figure 12c** and **d**). The surface morphologies of both calcium phosphates are well matched for bone tissue growth and osteointegration.

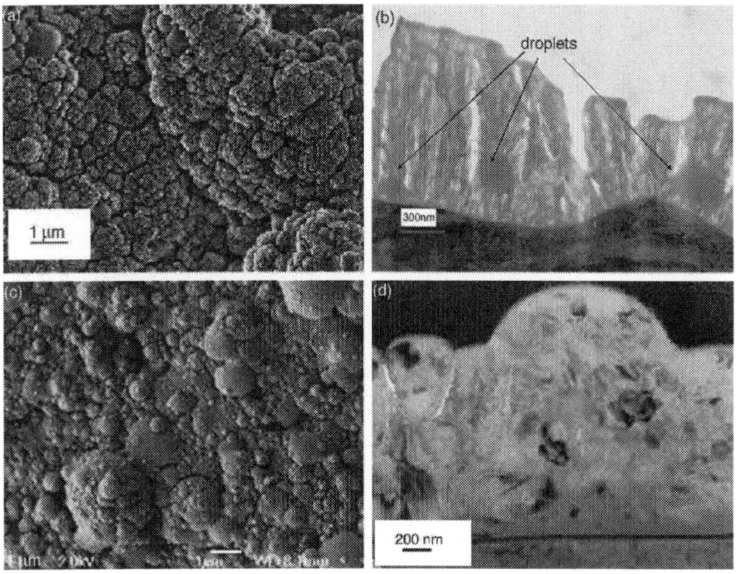

Figure 12. (a) Scanning and (b) Transmission Electron Microscopy (TEM) images of OCP coatings; (c) SEM and (d) TEM images of Mn-CHA coatings (reproduced with permission from Ref. [25]).

In case of OCP coatings, XPS measurements showed the total dissolution and disappearance of coating after 7 days of immersion in simulated body fluid (SBF) (**Figure 13a**). SEM investigations confirmed this advanced dissolution. As for Mn-CHA, the SEM and XPS investigations demonstrated that the coatings preserve the basic composition even the intensity of Ca and P peaks decreased (**Figure 13b**). After 7 days of immersion in SBF, the surface becomes slightly smoother.

Some interesting results were obtained by laser ablation of Mg:OCP and Sr:OCP compounds using the MAPLE technique [31]. The X-ray Diffraction (XRD) patterns revealed that all MAPLE coatings are constituted of OCP. This remark is sustained by the presence of the strong low angle reflection 2θ of 4.7° and the series of reflections in the range of 30–34°. Comparing these results with the previous one related to OCP deposition by PLD [70], one can remark that the gentle deposition conditions of MAPLE offer a higher degree of OCP crystallinity with respect to PLD [31]. The homogeneous distribution of magnesium and strontium on the thin film surface was evidenced by EDS analysis (**Figure 14**).

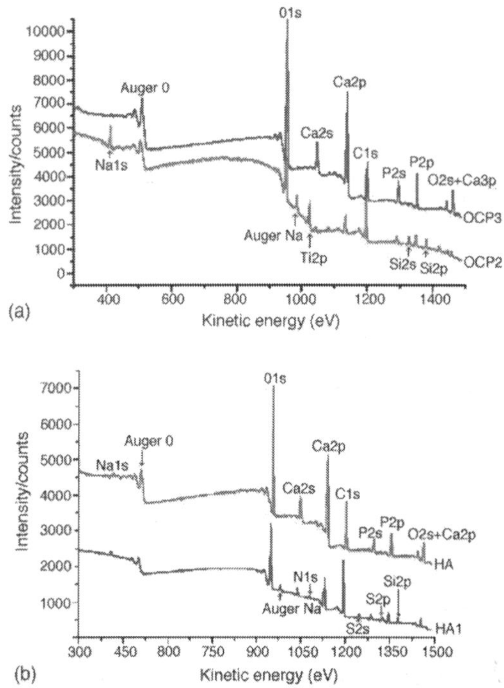

Figure 13. XPS spectra of (a) OCP and (b) Mn-CHA before and after degradation tests (reproduced with permission from Ref. [25]).

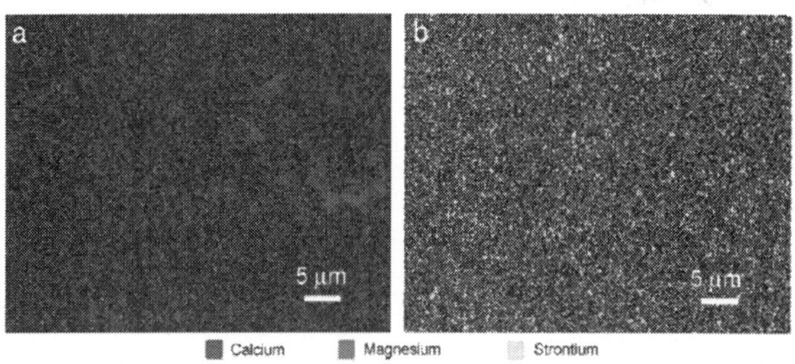

Figure 14. EDS maps recorded for (a) Mg:OCP and (b) Sr:OCP coatings (reproduced with permission from Ref. [31]).

To evaluate the proliferation and morphology of MG63 cells, one can also perform phalloidin staining on OCP, Mg:OCP, and Sr:OCP coatings deposited by MAPLE. The surface topography and chemical composition can influence cell behavior. Looking the images of MAPLE coating stain with phalloidin for 14 days, no visible differences were observed (**Figure 15**, left). Similar results are visible from SEM images of the same structures (**Figure 15**, right). Deeper studies showed that Mg:OCP and Sr:OCP coatings improved the proliferation and differentiation of MG63 cells.

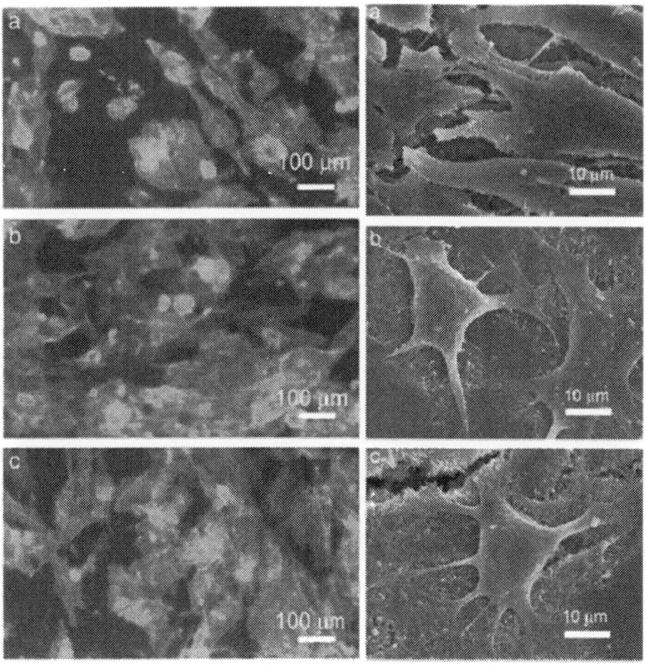

Figure 15. Phalloidin staining (left) and SEM images (right) of MG63 cells after 14 days of culture grown on (a) OCP, (b) Mg:OCP, and (c) Sr:OCP (reproduced with permission from Ref. [31]).

The biological analysis conducted on Mg:OCP and Sr:OCP coatings revealed no visible cytotoxic or inflammatory effects on osteoblast-like cells in all experimental times (**Table 1**).

Test MG63	3 days	7 days	14 days
WST1 (450/625 nm)	1.564 ± 0.125	3.132 ± 0.081	3.303 ± 0.047
Alkaline Phosphatase (ALP) (µg/mg proteins)	1.20 ± 0.17	2.33 ± 0.19	1.65 ± 0.02
Collagen Type I Production (CICP) (ng/mg proteins)	9.9 ± 1.2	11.3 ± 0.3	11.1 ± 0.3
Osteocalcin (OC) (ng/mg proteins)	2.00 ± 0.09	2.09 ± 0.18	2.63 ± 0.22
Interleukin-6 (IL-6) (pg/mg proteins)	37 ± 9	109 ± 5	97 ± 3
Transforming Growth Factor -β1 (TGF-β1) (pg/mg proteins)	454 ± 34	537 ± 67	370 ± 26

Table 1. Control values of proliferation and differentiation for MG63 osteoblast cells at 3, 7, and 14 days culture.

Mroz et al. also evaluated the performances of HA and OCP coatings deposited by PLD. The biological assays revealed that both layers are biocompatible with respect to human osteoblast cells, offering favorable conditions for their proliferation [71].

5. Conclusions and perspectives

This chapter highlighted the importance of laser ablation phenomenon that underlies the processing of biomaterials.

Laser ablation of biomaterials is an important and complex field. This domain was explored since long time, but there are a lot of natural or inorganic biomaterials, which wait to be developed and understood.

PLD and MAPLE techniques are of interest for thin film deposition, allowing varying some parameters to achieve the optimum conditions for the selected biomaterial. They permit the stoichiometric transfer. In case of PLD, the depositions are obtained from a solid target, as compared to MAPLE in which the target is a cryogenic mixture.

Sensitive biomaterials can be processed only by MAPLE, to avoid the chemical decomposition. The deposition of calcium phosphates (CaP) as thin films can be done by both techniques. Physical-chemical and biological analyses in the field of CaP recommend these coatings as potential biomaterial for the development of medical implants.

Acknowledgements

The authors acknowledge the support of UEFISCDI under the contracts 19_RO-FR/2014 and PCCA 244/2014.

Author details

Gianina-Florentina Popescu-Pelin[1,2], Carmen Ristoscu[1] and Ion N. Mihailescu[1*]

*Address all correspondence to: ion.mihailescu@inflpr.ro

1 National Institute for Lasers, Plasma and Radiation Physics, Magurele, Ilfov, Romania

2 Faculty of Physics, University of Bucharest, Magurele, Ilfov, Romania

References

[1] Ratner BD, Hoffman AS, Schoen FJ, Lemons J. Biomaterials science: a multidisciplinary endeavor. In: Ratner BD, Hoffman AS, Schoen FJ, Lemons J, editors. Biomaterials Science: An Introduction to Materials in Medicine. 2nd ed. San Diego: Elsevier Academic Press; 2004. p. 1–10.

[2] Desai TA, Hansford DJ, Kulinsky L, Nashat AH, Rasi G, Tu J, Wang Y, Zhang M, Ferrari M. Nanopore technology for biomedical applications. Biomedical Microdevices. 1999;2(1): 11–40. DOI: 10.1023/A:1009903215959.

[3] Brown MS, Arnold CB. Fundamentals of laser-material interaction and application to multiscale surface modification. In: Sugioka K, Meunier M, Piqué A, editors. Laser Precision Microfabrication. 1st ed. Springer Heidelberg Dordrecht London New York: Springer-Verlag Berlin Heidelberg; 2010. p. 91–120 (Chapter 4).

[4] Fadeeva E, Deiwick A, Chichkov B, Schlie-Wolter S. Impact of laser-structured biomaterial interfaces on guided cell responses. Interface Focus. 2014; 4(1): 20130048.

[5] Wang X. Overview on biocompatibilities of implantable biomaterials. In: Pignatello R. editor. Advances in Biomaterials Science and Biomedical Applications. Croatia: InTech; 2013. p. 111–155. DOI: 10.5772/53461.ch5.

[6] Saha N, Saarai A, Roy N, Kitano T, Saha P. Polymeric biomaterial based hydrogels for biomedical applications. Journal of Biomaterials and Nanobiotechnology. 2011; 2(1):85–90. DOI: 10.4236/jbnb.2011.21011.

[7] Ulery BD, Nair LS, Laurencin CT. Biomedical applications of biodegradable polymers. Journal of Polymer Science Part B: Polymer Physics. 2011;49:832–864. DOI: 10.1002/polb.22259.

[8] Chua PK, Chen JY, Wang LP, Huang N. Plasma-surface modification of biomaterials. Materials Science and Engineering R. 2002;36:143–206.

[9] Zhang W, Chung PH, Zhang A, Chen S. Laser processing of natural biomaterials. In: Schmidt V, Belegratis MR. editors. Laser Technology in Biomimetics: Basics and Applications. Springer Heidelberg New York Dordrecht London: Springer-Verlag Berlin Heidelberg; 2013. p. 237–257. DOI: 10.1007/978-3-642-41341-4_10.ch10.

[10] Zhu Y, Fu J, Zheng C, Ji Z. Effect of nanosecond pulse laser ablation on the surface morphology of Zr-based metallic glass. Optics & Laser Technology. 2016;83:21–27. DOI: 10.1016/j.optlastec.2016.03.021.

[11] Kulkarni M, Mazare A, Schmuki P, Iglič A. Biomaterial surface modification of titanium and titanium alloys for medical applications. In: Seifalian A, de Mel A, Kalaskar DM. editors. Nanomedicine. Manchester: One Central Press; 2014. p. 111–136 (Chapter 5).

[12] Shi D. Introduction to Biomaterials. 1st ed. Beijing: Tsinghua University Press and World Scientific Publishing Co. Pte. Ltd.; 2006. 253 p.

[13] Wang M. Bioactive materials and processing. In: Shi D. editor. Biomaterials and Tissue Engineering. 1st ed. Springer Heidelberg New York Dordrecht London: Springer-Verlag Berlin Heidelberg; 2004. p. 1–82. DOI: 10.1007/978-3-662-06104-6.ch1.

[14] Hazar Yoruç AB, Şener BC. Biomaterials. In: Kara S. editor. A Roadmap of Biomedical Engineers and Milestones. Croatia: InTech; 2012. p. 67–114, DOI: 10.5772/48057.ch3.

[15] Park JB, Bronzino JD. Biomaterials: Principles and Applications. Florida: CRC Press LLC; 2003. 250 p.

[16] Jokanovic V, Uskokovic D. Calcium hydroxyapatite thin films on titanium substrates prepared by ultrasonic spray pyrolysis. Materials Transactions. 2005;46(2):228–235.

[17] Gross UM, Lassner D. In vitro and in vivo evaluation of thin calcium phosphate coatings. In: Leon B, Jansen JA. editors. Thin Calcium Phosphate Coatings for Medical Implants. New York: Springer Science+Business Media LLC; 2009. p. 67–100. DOI: 10.1007/978-0-387-77718-4.ch4.

[18] Axente E, Sima F, Ristoscu C, Mihailescu N, Mihailescu IN. Biopolymer thin films synthesized by advanced pulsed laser techniques. In: Parveen FK. editor. Recent Advances in Biopolymers. Croatia: InTech; 2016. p. 73–104, DOI: 10.5772/61734.ch4.

[19] Berndt CC, Hasan F, Tietz U, Schmitz K-P. A review of hydroxyapatite coatings manufactured by thermal spray. In: Ben-Nissan B. editor. Advances in Calcium Phosphate Biomaterials. Springer Heidelberg New York Dordrecht London: Springer Berlin Heidelberg; 2014. p 267–329, DOI: 10.1007/978-3-642-53980-0.ch9.

[20] Padilla S, Vallet-Regi M, Ginebra MP, Gil FJ. Processing and mechanical properties of hydroxyapatite pieces obtained by the gelcasting method. Journal of the European Ceramic Society. 2005; 25:375–383.

[21] Rivera-Muñoz EM. Hydroxyapatite-based materials: synthesis and characterization. In: Fazel R. editor. Biomedical Engineering-Frontiers and Challenges. Croatia: InTech; 2011. p. 75–98, DOI: 10.5772/19123.ch4.

[22] Šupová M. Substituted hydroxyapatites for biomedical applications: a review. Ceramics International. 2015;41:9203–9231. DOI: 10.1016/j.ceramint.2015.03.316.

[23] Orlovskii VP, Komlev VS, Barinov SM. Hydroxyapatite and hydroxyapatite-based ceramics. Inorganic Materials. 2002;38(10):973–984. Translated from Neorganicheskie Materialy. 2002;38(10):1159–1172. DOI: 10.1023/A:1020585800572.

[24] Ridi F, Meazzini I, Castroflorio B, Bonini M, Berti D, Baglioni P. Functional calcium phosphate composites in nanomedicine. Advances in Colloid and Interface Science. 2016; DOI: 10.1016/j.cis.2016.03.006. Available online 7 April 2016, http://dx.doi.org/10.1016/j.cis.2016.03.006.

[25] Mihailescu IN, Torricelli P, Bigi A, Mayer I, Iliescu M, Werckmann J, Socol G, Miroiu F, Cuisinier F, Elkaim R, Hildebrand G. Calcium phosphate thin films synthesized by pulsed laser deposition: physico-chemical characterization and in vitro cell response. Applied Surface Science. 2005; 248:344–348. DOI: 10.1016/j.apsusc.2005.03.111.

[26] LeGeros RZ. Preparation of octacalcium phosphate (OCP): a direct fast method. Calcified Tissue International. 1985. 37:194–197. DOI: 10.1007/BF02554841.

[27] Komlev VS, Fadeeva IV, Fomin AS, Shvorneva LI, Ferro D. Synthesis of Octacalcium Phosphate by Precipitation from Solution. Doklady Chemistry. Part 2. 2010;432:178–182. DOI: 10.1134/S0012500810060066.

[28] Fedotov AY, Komlev VS, Teterina AY, Sirotinkin VP, Shamrai VF, Fadeeva IV, Barinov SM. Preparation of octacalcium phosphate from calcium carbonate powder. Inorganic Materials. 2013. 49(11):1148–1151. DOI: 10.1134/S0020168513110058.

[29] Honda Y, Anada T, Kamakura S, Suzuki O. Effect of topography of the octacalcium phosphate granule surfaces on its bone regenerative property. In: Sasano T, Suzuki O. editors. Interface Oral Health Science 2009. Springer Tokyo Berlin Heidelberg New York: Springer; 2010. p. 288–290. DOI: 10.1007/978-4-431-99644-6_82.

[30] Suzuki O. Chemical and physical factors affecting osteoconductivity of octacalcium phosphate bone substitute material. In: Sasano T, Suzuki O. editors. Interface Oral Health Science 2009. Springer Tokyo Berlin Heidelberg New York: Springer; 2010. p. 100–107. DOI: 10.1007/978-4-431-99644-6_82.

[31] Boanini E, Torricelli P, Fini M, Sima F, Serban N, Mihailescu IN, Bigi A. Magnesium and strontium doped octacalcium phosphate thin films by matrix assisted pulsed laser evaporation. Journal of Inorganic Biochemistry. 2012;107:65–72. DOI: 10.1016/j.jinorgbio.2011.11.003.

[32] Kolmas J, Groszyk E, Piotrowska U. Nanocrystalline hydroxyapatite enriched in selenite and manganese ions: physicochemical and antibacterial properties. Nanoscale Research Letters. 2015. 10(1):989. DOI: 10.1186/s11671-015-0989-x.

[33] Iliescu M, Nelea V, Werckmann J, Socol G, Mihailescu IN, Mayer I, Cuisinier F. Morphological and structural characterisation of osseointegrable Mn^{2+} and CO_3^{2-} doped hydroxylapatite thin films. Materials Science and Engineering C. 2007;27:105–109. DOI: 10.1016/j.msec.2006.03.006.

[34] György E, Toricelli P, Socol G, Iliescu M, Mayer I, Mihailescu IN, Bigi A, Werckman J. Biocompatible Mn^{2+}-doped carbonated hydroxyapatite thin films grown by pulsed laser deposition. Journal of Biomedical Materials Research Part A. 2004;71A(2):353–358. DOI: 10.1002/jbm.a.30172.

[35] Weingartner E, Wegener K, Dold C. Laser ablation. In: Laperrière L, Reinhart G. editors. CIRP Encyclopedia of Production Engineering. Springer Heidelberg New

York Dordrecht London: Springer-Verlag Berlin Heidelberg; 2014. p. 736–739. DOI: 10.1007/978-3-642-20617-7.

[36] Gamaly EG, Rode AV, Luther-Davies B, Tikhonchuk VT. Ablation of solids by femtosecond lasers: ablation mechanism and ablation thresholds for metals and dielectrics. Physics of Plasmas. 2002;9(3): 949. DOI: 10.1063/1.1447555.

[37] Dahotre NB, Harimkar SP. Laser Fabrication and Machining of Materials. New York: Springer US; 2008. 558 p. DOI: 10.1007/978-0-387-72344-0.

[38] Reif J. Processing with ultrashort laser pulses. In: Schaaf P. editor. Laser Processing of Materials Fundamentals, Applications and Developments. Springer Heidelberg Dordrecht London New York: Springer-Verlag Berlin Heidelberg; 2010. p. 113–126. DOI: 10.1007/978-3-642-13281-0.ch6.

[39] Schneider CW, Lippert T. Laser ablation and thin film deposition. In: Schaaf P. editor. Laser Processing of Materials Fundamentals, Applications and Developments. Springer Heidelberg Dordrecht London New York: Springer-Verlag Berlin Heidelberg; 2010. p. 89–109. DOI: 10.1007/978-3-642-13281-0.ch5.

[40] Ossi PM, Dinescu M. Creating nanostructures with lasers. In: Schaaf P. editor. Laser Processing of Materials Fundamentals, Applications and Developments. Springer Heidelberg Dordrecht London New York: Springer-Verlag Berlin Heidelberg; 2010. p. 131–165. DOI: 10.1007/978-3-642-13281-0.ch7

[41] Ristoscu C, Mihailescu IN. Biomimetic coatings by pulsed laser deposition. In: Schmidt V, Belegratis MR. Laser Technology in Biomimetics: Basics and Applications. Springer Heidelberg New York Dordrecht London: Springer-Verlag Berlin Heidelberg; 2013. p. 163–191. DOI: 10.1007/978-3-642-41341-4.ch7.

[42] Hansen SG, Robitaille TE. Laser ablation of a polysilane material. Journal of Applied Physics. 1987;62:1394. DOI: 10.1063/1.339643.

[43] Stafe M, Marcu A, Puscas NN. Pulsed Laser Ablation of Solids. Basics, Theory and Applications. 1st ed. Springer Heidelberg New York Dordrecht London: Springer-Verlag Berlin Heidelberg; 2014. 233 p. DOI: 10.1007/978-3-642-40978-3.

[44] Takeuchi I. Combinatorial pulsed laser deposition. In: Eason R. editor. Pulsed Laser Deposition of Thin Films: Applications-led growth of functional materials. Hoboken, New Jersey: John Wiley & Sons, Inc.; 2007. p. 161–175. DOI: 10.1002/0470052120.ch7.

[45] Urech L, Lippert T. Designed polymers for ablation. In: Phipps CR. editor. Laser Ablation and Its Applications. New York: Springer US; 2007. p. 281–297. DOI: 10.1007/978-0-387-30453-3.ch11.

[46] Lippert T. UV laser ablation of polymers: from structuring to thin film deposition. In: Miotello A, Ossi PM. editors. Laser-surface interactions for nw materials production. Tailoring structure and properties. Heidelberg Dordrecht London New York: Springer-Verlag Berlin Heidelberg; 2010. p. 141–175. DOI: 10.1007/978-3-642-03307-0.ch7.

[47] Autrique D, Alexiades V, Khanal H. Hydrodynamic modeling of ns-laser ablation. In: Ninth MSU-UAB Conference on Differential Equations and Computational Simulations, Electronic Journal of Differential Equations, Conference 20; October 2013; Texas State University-San Marcos; 2013. p. 1–14. https://www.researchgate.net/publication/258201771_Hydrodynamic_modeling_of_ns-laser_ablation (last access September 5, 2016)

[48] Gower M, Rizvi N. Applications of Laser Ablation to Microengineering. Available from: http://www.lasermicromachining.com/wp-content/uploads/2014/09/Applications-of-Laser-Ablation-to-Microengineering.pdf [Accessed: 2016-06-25]

[49] Haglund Jr. RF, Itoh N. Electronic processes in laser ablation of semiconductors and insulators. In: Miller JC. editor. Laser Ablation: Principles and Applications. Springer Berlin Heidelberg New York London Paris Tokyo Hong Kong Barcelona Budapest: Springer-Verlag Berlin Heidelberg; 1994. p. 11–52. DOI: 10.1007/978-3-642-78720-1.ch2.

[50] Ashfold MNR, Claeyssens F, Fuge GM, Henley SJ. Pulsed laser ablation and deposition of thin films. Chemical Society Reviews. 2004;33:23–31. DOI: 10.1039/B207644F.

[51] Popescu AC, Ulmeanu M, Ristoscu C, Mihailescu IN. Deposition and surface modification of thin solid structures by high-intensity pulsed laser irradiation. In: Lawrence JR, Dowding C, Waugh D, Griffiths JB. Laser Surface Engineering. AMSTERDAM • BOSTON • CAMBRIDGE • HEIDELBERG LONDON • NEW YORK • OXFORD • PARIS • SAN DIEGO SAN FRANCISCO • SINGAPORE • SYDNEY • TOKYO: Elsevier – Woodhead Publishing; 2015. p. 287–313 (Chapter 12).

[52] Sima F, Ristoscu C, Duta L, Gallet O, Anselme K, Mihailescu IN. Laser thin films deposition and characterization for biomedical applications. In: Vilar R. editor. Laser Surface Modification of Biomaterials. AMSTERDAM • BOSTON • CAMBRIDGE • HEIDELBERG LONDON • NEW YORK • OXFORD • PARIS • SAN DIEGO SAN FRANCISCO • SINGAPORE • SYDNEY • TOKYO: Elsevier – Woodhead Publishing;; 2016. p. 77–125 (Chapter 3).

[53] Norton DP. Pulsed laser deposition of complex materials: progress towards applications. In: Eason R. editor. Pulsed Laser Deposition of Thin Films: Applications-Led Growth of Functional Materials. Hoboken, New Jersey: John Wiley & Sons, Inc.; 2007. p. 3–27. DOI: 10.1002/0470052120.ch1.

[54] Duta L, Stan GE, Stroescu H, Gartner M, Anastasescu M, Fogarassy Zs, Mihailescu N, Szekeres A, Bakalova S, Mihailescu IN. Multi-stage pulsed laser deposition of aluminum nitride at different temperatures. Applied Surface Science. 2016;374:143–150. DOI: 10.1016/j.apsusc.2015.10.093.

[55] Erakovic S, Jankovic A, Ristoscu C, Duta L, Serban N, Visan A, Mihailescu IN, Stan GE, Socol M, Iordache O, Dumitrescu I, Luculescu CR, Janackovi Dj, Miskovic-Stankovi V. Antifungal activity of Ag:hydroxyapatite thin films synthesized by pulsed laser

deposition on Ti and Ti modified by TiO_2 nanotubes substrates. Applied Surface Science. 2014;293:37–45. DOI: 10.1016/j.apsusc.2013.12.029.

[56] Shepard KB, Priestley RD. MAPLE deposition of macromolecules. Macromolecular Chemistry and Physics. 2013;214(8):862–872. DOI: 10.1002/macp.201200621.

[57] Stamatin L, Cristescu R, Socol G, Moldovan A, Mihaiescu D, Stamatin I, Mihailescu IN, Chrisey DB. Laser deposition of fibrinogen blood proteins thin films by matrix assisted pulsed laser evaporation. Applied Surface Science. 2005;248:422–427. DOI: 10.1016/j.apsusc.2005.03.060.

[58] Piqué A. The Matrix-Assisted Pulsed Laser Evaporation (MAPLE) process: origins and future directions. Applied Physics A. 2011;105(3):517–528. DOI: 10.1007/s00339-011-6594-7.

[59] Sima F, Mihailescu IN. Biomimetic assemblies by matrix-assisted pulsed laser evaporation. In: Schmidt V, Belegratis MR. editors. Laser Technology in Biomimetics: Basics and Applications. Springer Heidelberg New York Dordrecht London: Springer-Verlag Berlin Heidelberg; 2014. p. 111–141. DOI: 10.1007/978-3-642-41341-4.ch5.

[60] Mihailescu IN, Ristoscu C, Bigi A, Mayer I. Advanced biomimetic implants based on nanostructured coatings synthesized by pulsed laser technologies. In: Miotello A, Ossi PM. editors. Laser-Surface Interactions for New Materials Production. Springer Heidelberg Dordrecht London New York: Springer-Verlag Berlin Heidelberg; 2010. p. 235–260. DOI: 10.1007/978-3-642-03307-0.ch10.

[61] Cristescu R, Doraiswamy A, Patz T, Socol G, Grigorescu S, Axente E, Sima F, Narayan RJ, Mihaiescu D, Moldovan A, Stamatin I, Mihailescu IN, Chisholm B, Chrisey DB. Matrix assisted pulsed laser evaporation of poly(d,l-lactide) thin films for controlled-release drug systems. Applied Surface Science. 2007;253:7702–7706. DOI: 10.1016/j.apsusc.2007.02.042.

[62] Oltra R, Yavas O, Cruz F, Boquillon JP, Sartori C. Modelling and diagnostic of pulsed laser cleaning of oxidized metalic surfaces. Applied Surface Science. 1996;96–98:484–490. DOI: 10.1016/0169-4332(95)00500-5.

[63] Zhou M, Zhang YK, Cai L. Adhesion measurement of thin films by a modified laser spallation technique: theoretical analysis and experimental investigation. Applied Physics A. 2002; 74(4):475–480. DOI: 10.1007/s003390101185.

[64] Gupta V, Argon AS, Parks DM, Cornie JA. Measurement of interface strength by a laser spallation technique. Journal of the Mechanics and Physics of Solids. 1992;40(1):141–180. DOI: 10.1016/0022-5096(92)90296-E.

[65] Zhigilei LV, Ivanov DS, Leveugle E. Computer modelling of laser melting and spallation of metal targets. Proceedings of SPIE. 2004;5448:14 p. DOI: 10.1117/12.548821.

[66] Wu C, Zhigilei LV. Microscopic mechanisms of laser spallation and ablation of metal targets from large-scale molecular dynamics simulations. Applied Physics A. 2014;114(1):11–32. DOI: 10.1007/s00339-013-8086-4.

[67] Du ZR, Palina N, Chen J, Aberle AG, Hoex B, Hong MH. Enhancement of laser-induced rear surface spallation by pyramid textured structures on silicon wafer solar cells. Optics Express. 2012; 20(23):A984–A990. DOI: 10.1364/OE.20.00A984.

[68] Zhu L, Ye X, Tang G, Zhao N, Gong Y, Zhao Y, Zhao J, Zhang X. Biomimetic coating of compound titania and hydroxyapatite on titanium. Journal of Biomedical Materials Research Part A. 2007;83(4):1165–1175. DOI: 10.1002/jbm.a.31401.

[69] Khor KA, Li H, Cheang P. Nanostructures and osteoblast behavior of thermal sprayed calcium phosphate splats. AZO Journal of Materials Online. 2006;2. Was also published in print form in Advances in Technology of Materials and Materials Processing. 2006;8(2):130–133. DOI: 10.2240/azojomo0215.

[70] Socol G, Torricelli P, Bracci B, Iliescu M, Miroiu F, Bigi A, Werckmann J, Mihailescu IN. Biocompatible nanocrystalline octacalcium phosphate thin films obtained by pulsed laser deposition. Biomaterials. 2004;25:2539–2545. DOI: 10.1016/j.biomaterials.2003.09.044.

[71] Mróz W, Bombalska A, Budner B, Burdynska S, Jedynski M, Prokopiuk A, Menaszek E, Scisłowska-Czarnecka A, Niedzielska A, Niedzielski K. Comparative study of hydroxyapatite and octacalcium phosphate coatings deposited on metallic implants by PLD method. Applied Physics A. 2010;101:713–716. DOI: 10.1007/s00339-010-5926-3.

Chapter 6

Concurrent Multi-Target Laser Ablation for Making Nano-Composite Films

Abdalla M. Darwish, Sergey S. Sarkisov and Darayas N. Patel

Additional information is available at the end of the chapter

http://dx.doi.org/10.5772/64816

Abstract

New method of using laser ablation for film deposition that can be called as concurrent multi-beam multi-target matrix-assisted pulsed laser evaporation and pulsed laser deposition (MBMT-MAPLE/PLD) is described. Practical MBMT-MAPLE/PLD system built at Dillard University has three separate laser beams, three targets and the remotely controlled plume overlapping mechanism that provides even mixing of the target materials during their deposition on the substrate. The system accommodates MAPLE targets in the form of polymer solutions frozen with flowing liquid nitrogen. The feasibility of the method was demonstrated when it was used for making polymer nano-composite films with two inorganic additives: upconversion fluorescent phosphor $NaYF_4:Yb^{3+}$, Er^{3+} and aluminum-doped ZnO (AZO). Three laser beams, an infrared 1064-nm beam for the MAPLE and two 532-nm beams for the PLD targets, were concurrently used in the process. The fabricated nano-composite films were characterized using X-ray diffraction, scanning electron microscopy (SEM), optical fluorescent spectroscopy, and the measurement of the quantum efficiency (QE) of the upconversion fluorescence. The size of the inorganic nanoparticles varied in the range 10–200 nm. The AZO additive increased QE by 1.6 times. The conclusion was made on the feasibility of MBMT-MAPLE/PLD method for making multi-component nano-composite films for various applications.

Keywords: laser ablation, pulsed laser deposition, matrix-assisted pulsed laser evaporation, polymer nano-composite films

1. Introduction

Pulsed laser deposition (PLD) is a popular application of laser ablation for making thin films. The technology in its variation of the matrix-assisted pulsed laser evaporation (MAPLE) has become a potent way of making polymer and other organic coatings. The chapter will describe a new method of making nano-composite films based on concurrent ablation/evaporation of more than one organic MAPLE and inorganic PLD targets with several laser beams. The advantage is an independent and optimized control of the deposition conditions for each target. The discussion will include the design of the multi-beam multi-target deposition system, examples of the deposited nano-composite films and their properties, the encountered problems/solutions, and the future trends.

For the last three decades, PLD technique has been intensively used for deposition of many kinds of oxides, nitrides, carbides, and metals and for fabrication of thin films including those composed of superconductors, electro-optic $BaTi0_3$, piezoelectric ZnO, electro-conductive TiO_2, rare earth (RE) doped phosphate glasses, etc [1–14]. Those thin films had many deficiencies that had to be addressed before making them suitable for commercial applications. The technique was used by Smith and Turner [2] in 1965 for the preparation of semiconductor and dielectric thin films. They demonstrated the stoichiometry transfer between the target and the deposited film and high deposition rates of about 0.1 nm per pulse. The occurrence of the droplets of the target material on the substrate surface has been observed in [3]. Despite some encouraging results for single-component films, the conventional single-beam single-target PLD technique remains to be poorly suitable for composite films made of the materials of different nature: one of the reasons that a single laser beam cannot be always suitable for a wide spectrum of materials. Further improvement is a new variant of PLD called the concurrent multi-beam multi-target PLD that deposits different materials simultaneously with different laser beams producing overlapping plumes. In case of polymer nano-composite films at least one of the laser beams and the target must be suitable for the process of deposition of the polymer host, such as MAPLE.

In MAPLE, a frozen solution of a polymer in a volatile solvent becomes a laser target [15–54]. The solvent and concentration of the solution are chosen in such a way that first, the polymer dissolves completely and forms a dilute solution free of particulate; second, the main portion of the laser energy is absorbed by the solvent and not by the solute; and third, no photochemical reaction occurs between the solvent and the solute. The interaction between the laser radiation and matter in MAPLE is a photothermal process. The energy of the laser beam absorbed by the solvent is converted to thermal energy that heats the polymer but causes the solvent to vaporize. As the solvent molecules escape into the gas phase, the polymer molecules gain sufficient kinetic energy via collisions with the evaporating solvent molecules and enter the gas phase as well. Selection of the optimal MAPLE conditions (laser energy and wavelength, pulse repetition rate, type of solvent, concentration of the polymer solution, temperature, and background gas and its pressure) can result in the deposition process with no significant degradation of the polymer. The MAPLE deposition is carried on layer-by-layer, while the concentration of the polymer solution in the ablated target remains constant. When a substrate is positioned directly in the path of the plume, a film starts forming from the evaporated polymer molecules on a substrate placed in the path of the plume. At the same time, the volatile solvent molecules are taken away from the chamber by the vacuum pump. In case of fabrication of polymer nano-composites, MAPLE targets are usually prepared as nano-colloids of the additives of interest in the initial polymer solutions. Combining the components of different nature, such as polymers and inorganic substances, in the same target and ablating them with the same laser beam rarely results in nano-composite films of fair quality. The laser beam energy and wavelength cannot fit all components in the mixture. Also, the proportion of the components in the film is dictated by the target and cannot be altered in the process. The improvements can be expected in the method of concurrent multi-beam and multi-target deposition using MAPLE polymer targets and inorganic PLD targets, each being concurrently ablated by laser beams of different wavelengths [55–64]. The method can be called as multi-beam multi-target MAPLE and PLD or with acronym MBMT-MAPLE/PLD that is described below.

2. Methods and materials

2.1. Multi-beam multi-target MAPLE/PLD system

The design of the MBMT-MAPLE/PLD system (it is a three-beam and three-target variant) is schematically presented in **Figure 1**. Three laser beams of different wavelength evaporate/ablate concurrently three targets made of different materials (organic or inorganic). The laser beams are forced to scan over the targets with oscillating (in X and Y directions) reflectors in order to eliminate pre-mature target erosion and cracking due to laser-induced target ablation in a single spot. The photograph of the 24″ vacuum chamber of the system at Dillard University and one of three laser beam guides is presented in **Figure 2**.

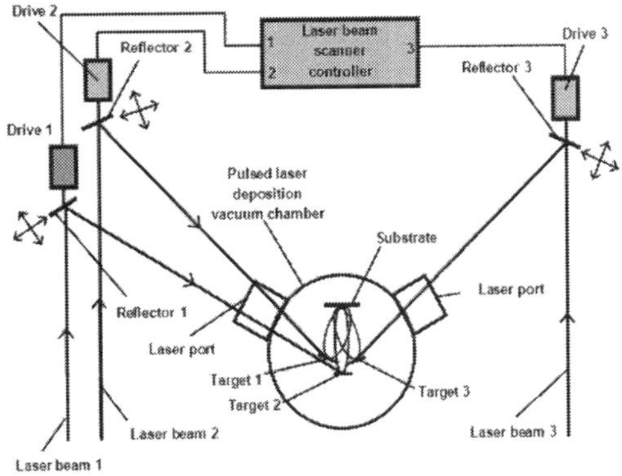

Figure 1. The schematic of the MBMT-MAPLE/PLD system [60].

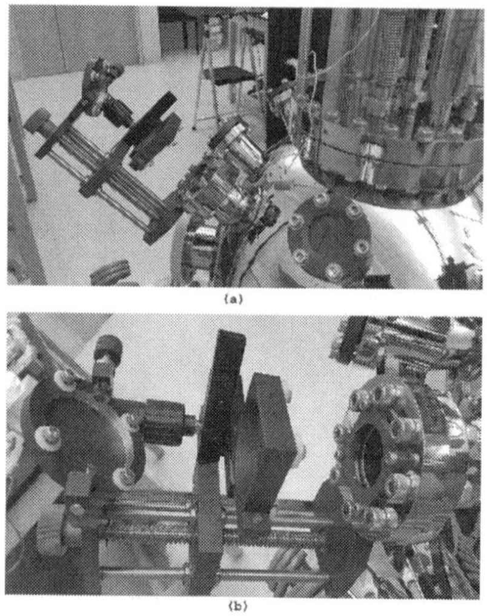

Figure 2. View of the 24″ vacuum chamber of the three-beam MBMT-MAPLE/PLD system at Dillard University (a). The side view of one of the laser beam guides including the holder for the laser beam scanner attached to the optical window of the chamber (b) [60].

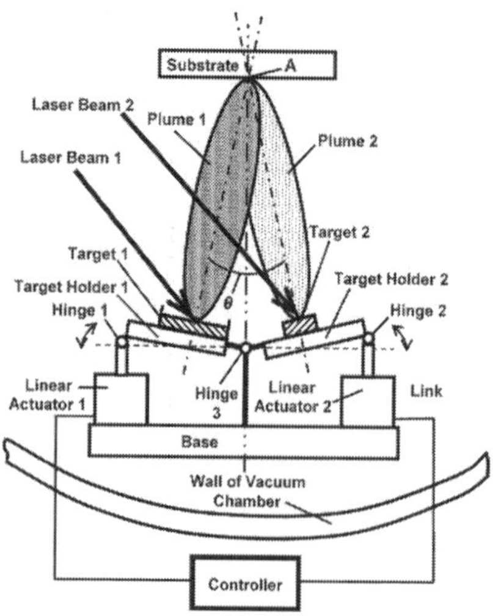

Figure 3. Schematic of the target tilting sub-system of the MBMT-MAPLE/PLD system with remote control of the directions of the plumes. Shown are two targets out of three. θ is the optimal angle between the plums at which they overlap in point A on the surface of the substrate.

The MBMT-MAPLE/PLD system has a sub-system of the plume direction control schematically presented (for two targets for the sake of simplicity) in **Figure 3**. Remotely controlled vacuum compatible linear actuators tilt the targets in order to achieve an optimal angle θ between the plumes (which are perpendicular to the target surfaces) at which the plumes overlap on the surface of the substrate. This secures the uniform mixing of the materials from the targets in the resulting composite film during the deposition process. The photographs of the tilt control sub-system for three targets are presented in **Figure 4**. One important feature of this sub-system is that the target holders are tilted around the axes in the horizontal plane instead of vertical plane, which reduces chances of dropping or spilling the target material.

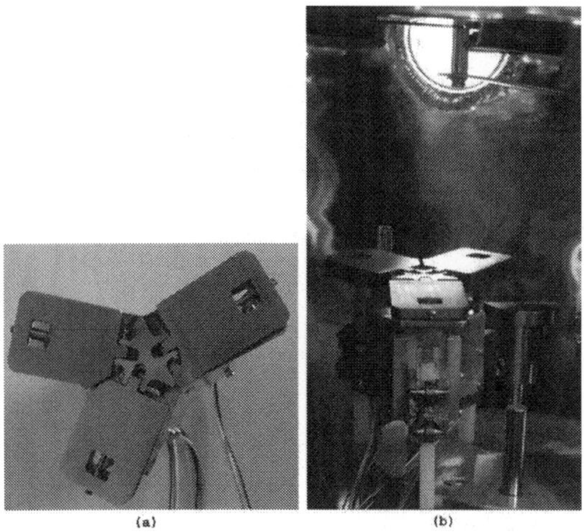

Figure 4. View of the three-target holder with remotely controlled tilt of each target (to achieve overlapping of the plumes on the substrate) (a) and the view of the three-target holder installed inside the vacuum chamber (b).

The target holders of the triple-target sub-system are designed to accommodate MAPLE targets cooled with flowing liquid nitrogen (LN) as presented in **Figure 5**. A copper container for a polymer solution (MAPLE target) is mounted on a copper container for LN (the cooler) that cools the polymer solution (the target) and keeps it frozen. The MAPLE target assembly is connected to the LN feeding and collecting lines (copper tubing) drawn through a flange to be attached to the vacuum chamber. The feeding line is connected to the LN feeding vessel external to the vacuum chamber. The collecting LN line is connected to another external vessel where the LN flows in after passing through the cooler. The MAPLE target assembly is mounted on the top of one of the tilting target holders (**Figure 4**) installed inside the vacuum chamber. Horizontal orientation of the MAPLE target in this design makes possible to conveniently install the empty copper cup in the chamber and fill it later with the liquid polymer solution without the risk of spilling it out before freezing.

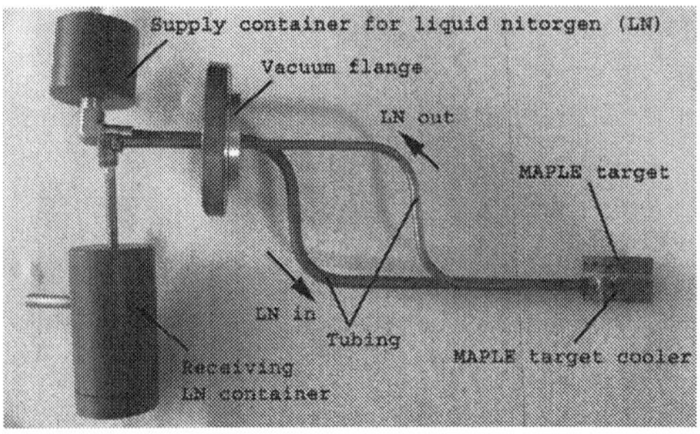

Figure 5. View of the MAPLE target assembly removed from the vacuum chamber. The target is cooled with liquid nitrogen (LN).

2.2. Target materials

The MBPT-MAPLE/PLD method was used to deposit thin films of a polymer nano-composite film with two inorganic additives: upconversion phosphor and the electro-optic enhancer of the upconversion emission.

2.2.1. *MAPLE target*

The MAPLE target material was the solution of poly(methyl methacrylate) known as PMMA in chlorobenzene at a proportion of 1.0 g solids per 10 mL. The solution was filtered through a 0.2-micron filter.

2.2.2. *Inorganic PLD targets*

2.2.2.1. *Upconversion phosphor*

The first inorganic PLD target was made by compressing a powder of efficient upconversion phosphor $NaYF_4$:Yb^{3+}, Er^{3+} with a 25-ton hydraulic press. Upconversion phosphor is a material that absorbs low-energy photons, such as infrared (IR) ones, and re-emits high-energy photons of visible or ultra-violet light. The compounds of the rare earth (RE) elements are particularly attractive as upconversion phosphors with high efficiency of converting IR radiation to upconversion emission. The efficiency depends significantly on the host material for the RE ions. There is a group of efficient upconversion phosphors based on the hexagonal crystalline fluoride $NaYF_4$ (β-$NaYF_4$) as a host. This is due to the low phonon energy of its crystalline lattice that keeps at minimum the rate of the non-radiative multi-phonon relaxation of the excited RE dopant ions. The powder of phosphor $NaYF_4$:Yb^{3+}, Er^{3+} with the doping rates of Yb^{3+} and Er^{3+} 10 and 2%, respectively (10 ytterbium ions and 2 erbium ions per 100 ions of sodium), was synthesized using the wet method [55] and baked for 1 h at 400°C in open air to convert host $NaYF_4$ to its hexagonal crystalline β-phase and maximize the upconversion efficiency. **Figure 6** presents the energy diagram of the phosphor and **Figure 7** its absorption/fluorescence spectra [60]. The energy of the laser photons is mostly absorbed by the ions of

Yb^{3+} acting as synthesizers and then transferred via the energy transfer process to the ions of Er^{3+}, which generated the upconversion, higher energy photons due to the two-photon mechanism. The phosphor powder produced intense visible upconversion emission with two green spectral peaks at 515 and 535 nm and one red spectral peak at 653 nm being pumped with infrared (980 nm) radiation from a laser diode as presented in **Figure 7**. **Figure 8** presents the X-ray diffraction spectrum of the phosphor powder taken with Bruker D2 Phaser X-ray diffractometer. All the observed diffraction peaks can be attributed to β-$NaYF_4$, thus indicating that this is the dominating crystalline phase of the powder.

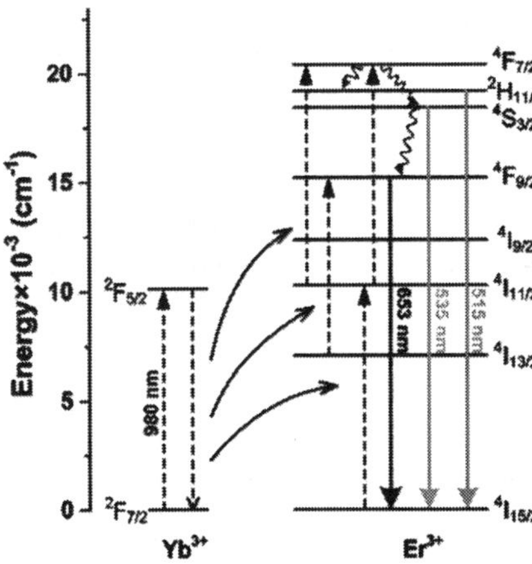

Figure 6. Energy diagram of upconversion phosphor $NaYF_4$: Yb^{3+}, Er^{3+} [60].

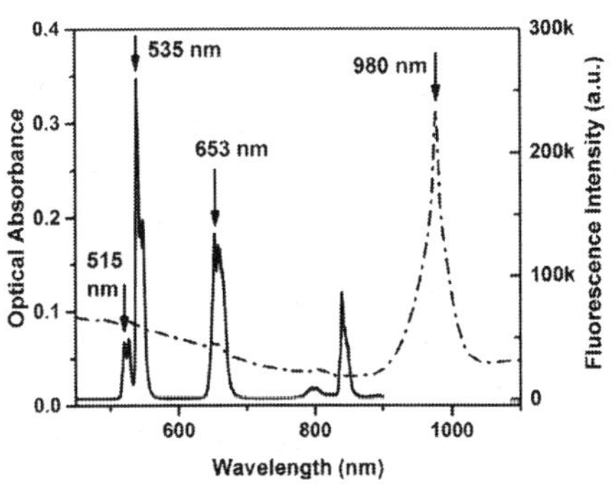

Figure 7. Spectra of optical absorbance (dash-dotted line) and upconversion fluorescence (solid line) of phosphor $NaYF_4$: Yb^{3+}, Er^{3+} [60].

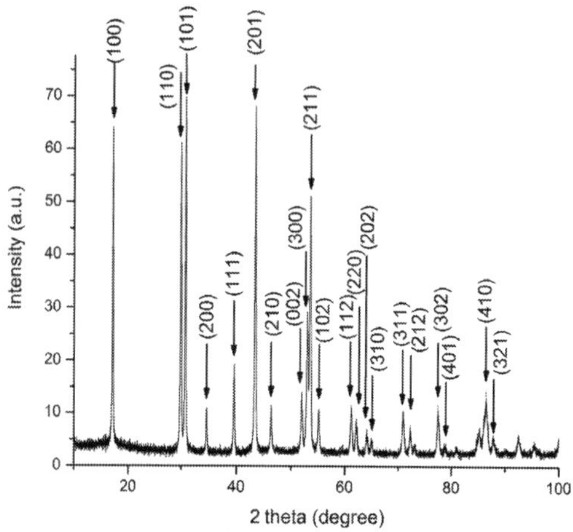

Figure 8. XRD spectrum of the upconversion phosphor powder baked for 1 h at 400°C with the diffraction peaks attributed to the hexagonal β-phase of $NaYF_4$.

2.2.2.2. *Electro-optic AZO compound*

Aluminum-doped ZnO (AZO) compound is an optically transparent electric conductor with the surface plasmon resonance (SPR) enhancement of local optical field similar to the noble metals, but without significant optical losses attributed to them [65–68]. It can thus be expected that adding AZO nanoparticles to the upconversion phosphor in a polymer nano-composite film will bring the local enhancement of the pumping IR optical field in the vicinity of the phosphor nanoparticles and consequently increase the intensity of the upconversion emission. PLD target was a pellet of $Zn_{0.98}Al_{0.02}O$ where the aluminum fraction was 2% of the total by weight as compared with zinc, not counting the oxygen. The AZO pellet had 20 mm in diameter and 3 mm in thickness. The pellet was prepared by spark plasma sintering (SPS), also referred to as pulsed electric current sintering (PECS). In SPS pulsed DC current passed through a graphite die, as well as the AZO powder compact. Joule heating is the key mechanism in the densification of the powder compact. The densification resulted in near theoretical density at a lower sintering temperature than in conventional sintering. The heat is produced internally. This differs from the regular hot pressing with the heat generated by external source. Due to high heating/cooling rate (up to 1000 K/min) sintering was very fast (within few minutes). High speed of the process ensured it densified the powder without coarsening, which accompanied standard densification routes. While the term "spark plasma sintering" is commonly used, a literal interpretation of the term may be misleading since neither a spark nor a plasma is present in the process. **Figure 9** presents the X-ray diffraction spectrum of the AZO target taken with Bruker D2 Phaser X-ray diffractometer. All the observed diffraction peaks can be attributed to ZnO.

http://dx.doi.org/10.5772/64816

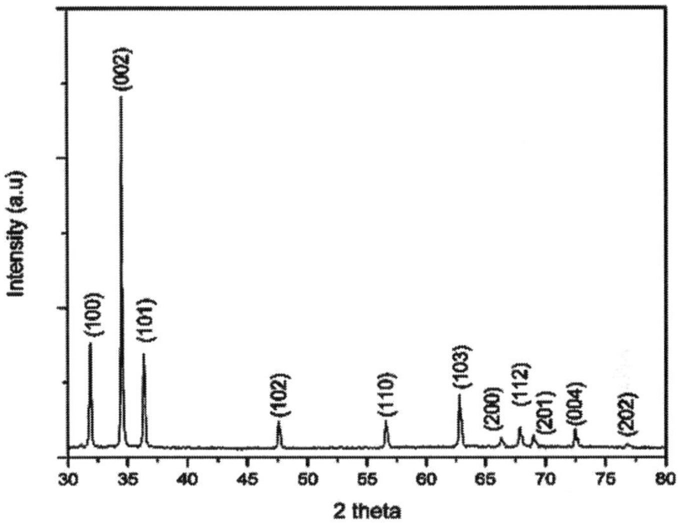

Figure 9. XRD spectrum of the prepared AZO PLD target with the diffraction peaks attributed to ZnO.

2.3. Deposition procedure

The PMMA solution in chlorobenzene was poured in a copper cup of the MAPLE target assembly (**Figure 5**) and frozen with liquid nitrogen. The first PLD target was the pellet of the upconversion phosphor. The AZO pellet was the second PLD target. Two pulsed laser sources were used. The first was a Spectra Physics Quanta Ray Nd:YAG Q-switched Pro-250-50 laser with a pulse repetition rate of 50 Hz, 750-mJ energy per pulse (1064-nm wavelength) and 400-mJ energy per pulse at the 532-nm second harmonic. The second laser source (synchronized with the first laser) was a Spectra Physics Quanta Ray Nd:YAG Q-switched Lab-170-10 laser with a pulse repetition rate of 10 Hz, 850-mJ energy per pulse (1064-nm wavelength) and 450-mJ energy per pulse at the 532-nm second harmonic. The MAPLE target was evaporated with the 1064-nm beam from the first laser source and exposed to the fluence ranging from 0.84 to 2.4 J/cm^2 per pulse. The first PLD target was concurrently ablated with the 532-nm frequency doubled Nd:YAG beam from the same source. The fluence was tuned up between 0.8 and 1.0 J/cm^2 to keep the proportion of the upconversion material in the polymer film at approximately 5% by weight. The second PLD target was concurrently ablated with the 532-nm frequency doubled Nd:YAG beam from the second source. The fluence was tuned up between 1.0 and 2.2 J/cm^2 to keep the proportion of the upconversion material in the polymer film at approximately 5% by weight. The polymer films were deposited on preoxidized Si (100) substrates with a SiO_2 layer thickness of 2.0 μm. The deposition time was ~3.0 min (until PLD targets started showing the signs of erosion). The thickness of the deposited films was between 180 and 200 nm as measured with an atomic force microscope.

3. Properties of the deposited films

3.1. Crystalline structure of the inorganic additives in the polymer nano-composite film

In order to conduct X-ray diffraction spectroscopy, the deposited films were separated from the substrates and placed in a Bruker D2 Phaser X-ray diffractometer. A reference sample of the PMMA nano-composite film containing only one upconversion phosphor additive was also made. The X-ray diffraction spectrum of this sample is presented in **Figure 10**. The spectrum has all the signatures of the hexagonal β-phase $NaYF_4$ that was initially present in the first PLD target made of the upconversion phosphor. **Figure 11** shows the X-ray diffraction spectrum of the polymer nano-composite film including nanoparticles of both inorganic additives: the upconversion phosphor and AZO. The observed spectral peaks include those that can be attributed to both β-phase $NaYF_4$ and AZO. It thus can be concluded that the two inorganic additives have been transferred to the polymer film without modification of their crystalline structure during the PLD process.

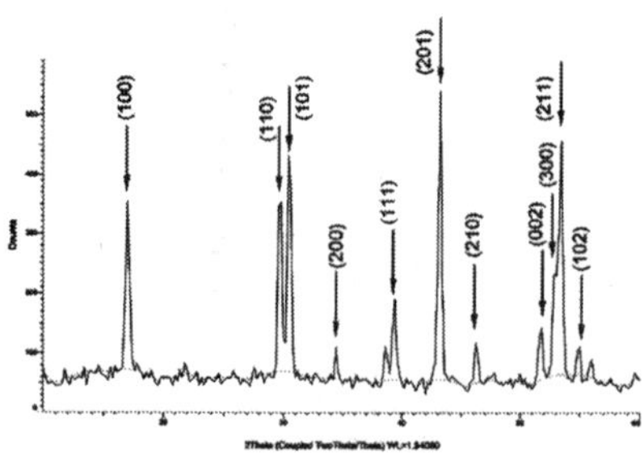

Figure 10. XRD spectrum of the two-component composite film made of PMMA and the nanoparticles of $NaYF_4$: Yb^{3+}, Er^{3+} with the diffraction peaks attributed to the hexagonal β-phase of $NaYF_4$.

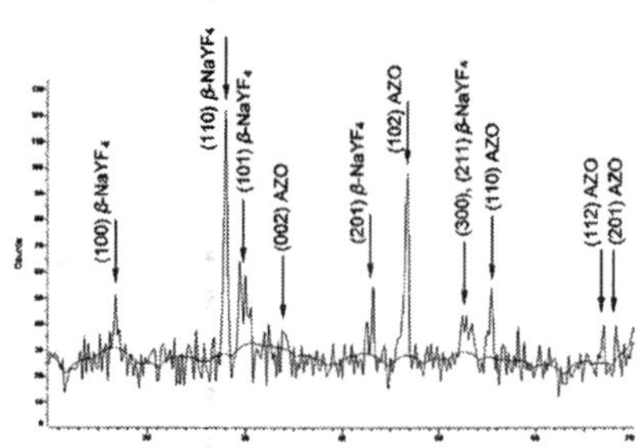

Figure 11. XRD spectrum of the three-component composite film made of PMMA and the nanoparticles of $NaYF_4$: Yb^{3+}, $Er^{3+}$and AZO.

3.2. Morphology

Figure 12 presents the high-resolution scanning electron microscope (SEM) image of the produced nano-composite film with a magnification of ×60 K. For the sake of convenience, the images of some exemplary nanoparticles are marked with arrows. The size of the nanoparticles varied widely in the range of the order of 10–200 nm. The nearly uniform distribution of the nanoparticles in the polymer film was occasionally disrupted by much larger particles or clusters (limited resolution of SEM prohibited distinguishing between large particles and the clusters of nanoparticles) of the order of 500–1000 nm.

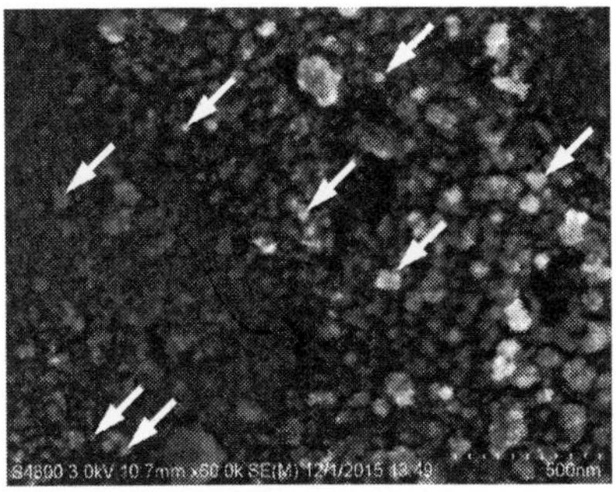

Figure 12. Scanning electron microscopy (SEM) image of the three-component composite film made of PMMA and the nanoparticles of $NaYF_4$: Yb^{3+}, $Er^{3+}$and AZO taken with magnification ×60,000. White arrows point to exemplary nanoparticles of various sizes (from ~10 to 200 nm) embedded in the polymer matrix.

3.3. Fluorescence

The deposited nano-composite films demonstrated visible upconversion fluorescence being pumped with a 980-nm IR radiation from a laser diode (PL980P330J from Thorlabs; 330-mW maximum power; quantum well laser chip, pigtailed with a wavelength stabilizing fiber Bragg grating). The spectrum of the upconversion emission was measured with a Princeton Instruments 500-mm focal-length Spectra Pro (SP–2500i) imaging spectrometer/monochromator equipped with 1200 gr/mm (blazed at 500 nm) holographic diffraction grating. The spectrum presented in **Figure 13** had peaks at 522, 540, and 656 nm and resembled the spectrum of the bulk phosphor target (**Figure 7**).

The upconversion fluorescence of the films was quantitatively characterized by the quantum efficiency (QE) η. QE is the ratio of the number of the photons of the upconversion radiation generated per unit of time n_{up} to the number of the photons of the infrared pump radiation $\eta = (n_{up}/n_{pump}) \times 100\%$ [69]. The quantum efficiencies of green (at 522- and 540-nm spectral peaks combined) upconversion emission of the deposited three-component films (PMMA-phosphor-AZO), two-component films (PMMA-phosphor), and the bulk phosphor target were measured to be of 0.072, 0.045, and 0.56% respectively. QE of the bulk phosphor PLD target compared well against the highest 3% quantum efficiency reported for similar upconversion phosphor $NaYF_4$: Yb^{3+}, Er^{3+} in the literature [69]. QE of the polymer nano-composite film containing only the nanoparticles of the upconversion phosphor was ~12 times less than that of the bulk phosphor. On the other side, the nano-composite film deposited under similar conditions, but also containing AZO nanoparticles, had QE ~1.6 times greater. This could be attributed to the plasmonic enhancement effect of the AZO nanoparticles on the local optical pump IR field. Since the upconversion emission is a two-photon process, QE could be increased proportional to approximately square of the pump power. In this experiment, the maximum pump power was limited to 150 mW, less than the damage threshold of the nano-composite film.

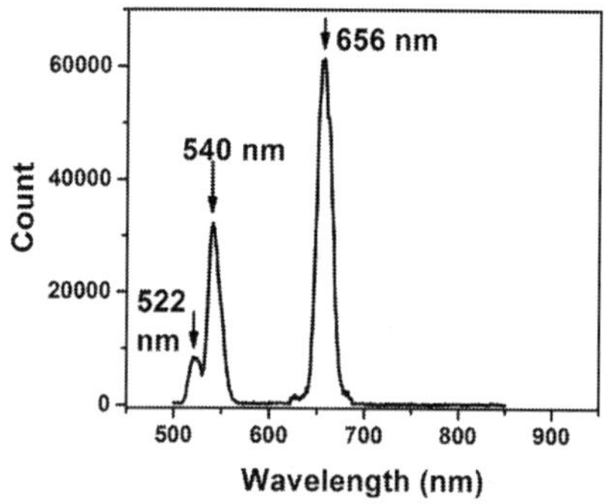

Figure 13. Spectrum of upconversion emission of the PMMA nano-composite film with $NaYF_4$: Yb^{3+}, Er^{3+} and AZO additives excited with a 980-nm laser diode.

The reason for the nano-composite film having an order of magnitude weaker upconversion fluorescence than that of the bulk phosphor powder, besides a limited concentration of the phosphor nanoparticles in the polymer host, could be related to the size effect. Based on the doping rate of the RE ions in the phosphor (see Sub-Section 2.2.2.1) and the computational approach in [70], the average number of the Er^{3+} and Yb^{3+} ions in the particles of an average diameter of 10 nm could be estimated as 128 and 628, respectively. The particles of 100-nm and 1-μm diameter would contain 10^3 and 10^6 times more RE ions. The upconversion emission involved two types of RE ions with the Yb^{3+} ion acting as a captor of the pump IR photons that later excited the Er^{3+} ion through the energy transfer process involving two IR photons, but not one. The more the RE ions contained in a phosphor particle, the stronger the upconversion emission would be. Accordingly, the nano-composite film including the nanoparticles of the upconversion phosphor of the size not exceeding ~200 nm should expectedly have upconversion QE less than that of the bulk powder with significant presence of 1-μm and greater particles. Adding the nanoparticles of AZO compound to the polymer nano-composite film helped to partially compensate the drop of upconversion QE due to the plasmon enhancement of the local pump IR optical field. As an illustration of possible applications for upconversion fiber illuminators, **Figure 14** presents the photograph of the tip of a single-mode fiber coated using the above-described MAPLE/PLD method with a nano-composite film of PMMA + $NaYF_4$: Yb^{3+}, Er^{3+} + AZO pumped with a 980-nm laser diode (125-mW power). The tip of the fiber illuminated the white back side of a business card with visible upconversion light. The picture was taken with an iPhone 6 digital camera at dimmed room light.

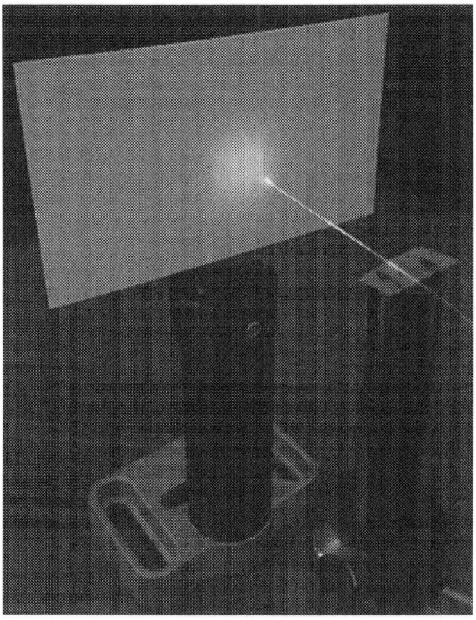

Figure 14. Photograph of the tip of a single-mode fiber coated with a nano-composite film of PMMA + $NaYF_4$: Yb^{3+}, Er^{3+} + AZO pumped with a 980-nm laser diode (125-mW power) illuminating white back side of a business card.

4. Conclusions

The multi-beam multi-target MAPLE/PLD method as a new variation of the film deposition via laser ablation makes possible to transfer concurrently several inorganic additives, such as an efficient upconversion RE inorganic phosphor and AZO compound, in nano-composite polymer films preserving crystalline structure of the additives and the upconversion emission properties. This is due to much better control of the deposition process of the materials of different nature from separate targets with different laser beams. The basic components of the new triple-beam triple-target MAPLE/PLD apparatus and the major process steps have been designed, built, and tested. The preliminary results indicated that adding AZO nanoparticles improved by a factor of 1.6 the upconversion quantum efficiency of the upconversion emission from the films possibly due to the plasmon enhancement of the local optical pump IR field in the vicinity of AZO nano-particles. The MBMT-MAPLE/PLD technique can find its applications in fabrication of multi-component polymer-inorganic composites with functionalities spanning emitting light, sensing bio- and chemical agents, energy harvesting, and other applications.

Acknowledgements

The authors would like to acknowledge the support from DoD grants FA 9550-12-1-0068, W911NF-15-0446, W911NF-14-1-0093, W911NF-11-1-0192 and NSF HRD# 1002541 DU LS-LAMP grants.

Author details

Abdalla M. Darwish[1*], Sergey S. Sarkisov[2] and Darayas N. Patel[3]

*Address all correspondence to: adarwish@bellsouth.net

1 Dillard University, New Orleans, LA, USA

2 SSS Optical Technologies LLC, Huntsville, AL, USA

3 Oakwood University, Huntsville, AL, USA

References

[1] Dijkkamp D, Venkatesan T, Wu XD, Shareen SA, Jiswari N, Min-Lee YH, et al. Pulsed laser evaporation form high Tc bulk materials. Appl. Phys. Lett. 1987;5(8):619–621.

[2] Turner AF, Croft M. Vacuum deposited thin films using a ruby laser. Appl. Optics. 1965;4(1):147.

[3] Chrisey DB, Hubler GK. Pulsed laser deposition of thin films. New York: Wiley-Interscience; 1st edition. 1994, 648p.

[4] Scarfone C, Norton MG, Carter CB, Li J, Mayer JW. Characterization of $BaTiO_3$ thin films deposited by pulsed-laser ablation. In: Volume 201 - Proceedings of the 1990 MRS Fall Meeting – Symposium A - Surface Chemistry and Beam-Solid Interactions; 1990; Boston. New York: Cambridge University Press; 1991. p. 183.

[5] Fahler S, Stormer M, Krebs HU. Origin and avoidance of droplets during laser ablation of materials. Appl. Surf. Science. 1997;109/110:433.

[6] Kelly R, Rotenberg JE. Laser sputtering. Part III. The mechanism of sputtering of metals low energy densities. Nuclear Instrum. Methods Phys. Res. 1985;B7–8:755–763.

[7] Singh K, Narajan J. Pulsed-laser evaporation technique for deposition of thin films: physics and theoretical model. Phys. Rev. 1990;B41(5):8843.

[8] Holzapfel B, Roas B, Schultz L, Bauer P, Saemann-Ischenko G. Off-axis deposition of $YBa_2Cu_3O_{7-\delta}$ thin films. Appl. Phys. Lett. 1992;61(26):3178–3180. doi: 0003-6951/92/513178-03

[9] Gorbunov AA, Pompe W, Sewing A, Gaponov SV, Akhsakhalyan AD, Zabrodin IG, et al. Ultrathin film deposition by pulsed laser ablation using crossed beams. Appl. Surf. Sci. 1996;96–98:649–655.

[10] Guarniery CR, Roy RA, Saenger KL, Shivashnkar SA, Yee DS, Cuomo. Thin-film Bi-Sr-Ca-Cu-O high-temperature superconductors using pulsed laser evaporation from sintered disks. Appl. Phys. Lett. 1988;53(6):532–533. doi:0003-6951/88/320532-02

[11] Foster CM, Voss KF, Hagler TW, Mihailovich D, Heeger AJ, Eddy MM, et al. Infrared reflection of epitaxial $Ti_2Ba_2CaCu_2O_8$ thin films in the normal and superconducting states. Solid State Comm. 1990;76:651.

[12] Shinde SR, Ogale SB, Greene RL, Venkatesan T, Canfield PC, Budko SL, et al. Superconducting MgB2 thin films by pulsed laser deposition. Appl. Phys. Lett. 2001;79(2): 227–229.

[13] Fogarassy E, Fuchs C, Slaoui A, Stoquert JP. SiO2 thin-film deposition by excimer laser ablation from SiO target in oxygen atmosphere. Appl. Phys. Lett. 1990;57(7):664–666. doi:0003-6951/90/330664-03

[14] Balooch M, Tench RJ, Siekhaus WJ, Allen MJ, Connor AL, Olander DR, et al. Deposition of SiC films by pulsed excimer laser ablation. Appl. Phys. Lett. 1990;57(15):1540–1542. doi:0003-6951/90/411540-03

[15] Mihaiescu DE, Cristescu R, Dorcioman G, Popescu CE, Nita C, Socol G, et al. Functionalized magnetite silica thin films fabricated by MAPLE with antibiofilm properties. Biofabrication. 2013;5(1):015007.

[16] Paun IA, Moldovan AM, Luculescu C, Dinescu M. Antibacterial polymeric coatings grown by matrix assisted pulsed laser evaporation. Appl. Phys. A: Mater. Sci. Process. 2013;110(4):895–902.

[17] Shepard KB, Yunlong Guo, Arnold CB, Priestley RD. Nanostructured morphology of polymer films prepared by matrix assisted pulsed laser evaporation. Appl. Phys A: Mater Sci Process. 2013;110(4):771–7.

[18] Shepard KB, Priestley RD. MAPLE deposition of macromolecules. Macromol. Chem. Phys. 2013;214(8):862–72.

[19] Iordache S, Cristescu R, Popescu AC, Popescu CE, Dorcioman G, Mihailescu IN, et al. Functionalized porphyrin conjugate thin films deposited by matrix assisted pulsed laser evaporation. Appl. Surf. Sci. 2013;8:207–10.

[20] Singaravelu S, Mayo DC, Park HK, Schriver KE, Haglund RF Jr. Anti-reflective polymer-nanocomposite coatings fabricated by RIR-MAPLE. Proc. SPIE. 2013;8607:860718.

[21] Shepard KB, Arnold CB, Priestley RD. Origins of nanostructure in amorphous polymer coatings via matrix assisted pulsed laser evaporation. Appl. Phys. Lett. 2013;103(12): 123105.

[22] Caricato AP, Arima V, Cesaria M, Martino M, Tunno T, Rinaldi R, et al. Solvent-related effects in MAPLE mechanism. Appl. Phys. B: Lasers Optics. 2013;113(3):463–71.

[23] Birjega R, Matei A, Mitu B, Ionita MD, Filipescu M, Stokker-Cheregi F, et al. Layered double hydroxides/polymer thin films grown by matrix assisted pulsed laser evaporation. Thin Solid Films. 2013;543:63–8.

[24] McCormick RD, Cline ED, Chadha AS, Zhou W, Stiff-Roberts AD. Tuning the refractive index of homopolymer blends by controlling nanoscale domain size via RIR-MAPLE deposition. Macromol. Chem. Phys. 2013;214(23):2643–50.

[25] Canulescu S, Schou J, Fester S, Hansen KV, Conseil H. Deposition of matrix-free fullerene films with improved morphology by matrix-assisted pulsed laser evaporation (MAPLE). Chem. Phys. Lett. 2013;588:119–23.

[26] Visan A, Grossin D, Stefan N, Duta L, Miroiu FM, Stan GE, et al. Biomimetic nanocrystalline apatite coatings synthesized by Matrix assisted pulsed laser evaporation for medical applications. Mat. Sci. Eng.: B (Advanced Functional Solid-State Materials). 2014;181:56–63.

[27] Stokker-Cheregi F, Matei A, Dinescu M, Secu CE, Secu M. Photoluminescence of Eu-doped $LiYF_4$ thin films grown by pulsed laser deposition and matrix-assisted pulsed laser evaporation. J. Phys. D: Appl. Phys. 2014;47(4):045304.

[28] Constantinescu C, Matei A, Ion V, Mitu B, Ionita I, Dinescu M, et al. Ferrocene carboxaldehyde thin films grown by matrix-assisted pulsed laser evaporation for nonlinear optical applications. Appl. Surf. Sci. 2014;302:83–6.

[29] Palla-Papavlu A, Rusen L, Dinca V, Filipescu M, Lippert T, Dinescu M. Characterization of ethylcellulose and hydroxypropyl methylcellulose thin films deposited by matrix-assisted pulsed laser evaporation. Appl. Surf. Sci. 2014;302:87–91.

[30] Grumezescu V, Socol G, Grumezescu AM, Holban AM, Ficai A, Bleotu C, et al. Functionalized antibiofilm thin coatings based on PLA-PVA microspheres loaded with usnic acid natural compounds fabricated by MAPLE. Appl. Surf. Sci. 2014;302:262–7.

[31] Rusen L, Dinca V, Mitu B, Mustaciosu C, Dinescu M. Temperature responsive functional polymeric thin films obtained by matrix assisted pulsed laser evaporation for cells attachment-detachment study. Appl. Surf. Sci. 2014;302:134–40.

[32] Caricato AP, Arima V, Catalano M, Cesaria M, Cozzoli PD, Martino M, et al. MAPLE deposition of nanomaterials. Appl. Surf. 2014;302:92–8.

[33] Socol M, Preda N, Vacareanu L, Grigoras M, Socol G, Mihailescu IN, et al. Organic heterostructures based on arylenevinylene oligomers deposited by MAPLE. Appl. Surf. Sci. 2014;302:216–22.

[34] Constantinescu C, Matei A, Ionita I, Ion V, Marascu V, Dinescu M, et al. Azo-derivatives thin films grown by matrix-assisted pulsed laser evaporation for non-linear optical applications. Appl. Surf. Sci. 2014;302:69–73.

[35] Grumezescu V, Holban AM, Grumezescu AM, Socol G, Ficai A, Vasile BS, et al. Usnic acid-loaded biocompatible magnetic PLGA-PVA microsphere thin films fabricated by MAPLE with increased resistance to staphylococcal colonization. Biofabrication. 2014;6(3):035002.

[36] Sima F, Axente E, Iordache I, Luculescu C, Gallet O, Anselme K, et al. Combinatorial matrix assisted pulsed laser evaporation of a biodegradable polymer and fibronectin for protein immobilization and controlled release. Appl. Surf. Sci. 2014;306:75–9.

[37] Grumezescu V, Holban AM, Iordache F, Socol G, Mogosanu GD, Grumezescu V, et al. MAPLE fabricated magnetite-eugenol and (3-hidroxybutyric acid-co-3-hidroxyvaleric acid)-polyvinyl alcohol microspheres coated surfaces with anti-microbial properties. Appl. Surf. Sci. 2014;306:16–22.

[38] Axente E, Sima F, Sima LE, Erginer M, Eroglu MS, Serban N, et al. Combinatorial MAPLE gradient thin film assemblies signaling to human osteoblasts. Biofabrication. 2014;6(3):035010.

[39] Dinca V, Florian PE, Sima LE, Rusen L, Constantinescu C, Evans RW, et al. MAPLE-based method to obtain biodegradable hybrid polymeric thin films with embedded antitumoral agents. Biomed. Microdevices. 2014;16(1):11–21.

[40] Wangyao Ge, Qian Yu, Lopez GP, Stiff-Roberts AD. Antimicrobial oligo(p-phenylene-ethynylene) film deposited by resonant infrared matrix-assisted pulsed laser evaporation. Colloids and Surf B: Biointerfaces. 2014;116:786–92.

[41] Aronne A, Ausanio G, Bloisi F, Calabria R, Califano V, Fanelli E, et al. Structural characterization of MAPLE deposited lipase biofilm. Appl. Surf. Sci. 2014;320:524–30.

[42] Califano V, Bloisi F, Aronne A, Federici S, Nasti L, Depero LE, et al. Biosensor applications of MAPLE deposited lipase. Biosensors. 2014;4(4):329–39.

[43] Qian Yu, Wangyao Ge, Atewologun A, Lopez GP, Stiff-Roberts AD. RIR-MAPLE deposition of multifunctional films combining biocidal and fouling release properties. J. Mat. Chem. B. 2014;2(27):4371–8.

[44] Constantinescu C, Rotaru A, Nedelcea A, Dinescu M. Thermal behavior and matrix-assisted pulsed laser evaporation deposition of functional polymeric materials thin films with potential use in optoelectronics. Mat. Sci. Semicond. Process. 2015;30:242–9.

[45] Caricato AP, Cesaria M, Leo C, Mazzeo M, Genco A, Carallo S, et al. Very low roughness MAPLE-deposited films of a light emitting polymer: an alternative to spin coating. J. Phys. D: Appl. Phys. 2015;48(13):135501.

[46] Aronne A, Bloisi F, Calabria R, Califano V, Depero LE, Fanelli E, et al. Lipase biofilm deposited by matrix assisted pulsed laser evaporation technique. Appl. Surf. Sci. 2015;336:196–9.

[47] Iordache F, Grumezescu V, Grumezescu AM, Curutiu C, Ditu LM, Socol M, et al. Gamma-cyclodextrin/usnic acid thin film fabricated by MAPLE for improving the resistance of medical surfaces to Staphylococcus aureus colonization. Appl. Surf. Sci. 2015;336:407–12.

[48] Matei A, Constantinescu C, Mitu B, Filipescu M, Ion V, Ionita I, et al. Laser printing of azo-derivative thin films for non-linear optical applications. Appl. Surf. Sci. 2015;336:200–5.

[49] Stiff-Roberts AD, McCormick RD, Wangyao Ge. Material properties and applications of blended organic thin films with nanoscale domains deposited by RIR-MAPLE. Proc. SPIE. 2015;9350:935007.

[50] Grumezescu V, Andronescu E, Holban AM, Grumezescu AM, Socol G, Iordache F, et al. MAPLE fabrication of thin films based on kanamycin functionalized magnetite nanoparticles with anti-pathogenic properties. Appl. Surf. Sci. 2015;336:188–95.

[51] Caricato AP, Anni M, Cesaria M, Lattante S, Leggieri G, Leo C, et al. MAPLE-deposited PFO films: influence of the laser fluence and repetition rate on the film emission and morphology. Appl. Phys. B. Lasers Optics. 2015;19(3):453–61.

[52] Wangyao Ge, Atewologun A, Stiff-Roberts AD. Hybrid nanocomposite thin films deposited by emulsion-based resonant infrared matrix-assisted pulsed laser evaporation for photovoltaic applications. Org. Electron.: Mater. Phys. Chem. Appl. 2015;22:98–107.

[53] Janković A, Eraković S, Ristoscu C, Mihailescu N, Duta L, Visan A, et al. Structural and biological evaluation of lignin addition to simple and silver-doped hydroxyapatite thin films synthesized by matrix-assisted pulsed laser evaporation. J. Mater. Sci: Mater. Med. 2015;26(1):17.

[54] Paun IA, Acasandrei AM, Luculescu CR, Mustaciosu CC, Ion V, et al. MAPLE deposition of polypyrrole-based composite layers for bone regeneration. Appl. Surf. Sci. 2015;357:975–84.

[55] Darwish AM, Sagapolutele MT, Sarkisov S, Patel D, Hui D, Koplitz B. Double beam pulsed laser deposition of composite films of poly(methyl methacrylate) and rare earth fluoride upconversion phosphors. Compos Part B: Eng. 2013;55:139–46.

[56] Darwish AM, Wilson S, Sarkisov S, Patel D. Double pulse laser deposition of polymer nanocomposite $NaYF_4:Tm^{3+}$, Yb^{3+} films for optical sensors and light emitting applications. Proc. SPIE. 2014;8847:884702.

[57] Darwish AM, Burkett A, Blackwell A, Taylor K, Sarkisov S, Patel D, et al. Polymer-inorganic nano-composite thin film upconversion light emitters prepared by double-beam matrix assisted pulsed laser evaporation (DB-MAPLE) method. Compos Part B: Eng. 2015;68:355–64.

[58] Darwish AM, Burkett A, Blackwell A, Taylor K, Walker V, Sarkisov S, et al. Efficient upconversion polymer-inorganic nanocomposite emitters prepared by the double beam matrix assisted pulsed laser evaporation (DB-MAPLE), Proc. SPIE. 2014;9200:92000C.

[59] Darwish AM, Wilson S, Blackwell A, Taylor K, Sarkisov SS, Patel DN, et al. Ammonia Sensor based on polymer-inorganic nano-composite thin film upconversion light emitter prepared by double-beam pulsed laser deposition. Am. J. Mater. Sci. 2015;5(3A): 8–15.

[60] Darwish AM, Wilson S, Blackwell A, Taylor K, Sarkisov S, Patel D, et al. Polymer-inorganic nanocomposite thin film emitters, optoelectronic chemical sensors, and energy harvesters produced by multiple-beam pulsed laser deposition. Proc. SPIE. 2015;9586:958602.

[61] Darwish AM, Wilson S, Blackwell A, Taylor K, Sarkisov S, Patel D, et al. Multi-beam pulsed laser deposition: new method of making nanocomposite coatings. Proc. SPIE. 2015;9586:958605.

[62] Darwish A, Sarkisov S. Multiple beam pulsed laser deposition of composite films, US Patent Application No. 14/158,567, Filed 1/17/2014, Publication No. US 2014/0227461 A1.

[63] Darwish A, Sarkisov S. Multiple beam pulsed laser deposition of composite films, US Patent Application No. 14/506,685, Filed 10/15/2014.

[64] Darwish A, Mele P, Sarkisov S. Nano-composite thermo-electric energy converter and fabrication method thereof. US Patent Application No. 14/853,674, Filed 9/14/2015.

[65] West PR, Ishii S, Naik GV, Emani NK, Shalaev VM, Boltasseva A. Searching for better plasmonic materials. Laser Photonics Rev. 2010;4(6):795–808.

[66] Kim Y, Lee W, Jung D-R, Kim J, Nam S, Kim H, et al. Optical and electronic properties of post-annealed ZnO:Al thin films. Appl. Phys. Lett. 2010;96(17):171902.

[67] Tong H, Deng Z, Liu Z, Huang C, Huang J, Lan H, et al. Effects of post-annealing on structural, optical and electrical properties of Aldoped ZnO thin films. Appl. Surf. Sci. 2011;257(11):4906–11.

[68] Lin J-Y, Zhong K-D, Lee P-T. Plasmonic behaviors of metallic AZO thin films and AZO nanodisk arrays. Opt. Express. 2016;24(5):5125.

[69] Boyer JCh, van Veggel FCJM. Absolute quantum yield measurements of colloidal $NaYF_4$: Er^{3+}, Yb^{3+} upconverting nanoparticles. Nanoscale. 2010;2:1417–19.

[70] Liu H, Xu CT, Dumlupinar G, Jensen OB, Andersen PE, Andersson-Engels S. Deep tissue optical imaging of upconverting nanoparticles enabled by exploiting higher intrinsic quantum yield through use of millisecond single pulse excitation with high peak power. Nanoscale. 2013;20(5):10034–40.

Chapter 7

Pulsed Laser Deposition of Large-Area Thin Films and Coatings

Armen S. Kuzanyan and Astghik A. Kuzanyan

Additional information is available at the end of the chapter

http://dx.doi.org/10.5772/64978

Abstract

Simple and inexpensive methods of obtaining large-area uniform in thickness and composition thin films on rotating substrates and moving ribbons through pulsed laser deposition have been proposed. Thin films of different compositions were prepared using these methods. The thickness uniformity of obtained films was preserved within the limits of ±3% on up to 300 mm diameter substrates. Also, a method of creating a laser spot with a certain configuration on the target is proposed allowing almost full utilization of the target material.

Keywords: pulsed laser deposition, large-area films

1. Introduction

The wide application of thin films in newest technologies has revealed the need of developing simple methods of obtaining large-area thin films. One of these methods is provided by pulsed laser deposition (PLD). It equally provides high speed of deposition, correspondence of the film composition to that of the target, and possibilities to vary the gas pressure in the deposition chamber in a wide range.

PLD started developing rapidly after the discovery of high-temperature superconductivity. Scientists needed samples of new multicomponent materials, superior in their characteristic to the ceramic ones, for identification of their physical properties and disclosure of high-temperature superconductivity mechanism. For solution of these issues, the Laboratory of High-Temperature Superconductivity was founded in the Institute for Physical Research NAS of Armenia in September 1987. Here, we present the research conducted in this laboratory to develop new approaches and solutions for the preparation of large-area homogeneous in composition and thickness thin films and coatings of complex compounds. PLD turned out to be a very convenient tool to obtain thin films of high-temperature superconductors.

PLD is used to grow thin films of metals, oxides, polymers, biocompatible materials, and so on. It allows the fabrication of ultra-thin epitaxial single-crystalline, polycrystalline and amorphous films, heterostructures and nanocrystalline coatings [1, 2]. PLD is simple in application and, therefore, is widely used in research laboratories. However, it is also promising for various commercial applications, in particular, growth of large-area films. Films of uniform thickness on large-diameter substrates are necessary for many applications in microelectronics, optical industry, and other modern technologies.

Wide use of PLD in the growth of large-area films is impeded by the following circumstance: the angular distribution of the mass-transfer rate in the plasma plume formed by laser radiation is nonuniform. Therefore, using conventional laser deposition, one cannot obtain films of uniform thickness on substrates larger than 10 mm in diameter. In this chapter, we describe some main solutions to this problem and propose a new technique for depositing thin films of uniform thickness on large-area substrates the size of which is limited by the deposition chamber dimensions only.

All versions of PLD of large-area films are based on the fact that the angular distribution of the mass-transfer rate in a plasma plume is set by the function $F(\theta) = A\cos^m \theta$ [1], where θ is the angle of deviation from the perpendicular to the target plane. The plasma plume axis, that is, the direction in which the mass-transfer rate is maximal, is perpendicular to the target surface in a wide range of variation in the angle of incidence of laser beam on the target. Knowing the angular distribution of the mass-transfer rate of the material evaporated from the target, one can arrange the mutual position and motion of the target and substrate to provide identical amount of evaporated material per substrate unit area over the substrate surface and thus grow films uniform in thickness.

2. Main solutions for depositing large-area films

Let us briefly consider the main existing ways for large area films deposition [1]. In one of them, a laser beam is incident on a rotating target, before which a rotating substrate is located parallel to the target. The substrate axis is shifted with respect to the plasma plume axis by some distance (**Figure 1a**). This distance is determined by the width of the gap between the target and the substrate, the substrate diameter, and the angular distribution of evaporated particles in the plasma plume. Thus, the part of the plasma plume that is characterized by a higher mass transfer of evaporated material arrives at the substrate edge, where a larger area must be coated for the same time; as a result, a film of uniform thickness grows. A modification of this technique is the version shown in **Figure 1b**. Here, the substrate simultaneously rotates and moves in the horizontal plane. A computer controls the horizontal displacement velocity of the substrate in such a way that the plasma plume axis is directed toward the substrate edge for a longer period as compared to the center.

http://dx.doi.org/10.5772/64978

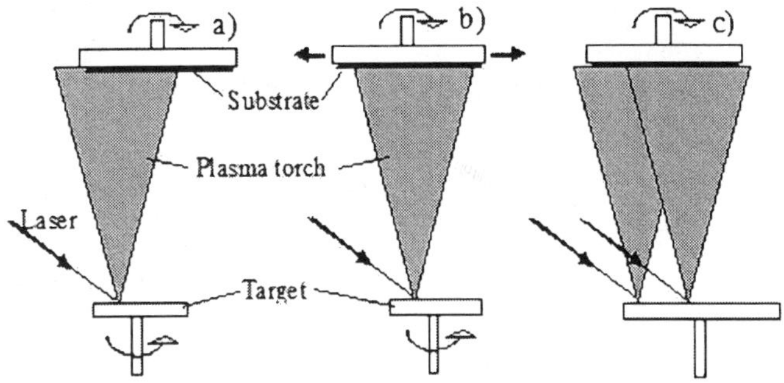

Figure 1. Schematics of the main techniques for depositing large-area films: (a) "off-axis" deposition, (b) deposition on substrate that simultaneously rotates and moves in the horizontal direction, and (c) scanning the laser beam on the surface of a large target.

The drawback of the first version (**Figure 1a**) is the spatial confinement of the ablation plasma plume. Substrates whose diameter exceeds some limiting size are not overlapped completely by the plume region where the mass-transfer velocity is sufficiently high. This drawback can be compensated for by mounting the substrate at a larger distance from the target; however, the larger the substrate, target distance, the lower the deposition rate. In addition, the stoichiometry of multicomponent films can be preserved only in certain range of variation in the substrate—target distance.

The drawback of the second version (**Figure 1b**) is the necessity of preliminary analysis of the angular distribution of the mass-transfer rate in the plasma plume in specific geometry and under specific deposition conditions. In turn, the angular distribution of the mass transfer may vary during deposition, because it depends on several parameters, which may also vary during long-term deposition. There are many modifications of the above-described techniques. Some achievements in growing large area films by PLD were described in the study of Eason [2]. As previously, the main technique is based on scanning the laser beam on the surface of a large target (**Figure 1c**). However, the manufacture of a large target for some of the compounds can be very expensive.

3. Proposed methods of deposition on rotating substrate

Three relatively simple methods of laser deposition of large area thin films are proposed [3, 4]. The first method is based on the target tilt control relative to the laser beam and its focal spot while the substrate is maintained in the same position. Specifics of other two methods include the deposition of a compound on a substrate through a mask with consideration of various slits on it. In one of the options, the mask has a slit in the form of a sector symmetrical to the substrate radius and with variation of angular sizes at different values of the radius. Another option uses a mask with two bent sector-shaped slits located symmetrically relative to the line of equal velocity of the compound mass transfer path.

Conditions of films deposition at experimental verification of the proposed methods were the following. The deposition was performed using the third harmonic of an Nd^{3+}:YAG laser (λ = 355 nm) with the pulse characteristics: energy, 15 mJ; width, 10 ns; and repetition rate, 20 Hz. The target and substrate rotation speeds were 37 and 2 rpm, respectively. AMBIOS XP-1 profiler was used to control the thickness of the obtained films.

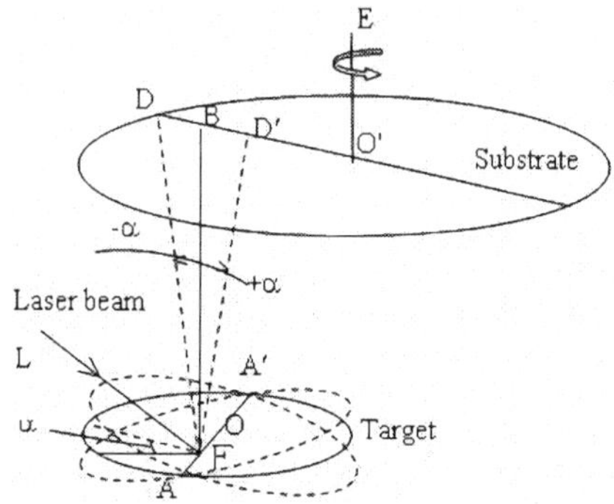

Figure 2. Geometry of tilting target method.

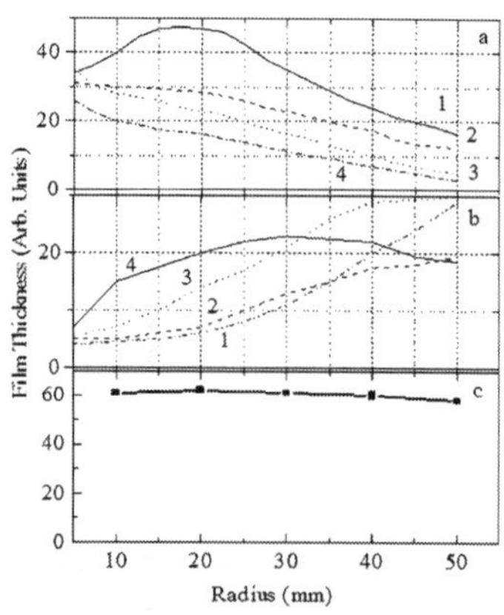

Figure 3. Dependence of film thickness on substrate radius. (a) α = +18°: d = 0.8 mm (1), 0.65 mm (2), 0.45 mm (3), 0.3 mm (4); (b) α = −30°: d = 1.05 mm (1), α = −24°: d = 1.05 mm (2), α = −30°: d = 0.8 mm (3), α = −24°: d = 0.65 mm (4); and (c) α = −24°: d =1.05 mm and α = +18°: d =0.45 mm.

3.1. The method of tilting target

The first method directing the plume flux to various substrate areas is illustrated in **Figure 2**. Here, the target with a center in point O is tilted around the axis passing through the center and perpendicular to the laser beam. This AOA′ axis is parallel to the substrate plane rotating around O′E. In original substrate position, the plume axis coinciding orthogonal to the target plane is crossing the substrate at point B. With tilted substrate, this intersection point is traveling from point D to point D′ as it is marked by the dashed lines.

Angular distribution of the material ablated by the laser beam has a number of parameters it depends on. We have narrowed the variability to the slant angle (α) and the laser beam diameter (d). **Figure 3** shows the dependence of deposited film thickness on mentioned parameters as a function of substrate radius. It is clear from the charts in **Figure 3** that the proposed target tilting method allows achievement of a thicker deposition on the center and edges of the substrate. So it is possible to obtain uniform film thickness by superposing two tilts as it is shown in **Figure 3c**. Variation of the thickness of the film deposited by such a technique did not exceed ±3.3%.

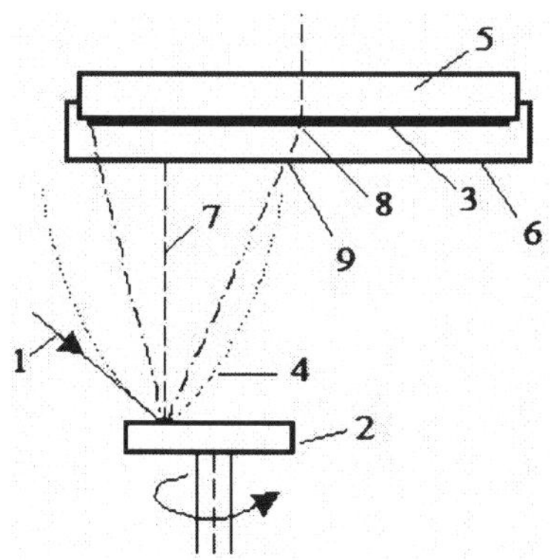

Figure 4. Geometry of mask method of deposition on a rotating substrate. 1, laser beam; 2, target; 3, substrate; 4, plasma torch; 5, substrate holder; 6, mask; 7, axis of the plasma plume; 8, center of rotation of the substrate; 9, point of intersection of the line connecting the focal spot and the center of rotation of the substrate with the mask plane.

3.2. Mask method of deposition

We have proposed two essentially different solutions of the problem of obtaining large-size films on a rotating substrate. The first method is based on the use of a mask with a slit in the form of a sector of different angular sizes at different distances from the substrate rotation axis. The slit configuration is determined on the basis of the angular dependence of the mass-transfer rate so that the thickness uniformity of the deposited film is provided. The second method uses the slits in the form of a sector with a curved symmetry axis coinciding with the equal thickness line of the film deposited on a stationary substrate. **Figure 4** shows schematically the geometry of film deposition on a rotating substrate. It is seen that it differs from the conventional off-axis deposition (**Figure 1a**), in that the deposition is performed through a mask placed in the immediate vicinity of the substrate.

3.2.1. Sectorial slit with varying angular sizes

This technique supposes deposition through a mask with a slit in the form of a sector symmetric with respect to the substrate radius and having different angular sizes at different distances from the axis of substrate rotation. The sector vertex must be in the point of intersection of the straight line connecting the focal spot and the center of the rotation of the substrate with the mask plane. **Figure 5a** shows the variation of the thickness of films along the substrate radius in different deposition processes. Curve 1 presents the process with no mask and rotating substrate. Curve 2 corresponds to deposition on the rotating substrate through a mask with a slit in the form of a sector. As expected, in this case, we obtain films of nonuniform thickness. However, curve 2 indicates where the sector should be narrowed and where broadened. Performing several depositions with different masks, we succeeded to reveal a mask configuration (**Figure 5b**) providing the thickness uniformity of the films (curve 3).

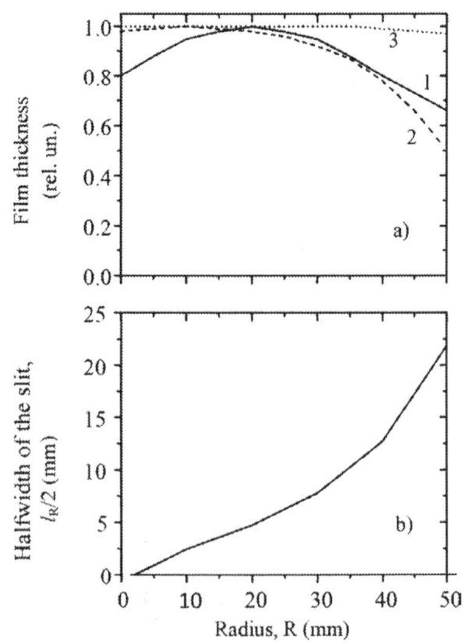

Figure 5. Radial distribution of film thickness (a) slit configuration which provided film thickness uniformity (b).

3.2.2. Calculation of the configuration of slit in the mask

In order to exclude experimental optimization of the mask-slit configuration, a method has been designed to calculate the slit form based on the data from one deposition process onto a stationary substrate. With this purpose, the deposition is performed onto a non-rotating substrate without a mask. **Figure 6** shows the relative thickness of a CuO film obtained under indicated conditions. After choosing the position of the plume axis with respect to the axis of rotation of the substrate and taking the center of rotation for the origin, we determine the

http://dx.doi.org/10.5772/64978

variation of the film thickness along an arc of circumference for different values of R inside a sector with an angle δ.

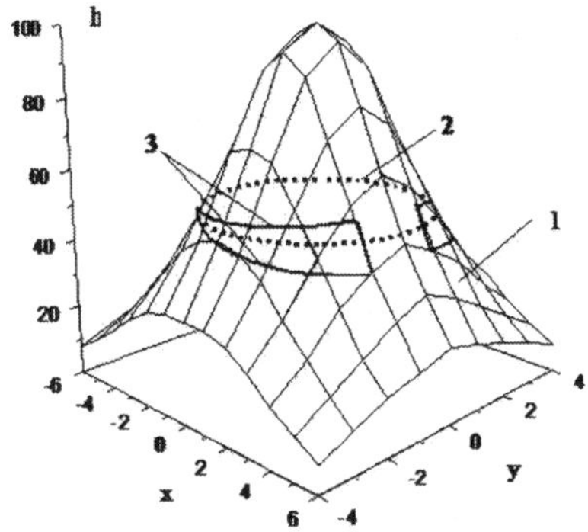

Figure 6. Relative thickness of the film deposited from CuO target on a resting substrate (1), ellipse of equal thickness (2), and the part of the plasma plume passing through the mask (3).

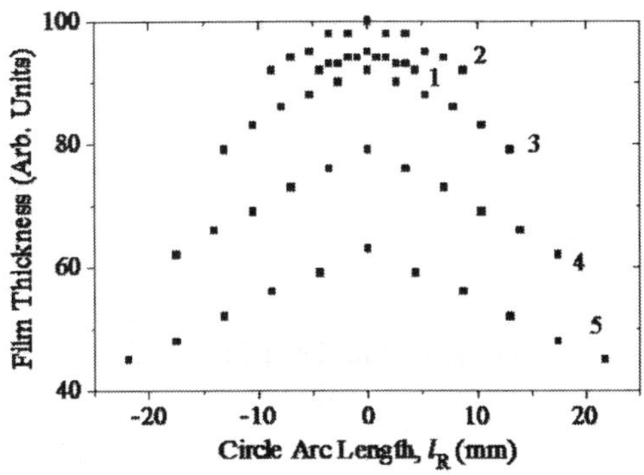

Figure 7. Distribution of film thickness along the arc at different radii R: 1, 10 mm; 2, 20 mm; 3, 30 mm; 4, 40 mm; and 5, 50 mm. The pattern is symmetric and the maximal thickness is associated with the origin of the l_R coordinate.

For given (parallel) position of substrate and target rotation axes, the film thickness on various R from the center of substrate inside the sector with δ angle is determined. **Figure 7** shows the case corresponding to $\delta = 50°$ (the distance between two rotation axes is 2 cm). Using these data, the average film thickness $\bar{h}_R$ on various R is calculated in the first approximation. Then, the optimal length of the circle arc l_R is determined (i.e., the slit width on various R). This provides obtaining films with the same thickness on various R, equal to the average thickness $\bar{h}_{Rm}$ on the maximal radius R_m, according to the formula $l_R = \bar{h}_{Rm} l_R(\delta)/\bar{h}_R$. Here, $l_R(\delta) = \pi R\delta/180°$ is the length of the arc of the radius R in the sector with an angle δ. More accurate values, l_{R}^{2}, may be obtained if we repeat the calculation using the average values of the film thickness $\bar{h}_{R}^{2}$ for the slit corrected in the first approximation. After this step, a test of the obtained results is performed.

Both the calculation data and the experiments have shown that the described computation methodology provides the necessary precision for obtaining films with excellent thickness uniformity.

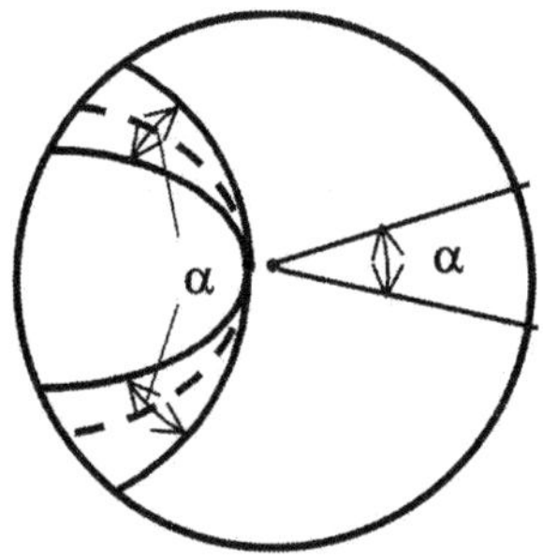

Figure 8. Distortion of sectorial slit. The mask slit curved along the uniform thickness ellipse (dashed line) provides both thickness and composition uniformity of deposited films.

3.2.3. Slit in the form of a sector with the curved symmetry axis

The method of the mask with a slit in the form of a sector with the curved symmetry axis is realized also with the use of geometry presented in **Figure 4**. This possibility to obtain thickness-uniform large-area films is due to the following circumstances. For obtaining a thickness-uniform film on rotating substrate, the slit in the mask must be a sector with the vertex coinciding with the rotation axis of the substrate. This is valid for the case of uniform flow of deposited substrate. Consider now the realistic pattern shown in **Figure 6**. If we intersect the three-dimensional surface outlining the thickness of the film, by a plane parallel to the xy-plane, we obtain the line of equal thickness of the film. This line is usually an ellipse. Intersecting the surface above by planes parallel to the xy-plane at different heights, we obtain a set of concentric ellipses. **Figure 8** shows the ellipse passing through point 9 in **Figure 4**. In every point of this ellipse, we have the same thickness of the film. The reason for this is the spatial uniformity of the flow of deposited material along this line. Hence, the slit in the mask in the form of a curved sector having the middle line coinciding with equal thickness ellipse (**Figure 8**) should provide the uniformity of the film thickness. The choice of the ellipse is caused by striving to obtain a uniform film over the entire surface of the substrate. Symmetry of the pattern with respect to the substrate radius causes the possibility to use two slits which doubles the deposition rate. The deposition rate depends as well on the value of angle α.

http://dx.doi.org/10.5772/64978

Figure 6 shows the domain corresponding to the part of the plasma torch passing through the mask. It is obvious that if in this domain the three-dimensional surface outlining the film thickness has no abrupt changes in curvature, inside this domain the average thickness of the film along the radius of substrate rotation will be the same for different radii and equal to the thickness corresponding to the chosen ellipse. The smaller the angle α, the better this condition is met. On the other hand, the smaller is the angle α, the lower is the rate of film deposition. Hence, it is necessary to choose α in each case depending on the specific problem.

Implementation of both mask methods of deposition onto a rotating substrate has revealed a possibility of providing for thickness uniformity for both CuO and $YBa_2Cu_3O_{7-\delta}$ films within ±3.3% on substrates with the diameter of 100 mm. The $YBa_2Cu_3O_{7-\delta}$ films had a high value of temperature of a superconducting transition (T_c= 90 K) independent of the distance between the chosen domain and the center of substrate rotation which indicates radial uniformity of the composition of deposited films. This refers more to especially the method of mask in the form of a sector with curved symmetry axis, since the rate of mass transfer for components of complex compound may vary depending on the solid angle inside the plasma torch. Another advantage of such a mask is that the central part of the plasma torch is cut off. Thereby, the density of micron-size particles inherent to laser deposition decreases.

We also point out the possibility to combine the principles underlying the two different types of slit configurations, when requirements to the film quality are especially rigid. This may be done if we employ a mask with slits having curved symmetry axis and different angular sizes at different distances from the axis of rotation of the substrate.

3.3. Deposition of large-size films

Figure 9 shows schematically the deposition geometry for large-area thin uniform films, whose transverse sizes are limited by only the deposition chamber dimensions [4]. Its distinctive features are as follows:

- the presence of a diaphragm, partially transmitting the evaporated material, between the target and the substrate.
- the translatory motion of the rotating substrate with respect to the target at a certain velocity and selecting beams with identical and maximum mass-transfer rate from the plasma torch.

It can be seen in **Figure 9** that the substrate undergoes translatory motion with respect to the target. Obviously, in this geometry the substrate motion can be replaced by the joint motion of the target and the diaphragm.

To grow a film of uniform thickness over the entire substrate surface, it is necessary to determine the motion law for the substrate. The calculation results for substrate transferred velocity V were experimentally checked by depositing thin CuO films on silicon substrates 30 mm long, located over the radius of a disk 300 mm in diameter. The angles between the target plane, laser beam axis, and diaphragm plane were chosen so as to make the diaphragm select a particle beam oriented perpendicular to the target from the plasma plume. At any changes in the deposition parameters during a long-term process in the chosen geometry, the mass transfer rate of the target material to the substrate will be maximum and the deposition time of a film of specified thickness will be minimum.

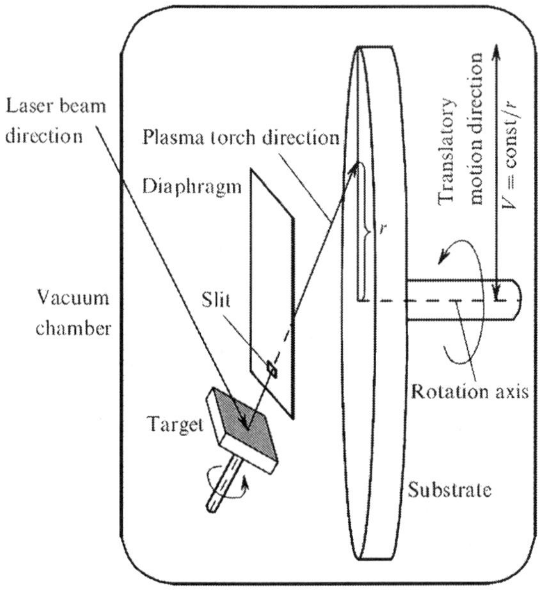

Figure 9. Schematic of the new technique for depositing thin films of arbitrary sizes.

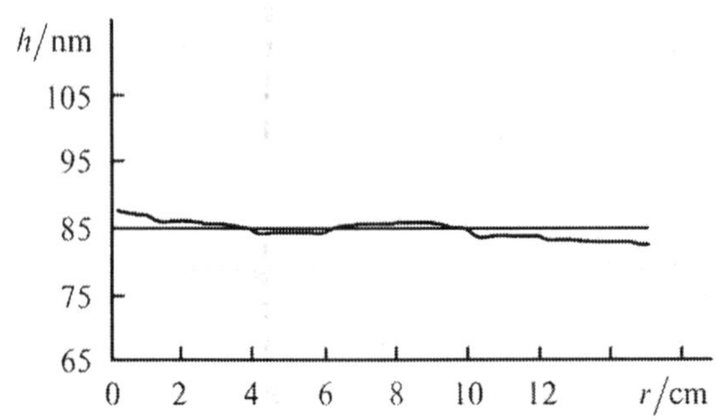

Figure 10. Surface profile of a CuO film deposited on a substrate transferred at a velocity $V = const/r$.

The surface profile of a CuO film deposited on a substrate transferred with a velocity $V = const/r$ is shown in **Figure 10**. The average film thickness is 85 nm. The thickness deviation from the average value did not exceed ±3% over the entire substrate surface. The method proposed

http://dx.doi.org/10.5772/64978

makes it possible to obtain thin films of uniform thickness on substrates with sizes limited by only the deposition chamber size.

4. Effective utilization of the target material

A commercial of PLD is hindered by its few drawbacks, main among those are a nonuniform thickness of the deposited film and a low coefficient of effective utilization of the target material. Consider the second drawback in more detail. In this method, the films are deposited due to ablation of the target material by a laser beam. Obviously, if both the target and the laser beam are immovable, then a crater arises on the target in a certain time interval. The crater affects the angular distribution of the evaporated material, that is, the thickness of the film deposited per unit time on a certain area of the substrate.

One more undesirable result is overheating of the target and distortion of its composition. The target is also overheated in the case when the laser beam is focused onto the entire surface of the target. In this geometry, no crater is formed in the target; however, the target undergoes overheating. Thus, in the case of PLD with both immovable target and the laser beam, the thickness and composition of the film deposited would vary in time. There is a simple solution to the problem—rotation of the target, which substantially facilitates the situation preventing the target from overheating and prolonging the duration of deposition without disturbing the film composition and angular distribution of the evaporated material.

However, such geometry does not solve the problem yet. A groove arises in the rotating target (**Figure 11**) and the angular distribution of the evaporated material changes with time. After several deposition cycles with a variation in the distance of the laser beam from the target center, concentric grooves arise on the surface of the target and the latter becomes unsuitable for further employment. In the conventional geometry, utilization efficiency of PLD (the fraction of its evaporated volume) is very poor (0.01–0.02). It is not important if the target material is cheap. However, if the films of rare metals and its alloys are deposited or the target is made of very high purity chemicals or isotopes, then such a low efficiency is inappropriate.

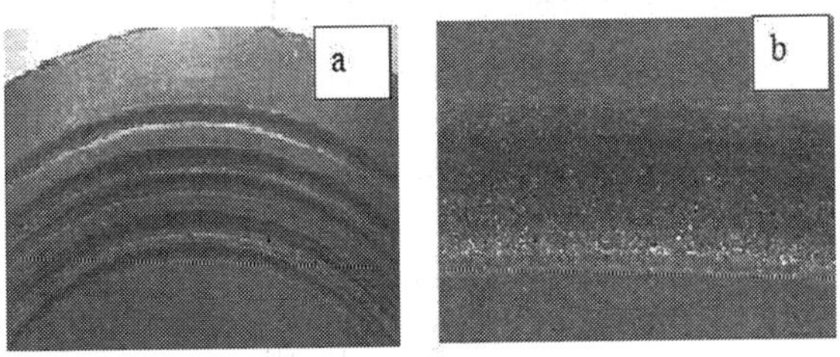

Figure 11. Target made of $YBa_2Cu_3O_{7-\delta}$ of diameter 50 mm after five deposition cycles performed by the conventional PLD method at magnification (a) 10× and (b) 60×.

4.1. Ways of increasing the target utilization efficiency

There are several known solutions to the problem for improving the target utilization efficiency in PLD of thin films. The targets employed for deposition may have polished surfaces. In this case, up to 75% of the target material may be lost [5]. There are cardinal approaches, which, however, substantially complicate the deposition installation suggesting the computer-controlled scanning of the large target surface by a laser beam (**Figure 12a**) or moving the target in two perpendicular directions relative to the fixed laser beam (**Figure 12b**). In this case, the material ablation occurs from 90% of the target surface providing in this way a noticeable improvement in the utilization efficiency.

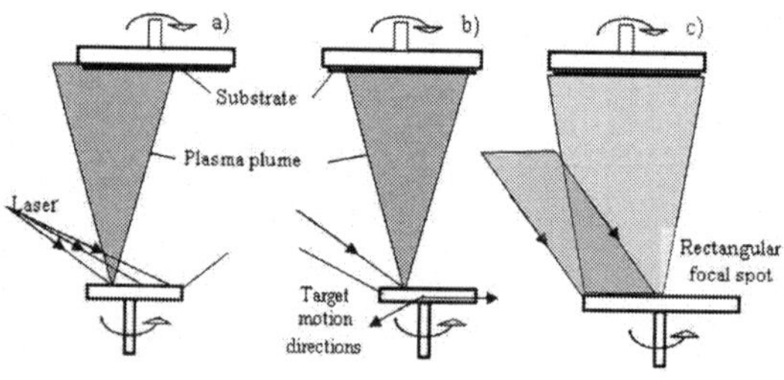

Figure 12. Geometry of the PLD method with: (a) laser beam screening the surface of target; (b) target motion relative to the laser beam, (c) rectangular focal spot along the radius of the target.

The suggestion in [6] is simple to realize: it is assumed that the laser beam is split into four parts, each of them performing material ablation from a certain area of the rotating target. Thus, a wider plasma jet of the evaporated material is formed (which facilitates the uniformity of the film thickness) and ablation occurs from a larger surface of the target. From the viewpoint of enhancing the target utilization efficiency, this approach is simply an increase in the focal spot dimensions. However, the grooves on the rotating target will still be produced with the following negative consequences. A more reasonable suggestion was patented [7, 8]. The authors complicated the mechanism of target rotation in such a way that the target rotates around two parallel axes. In this case, the laser beam circumscribes a cycloid rather than a circle over the target surface, which will increase the target utilization efficiency. Unfortunately, this approach has also a serious disadvantage (see below), which hinders obtaining high values of target utilization.

A most simple solution is, evidently, a uniform distribution of laser radiation over the entire surface of a fixed target. Such a distribution can be obtained, for example, with raster-focusing systems [9]. Nevertheless, as mentioned above, such a scheme cannot be used for depositing multicomponent compounds, because overheating of the target will affect the composition of the latter. Thus, the known methods of enhancing a PLD process aimed at increasing the target utilization efficiency either noticeably complicate the installation or do not solve the problem completely. In the present work, we suggest a simple solution to the problem for attaining a

http://dx.doi.org/10.5772/64978

maximal utilization of the target material, capable of increasing the utilization efficiency actually to unity.

4.2. Rectangular focal spot arranged along the target radius

One possible solution to the problem for increasing the target utilization efficiency is a rectangular spot of the laser beam arranged along the radius of the rotating target, where the focal spot of width L has the length of at least the target radius (**Figure 12c**). The center of one focal spot side coincides with the target rotation center O, and the axis of symmetry coincides with the target radius (**Figure 13**). At a first glance, it may appear that it is a simple and effective solution, because ablation of the target material will occur from the entire surface. But the quantity of the substance evaporated from a unit area is proportional to the energy passed on it. The sites of focal spot closer to the target center will affect (per single round) a smaller area than those residing far from the center. Consequently, the target material at its center will be consumed faster than at periphery.

Let a laser beam fall onto the rotating target, which has the form of a disk with radius R and thickness h_0, forming the rectangular focal spot of width L arranged along the target radius. In time t_0, the target executes N revolutions and the laser burns a hole at the center of the target. How much of target volume is ablated in this case?

According to **Figure 13**, one should consider two domains of the target surface: inside the circle of radius $L/2$ and outside it [10]. When the target rotates at a constant angular velocity ω, all the points inside the circle of radius OA = $L/2$ are exposed to laser radiation during half the process duration ($t_0/2$), because for these points the focal laser spot is a semicircle. In other words, inside the rectangular focal spot of width L the trajectory of any point residing closer than $L/2$ to the center of rotation (point O) is a semicircle. For example, point D on the target surface is subjected to laser radiation as long as it follows semicircle DF (**Figure 13**). Point B residing at a distance longer than $L/2$ is subjected to laser radiation until it reaches point C having passed the arced path BC.

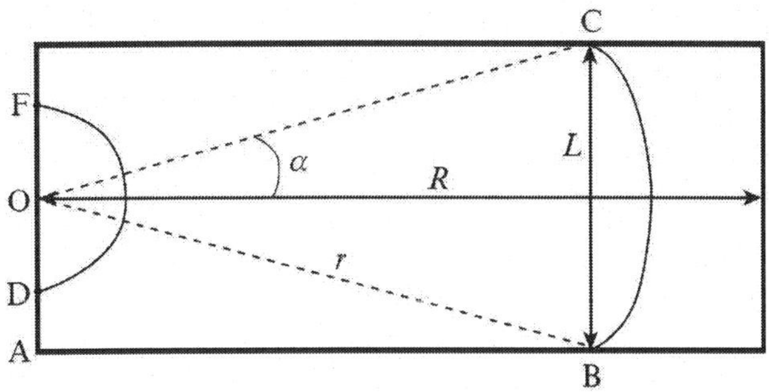

Figure 13. Scheme of the rectangular focal spot directed along the radius of the rotating target [10].

By denoting the rate of target evaporation σ (the thickness evaporated per unit time), from the condition of burning the target to a throughout hole, we may write

$$h_0 = \sigma t_0 / 2, \sigma = 2h_0 / t_0 \tag{1}$$

The points of the target surface outside the circle of radius $L/2$ at a distance r from the center are exposed to laser radiation during the time interval Δt (a single revolution of target) so that after N revolutions the target thickness reduces by the value ,

$$h = \sigma \Delta t N = \sigma N|\breve{BC}|/v = \sigma N|\breve{BC}|/(\omega r), \tag{2}$$

where v is the linear speed of point B, $|\breve{BC}| = 2ra = 2r \times \arcsin(L/2r)$ is the length of arc BC, and $\omega = 2\pi N/t_0$.

Then at $r \geq L/2$, the reduction of the target thickness will be

$$h = \frac{2h_0 \arcsin\left(L/2r\right)}{\pi} \tag{3}$$

Dependences of $h(r)$ for various L are given in **Figure 14**. For the points $0 < r \leq L/2$, we obtain $h = h_0$. The volume of the ablated target material is determined by the rotation of curve $h(r)$ around axis h:

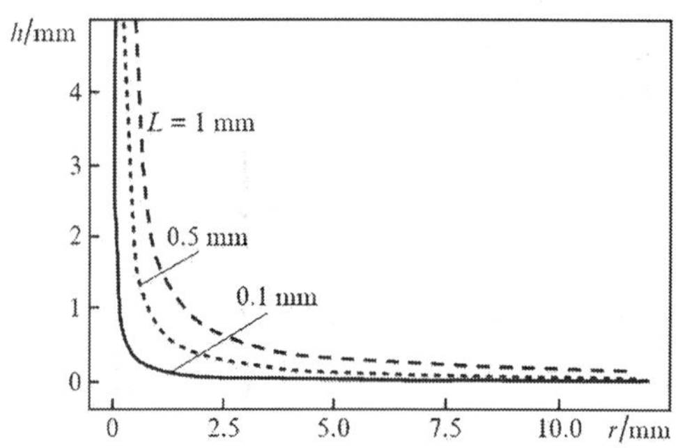

Figure 14. Target thickness h versus radius r under PLD in the rectangular focal spot geometry [10].

$$V = \pi h_1 R^2 + \pi \int_{h_1}^{h_2} \left[\frac{L}{2\sin\left(\pi h / 2h_0\right)} \right]^2 dh, \tag{4}$$

http://dx.doi.org/10.5772/64978

where $h_1 = [2h_0 \arcsin L/2R]/\pi$ and $h_2 = h_0$. This entails

$$V = 2h_0R^2 \arcsin\left(\frac{L}{2R}\right) + h_0LR\left(1 - \frac{L^2}{4R^2}\right)^{1/2}. \quad (5)$$

One can see that the volume depends on three parameters: the target height h_0, the target radius R, and the width of the laser beam L. Because the value of the target material utilization efficiency is $\eta = V/V_0$, where $V_0 = \pi h_0 R^2$ is the target volume prior to ablation, in view of Eq. (5), we obtain

$$\eta = \left[2\arcsin\left(\frac{L}{2R}\right) + LR^{-1}\left(1 - \frac{L^2}{4R^2}\right)^{1/2}\right]\pi^{-1}. \quad (6)$$

The value of η depends only on two parameters L and R. Dependences of $\eta(R)$ for various values of L are shown in **Figure 15**. One can see that at a fixed parameter R, the volume of the ablated part of the target is greater at longer L.

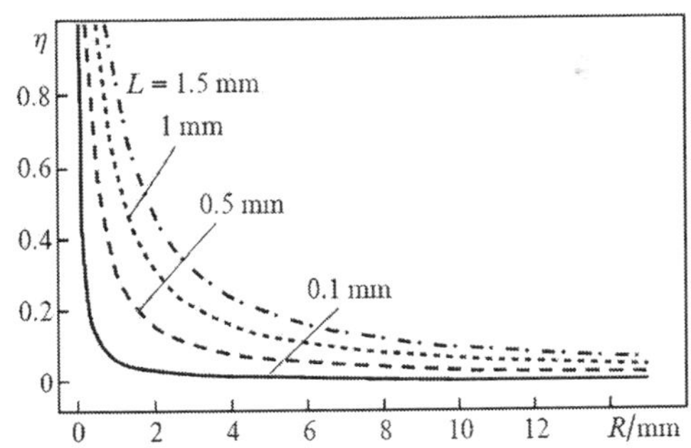

Figure 15. Target utilization efficiency versus target radius at various laser beam spot widths [10].

The expression for the target material utilization efficiency may be simplified. By introducing the parameter $k = L/R$, we may write Eq. (6) in the form

$$\eta = \left[2\arcsin(k/2) + k\left(1 - k^2/4\right)^{1/2}\right]\pi^{-1} \quad (7)$$

Recall that the calculations are performed for the case $R \geq L/2$. At $0 < R < L/2$, the equality $\eta = 1$. **Figure 16** presents the dependence $\eta(k)$.

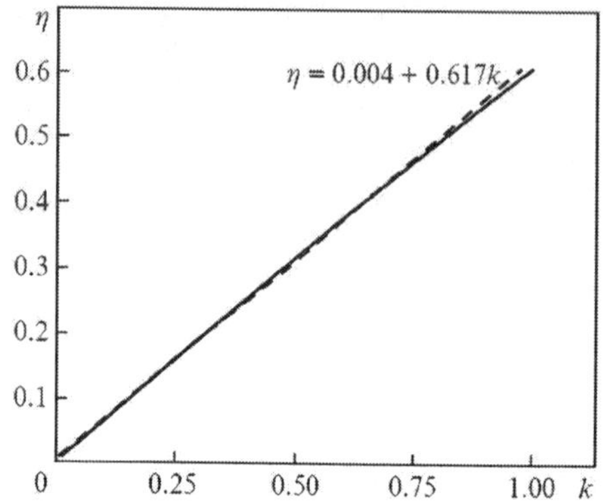

Figure 16. Target utilization efficiency η versus k. The dashed line refers to approximation [10].

One can see that at $k < 1$, it is well approximated by the straight line according to the formula

$$\eta = 0.004 + 0.617k \approx 0.6k \tag{8}$$

Thus, we have the simple expression for target utilization efficiency with a sufficiently good approximation. Obviously, for experimentally actual values of L and R where $k < 1$, the value of η will not be greater than 0.5.

Note that the method suggested in patents [10, 11] has a similar drawback. The cycloidal trajectory described by the focal laser spot on the surface of the target arising due to the rotation around two parallel axes will also result in more intense material evaporation from the central part of the target and reduced utilization efficiency.

4.3. Focal spot in the form of a sector

The consideration of the problem stated above suggests its cardinal solution. In the range $0 < R < L/2$ where the focal spot is a semicircle, the equality $\eta = 1$ holds. But a semicircle is the particular case of a sector with the angle of 180°. The scheme of the modified PLD method is shown in **Figure 17**, which simply and cardinally solves the problem on the maximal utilization of the target material. The laser deposition installation is suggested, which differs from ordinary devices by a simple optical system placed outside the deposition chamber. It comprises two lenses and a diaphragm and provides the focal spot in the form of a sector on the target surface.

If such a focal spot coincides with a sector-shape area on the target surface and the density of energy is uniform over the focal spot, then the surface of the uniformly rotating target will be uniformly irradiated, which will provide a uniform material ablation from the surface.

http://dx.doi.org/10.5772/64978

The device suggested was employed for depositing CuO and $YBa_2Cu_3O_{7-\delta}$ films from the targets 10 mm in diameter by the pulses of second harmonic radiation of the Nd^{3+}:YAG laser with the repetition rate of 20 Hz. The optical system provided the focal spot of laser radiation on the target in the form of the sector with an angle of 60° and the energy density of 4 J cm^{-2}. Variations in the target thickness were within ±2% both before deposition and after five deposition cycles lasting for 45 min. One may assert that the device suggested enhances the target material utilization efficiency up to $\eta = 1$.

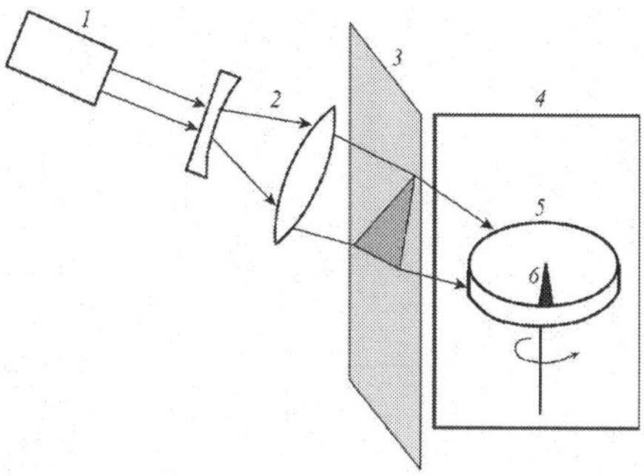

Figure 17. Geometry of the PLD method with the focal spot in the form of a sector: (1) laser, (2) optical system, (3) diaphragm with a hole, (4) deposition chamber, (5) target, and (6) laser focal spot in the form of the sector coinciding with the sector of the rotating target surface [10].

5. Deposition onto ribbon

A method of PLD of thickness- and composition-homogeneous thin films on a translationally moving ribbon of up to 100 mm-width is proposed [11]. The peculiarity of the method is in the matter deposition via a mask placed in vicinity of the substrate. Two configurations of mask slits providing for the thickness homogeneity of films are considered. Exact dimensions of the mask slits are calculated using the data of an angular distribution of mass transfer in a plasma torch.

5.1. The geometry of the proposed technique

The geometry is shown in **Figure 18**. The laser beam (2) is incident at the target (1). The mask (4) and translationally moving substrate (5) are placed perpendicularly to the axis of the plasma torch. The arrow shows the direction of the substrate displacement.

In order to provide an equal thickness of the film deposited onto the uniformly moving substrate, the following condition should be met: equal amounts of the substance must be deposited per unit time on a unit length section placed along the direction of the substrate displacement and this condition should be satisfied over the entire width of the substrate.

Taking into account the peculiarities of angular distribution of the mass-transfer rate in the plasma torch [6], we propose two configurations of the slit in the mask (**Figure 19**). In both cases, the plasma torch axis passes through the geometrical center of the mask. The displacement of substrate occurs in the x-direction. The dashed line in **Figure 19a** indicates the line along which the mass-transfer rate of the deposited matter is constant. A rectangular slit (or slits) curved along dashed line provides, obviously, the uniformity of the film thickness. The second configuration of the mask (**Figure 19b**) takes into account the fact that the angular distribution of the mass-transfer rate in the plasma plume is given by the function $F(\theta) = A \cdot \cos^m(\theta)$ [6], that is, the greater is the angle θ (**Figure 18**), the lower is the mass-transfer rate; thus, in order to provide the uniformity of the film thickness over the substrate width, the greater should be the width of the slit in the mask.

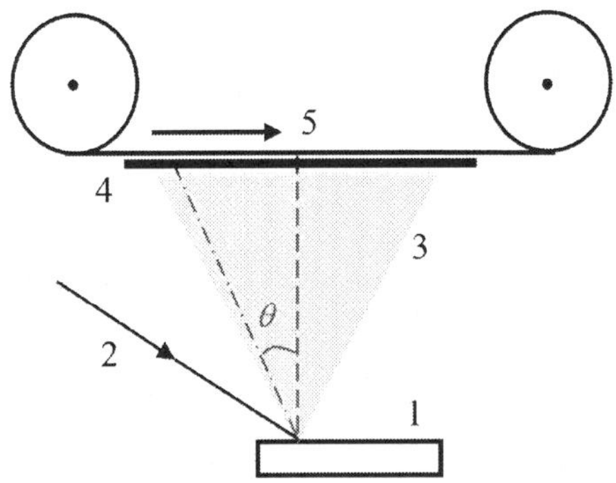

Figure 18. Geometry of deposition onto a moving ribbon.

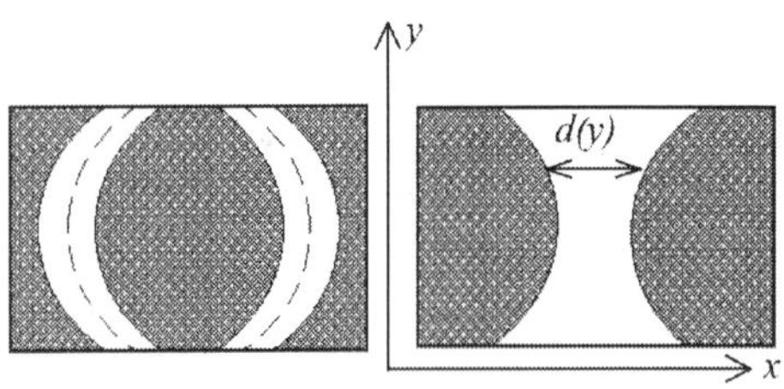

Figure 19. Configurations of slits in the mask.

5.2. Calculation of the configuration of slit in the mask

We now give a method for calculation of the configuration of the slit in the mask (**Figure 19b**) providing the thickness homogeneity of the film deposited onto the substrate moving trans-

http://dx.doi.org/10.5772/64978

lationally at a constant velocity. The width of the slit $d(y)$ is determined from the relation $d(y)\bar{A}(y)$ = const, where $\bar{A}(y)$ is the average thickness on the section $d(y)$ of the film deposited on the resting substrate per unit time. The quantities $\bar{A}(y)$ are determined from the experimentally determined function $D(x,y)$. The angular distribution of the ablated material depends on many factors. In our experiments, we tried to change only one parameter, leaving the others unchanged. We chose the laser spot dimensions (S) and the laser fluence (F) as variable parameters. The values of these parameters and the obtained films relative thickness are listed in **Table 1**. The data of films thickness are fitted by the function $D(x, y) = A\cos^{Px+3}(\theta_x)\cos^{Py+3}(\theta_y)$. The used designations are clear from **Figure 20**. The obtained values of parameters A, p_x and p_y are presented in **Table 2**. Now, it is possible to initiate calculation of the configuration of the slit.

No.	F, J/cm²	Spot dimension, mm²	D(θ = 0)[a], a. u.
1	53	0.205	68
2	76	0.205	83
3	97	0.205	84
4	55	0.361	103
5	18	1.11	157
6	9.9	2.02	127
7	6.2	3.2	76

[a]Obtained films relative thickness at $\theta = 0$.

Table 1. Variable parameters of a laser deposition.

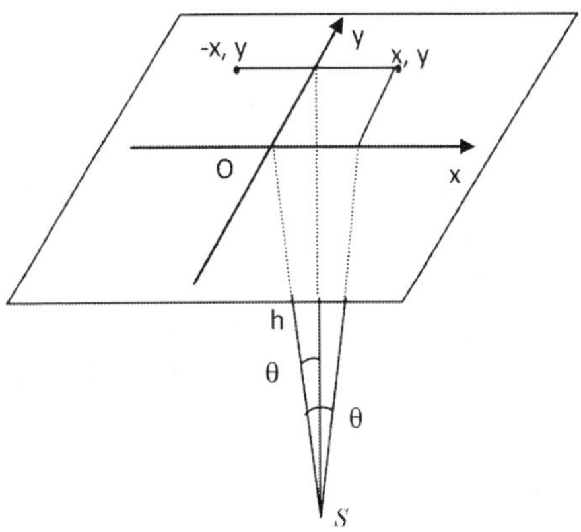

Figure 20. Schematic of deposition geometry. S is the laser spot.

No.	A	$p_x + 3$	$p_y + 3$	χ^{2a}	Area of a slit, cm^2	V^b, a.u.
1	66.83	8.45	5.8	2.58	24.76	713
2	80.13	8.2	5.3	4.46	26	925
3	80.28	7.09	4.31	5.7	30.39	1220
4	107	7.8	4.5	6.71	28.11	1403
5	147.33	10.03	5.23	6.67	22.24	1219
6	117.51	11.55	7.69	4.99	18.25	700
7	76.22	12.84	6.86	0.96	16.56	340

[a] $\chi^2 = \frac{1}{N}\sum_{i=1}^{N}\left(f\left(x_i, y_i\right) - z_i\right)^2$.

[b] Volume of a matter deposited on the ribbon through the mask in a unit time.

Table 2. Parameters of a function $D(x, y)$ and corresponding uniform deposition slit.

If the origin of the coordinates is located at the point that corresponds to $\theta_x = \theta_y = 0$, and a value x_o, y_o is chosen (where y_o stands for the half-width of the ribbon and x_o is determined so that the film thickness is not too small), then the amount of the matter deposited on the segment $[(-x_o, y_o), (x_o, y_o)]$ in a unit time will be given by the expression

$$m = \rho dy \int_{-x_0}^{+x_1} D(x,y)\,dx \qquad (9)$$

where $D(x, y)$ is the thickness profile of the film, ρ is the matter density, and dy is the width of the section. The thickness profile of the film deposited from a point source on a stationary substrate is usually determined by expression $D(\theta) = A\cos^{p+3}(\theta)$ [8]. However, when the focal spot has real sizes that differ in different directions, an asymmetry of function $D(\theta)$ is observed. In this case, the profile of the thickness of the film can be described by the expression

$$D(\theta) = A\cos^{p_x+3}(\theta_x)\cos^{p_y+3}(\theta_y) = A\left(\frac{h}{\sqrt{h^2+x^2}}\right)^{p_x+3}\left(\frac{h}{\sqrt{h^2+y^2}}\right)^{p_y+3} \qquad (10)$$

Upon substitution of Eq. (10) into Eq. (9), we shall receive an equation for the amount of matter $m(y)$ deposited in unit time on the segment $[(-x, y), (x, y)]$ for any value of y. From the requirement that $m(y)$ is constant for all y from $-y_o$ to y_o and equals m, it is possible to determine the values $x(y)$ at which this condition is fulfilled:

$$m_y = m = \rho dy \int_{-x_0}^{+x_0} \left(\frac{h}{\sqrt{h^2 + x^2}} \right)^{p_x + 3} \left(\frac{h}{\sqrt{h^2 + y^2}} \right)^{p_y + 3} dx \tag{11}$$

The function $x(y)$ will determine the profile of a slit that provides uniform deposition of matter on the ribbon. At uniform translation of the ribbon with speed v, the thickness of the film D will be determined by the expression

$$D = \frac{A}{2y_o v} \int_{-y_0}^{+y_0} \int_{-x_0}^{+x_0} \left(\frac{h}{\sqrt{h^2 + x^2}} \right)^{p_x + 3} \left(\frac{h}{\sqrt{h^2 + y^2}} \right)^{p_y + 3} dx dy \tag{12}$$

The computer calculation of the configuration of the slit has been performed. The results are presented in **Figure 21**.

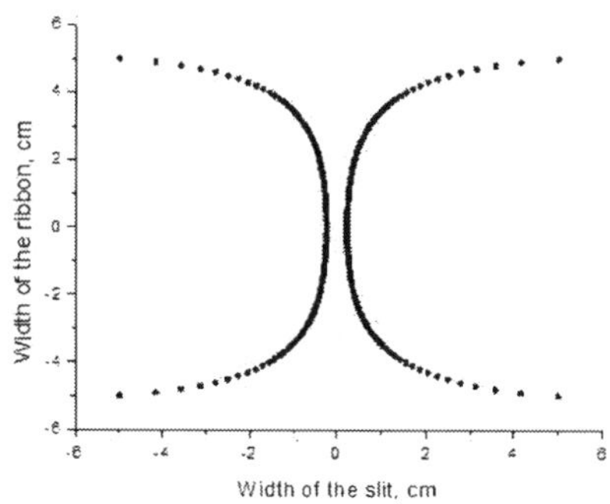

Figure 21. The calculated slit configuration for deposition No 5.

6. Deposition of very large-size films

We have developed special techniques for depositing thin films on rotating disks and translationally moving ribbons. In this paragraph, we propose new solutions, which imply more than one target for deposition, allowing in case of a rotating substrate to shorten the deposition duration and in case of a ribbon to increase the ribbon width or the speed of depositing a film of the same thickness [12].

To reduce the deposition time for films on substrates of very large size, one can modify the above-described way (see paragraph 2.3.) as follows: use several targets instead of one and scan a laser beam over them, applying target after target for deposition. If the laser pulse repetition rate increases in correspondence with the number of targets, the deposition from each target will be performed at the same frequency as in the case of one target. Each target must have its own diaphragm, moving simultaneously with the target at a velocity *V*=const/*r*.

Suppose we are going to use *N* targets to deposit a film on a substrate with a radius *R*. Obviously, the targets must be located with respect to each other and with respect to the substrate in such a way as, having started deposition from all targets simultaneously and moving each target at a velocity *V* = const/*r*, to deposit a film on the entire substrate surface and finish deposition from all targets simultaneously. It can be shown that, to deposit a film of uniform thickness on a substrate of radius *R*, the *N* targets must be initially located so as to provide a distance $X_l(t_0) = R(l\text{-}1)^{1/2}(N)^{-1/2}$ from the substrate center (**Figure 22**) to the point of intersection of the perpendicular dropped from the first target with the substrate plane.

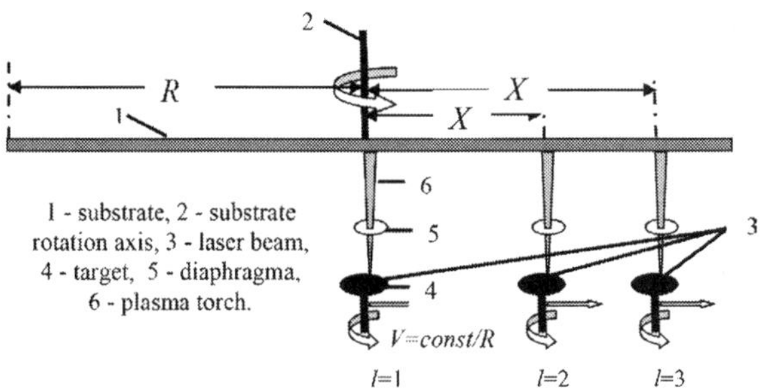

Figure 22. Moving *N* targets.

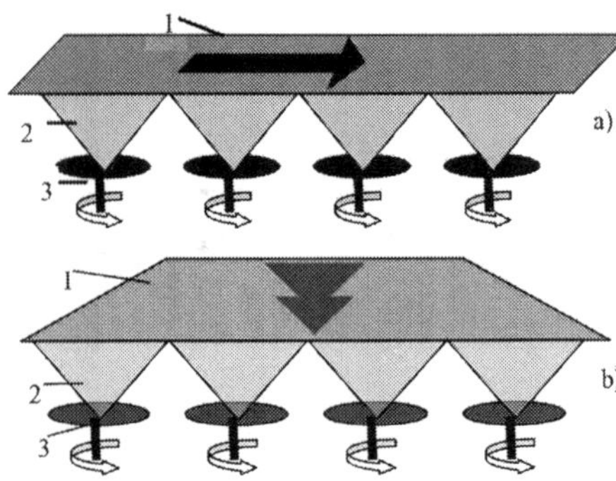

Figure 23. Moving ribbon, *N* targets: 1, ribbon; 2, plasma torch; 3, target. (a) N times acceleration of sputtering process, (b) deposition on N time wider tape.

We can use more than one target also in the mask method (paragraph 3.2.3). Certainly, we can use more than one target as described in paragraph 5 designs for deposition on a moving ribbon. Using *N* targets, we can accelerate sputtering process *N* times (**Figure 23a**), or obtain a film on the tape which is *N* times wider (**Figure 23b**).

7. Conclusions

We believe that the main result of the present work is the proposal and realization of a simple and reliable method of obtaining thickness- and composition-uniform thin films on large areas, with the use of laser deposition. The mask method of deposition on a rotating substrate can be used at any standard arrangement. The size of the obtained films and achieved degree of the thickness uniformity are not limiting. In any specific case where the permissible degree of the thickness nonuniformity is given, one may calculate the slit configuration providing the maximal rate of the deposition.

The new device for laser deposition of thin films is suggested, substantiated, and experimentally tested, which simply and cardinally overcomes the main drawback of the PLD method, namely a low coefficient of the target material utilization efficiency. Taking into account the advantages of laser deposition as compared to other methods of the manufacture of thin films, we hope that the proposed methods will be used in elaboration of new technologies in microelectronics and optical industry.

A more detailed description of the proposed methods can be found in patents [13–20]. Apart from PLD, the proposed methods are applicable to all methods of deposition from a point source.

Acknowledgements

The authors wish to thank all those who have contributed to this area of research, the members of the High Temperature Superconductivity laboratory at the Institute for Physical Research, Armenian National Academy of Sciences. This work was made possible in part by a research grant from the Armenian National Science and Education Fund (ANSEF) based in New York, USA, and was supported by the RA MES State Committee of Science and Russian Foundation for Basic Research (RF) in the frames of the joint research projects SCS 15RF-018 and RFBR 15-53-05047 accordingly.

Author details

Armen S. Kuzanyan* and Astghik A. Kuzanyan

*Address all correspondence to: akuzanyan@yahoo.com

Institute for Physical Research, Ashtarak, Republic of Armenia

References

[1] Chrisey D, Hubler G, editors. Pulsed Laser Deposition of Thin Films, NY: Wiley; 1994. 613 p.

[2] Eason R., editor. Pulsed Laser Deposition of Thin Films: Applications-Led Growth of Functional Materials, New Jersey: Wiley; 2006. 682 p. DOI: 10.1002/0470052120.

[3] Kuzanyan A, Badalyan G, Karapetyan V, Gyulamiryan A, Gulian A. three simple methods to obtain large area thin films by pulsed laser deposition. IEEE Trans. Appl. Supercond. 2001; 11(1), 3852–3855. DOI: 10.1109/77.919905.

[4] Kuzanyan AS, Petrosyan VA, Pilosyan S.Kh, Nesterov VM. Laser deposition of large-area thin films. Quantum Electron. 2011, 41(3), 253–256. DOI: 10.1070/QE2011v041n03ABEH014417.

[5] Greer JA. Commercial scale-up of pulsed laser deposition. In Chrisey D, Hubler G, editors. Pulsed Laser Deposition of Thin Films, NY: Wiley; 1994. p. 293–311.

[6] Nagaishi R. Formation of large surface area thin film by laser deposition method and device therefore. Japanese paten JP83333674 (A), 12.17 (1996).

[7] Takano S. Film forming device by laser vapor deposition. Japanese paten JP2305957 (A), 12.19 (1990).

[8] Iwamoto D, Nagao M. Laser ablation mechanism. Japanese paten JP2006124821(A), 05.18 (2006).

[9] Grasyuk AZ, Zubarev IG, Efimkov VF, Smirnov VG. High-power SRS lasers – coherent summators (the way it was). Quantum Electron. 2012; 42(12): 1064-1072. DOI: 10.1070/QE2012v042n12ABEH015060.

[10] Kuzanyan AS, Kuzanyan AA, Petrosyan VA, Pilosyan S.Kh, Grasiuk AZ. A simple solution to the problem of effective utilization of the target material for pulsed laser deposition of thin films. Quantum Electron. 2013;43(12):1170–1174. DOI: 10.1070/QE2013v043n12ABEH015194.

[11] Kuzanyan AS. PLD of large area films onto substrate undergoing translational motion by mask method. IEEE Trans. Appl. Supercond. 2003; 13(2): 2868-2870. DOI: 10.1109/TASC.2003.812027A.A.

[12] Kuzanyan VA, Petrosyan AS, Kuzanyan. Methods for pulsed laserdeposition of large-area films using more than one target. Int. J. Mod. Phys. Conf. Ser. 2012;15:170–178. DOI: 10.1142/S2010194512007118.

[13] Kuzanyan A, Karapetyan V, Gyulamiryan A, Badalyan G. Film production method. Patent of Armenia No. 692, 15. 12. 1999.

[14] Kuzanyan A, Karapetyan V, Gulian A, Badalyan G. Film production method. Patent of Armenia No. 962, 10. 06. 2001.

[15] Kuzanyan A, Karapetyan V, Gyulamiryan A, Badalyan G. Apparatus for films production. Patent of Armenia No. 963, 10. 06. 2001.

[16] Kuzanyan A. Method for producing of thin films. Patent of Armenia No. 1432, 21.04.2003.

[17] Petrosyan V, Kuzanyan A. Method for producing of thin films. Patent of Armenia No. 2048, 17. 12. 2007.

[18] Petrosyan V, Kuzanyan A. Method of thin films production. Patent of Armenia No. 2530, 27. 06. 2011.

[19] Kuzanyan AS, Petrosyan VA, Kuzanyan AA. Device to produce thin films. Patent of Armenia No. 2531, 27. 06. 2011.

[20] Kuzanyan A, Kuzanyan A, Petrosyan V. System for production of thin films. Patent of Armenia No. 2737, 26. 11. 2012.

Section 2

Nanoparticle and Other Nanomaterial Fabrication

Chapter 8

Laser Ablation in Different Environments and Generation of Nanoparticles

Abubaker Hassan Hamad,
Khawla Salah Khashan and Aseel Abdulkreem Hadi

Additional information is available at the end of the chapter

http://dx.doi.org/10.5772/65241

Abstract

In the last two decades, pulsed-laser ablation has received attention from researchers in micro- and nanotechnology. During the development of laser ablation in materials processing, several media, such as vacuum, air, gases and liquids, have been used to improve the quality and quantity of laser machining and production of nanoparticles. The laser-ablation environment is important in order to control the average size and chemical compositions of nanoparticles. Conducting the laser-ablation process in liquid environments has become of increasing relevance for the production of precise and pure micromachining and nanomaterials. In addition, deionised water has been found to be the optimal environment to produce nanoparticles for bioapplications.

Keywords: laser ablation, vacuum, air, liquids, gases, nanoparticles, ablation mechanism, productivity

1. Introduction

Different pulsed-laser ablations of the solid target materials have shown great potential in the fields of laser-material microprocessing, nanotechnology and device fabrication. To develop the quality and quantity of micro- and nanomachining, laser ablation of materials has been carried out in different environments, such as in a vacuum [1], in air [2], in gases [3] and in liquids [4] for different applications such as welding [5], cladding [6], cutting [7], cleaning [8] and generation of nanoparticles.

Although some laser beam parameters such as laser fluence, wavelength and pulse duration are important for controlling material processing, the ablation environment is also an important factor in laser-material interaction. For example, Zhu et al. [9] concluded that the ablation rate of a Si target material is greatly enhanced by using a water confinement regime (WCR) at laser fluence ranges from 2.0 to 5.0 J/cm^2. They also found that in water, the first peak-to-peak amplitude of the acoustic waves is approximately 25% higher than that in an ambient environment. Besner et al. [10] used a femtosecond laser for ablation in a vacuum, air and water for surface modifications; they showed that the threshold values of the Si and Au target materials were almost identical in all three environments. The values of Si in the single and multi-pulse irradiation regime were 0.4 and 0.2 J/cm^2; for Au, their values were 0.9 and 0.3 J/cm^2, respectively. Patel et al. [11] showed that the laser-ablation process is more efficient in water than in air, which depends on the thermal properties of the materials. It was also concluded that laser-ablation production in water is more suitable for the production of uniform nanoparticles and the mass production of nanoparticles. In addition to the aforementioned environments for laser ablation, Lindley et al. [12] studied the laser-ablation plume of an aluminium target material in a plasma environment, as well as in a vacuum and argon gas. It was concluded in this study that the laser-ablation plumes in the plasma expand and dissipate slightly faster than in the gas and the vacuum. Iqbal et al. [13] showed the effects of the laser fluence and ablation environments (vacuum and hydrogen) on the micro- and nanostructure of a Ge target material. It was shown in their study that the formation and growth of laser-induced periodic surface structures (LIPSS), cones and micro-bumps strongly depended on the laser fluence and environmental conditions. Hence, the growth, size and shape of these structures strongly depended on the laser fluence.

In this chapter, a critical comparison of laser ablation in different environments such as a vacuum, ambient air, different liquid environments and different background gases is presented. The optimal medium and laser-beam parameters for laser ablation will be designated.

2. Laser-material interaction

During laser-material interaction, the product materialises directly after laser irradiation onto the surface of the solid target material by condensing the plasma plume. After laser-pulse deposition onto the surface of the target material, the laser-pulse energy will heat the target, leading to an increase in the temperature of the materials. The temperature propagates in the axial and radial directions in a specific area. In the case of a perpendicular laser beam on a flat target material, the temperature propagation as a function of time (t) and depth (x) can be written as follows [14]:

$$\Delta T(x,t)=2(1-R)\alpha I_o\left(\frac{t}{\pi kpc}\right)ierfc\left(\frac{x}{2\sqrt{kt/\rho c}}\right) \tag{1}$$

where t, R, α, I_o, k, ρ and c are irradiation time, reflectivity, absorptivity, spatial distribution of laser intensity, thermal conductivity, target material density and velocity of light, respectively. When $x > 2\sqrt{kt/\rho c}$ the surface temperature can be simplified as follows:

$$\Delta T(t) = \frac{2\alpha I_o \sqrt{t}}{\sqrt{\pi k \rho c}} \tag{2}$$

Two main things which have an effect on the laser ablation of materials are: laser-beam parameters such as laser fluence, laser wavelengths and laser-pulse duration, and factors in the experimental set-up, such as the type of environment and the solution. Laser-ablation mechanisms and their products are different depending on whether the laser-material interaction is produced by nanosecond, picosecond or femtosecond lasers [4]. In the laser-ablation process, groove width and depth are two important factors with which to characterise the results (see **Figure 1**) [15].

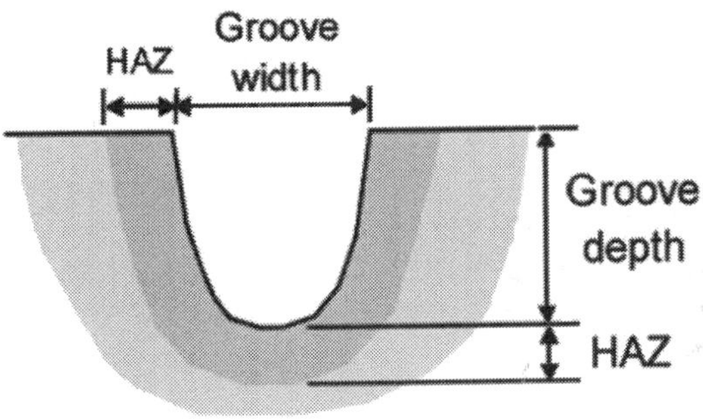

Figure 1. Groove width, groove depth and heat-affected zone of a laser-ablated material target.

Laser ablation in the target material occurs when the laser intensity reaches its threshold value. The laser-intensity threshold for material removal (I_{th}) can be written as follows [15]:

$$I_{th} = I_o \exp\left(\frac{-2G_w^2}{d_b^2}\right) \tag{3}$$

where I_o, G_w, and d_b are laser-beam intensity, groove width and laser-beam diameter, respectively. This equation can be written as follows:

$$\ln\left(\frac{I_o}{I_{th}}\right) = \left(\frac{2G_w^2}{d_b^2}\right)^{\beta} \tag{4}$$

where β is an empirical coefficient which may be added to the formula due to some effects, such as recoil pressure, plasma-shielding effect, dynamics of vaporisation and multiple reflection of the laser beam in the cut channel. Thus, the groove width (G_w) equation can be written as a general equation as follows:

$$G_w = \sqrt{\frac{d_b^2}{2}\left[\ln\left(\frac{I_o}{I_{th}}\right)\right]^{1/\beta}} \tag{5}$$

As shown in the following equation, the predictive model for groove depth (G_d) can be calculated based on 'Furzikov's study and the b coefficient applied to the logarithmic function of the laser intensity ratio' [15]:

$$G_d = \left[\gamma\frac{\lambda}{\sqrt{n}}\left(\frac{1}{\alpha} + \sqrt{\frac{4kd_b}{v}}\right)\right]^{1/2}\left[\ln\frac{I_o}{I_{th}}\right]^{1/\beta} \tag{6}$$

where γ is another empirical coefficient, λ is the laser wavelength, n is the refractive index of the target material, α is the absorption coefficient of the target material, k is the thermal diffusivity of the target material and v is the laser-traverse speed. The last empirical coefficient (γ) is related to the degree of laser absorption and plasma formation in different environments such as air, water and ethanol, in which their penetration depths to produce a cut are different. In polymethyl methacrylate (PMMA) the approximated values of the refractive index n, absorption coefficient α and the thermal diffusivity k are 1.4827, 500 cm^{-1}, and 1.073×0^{-7} m^2/s. The HAZ formula can be written as [15]:

$$G_{HAZ} = \sqrt{\frac{d_b^2}{8}\left[\ln\left(\frac{I_o}{I_{th}}\right)\right]^{1/\beta}} - \frac{G_w}{2} \tag{7}$$

The groove formed by laser ablation of a Si target material in air and water is quite different. In water, the hole edge is very smooth, but in an ambient environment, the hole has bumps with a height of 3.6 mm. In addition, the hole diameter in water is larger than that in air, and the hole depth in water is several times greater than in air [8]. **Figure 2** shows the laser-ablated groove profile of a Silicon target material ablated in air and in water [8].

The ablation depth increases with increasing laser power, reducing spot size and decreasing scan speed [7]. In addition, the ablation rate significantly decreases with the depth of the hole [16].

During laser ablation, some of the laser energy will be lost in the ablation environment before it reaches the target material. The ratio of loss is higher in water than in air and in air, it is higher than in a vacuum. As shown in **Figure 3**, under the same laser-beam parameters the energy loss in air is higher than that in a vacuum. This is because for laser ablation in a vacuum, laser-induced air breakdown and ionisation do not exist; in addition, the ejected energetic electrons freely diffuse in the vacuum. The absorption of the laser beam which causes laser energy loss in the vacuum may only be caused by hot electron. Furthermore, absorption rate in a vacuum is very low because the collision frequency is low and the electron density is small [17]. In contrast, in air, laser-induced air breakdown does exist; as a result, laser absorption is increased. In water, laser absorption will be increased considerably due to the water level above the sample.

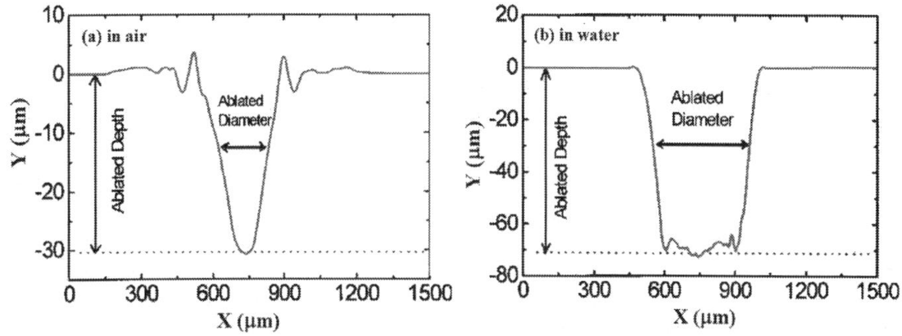

Figure 2. Laser-ablated region profile of a Si target after 1000-pulse irradiation in air (a) and water (b).

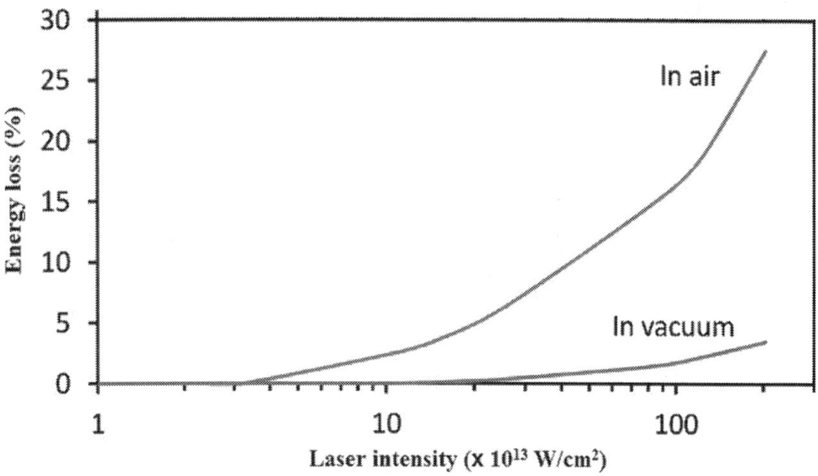

Figure 3. Pulsed laser-beam energy loss during laser ablation of aluminium target materials in air and in a vacuum. Laser wavelength (λ) was 800 nm and laser pulse duration (τ) was 100 fs.

3. Laser ablation in a vacuum

Laser ablation has been used in a vacuum using a vacuum chamber. A vacuum chamber is an empty rigid enclosure free from air and any gases, which are removed using a vacuum pump. As a result, a low-pressure environment is produced in the chamber. Laser ablation in a vacuum has been used to avoid any contamination during laser-material processing.

Laser-material interaction in a vacuum produces plasma at the surface of the target material, which expands considerably in the vacuum chamber [18]. An important factor in pulsed-laser ablation in different environments is the surface temperature of the target material. **Figure 4** shows the temporal variations of the surface temperature (T_w) and equilibrium vapour density (n_s) above the surface of a bulk flat material of pure niobium (Nb) in a vacuum chamber at a base pressure of 10^{-5} Pa. The target material was irradiated using a Nd:YAG laser pulse at the 1064 nm wavelength. It can be noted that the maximum values of both T_w and n_s were reached at 0.3 τ_L (4 ns) at the maximum laser pulse [19]. It is worth mentioning that the temperature profile is not only a function of the laser-beam parameters but is also a function of the type of target material.

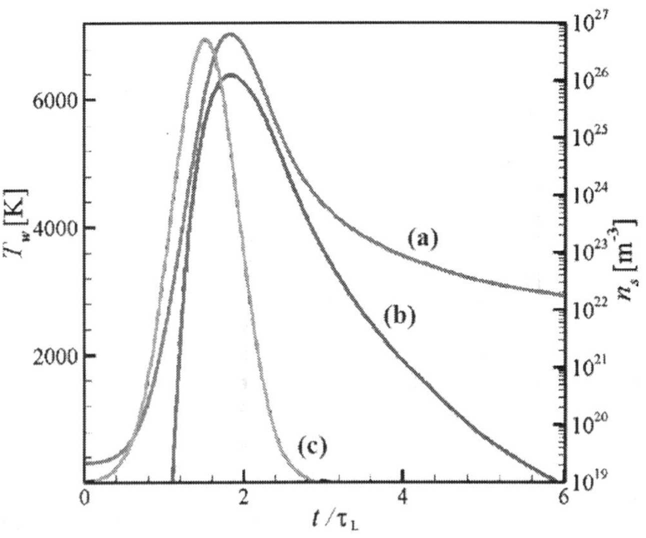

Figure 4. (a) The surface temperature, T_w, (b) equilibrium vapour density, n_s, as a function of time and (c) irradiated pulsed laser.

The temperature profile of laser-material interaction in the vacuum can also be shown as a function of plume density. **Figure 5** shows the temperature-density phase diagram of an aluminium (Al) foil target material irradiated by a femtosecond laser (wavelength: 800 nm and pulse duration: 100 fs) in a vacuum chamber at a backing pressure of 10^{-7} mbar [20]. The

http://dx.doi.org/10.5772/65241

corresponding temperature (T) of the plasma from which the ions originated, which were estimated for the IR and UV laser wavelengths, is about 10^6 and 10^5 K, respectively [18].

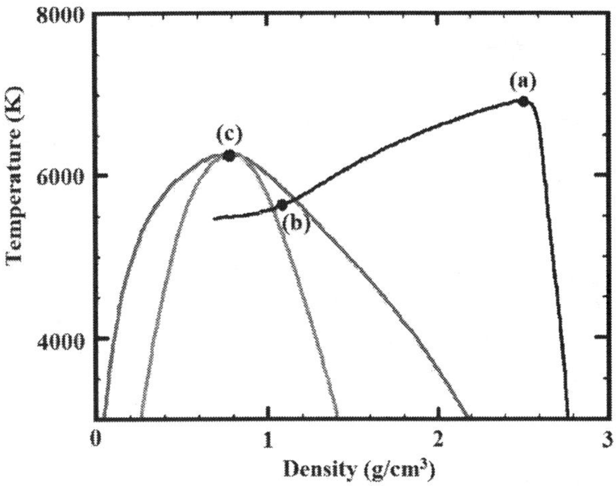

Figure 5. Diagram showing temperature as a function of plume density of Al foil target material irradiated by a femtosecond laser in a vacuum. The solid line represents the thermodynamic path of the Al absorbing volume (at a laser fluence of 400 mJ/cm^2). The dashed line (binodal) is the location of liquid-gas equilibrium states and the dotted line (spinodal) curve defines the limit of the homogenous phase. (a) and (b) are a particular case representing the physical conditions reached by the Al plume about 5 and 7 ps after the 100-fs beings laser pumping, respectively. (c) indicates the critical point.

After laser-material interaction in the vacuum, the debris materials from the expanding plume have different speeds. The temporal profile of the emission intensity for nanoparticles and atoms, measured 5 mm from the surface of the target material at a laser fluence of 0.8 J/cm^2, shows that the light atoms fly faster than the nanoclusters or nanoparticles. It is shown that the average velocities of the nanoparticles and atoms are about 12×10^3 and 1×10^3 m/s, respectively [20]. Time-of-flight (TOF) measurements showed that the average velocity of the ion emission from a Cu material target in a vacuum by infrared (IR) (1064 nm wavelength) and ultraviolet (UV) (308 nm) are different and about 4.7×10^4 and 2.3×10^4 m/s, respectively [18].

The ablation depth (ablation depth per pulse) and emission yield as a function of the laser fluence at low and moderate laser fluence (about 150–500 mJ/cm^2) show logarithmic dependence. On the other hand, at higher laser fluence (>500 mJ/cm^2) the laser ablation sharply increased [20]. The laser ablation threshold in terms of the laser power density at 1064 and 308 nm laser wavelengths are about 7 and 3 J/cm^2, respectively [18].

In the vacuum, it was shown that the nanoparticles are directly generated from the target material by phase explosion. The condensation processes in the gas phase in the first stages of the plume expansion is not a means of producing nanoparticles in the vacuum [20]. It has been shown that during femtosecond laser interaction with tungsten to produce nanoparticles in a vacuum, the atomic plume emission acquired about 400 ns, but the nanoparticle plume acquired about 100 μs after the ablating laser pulse [21].

4. Laser ablation in air

Another environment that has been used in laser-material processing is air. In general, air is the name given to the Earth's atmosphere. In the case of these experiments, it means that laser ablation will be carried out in an environment in which about 78% is Nitrogen (N) and 21% is Oxygen (O). Laser ablation in air plays a substantial role in deep holes but has a low effect on initial surface ablation rates [16].

In general, ablation rate depends on laser fluence, repetition rate and the number of laser pulses [22]. It has been shown that the ablation rate in air at a laser fluence of less than 5.9 J cm^{-2} sharply increases and then slowly increases up to 40.7 J cm^{-2}; after this point, the ablation rate decreases. The ablation rate also drops at a high repetition rate. The drop in ablation rate in both cases is thought to be due to attenuation of the laser energy caused by particle and plasma shielding, produced due to interactions with the remaining laser-generated particles on the ablated crater [22]. The ablation rate at reduced pressure in air shows that the averaged ablation rate increases with decreasing pressure from 1000 to about 250 mbar, regardless of the laser fluence and type of target material [23].

5. Laser ablation in background gases

Laser ablation in different background gases such as He, Ne, Ar, Kr, Xe and N_2 has been used to improve the cutting-edge quality of laser ablation [24]. Heavier background gases produce a slower expansion, or more confinement of the vapour plume [3]. The gas environment has an effect on ablation efficiency. The improvement in laser-machining in different gases has been found to correlate to their potential ionisation [25].

As shown in **Figure 6**, the surface temperature increases due to the laser (photon) influence to a maximum of about 7000 K at 8 ns. This corresponds to the maximum laser irradiance time profile. When the laser pulse is finished, the surface temperature reduces sharply to about 3000 K after 20 ns, after which point, it gradually drops to about 1100 K at 100 ns. It has been shown that the maximum surface temperatures for He, Ne, Ar, Kr and N_2 are 7088, 7062, 7036, 7025 and 7037 K, respectively. It can be concluded that the surface temperature increases slightly with decreasing mass and increasing ionisation potential of the background gas [3].

Concerning the generation of nanoparticles in a background gas, Nichols et al. [26] produced Ag nanoparticles by laser ablation in argon, nitrogen and helium at a variety of gas pressures. It was concluded that by selecting an appropriate gas type and pressure, Ag nanoparticles can be produced and controlled in the range of 4–20 nm. In addition, the smallest Ag nanoparticles (with a mean diameter of 5 nm) were produced in helium gas at 1 atm and below, and the

http://dx.doi.org/10.5772/65241

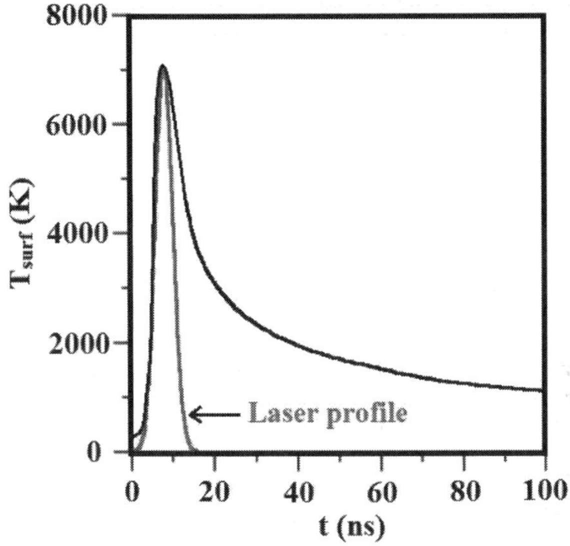

Figure 6. Calculated surface temperature (T_{surf}) and laser irradiance-time profile as a function of time (t) for a laser of 266 nm, with 5 ns, and an irradiance of 1 GW/cm^2.

largest nanoparticles (with a mean diameter of 19 nm) were produced in nitrogen gas. Furthermore, the average size of the nanoparticles increases with increasing molecular weight of the gas. The kinetic energy of the ablated atoms in a gas background and the width of their angular distribution has also been found to decrease with increasing ambient pressure [27].

6. Laser ablation in liquid environments

The use of pulsed-laser ablation at the solid–liquid interface was first reported by Patil and co-workers in 1987 to produce a metastable form of iron oxide from a pure iron target material [28]. Laser ablation in liquids has been used to produce nanoparticles as an alternative to chemicals because ablation in liquid is considered a cleaner environment in which to produce nanoparticles. Different liquids have different effects on the production of nanoparticles. For example, laser ablation of a Tin (Sn) target in water produces polycrystalline tin dioxide (SnO_2) nanoparticles, while ablation in ethanol produces single crystals of tin coated with tin hydroxide ($Sn(OH)_2$) nanoparticles [29].

The generation of nanoparticles in different solutions, particularly in pure water or deionised water, has received much attention from researchers in the field of nanoparticle generation because pure water is a suitable environment for the synthesis of nanoparticles and is free from any contamination.

Many number of research studies have been published on the production of several types of nanoparticle using laser ablation in deionised water [30–36], sodium dodecyl sulfate (SDS) [35, 37–40], acetone [41, 42], ethanol and ethylene chloride [33, 43, 44], polyvinylpyrrolidone (PVP) [45, 46] and liquid nitrogen solutions (LN) [47]. Both pulsed-laser beams [31, 33, 36–39, 47] and continuous-wave (CW) laser beams [30, 48–56] were used to produce nanoparticles.

It has been shown that the ablation rate of Si varies with the water level above the target material. It has been concluded that the laser-ablation rate can be considerably enhanced by using a water level of 1.1 mm [57].

Laser ablation of solids in liquids is an effective technique with considerable potential in the generation of nanocrystals, which allows multilateral design through choosing appropriate solid target materials and confining liquids [58]. The response of different liquid solutions in the generation of nanoparticles varies considerably. Considering the generation of ZnO nanoparticles by laser ablation, in a cetrimonium bromide (CTAB) solution, the highest ablation rate and highest crystallinity of nanoparticles were observed. The largest nanoparticles were produced in acetone. In addition, in a CTAB solution, the morphology of ZnO nanoparticles changed. However, the number of Zn@ZnO core-shell nanoparticles was found to be larger in CTAB and SDS. In general, cationic surfactants have been found to considerably change the shape of ZnO nanoparticles. The morphology of ZnO nanoparticles in SDS, acetone and water was found to be spherical, while in CTAB was spindle-like [59].

7. Comparison of laser ablation and nanoparticle production in different environments

Conducting laser ablation in different environments leads to different ablation mechanisms, ablation rate, ablation profile and efficiency. In addition, laser-generation of nanoparticles in various media produces different nanoparticles in terms of their morphology, size and size distribution. In the following sections, the compassion of laser ablation and nanoparticle production in different environments in terms of ablation mechanism, ablation rate, ablation efficiency and their differences and advantages will be presented.

7.1. Ablation mechanisms in different environments

Laser ablation is an important process in laser-material processing to obtain precision machining or to produce controllable nanoparticles in terms of size, size distribution and morphology. The laser-ablation mechanism various in different environments; for example, in the case of laser ablation in a liquid environment, when a target material is irradiated in a liquid environment by a pulsed laser, the laser energy will be absorbed by the target, which heats it, causing the target to reach melting temperature and evaporation. As a result of this, the material will be ablated from the surface of the target into the liquid solution, and finally nanoparticles are produced as a result of condensation and nucleation. The water level above the sample has different effects on the laser ablation: the main effect is decreasing the laser

http://dx.doi.org/10.5772/65241

power for some laser wavelengths, such as at 1064 nm. Furthermore, the focal length will change as it passes through the liquid medium.

In femtosecond-laser ablation of an aluminium target material, it has been shown that the ablation (material ejection) mechanism and shock-wave evolution are different in water and in air. In the case of a liquid environment, the ejected materials are confined to a small area near the surface of the target material. Furthermore, the liquid media significantly decreases the velocity of the shock waves close to the velocity of sound within nanoseconds, decelerating slowly afterwards. However, in the case of air, the ejected material can propagate away from the target material with a low resistance within the shock waves [60]. Different liquid environments such as deionised water and methanol produce different sizes of nanoparticles in laser generation of nanoparticles [61]. Small Au nanoparticles were produced by pulsed-laser ablation in sodium dodecylbenzene sulfonate (SDBS) in comparison with those produced in sodium dodecyl sulfate (SDS) surfactant due to SDS having smaller molecules than SDBS [62].

There is a considerable difference in the ablation mechanism and thermal damage of laser ablation in air and in water. No thermal damage has been observed in a water environment, whereas considerable thermal damage has been documented in air. Thermal damage has been found to increase with an increase in the number of laser pulses. In addition, in the case of machining in air, the ablated surface is smooth, while in the case of underwater machining, the surface is rough. In the later case, the surface shows molten materials rapidly solidified with a dendritic structure. Consequently, the ablation depth is more affected by laser fluence in the case of underwater machining than in ambient air. In contrast, at low laser fluence, the ablation depth is significantly higher in ambient air than in water [63]. All these differences are due to the use of different ablation mechanisms in different environments.

In a vacuum, the time taken for energy to be transferred from the bulk target material to the surface, which produces the high-energy tail, exceeds other characteristic timescales such as surface cooling time and the electron-ion temperature equilibration time. In air, the interaction between the gas atoms and the surface significantly reduces the lifetime of this non-equilibrium surface state, which allows thermal evaporation to take place before the surface cools. The ablation threshold in air is lower by half than that in a vacuum. This is because in air, after the laser pulse, thermal evaporation is responsible for the lower ablation threshold, while in a vacuum, the threshold corresponds to non-equilibrium ablation during the laser pulse [64]. The plasma produced in air was found to be much hotter and denser than that produced in a vacuum [65]. It was shown that the breakthrough times of laser ablation for drilling in a vacuum are slightly lower for stainless steel and much lower for aluminium than in air. In addition, shorter laser pulses in the range of 150 fs to 20 ps drill faster than 500 ps pulses [16].

It has also been shown that the dynamic process of femtosecond laser ablation of silicon (Si) in a vacuum is considerably different from that in air. In the case of the laser ablation in air at under 1×10^5 Pa, all the expected processes, such as the shock-wave front, the nearly concentric semicircular stripes and the contact front were observed, while in the case of laser ablation in a vacuum, these phenomena were not observed. In addition, another significant factor in the dynamic process of femtosecond laser ablation is the ambient air around the target, which acts as an effective heating dissipation factor and a stagnation force [66].

7.2. Ablation rate in different environments

The rate at which materials are ablated is also different in different environments. This may be due to different ablation mechanisms and ablation thresholds. Ben-Yakar and Byer [67] showed that in the case of single-shot laser ablation, the ablation threshold in air is lower than that in a vacuum. The reason for this may be due to the modified absorption process because air may accelerate the ablation chemistry; due to air resistance, the hot plasma expands at a slower speed in comparison with that in a vacuum. On the other hand, in the case of multiple laser pulses overlapping, the resulting incubation effect reduces dependence on the processing environment.

In air, ablation rate decreases with ablating deep holes because of the non-linear effect. The non-linear effect of laser ablation in air is due to localising the laser beam. This means that ablation products cannot escape the holes, which causes self-focusing of the laser pulses even at a power less than the critical power for self-focusing in ambient air. Based on these effects, it has been shown that the ablation efficiency in air is slightly higher than that in a vacuum (0.51 and 0.43%, respectively) [67].

The effects of using a buffer gas on the ablation rate of Ti, W, Fe, Cu and Mo target materials at different fluencies have been studied. In general, the effect of He gas at 1000 mbar and air at 1 mbar is quite similar, so in both cases, the averaged ablation rate slightly increased; this effect is more acute with increasing laser fluence. In the case of Ar gas, the average ablation rate decreased because the absorption of the laser beam by the plasma is more considerable in comparison with this in He or in ambient air. This effect was found to be independent of the laser fluence. As a result, 'more absorption of the laser energy reaching the sample necessarily implies that less sample is vaporized' [23].

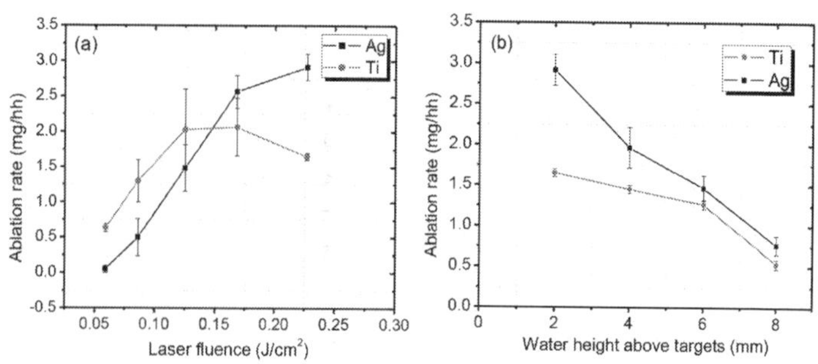

Figure 7. (a) The ablation rate of Ag and Ti target materials as a function of laser fluence at a fixed water level (2 mm), and (b) as a function of water level at a fixed laser fluence (0.22 J/cm²).

Figure 7a shows the ablation rate of Ag and Ti target materials as a function of laser fluence in deionised water at a fixed water level above the targets. A picosecond laser at $\lambda = 1064$ nm, $f = 200$ kHz, $v = 30$ mm/s, $t = 1/2$ h and a laser spot size = 125 μm was used to ablate the materials in deionised water. At lower laser fluencies, the ablation rates of both samples increased slightly, but at higher laser fluencies the ablation rates decreased. It can be seen that the ablation rate of the Ti target decreases significantly when the laser fluence increases above 0.15 J/cm^2. This is because of the reduction in laser pulse energy reaching the target material, which occurs after increasing laser-beam absorption and scattering due to higher plasma/plume density [4]. The ablation rate of the nanoparticles in liquid environments is also a function of the scan speed of the laser beam. This decreased with an increase in scan speed because at the beginning, higher scan speeds produce a high amount of nanoparticles, leading more nanoparticles to disperse in the solution, which rapidly prevent the laser pulse energy from reaching the target material. In other words, at high scan speeds, there is not enough time for nanoparticles to disperse in the solution, but at low scan speeds, there is enough time for nanoparticles to disperse, allowing the laser energy to reach the target almost completely. It has been shown that nanoparticles produced in deionised water would travel towards the laser beam which causes the laser beam to scatter and prevent it from reaching the target material [68].

In the case of laser ablation in deionised water, the ablation rate of materials (Ag and Ti targets) decreased as the water level above the samples increased at fixed laser fluence. Broadly speaking, the ablation rate decreased with increasing water level and the rate of decrease of Ag was found to be faster than that of Ti. It was also shown that the amount of Ag materials or nanoparticles ejected was greater than the amount of Ti nanoparticles (see **Figure 7b**) [4].

Recently, Hamad et al. [69] produced nanoparticles in ice water by picosecond laser ablation; it was shown that the ablation rate in ice water is lower than that produced in deionised water because of the higher attenuation coefficient and absorptivity of the ice water (frozen deionised water) to the laser-beam wavelength in comparison with unfrozen deionised water. The ratio of the ablation rate in ice water and deionised water was about 1:3.

7.3. Ablation efficiency in different environments

It was expected that the ablation efficiency would be lower in a vacuum than that in air because in a vacuum less work is required for the hydrodynamic expansion of plasma in a non-resistant environment. However, a larger groove diameter can be obtained in air because the single-shot ablation threshold is lower in air. In addition, the ablation volume is slightly higher in air than in a vacuum because the ablation depth is determined by the optical penetration depth, which is independent of the processing environment [67]. The non-linear dependence of the ablation rate or productivity on the laser fluence in a vacuum with femtosecond-laser ablation of nickel was also observed [70].

It has been reported that the laser-ablation efficiency of nanosecond-laser pulses is similar in both a liquid environment and ambient air, but in the case of laser ablation by femtosecond-laser pulses, the ablation efficiency was higher in air in comparison with that in water [32]. The mass rate of nanoparticle generation in air (aerosol) was observed to be 100 times higher than in water (colloid) [71]. The laser-ablation efficiency was also influenced by the laser-spot size and the irradiation time [72]. It was also reported that the laser-ablation efficiency in the vacuum decreased with increasing laser fluence, caused by increasing the duration of the atomisation stage [73, 74]. Although laser ablation of materials depends upon their thermal properties, laser ablation is more effective in a liquid environment that in air. In addition, in air, the microparticles were notably distinct both near and far from the crater on the brass target. However, in water, the narrow distribution of nanoparticles away from the crater was also sound. It was concluded that water is a more suitable environment in which to produce uniform and large quantities of nanoparticles by laser ablation [11]. Hermann et al. [74] determined that the ablation depth of copper and gold in a vacuum chamber increased gradually at low laser fluence up to 0.5 J/cm^2, but that this figure increased considerably at high laser fluence.

The laser-ablation process is also used to clean the surfaces of materials. The effectiveness of laser-ablation cleaning methods on silver artefacts in different media (air, water and a vacuum) was characterised. Before cleaning the samples, they were treated with HCl (37%) solutions for several hours to simulate the formation of a chloride patina (AgCl) on marine archeological silver artefacts. It was shown that the patina was removed from a few to 300 μm in air and in water and removed completely under vacuum conditions. However, the formation of a white patina in air and water and an increasing amount of oxygen on the surface of the samples were observed during cleaning process. This situation shows that 'in ambient reach in oxygen laser ablation favourite the oxygen absorbance on surface and the formation of a thin layer of AgO which dulls the surface' [75]. This hypothesis is proven by unobserving surface tarnishing in a vacuum during laser ablation. It can be concluded that the optimal conditions in which to clean silver artefacts is under vacuum conditions [75].

7.4. Differences and advantages of different environments to laser ablation and generation of nanoparticles

Many important differences have been observed between laser ablation in a vacuum, air, gas and liquid environments. These differences can be summarised as follows: first, in water, at the liquid-solid interface, a plasma plume is created. In a liquid solution, the ablated materials and plasma plume expansion are confined, which leads to the production of a high temperature and pressure caused by the mechanical effects. Second, chemical reactions are produced at the interfaces between the laser-induced plasma and the liquid and in the laser-induced plasma at high temperature and pressure. Third, in water, the quenching time of the plasma plume is much shorter than in gas and a vacuum [57].

The advantages of using laser ablation in liquid environments to produce nanomaterials over other methods such as chemicals are as follows: (i) simple and clean production because the process does not need a catalyst and no byproducts are formed; (ii) the process does not require extreme temperature and pressure; (iii) the method leads to the production of a new phase of nanocrystals which may occur in both liquids and solids, which presents more options to choose and combine interesting and desired solid targets and liquids to produce nanocrystals and nanostructures of new compounds or bimodal [57]. Another advantage of using liquid environments in the production of nanoparticles is the ability to use the liquid as a medium in which to collect the nanoparticles. In addition, it has also been shown that the application advantages of nanoparticles produced in liquid environments in comparison with conventionally generated nanoparticles are the following: (i) the number of bimolecules which can be conjugated on the surface of laser-produced nanoparticles in water is three to five times higher than their chemically produced counterparts; (ii) pulsed-laser ablation in a polymer solution allows doping or embedding of nanoparticles into matrices of the polymer for quick nanomaterial prototyping; (iii) production of nanoparticles in liquid environments via pulsed-laser ablation generates pure and aggregated nanoparticles. Using aggregated nanoparticles as a substrate in surface-enhanced Raman scattering (SERS) supplies higher signal-to-noise ratio [76].

The advantages of performing laser ablation in liquid environments over vacuum chambers are simple experimental set-up; low price of liquids and higher availability; easy transportation and storage; no chemical precursors; no chemical contamination resulting in 100% pure nanoparticles. In other words, more nanoparticle sites or surface atoms are available for drug loading, fictionalisations and environmental sensing. Furthermore, laser ablation in liquids produces charged nanoparticles which form high-stability colloidal nanoparticles. In addition, nanoparticles produced in liquid solution are not inhalable, unlike the dry nanopowders generated in gas-phase pulsed-laser ablation [76].

Yang [57] showed that laser ablation of solids in confined liquid environments is more effective than in a vacuum and diluted gas. This is because in liquids, the solid target material continuously etches at the plasma-solid interface due to the high temperature, high pressure and high density of plasma, as a result, promoting the total ablation rate.

8. Conclusions

Laser ablation in laser-material processing has been carried out in different environments such as a vacuum, ambient air, different liquid environments and different background gases. Each medium presents different responses to laser ablation; in other words, different ablation mechanisms are observed in each environment. Although, generation of nanoparticles via pulsed-laser ablation of a solid target material in a vacuum chamber, air and gas have been widely developed, but the advancement of techniques performed in liquid environments has been of the greatest interest to researchers. The optimal medium for generation of nanoparticles is deionised water.

Author details

Abubaker Hassan Hamad[1*], Khawla Salah Khashan[2] and Aseel Abdulkreem Hadi[2]

*Address all correspondence to: abubaker.hamad75@yahoo.co.uk

1 General Science Department, Faculty of Education, Soran University, Soran City, Erbil, Kurdistan Region, Iraq

2 Division of Laser physics, Department of Applied Science, University of Technology, Baghdad, Iraq

References

[1] Siraj, K., Y. Sohail and A. Tabassum, Turkish Journal of Physics, 2011. 35(2): p. 179–183.

[2] Nakamura, D., T. Tanaka, T. Ikebuchi, T. Ueyama, F. Nagasaki, Y. Fujiwara, M. Higashihata, H. Ikenoue and T. Okada. Ultraviolet lasing from spherical ZnO microcrystal produced by laser ablation in air. in 2015 20th Microoptics Conference (MOC). 2015: IEEE.

[3] Bogaerts, A., Z. Chen and Bleiner, D. Journal of Analytical Atomic Spectrometry, 2006. 21(4): p. 384–395.

[4] Hamad, A., L. Li and Liu, Z. Applied Physics A, 2015. 120(4): p. 1247–1260.

[5] Howard, K., S. Lawson and Y. Zhou, Welding Journal New York, 2006. 85(5): p. 101.

[6] Santo, L., International Journal of Surface Science and Engineering, 2008. 2(5): p. 327–336.

[7] Ullmann, F., U. Loeschner, L. Hartwig, D. Szczepanski, J. Schille, S. Gronau, T. Knebel, J. Drechsel, R. Ebert and H. Exner. Highspeed laser ablation cutting of metal. in SPIE LASE. 2013: International Society for Optics and Photonics.

[8] Zhou, X., K. Imasaki, H. Furukawa, H. Umino, K. Sakagishi, S. Nakai and Yamanaka, C. Optics & Laser Technology, 2001. 33(3): p. 189–194.

[9] Zhu, S., Y. Lu, M. Hong and Chen, X. Journal of Applied Physics, 2001. 89(4): p. 2400–2403.

[10] Besner, S., J.-Y. Degorce, A.V. Kabashin and M. Meunier. Surface modifications during femtosecond laser ablation in vacuum, air, and water. in Photonics North. 2004: International Society for Optics and Photonics.

[11] Patel, D., R.P. Singh and R.K. Thareja, Applied Surface Science, 2014. 288: p. 550–557.

[12] Lindley, R., R. Gilgenbach, C. Ching, J. Lash and Doll, G. Journal of Applied Physics, 1994. 76(9): p. 5457–5472.

[13] Iqbal, M.H., S. Bashir, M.S. Rafique, A. Dawood, M. Akram, K. Mahmood, A. Hayat, R. Ahmad, T. Hussain and A. Mahmood, Applied Surface Science, 2015. 344: p. 146–158.

[14] Jinghua Han and Yaguo Li (2011). Interaction Between Pulsed Laser and Materials, Lasers - Applications in Science and Industry, Dr Krzysztof Jakubczak (Ed.), ISBN: 978-953-307-755-0, InTech, Available from: http://www.intechopen.com/books/lasers-applications-in-science-and-industry/interaction-between-pulsedlaser- and-materials.

[15] Tangwarodomnukun, V. and Chen, H.-Y. Materials and Manufacturing Processes, 2015. 30(5): p. 685–691.

[16] Wynne, A. and Stuart, B. Applied Physics A, 2003. 76(3): p. 373–378.

[17] Zhao, X. and Y.C. Shin, Applied Surface Science, 2013. 283: p. 94–99.

[18] Torrisi, L., S. Gammino, L. Andò, V. Nassisi, D. Doria and Pedone, A. Applied Surface Science, 2003. 210(3): p. 262–273.

[19] Bykov, N.Y., N. Bulgakova, A. Bulgakov and G. Loukianov, Applied Physics A, 2004. 79(4–6): p. 1097–1100.

[20] Amoruso, S., R. Bruzzese, M. Vitiello, N. Nedialkov and P. Atanasov, Journal of Applied Physics, 2005. 98(4): p. 044907.

[21] Harilal, S., N. Farid, A. Hassanein and V. Kozhevin, Journal of Applied Physics, 2013. 114(20): p. 203302.

[22] Shaheen, M., J. Gagnon and B. Fryer, Laser Physics Letters, 2015. 12(6): p. 066103.

[23] Vadillo, J., J. Fernandez Romero, C. Rodriguez and J. Laserna, Surface and Interface Analysis, 1999. 27(11): p. 1009–1015.

[24] Guo, H. and Q. Lou, Thin Solid Films, 1992. 218(1): p. 274–276.

[25] Chan, W., A.P. Leung, X. Mao and R. Russo, Applied Surface Science, 1998. 127: p. 269–273.

[26] Nichols, W.T., G. Malyavanatham, D.E. Henneke, J.R. Brock, M.F. Becker, J.W. Keto and H.D. Glicksman. Journal of Nanoparticle Research, 2000. 2(2): p. 141–145.

[27] Gusarov, A., A. Gnedovets and I. Smurov, Applied Surface Science, 2000. 154: p. 66–72.

[28] Patil, P., D. Phase, S. Kulkarni, S. Ghaisas, S. Kulkarni, S. Kanetkar, S. Ogale and V. Bhide, Physical Review Letters, 1987. 58(3): p. 238.

[29] Musaev, O., M. Driver, E. Sutter, A. Caruso, J. Wrobel and M. Kruger, Applied Physics A, 2013. 113(2): p. 355–359.

[30] Tsuji, T., Y. Okazaki, T. Higuchi and M. Tsuji, Journal of Photochemistry and Photobiology A: Chemistry, 2006. 183(3): p. 297–303.

[31] Simakin, A.V., V.V. Voronov, G.A. Shafeev, R. Brayner and F. Bozon-Verduraz, Chemical Physics Letters, 2001. 348(3–4): p. 182–186.

[32] Tsuji, T., T. Kakita and M. Tsuji, Applied Surface Science, 2003. 206(1–4): p. 314–320.

[33] Dolgaev, S.I., A.V. Simakin, V.V. Voronov, G.A. Shafeev and F. Bozon-Verduraz, Applied Surface Science, 2002. 186(1–4): p. 546–551.

[34] Kazakevich, P.V., A.V. Simakin, V.V. Voronov and G.A. Shafeev, Applied Surface Science, 2006. 252(13): p. 4373–4380.

[35] Liang, C.H., Y. Shimizu, T. Sasaki and N. Koshizaki, Applied Physics A, 2005. 80(4): p. 819–822.

[36] Tsuji, T., K. Iryo, Y. Nishimura and M. Tsuji, Journal of Photochemistry and Photobiology A: Chemistry, 2001. 145(3): p. 201–207.

[37] Mafuné, F., J.-y. Kohno, Y. Takeda, T. Kondow and H. Sawabe, The Journal of Physical Chemistry B, 2000. 104(35): p. 8333–8337.

[38] Mafuné, F., J.-y. Kohno, Y. Takeda, T. Kondow and H. Sawabe, The Journal of Physical Chemistry B, 2000. 104(39): p. 9111–9117.

[39] Chen, Y.-H. and C.-S. Yeh, Colloids and Surfaces A: Physicochemical and Engineering Aspects, 2002. 197(1–3): p. 133–139.

[40] Sasaki, T., C. Liang, W.T. Nichols, Y. Shimizu and N. Koshizaki, Applied Physics A, 2004. 79(4–6): p. 1489–1492.

[41] Tarasenko, N.V., A.V. Butsen and E.A. Nevar, Applied Surface Science, 2005. 247(1–4): p. 418–422.

[42] Menéndez-Manjón, A., J. Jakobi, K. Schwabe, J.K. Krauss and S. Barcikowski, JLMN-Journal of Laser Micro/Nanoengineering, 2009. 4(2): p. 95–99.

[43] Ganeev, R.A., M. Baba, A.I. Ryasnyansky, M. Suzuki and H. Kuroda, Optics Communications, 2004. 240(4–6): p. 437–448.

[44] Ohshima, T., S. Nakashima, T. Ueda, H. Kawasaki, Y. Suda and K. Ebihara, Thin Solid Films, 2006. 506–507(0): p. 106–110.

[45] Tsuji, T., D.H. Thang, Y. Okazaki, M. Nakanishi, Y. Tsuboi and M. Tsuji, Applied Surface Science, 2008. 254(16): p. 5224–5230.

[46] Tsuji, T., T. Mizuki, S. Ozono and M. Tsuji, Journal of Photochemistry and Photobiology A: Chemistry, 2009. 206(2–3): p. 134–139.

[47] Sasaki, K. and N. Takada, Pure and Applied Chemistry, 2010. 82(6): p. 1317.

[48] Liu, Z., Y. Yuan, S. Khan, A. Abdolvand, D. Whitehead, M. Schmidt and L. Li, Journal of Micromechanics and Microengineering, 2009. 19(5): p. 054008.

[49] Phuoc, T.X., Y. Soong and M.K. Chyu, Optics and Lasers in Engineering, 2007. 45(12): p. 1099–1106.

[50] Ramesh, P., S. Muthukkumarasamy, K. Dhanabalan, T. Sadhasivam and K. Gurunathan, Digest Journal of Nanomaterials and Biostructures, 2012. 7(4): p. 8.

[51] Zhang, Q., C. Sun, Y. Zhao, S. Zhou, X. Hu and P. Chen, Environmental Science & Technology, 2010. 44(21): p. 8270–8275.

[52] Cozzoli, P.D., R. Comparelli, E. Fanizza, M.L. Curri, A. Agostiano and D. Laub, Journal of the American Chemical Society, 2004. 126(12): p. 3868–3879.

[53] Yu, B., K.M. Leung, Q. Guo, W.M. Lau and J. Yang, Nanotechnology, 2011. 22(11): p. 115603.

[54] Nikov, R.G., A.S. Nikolov, N.N. Nedyalkov, I.G. Dimitrov, P.A. Atanasov and M.T. Alexandrov, Applied Surface Science, 2012. 258(23): p. 9318–9322.

[55] Liu, C., M. Hong, Y. Zhou, G. Chen, M. Saw and A. Hor, Physica Scripta, 2007. 2007(T129): p. 326.

[56] Khan, S., Y. Yuan, A. Abdolvand, M. Schmidt, P. Crouse, L. Li, Z. Liu, M. Sharp and K.G. Watkins, Journal of Nanoparticle Research, 2009. 11(6): p. 1421–1427.

[57] Yang, G., Progress in Materials Science, 2007. 52(4): p. 648–698.

[58] Wafaa Mustafa Aly Soliman, Fundamental Studies on the Synthesis Dynamics of Nanoparticles by Laser Ablation in Pressurized Water, a dissertation submitted in partial fulfillment of the requirements for the degree of doctor of engineering, department of electrical engineering and computer science, Nagoya Univeristy, Nagoya, Japan, March 2011.

[59] Moradi, M., E. Solati, S. Darvishi and D. Dorranian, Journal of Cluster Science, 2016. 27(1): p. 127–138.

[60] Hu, H., T. Liu and H. Zhai, Optics Express, 2015. 23(2): p. 628–635.

[61] Farahani, S.V., A. Mahmoodi and M. Goranneviss, International Nano Letters, 2016. 6: p. 5.

[62] Essaidi, A., B. Schöps, A. Aumman, S. Xiao, C. Esen and A. Ostendorf, Journal of Laser Micro Nanoengineering, 2013. 8(2): p. 131.

[63] Choo, K., Y. Ogawa, G. Kanbargi, V. Otra, L. Raff and R. Komanduri, Materials Science and Engineering: A, 2004. 372(1): p. 145–162.

[64] Gamaly, E.G., N. Madsen, M. Duering, A.V. Rode, V.Z. Kolev and B. Luther-Davies, Physical Review B, 2005. 71(17): pp. 174405.

[65] Liu, Y.H., X.D. Liu, M. Chen and M.W. Zhao. Laser ablation of Ti-Al alloy in vacuum and air environments. in Applied Mechanics and Materials. 2012: Trans Tech Publ.

[66] Wu, Z., N. Zhang, M. Wang and X. Zhu, Chinese Optics Letters, 2011. 9(9): p. 093201.

[67] Ben-Yakar, A. and R.L. Byer, Journal of Applied Physics, 2004. 96(9): p. 5316–5323.

[68] Hamad, A., L. Li, Z. Liu, X.L. Zhong, G. Burke and T. Wang, Applied Physics A, 2016. 122(4): p. 1–12.

[69] Hamad, A., L. Li, Z. Liu, X.L. Zhong and T. Wang, Applied Physics A, 2015. 119(4): p. 1387–1396.

[70] Amoruso, S., R. Bruzzese, X. Wang, N. Nedialkov and P. Atanasov, Journal of Physics D: Applied Physics, 2007. 40(2): p. 331.

[71] Hahn, A., S. Barcikowski and B.N. Chichkov, Pulse, 2008. 40(45): p. 50.

[72] Pyatenko, A., K. Shimokawa, M. Yamaguchi, O. Nishimura and M. Suzuki, Applied Physics A, 2004. 79(4–6): p. 803–806.

[73] Noël, S., J. Hermann and T. Itina, Applied Surface Science, 2007. 253(15): p. 6310–6315.

[74] Hermann, J., S. Noël, T.E. Itina, E. Axente and M.E. Povarnitsyn, Laser Physics, 2008. 18(4): p. 374–379.

[75] Stranges, F., M. Barberio, P. Barone, V. Pingitore, F. Xu and A. Bonanno, Journal of Earth Science and Engineering, 2012. 2(12): p. 697.

[76] Subhash Chandra Singh, Haibo Zeng, Chunlei Guo and Weiping Cai (2012).Lasers: Fundamentals, Types, and Operations. In, Nanomaterials: Processing and Characterization with Lasers (pp. 1-34), DOI: 10.1002/9783527646821.ch1. Copyright © 2012 Wiley-VCH Verlag & Co. KGaA.

Chapter 9

Laser Ablation in Liquid: An Unconventional, Fast, Clean and Straightforward Technique for Material Preparation

Walter Mendes de Azevedo ,
Sérgio de Lemos Campello , Diego Leite da Cunha ,
Leonardo Tadeu Boaes de Mendonça and
Ohanna Maria Menezes Madeiro da Costa

Additional information is available at the end of the chapter

http://dx.doi.org/10.5772/66245

Abstract

The laser ablation in liquid environment (LALE) technique is a straightforward experimental technique with few controllable parameters, capable to provide extreme pressure and temperature conditions during target ablation without the need for dedicated systems to provide those variables. Additionally, we can state that LALE can be considered a low-cost experimental technique, with few steps and a clean synthesis method, by which a wide variety of materials can be synthesized with high yield. The majority of studies published in the literature using this technique seem to be limited only to the synthesis of metal nanoparticles, metal oxides, nitrates and semiconducting. However, in order to extend the synthesis potential of this technique, in this chapter we are going to demonstrate that with the appropriate choice of reactants, solvent, target materials and the solid-liquid interface interactions we will be able to prepare more complex molecules such as carbonate compound $Pb_3(CO_3)_2(OH)_2$, metal-organic frameworks (MOFs), luminescent metal-organic frameworks (LMOFs), highly dispersed CdS quantum dots and magnetic materials. Also for each material synthesized, we are going to propose a mechanism to explain its preparation using the LALE technique.

Keywords: laser ablation in liquid, nonconventional synthesis, materials preparation, complex compounds, inorganic compound

1. Introduction

Searching the literature we can observe that there is a huge amount of techniques that can be used for material preparation. As a matter of fact, for each group of materials there is a specific route more suitable to be used. An interesting conclusion we can make is the fact that unusual properties obtained in a specific material is sometimes related to the selected preparation method. To be more specific, not only the method used but also the solvent, the concentration, the manner, or the sequence in which the reaction components are mixed make these characteristics more pronounced when we are taking into consideration the preparation of nanomaterials.

On the other hand, among several methods developed in the last decade for material preparation, it can be observed that one stands apart: the pulsed laser ablation technique (PLA). The reason seems to be the high potential presented by the laser-based material processing for the thin solid film preparation, nanocrystal growth, surface cleaning and microelectronic device fabrication [1–5]. The pulsed laser ablation of solid materials was initially carried out in a conventionally evacuated deposition chamber and the majority of the results have focused on pulsed laser deposition (PLD) of solid targets in vacuum and in a diluted gaseous ambient [6, 7].

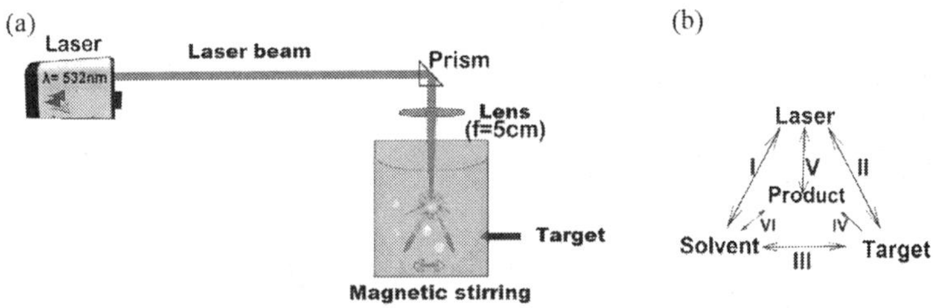

Figure 1. (a) Experimental set up for the LALE experiments (b) interaction diagram.

Moreover, a slight modification of the PLD technique for materials preparation is the laser ablation in liquid environment (LALE). This technique has been extensively used at the beginning for the preparation of colloidal solution of metallic nanoparticles and subsequently as a successful technique for the fabrication of nanostructured metals oxide, nitrate and other materials of various compositions, morphologies and phases [8, 9]. Additionally, this technique nowadays can be considered a chemically clean and one-pot synthetic route by which a variety of functionalized new nanostructures can be prepared with a high yield of the final product without any by-products and, mostly, there is no need for further purification.

Compared with others techniques, LALE can be considered as a low cost experimental technique, which presents few controllable parameters and provides extreme experimental conditions of high temperature and pressure that favors the formation of unusual metastable phases. The technique consists of the interaction of a high-intense Laser pulse with a solid target immersed in a liquid solution as shown in **Figure 1(a)**. In this case, both the ablation of the target material and the interaction of the laser beam with the solvent (**Figure 1b**) occur simultaneously. As a consequence, the plasma plume formed by the target ablation expands and is confined by the liquid. Some works indicated that the density of the ablated species is of the order of 10^{22}–10^{23} cm^3, the temperature could reach 4000–5000 K and the pressure is of the order of 10 GPa inside the plasma plume generated by the ablation in liquid [8]. According to Ogata and coworkers [10], interesting chemical reactions could take place at the interface between the plasma plume and the liquid, allowing the synthesis of a new chemical compound.

Although this technique has proved to be an excellent route for material preparation, this methodology has been used mainly to obtain metal oxide, metals colloids, semiconductor and nitrites [11–17]. Therefore, it will be presented, in detail, the synthesis of five chemical compounds prepared by the LALE technique that had been prepared at the Solid State Laboratory of Federal University of Pernambuco in order to demonstrate the versatility of the laser ablation technique and also to convince the scientific community that this is a very powerful tool for the preparation of complex materials.

2. Experimental

Figure 1 shows the experimental set up used for all compounds preparation. For the target ablation, a second harmonic of a pulsed Nd:YAG laser (Spectra-Physics Quanta-Ray GCR-170 or a Quantel, model Brilliant B) is used, operating at 10 Hz with 8 ns of pulse width, with the beam focused on the target with a spot size of about 1 mm in diameter using a lens with a focal length of 50 mm. The synthesis processes differ from each other only on the Laser fluency used, ablation time, kind or type of target, solvent used and the proposed mechanism that explains its formation.

2.1. Synthesis of $Pb_3(CO_3)_2(OH)_2$

For the synthesis of the carbonate compound $Pb_3(CO_3)_2(OH)_2$, a target of solid Pb metal (99.5%, from America Elements) was used and ablated for 10 minutes with a pulsed Nd:YAG laser as shown in **Figure 1**. Finally, the products were centrifuged and collected. The solutions obtained, including sediments, were dropped and dried on glass substrates for sample characterization. For the purpose of comparing the results, three different laser fluencies on the solid target were used: 0.5, 1.0 and 2.0 J/cm^2 and in order to establish the mechanism for sample preparation, the ablation reaction was performed in methanol (Merck PA), ethanol (Kinetics PA), 1-propanol (Merck PA) and 1-butanol (Merck PA) [17, 18].

2.2. Synthesis of CdS quantum dots

In this case, the synthesis of the nanostructured CdS quantum dots was performed using a metal cadmium foil, which is 0.5 mm thick (99.99%) from Aldrich Company as a target for the Cd ion source production and thiosulfate and thioglycerol from Merck company as a sulfur source. The cadmium metal foil was washed with a mixture of H_2O:HNO_3 1:1 (v/v) to remove the oxide layer, followed by distilled water and acetone washing under ultrasound bath for 5 min.

The Cd foil was immersed in a Beaker (50 ml) filled with aqueous solution of thiosulfate (5, 7, 9, 12 and 15 mmols) and thioglycerol (5 mmols) and ablated for 5 min with the second harmonic order light pulse of a Nd:YAG laser. Soon after the ablation starts, the solution changes its color from transparent to yellow as an indicator that the reaction is taking place. Finally, the products were centrifuged and washed several times with acetone and the material was collected for characterization [19].

2.3. Synthesis of magnetic iron compounds

For the synthesis of magnetic nano– and micro–iron oxide particles, the solid iron targets were ablated immersed in an aqueous solution of basic pH. As a result of the iron ablation, we observed the formation of ions ferrate (VI), α-iron and the unusual form of iron oxides FeO. The aim of this study is to evaluate how the material contained in the plasma plume formed by the laser ablation of an iron target evolves into a medium of high pH and chemically reacts with the solution to form iron oxides. We also examined how the fluency of laser pulses applied to solid iron target surface influences the preparation process.

For this experiment, we used Fe plates that were chemically cleaned, washed with water and then subjected to ultrasonic bath in ethanol for 5 min. The ambient liquid used was a solution with a concentration of 1 molar sodium hydroxide in distilled water and measurements indicated that the pH of the solution was 14. The ALAL was performed with three laser beam fluency values, the materials prepared under these conditions were designated as Type1, Type2 and Type3 and **Table 1** shows the synthesis parameters used for the preparation of these samples. All experiments were conducted at a temperature of 293 K and pressure of 1 atm. To minimize the number of variables in the preparation process, the same values for repetition rate of the laser pulses and the height of the liquid layer on the solid target (10 mm) were used, both for the volume to the liquid environment and for the laser ablation time at which the target has been subjected.

Sample	Solid target	Liquid environment	Fluency (J/cm^2)	Frequency (Hz)	Ablation time (min)
1	Iron (bulk)	20 ml of NaOH 1.0 M	0.5	10	10
2	Iron (bulk)	20 ml of NaOH 1.0 M	2.0	10	10
3	Iron (bulk)	20 ml of NaOH 1.0 M	4.0	10	10

Table 1. Experimental parameters for ALAL of iron plates.

2.4. Synthesis of $Cu_3(BTC)_2(H_2O)_3$

For the synthesis of the $Cu_3(BTC)_2$ MOF, a Cu powder metal (99.5%) was used as a target for the ablation dispersed in a solution of a DMF/H_2O (1:1) containing the organic linker benzene 1,3,5, benzene tricarboxilic acid and $NaNO_3$ as a hydroxyl radicals (•OH) precursor. The target was ablated for 60 min under magnet stirring with the second harmonic of a pulsed Nd:YAG laser. Soon after the ablation starts, the solution changes its color and a blue precipitate occurs. Finally, the products were centrifuged and collected for sample characterization [20].

2.5. Synthesis of $Eu(TMA)(H_2O)_4$

For the synthesis of the microstructured luminescent MOF $Eu(TMA)(H_2O)_4$, instead of using metal or metal powder as a target, europium oxide powder (Eu_2O_3) (99.99%, Sigma Aldrich) was used as a target. In this case, the powder was dispersed in a solution of DMF/H_2O (1:1), containing the organic linker Trimesic acid (TMA) and sodium nitrate ($NaNO_3$) under a magnetic stirring. As it is shown in **Figure 1**, the target was irradiated for 60 min with the second harmonic (λ = 532 nm) of a pulsed Nd:YAG laser. It is interesting to notice that as the ablation starts, the solution changes its optical characteristic, with the initial transparent solution becoming milk-like white as the reaction continues and finally a white precipitate occurs. In order to characterize the materials, the products were centrifuged, washed with hot water several times and dried under vacuum. Sample characterization allows us to conclude that the reaction yield is of the order of 70%. Compared with the literature results for different synthesis method we can conclude that it is a good yield [21–23].

3. Material characterization

Spectroscopic characterization in the UV-visible and infrared region was performed with an Ocean Optics model 2000 spectrophotometer and with a FTIR/FT-NIR Spectrometer model Spectrum 400 Perkin Elmer, respectively, while room temperature photoluminescence (PL) was measured in Ocean optics model QE65000 spectrophotometer with the sample been exited with a N_2 laser or in a Horiba Jobin Yvon spectrofluorometer model Fluorolog-3 FL3-22 using a 450-W xenon short-arc lamp light as the excitation source. The luminescence quantum yields measurements were recorded with a Horiba Jobin Yvon spectrofluorometer using an integrating sphere model F-3018.

The particle morphology and structure was characterized by transmission electron microscope (TEM) and X-ray diffraction measurement, using a Tecnai 20 transmission electron microscope operating with an accelerating voltage of 200 KV and a Rigaku diffractometer model DMAX 2400, using Cu K_ radiation (λ = 0.154 nm) at a scanning speed of 0.1° s^{-1} in the range of 2θ = 10–60° with a step 0.02°, respectively. The sample for the TEM measurements was diluted and dropped onto a carbon-covered 200-mesh copper grid followed by natural evaporation of the solvent.

4. Results and discussion

The first example of a compound synthesized by LALE technique shown in this chapter is the synthesis of hydrocerussite $Pb_3(CO_3)_2(OH)_2$, a quite rare mineral in nature that is found in soil only in small quantities. It consists of basic lead carbonate with a trigonal crystalline structure [24]. Although it is rare in nature, it was extensively synthesized and used not only as a cosmetic or in religious and medical practices [25] from Ancient Egypt to the Greco-Roman period [26] but also as a white pigment in easel painting [27] from the Renaissance to the end of the nineteenth century. More recently, it has been playing a significant role in geology, in environmental issues [28, 29], applied as a polymer stabilizer [30] and has been extensively investigated in the lead acid battery industry because its presence in battery plates increases the resistivity of the electrolyte [31].

The ablation of a Pb target in methanol or ethanol results in a whitish powder precipitation while a grayish color solution was obtained when 1-propanol or 1-butanol was used. The mechanism proposed to explain the whitish precipitated material is demonstrated in the scheme below, in which the ablation of the Pb target in a solution produces Pb atoms, ions and Pb particles that become dispersed in the liquid environment and subsequently interact with absorbed O_2 to form the lead oxide PbO.

The decomposition of methanol into CO_2 and H_2O is catalyzed by the ablated metal that remains in solution [32–34]. The remaining CO_2 reacts with PbO to form the hydrocerussite compound.

$$\mathrm{Pb}_{(\mathrm{solid})} + h\nu + \mathrm{O}_2 \rightarrow n\mathrm{Pb} + \mathrm{Pb}^{+m} + \mathrm{PbO} \tag{1}$$

$$\mathrm{CH_3OH} + \mathrm{O}_2 \xrightarrow{hn\ +\ n\mathrm{Pb}^0} \mathrm{CO}_2 + \mathrm{H}_2 + \mathrm{H_2O} \tag{2}$$

$$3\mathrm{PbO} + \mathrm{CO}_2 + \mathrm{H}_2 \rightarrow \mathrm{Pb}_3\left(\mathrm{CO}_3\right)\left(\mathrm{OH}\right)_2 \tag{3}$$

In order to prove the efficacy of the mechanism proposed in **Figure 2**, we present the X-ray diffraction analysis of the samples prepared in methanol, ethanol 1-propanol and 1-butanol. For the samples prepared in methanol and ethanol the diffraction peaks correspond to the crystalline planes of the trigonal crystalline structure of the hydrocerussite reported on JCPDS card 130131. While for the Pb target ablated in 1-propanol and 1-butanol it is observed that diffraction peaks refer to the hydrocerussite and the crystalline planes of the cubic FCC of metallic phase lead to a structure in agreement with the reported on JCPDS card 040686. It is interesting to notice that the metal became the major phase for the Pb target ablated in 1-butanol.

In conclusion, we may say that if we carefully choose the solvent, we are able to decide which material will be synthesized. Hydrocerussite is obtained when the Pb target is ablated in ethanol and methanol solution whereas hydrocerussite and Pb metal particles are obtained from the solvent 1-propanol and 1-butanol and the metallic phase is more abundant in 1-butanol.

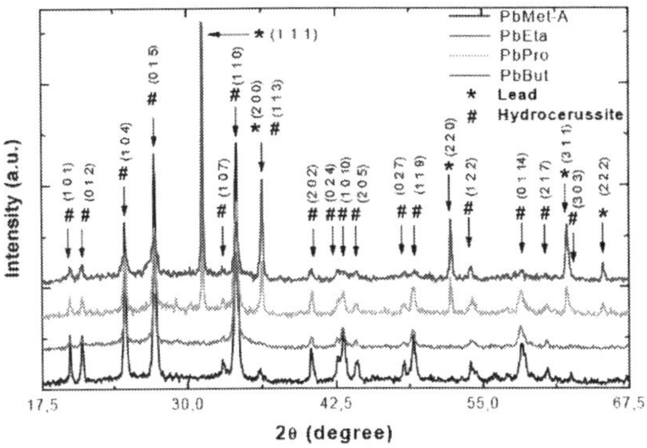

Figure 2. XRD patterns recorded for the samples PbMet-A, PbEtha, PbPro and PbBut. The crystalline phases and main crystalline planes associated with each phase are also shown.

As a matter of fact, several metal nanoparticles such as cobalt, nickel, silver, copper, palladium and semiconductor quantum dots (QDots) [35–39] have been obtained using the reducing power of alcohols with large carbon chains.

Figure 3 shows the SEM characterization for the sample prepared in methanol and it is observed that a hexagonal structure was obtained with average widths of 1.0 ± 0.3 μm and an average thickness of 80 ± 20 nm. This result is similar to the morphology results found in the literature for hydrocerussite [24, 40].

Figure 3. SEM image for PbMet-A sample as made.

The second example of a chemical compound prepared by LALE technique is the synthesis of II–VI semiconductor quantum dots (QDots). The reason for that interest seems to be the high technological potential that these materials present in several areas such as optic, electronic, catalyst application, solar photovoltaic cells, nanobar codes, field effect transistors, light emitting diodes and *in vivo* biomedical detection fluorescent tags in biology and the development of chemical and biological sensors [41–47]. Concerning to the CdS preparation, the synthesis consists of the ablation of cadmium foil in the thiosulfate and thioglycerol solution for 5 minutes. As a consequence, a highly stable colloidal solution is obtained and presents a strong yellow fluorescence peak at 590 nm that can be seen by the naked eye when the sample is excited with UV light at 365 nm as shown in **Figure 4**.

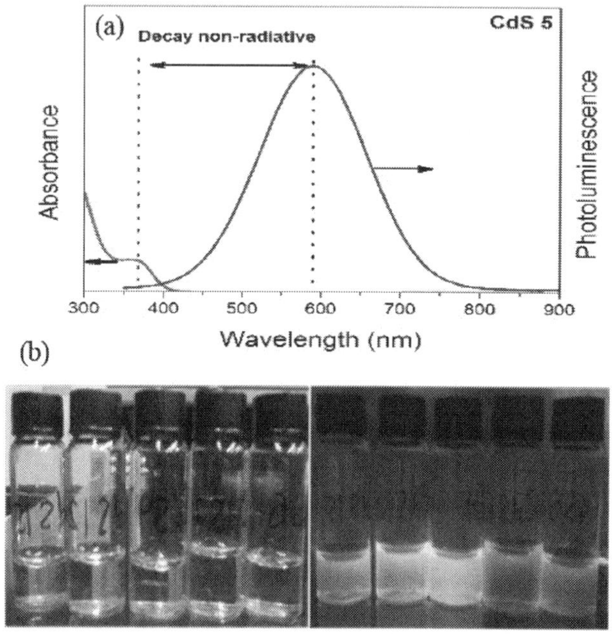

Figure 4. (a) The absorption luminescence spectra and (b) photograph of the CdS quantum dots solutions before and after excitation with UV lamp.

These results allow us to conclude that quantum confinement is observed since the absorption bands of the colloidal solution suffer a blue shift to 360 nm when compared with CdS bulk emission at 515 nm [47].

The mechanism proposed to explain the CdS quantum dots formation is shown in Eqs. (4)–(6), where the Cd target is ablated-producing ionizing species together with the action of thioglycerol as a catalyst for the thiosulfate hydrolysis [48, 49], which results *in situ* formation of sulfide ion (S^{2-}) and consequently the formation of CdS particle.

$$S_2O_3^{-2} + H_2O \xrightarrow{TG} SO_4^{-2} + H_2S \tag{4}$$

$$(m+n)Cd_{(s)} \xrightarrow{hv} nCd_{(s)} + mCd^{+2} + m2e \tag{5}$$

$$Cd^{+2} + S^{-2} \rightarrow CdS \tag{6}$$

Additionally, it can be seen that, in this process, thioglycerol is used as a hydrolyzing agent for thiosulfate (1) and as stabilizing agent for the nanoparticles.

From the analysis of the absorption spectra we found that the band gap of the CdS semiconductor is of the order of 3.1 eV and using the effective mass approximation method developed by Burs [50], we can estimate that the optical particle size is of the order of 3.2 nm. From the analysis of X-ray diffraction measurement of **Figure 5** and the comparison with the JCPDS card file number 10-454, we conclude that the CdS particle crystallize in a zinc blende structure and from the broadening of the diffraction peak we can see that the peaks present a very broad distribution indicating the presence of a very small crystallite size.

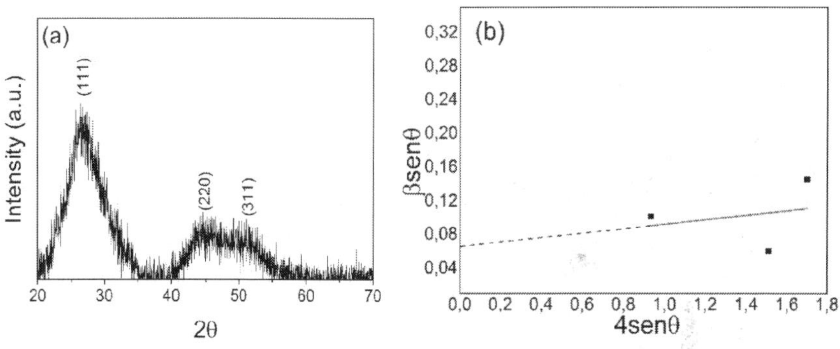

Figure 5. (a) X-ray diffraction of quantum dot CdS and (b) plot Willianson-Hall for crystallite size determination.

Using the Williamson-Hall method for the diffraction peaks analysis, we found that the crystallite size is of the order of 2.2 nm. This result is slightly smaller than those calculated by optics method. However, it indicates that we were successful in preparing very small CdS quantum dots particle. It is interesting to mention that the CdS nanocrystals possess the cubic space group, instead of the hexagonal wurtzite structure which is thermodynamically more stable. This probably could be attributed to the nonconventional chemical reaction condition of high pressures and temperatures which favors the occurrence of chemical processes in nonequilibrium conditions and the formation of materials with metastable configurations [51, 52].

The third example is the synthesis of magnetic iron samples. In this study, all prepared materials were dispersed or stabilized in liquid environment after synthesis. The samples were further characterized in a liquid phase by absorption spectroscopy in the ultraviolet and visible region. The materials of samples Type1, Type2 and Type3 were separated from the NaOH precursor solution by centrifugation and also washed in distilled water. A fraction of the prepared samples was precipitated on glass plates for the scanning electron microscopy characterization, while another fraction of the samples was pulverized and subjected to X-ray diffraction, infrared spectroscopy and magnetic measurement. **Figure 6** shows UV-Vis spectra for the precursor solution of sodium hydroxide and for the resulting solutions of the preparation process for samples 1, 2 and 3.

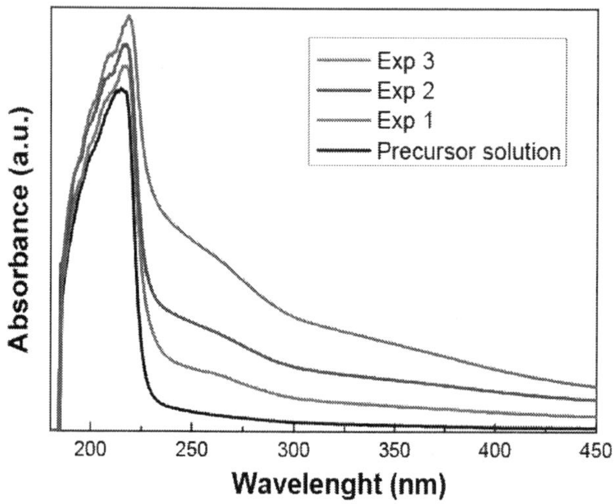

Figure 6. UV-Vis spectra of samples Type1, Type2 and Type3 and of precursor solution.

It is possible to see in **Figure 6** that the spectra of samples 1, 2 and 3 have three additional absorption bands when compared with the spectrum of the precursor solution of NaOH. These absorption bands are related to the presence of the dispersed material prepared in the solution resulting from laser ablation. It is also observed that the absorption bands were more intense for the materials prepared with the application of higher fluencies on the solid target. This fact indicates that the higher the fluency used in experiments, the higher is the amount of material obtained and dispersed in solution per unit time. This result is consistent with the UV-Vis spectrum for iron and iron oxide nanoparticles in aqueous solution. The absorption bands with maxima at 216 and 268 nm may be associated with iron nanoparticles, while the band located at 350 nm is likely due to other phases present in the samples [53].

From the analysis of the results obtained by ultraviolet and visible spectroscopy, FTIR, scanning electron microscopy, X-ray diffraction and magnetometry, it allowed us to conclude that all samples (named as Type1, Type2 and Type3) prepared, under three different experimental conditions, are nanostructured and formed respectively by: Type1 sample crystalline phases regarding the presence iron (Fe), iron oxide (FeO), sodium ferrate (V) ($NaFeO_3$), sodium ferrate (VI) (Na_2FeO_4) and sodium hydroxide (NaOH); Type2 sample, Fe, FeO, Na_2FeO_4 and NaOH; Type3 sample, Fe and FeO. The metastable FeO phase is especially interesting because it is generally prepared only at high temperatures and by rapid cooling process [51].

The prepared samples show magnetic properties, with the Fe and FeO (Type3) having the most intense magnetic response, whereas the Type1 sample shows intermediate intensity response and Type2 material shows the less intense magnetization response. Some modification of the synthesis conditions used in these experiments resulted in the preparation of ferrate sodium compound (V) ($NaFeO_3$) and sodium ferrate (VI) (Na_2FeO_4). In these compounds the iron atoms have a high oxidation number lying the state Fe^{+5} and Fe^{+6}. The FeO_3^{-1} and FeO_4^{-2} ferrate ions present high potential for technological application in several scientific areas, such as in the degradation of environmental pollutants and as disinfectants [54]. Furthermore, recent studies indicate that FeO_4^{-2} ions have promising applications as electrolytes materials in rechargeable batteries [55].

Thus, the laser ablation of iron targets in sodium hydroxide solution presents itself as a new method for the preparation of sodium ferrate, enabling control of the properties of materials prepared by tuning the fluency of laser pulses. To the best of our knowledge, obtaining sodium ferrate by ALAL and obtaining results regarding the control of crystallinity and magnetization of such samples by ALAL are unprecedented in the literature.

The scanning electron microscopy performed on samples Type1, Type2 and Type3 shows that these materials consist of particles of approximate spherical shape and with different size distribution for each type of material. **Figure 7(a)** shows the Type1 material consists of spherical particles with an average size of 170 nm that are both dispersed and isolated on the substrate or agglomerated as structures formed by larger particle clusters. **Figure 7(b)** shows that the Type2 material has particles with spherical shape and average size of the order of 0.7 μm and it seems that the distribution of particle sizes in this sample is wider than that presented by the sample Type1. **Figure 7(c)** shows that the SEM image is composed of agglomerated particles of 50 nm in size, where it is also possible to visualize a few micron size particles.

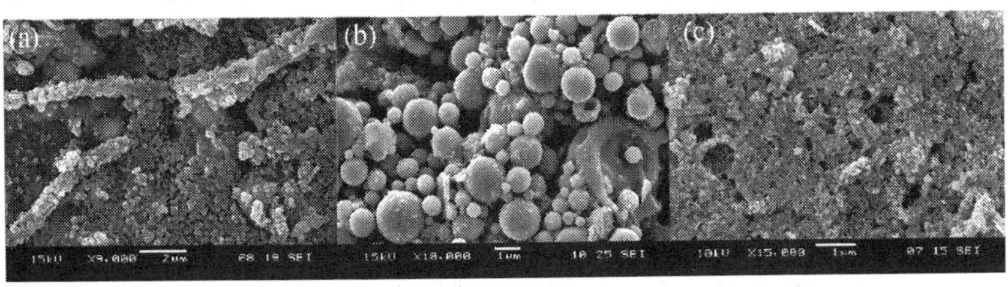

Figure 7. Image of SEM for: (a) Type1 sample; (b) Type2 sample; (c) Type3 sample.

The SEM shown in **Figure 7(a)–(c)** allowed to estimate the average size and average standard deviation for the size distribution of the particles in the samples Type1, Type2 and Type3. The results of these measurements are shown in **Table 2**.

These results indicate that the structural properties of the obtained materials are susceptible to the fluency of the laser beam used in the preparation procedure and therefore it is possible to control the size and size distribution of the sample by controlling the fluency. This behavior may be related to the photolysis promoted by the laser beam onto the prepared particles and increases the ablated material density in the plasma plume. These two effects increase when the fluency is increased and their joint action could explain the observed results: photolysis tends to decrease the size of the particles while high density plume favors to obtain larger particles.

Sample	Average size	Standard deviation
1	170 nm	50 nm
2	0.7 μm	0.3 μm
3	50 nm	30 nm

Table 2. Estimate of the average size for Type1, Type2 and Type3 samples.

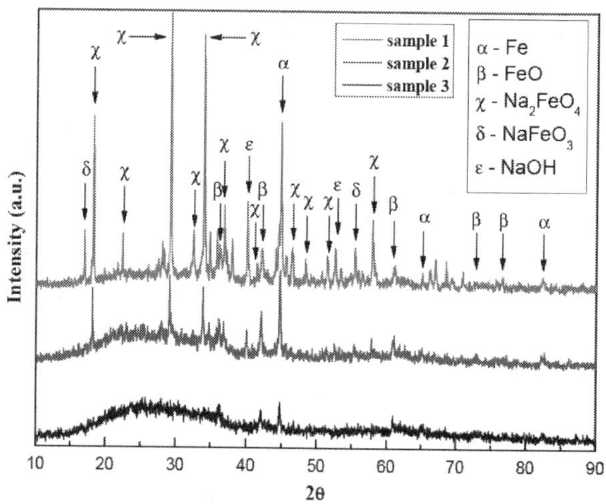

Figure 8. X-ray diffraction patterns of samples.

The X-ray diffraction was used for the identification and structural characterization of the prepared materials. The XRD patterns obtained for Type1, Type2 and Type3 samples are shown in **Figure 8**, where they are also indicating crystalline phases present in the samples that were identified by comparison with the software database PCPDF WIN of JCPDS-ICCD. Analyzing the XRD patterns of the samples and correlating the maximum diffracted intensity with the major peaks of diffraction patterns in JCPDS database is verified that prepared under Type1 experimental conditions is formed by crystalline phases regarding the presence of iron (Fe), iron oxide (FeO), sodium ferrate (V) ($NaFeO_3$), sodium ferrate (VI) (Na_2FeO_4) and sodium hydroxide (NaOH). The Type2 sample shows evidence of the presence of Fe, FeO, Na_2FeO_4,and NaOH, while in Type3 sample, the presence of Fe and FeO phases was only found. It is interesting to notice that the FeO phase is metastable and can be prepared only at high temperatures and by a fast cooling process [51].

Figure 7 shows that when the fluency applied to the target solid surface increases, it is possible to observe progressively reducing on the crystallinity of the prepared samples. Clearly, the Type1 sample with a fluency of 0.5 J/cm^2 has well-defined crystalline phases, while Type3 sample prepared using 4.0 J/cm^2 fluency is majorly amorphous. This behavior can be explained by considering the rapid cooling effect of the liquid environment of the plasma plume. The temperature of the ablated material in the plume plasma should increase proportionally to the fluency of laser pulses; on the other hand, the cooling capacity of the liquid environment basically depends on the temperature and the volume of liquid used. In this way, the higher the fluency used, the higher the cooling rate due to liquid environment, because greater will be the temperature variation during the short time cooling.

The proposed cooling model can explain the crystalline dependence of the material obtained with the fluency of the laser beam used in the preparation process. The fluency values used during the preparation of the samples Type1, Type2 and Type3 correspond to cooling rates T_1, T_2 and T_3 and $T_3 > T_2 > T_1$. Although the specific time (time per mass) available for crystallization of the ablated material is higher in Type1 than in Type2, this in turn has the time available for crystallization higher than the Type3 sample. The short crystallization time for Type2 and Type3 samples inhibits the formation of well-defined crystalline phases, leaving the samples Type2 and Type3 less crystalline and substantially amorphous.

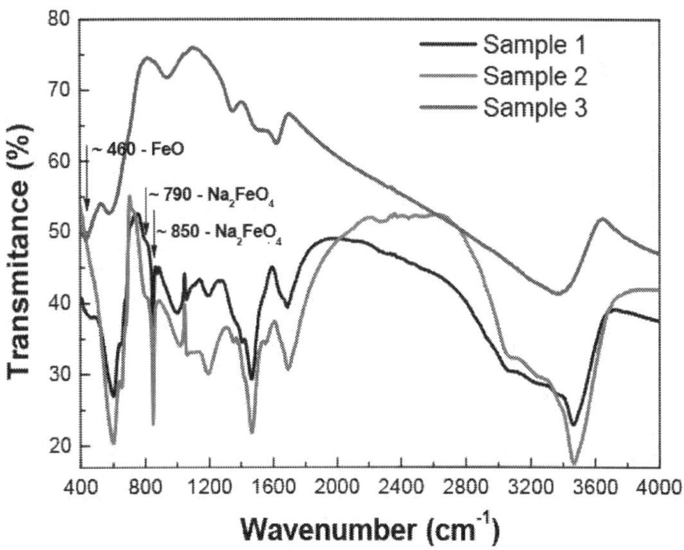

Figure 9. IR transmittance spectrum.

Figure 9 shows the infrared transmission spectra (IR spectra) for the materials prepared. It is observed that most of the absorption modes of Type1 and Type2 materials coincide, although it does not happen for sample Type3. This indicates that the composition and chemical structure of Type1 and Type2 samples are similar, while they differ from the composition and structure presented by the sample Type3. Comparing the IR spectra obtained from data available in [56] literature it was possible to verify the presence of Na_2FeO_4 in Type1 and Type2 samples by identifying active stretching and vibration modes relating to the tetrahedral distribution of four oxygens around the Fe ion (VI). The modes for the Na_2FeO_4 are not displayed in the IR spectrum of Type3 sample confirming the assumption that Type1 and Type2 samples are chemically similar and that the composition of the sample differs from that presented by Type3. The presence of FeO in all samples was attributed comparing it with data from the literature [57]. The IR spectroscopic analysis results corroborate the X-ray diffraction.

Figure 10 shows the magnetization curves measured as a function of applied magnetic field for the samples Type1, Type2 and Type3. The analysis of magnetization curves indicates that all samples exhibit hysteresis, thus being classified as ferromagnetic, although it is noticed that the magnetic response is different for each sample. When the materials are subjected to a field of the same intensity, Type3 material has higher magnetization, this value is higher than the Type1 and Type2 samples and the Type2 sample has the lowest value of magnetization.

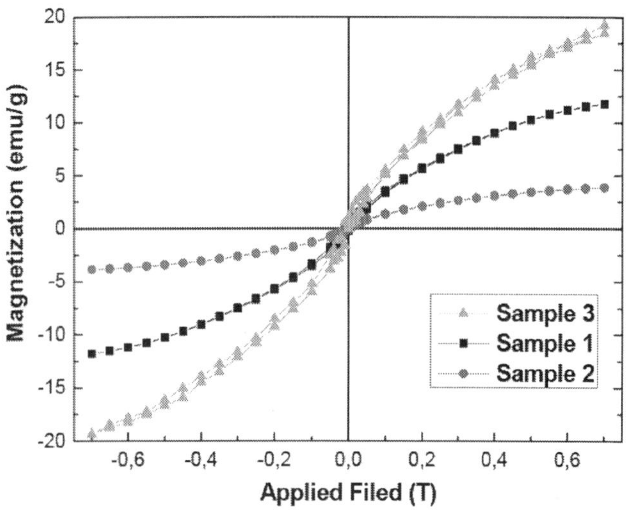

Figure 10. Magnetization per unit mass as a function of applied magnetic field.

The magnetic response of samples can be related to the size of their structures. It was observed in this experiment that for the same value of applied magnetic field, the magnetization of the module M (H) of Type1, Type2 and Type3 samples increases with decreasing diameter, D, of the particles in the sample. In other words, it was found that $M(H)_{Type3} > M(H)_{Type1} > M(H)_{Type2}$

while $D_{Type3} > D_{Type1} > D_{Type2}$. We observed size effects on the magnetic response of the samples, possibly related to the size and presence of magnetic monodomain within the particles of a smaller size, while larger particles should contain several oriented monodomain randomly causing a less effective magnetic response.

We can also qualitatively relate the magnetic behavior of the samples to their chemical composition: The Type1 sample is formed by phases Fe, FeO, $NaFeO_3$, Na_2FeO_4,and NaOH, while Type2 sample is formed by Fe, FeO, Na_2FeO_4,and NaOH and Type3 by Fe and FeO. It is known that unpaired electrons in the atomic orbitals of a material are responsible for their magnetic properties and the greater the number of unpaired electrons will be, the higher its magnetization per unit mass. Iron atoms in Fe and FeO phases are respectively 4 and 4 unpaired electrons. However, the phase $NaFeO_3$ the iron atoms present only 3 unpaired electrons and Na_2FeO_4 phase 2 has unpaired electrons. Since the magnetization was measured by mass, this argument justified the reason that Type3 sample, composed of Fe and FeO, is the most intense magnetic response. The Type1 material its intermediate intensity response and the Type2 material is the less intense magnetization ($M(H)_{Type3} > M(H)_{Type1} > M(H)_{Type2}$).

We can summarize that the preparation of micro- and nanostructured magnetic particles through ALAL iron targets of sodium hydroxide 1 M was successfully obtained. The results demonstrate that structural and magnetic properties of the materials prepared are influenced by the applied laser beam fluency during the ablation process. Specifically it was found that the fluency can be used to control the size distribution of structures present in the sample as well as controlling the magnetization and the crystallinity of the material prepared.

The fourth and fifth examples are the synthesis of the MOFs and LMOFS. For the synthesis of metal organic framework, we have developed two procedures: for the $Cu_3(BTC)_2(H_2O)_3$ preparation, a cupper metal powder was used as a target, while for the synthesis of the luminescent MOF $Eu(TMA)(H_2O)_4$, a metal oxide powder was used as a target instead.

We chose the metal organic frameworks (MOFs) materials as the next example for the fact that this structure are quite complex and consist of a three-dimensional (3D) metal coordination network formed by metal ions or clusters linked together by organic bridging ligands. Consequently, this material presents high surface area, uniform small porous architecture [58], open metal sites [59], possibility of structure modification through postsynthesis functionalization routes [60], that together with luminescent proprieties make MOFs materials promising for several applications such as gas storage, gas separation, heterogeneous catalysis [61, 62], sensing devices [63, 64], ion exchange [65, 66], optoelectronics [67, 68] and efficient phosphors for displays and lighting [69]. Regarding to the sample preparation, almost all methods used for MOFs preparation are time consuming, except for the microwave-assisted processes and therefore new methods should be developed in order to overcome this advantage.

The main difference between the methods applied before for the materials preparation is the fact that for the $Cu_2(BTC)_2(H_2O)_3$ preparation, Cupper metal powder dispersed in liquid medium was used as a target, whereas to synthesize $Eu(BTC)(H_2O)_4$, europium oxide was used as a target. In both syntheses, magnetic stirring was used to homogenize the suspension in a solution of trimesic acid dissolved in water and DMF. In $Cu_2(BTC)_2(H_2O)_3$ case, $NaNO_3$ was also used as oxidizing agent.

To explain the formation of $Cu_2(BTC)_2(H_2O)_{3,}$ the mechanism proposed consists of a combination of two simultaneous processes. The first step involves the ablation of the metal surface, producing the ejection of the plume containing Cu^0 and metal ion Cu^{+1} and Cu^{+2} and also the photochemical decomposition of nitrate ions, that results in hydroxyl radicals (•OH) production, which in turn results in the increase of the amount of Cu^{+2}. Finally, the Cu^{+2} reacts with the BTC to form the $Cu_3(BTC)_2.(H_2O)_2$ framework, as shown in Eq. (7)–(12).

$$Cu^{0}_{(s)} + h\upsilon \xrightarrow{HDMF^{+}\ BTC^{3-}\ NO_3^{-}} Cu^{0}_{(plasma)} + Cu^{1+}_{(plasma)} + Cu^{2+}_{(plasma)} \tag{7}$$

$$NO^{-}_{3(aq)} + h\upsilon \xrightarrow{HDMF^{+}\ BTC^{3-}\ NO_3^{-}} NO_{(g)} + O^{-}_{2(aq)} \tag{8}$$

$$O^{-}_{2(aq)} + 2H_2O_{(l)} \xrightarrow{HDMF^{+}\ BTC^{3-}\ NO_3^{-}} 2OH^{o} + 2H_2O_{(l)} \tag{9}$$

$$2OH^{o} \xrightarrow{DMF\ BTC^{3-}\ NO_3^{-}} H_2O_{(l)} + [O] \tag{10}$$

$$Cu^{0}_{(plasma)} + [O] \xrightarrow{DMF\ BTC^{3-}\ NO_3^{-}} 3Cu^{2+}_{(aq)} + H_2O_{(l)} \tag{11}$$

$$3Cu^{2+}_{(aq)} + 2BTC^{3-} \xrightarrow{DMF\ O_3^{-}} Cu_3(BTC)_2(H_2O)_3 + 4H_2O_{(l)} \tag{12}$$

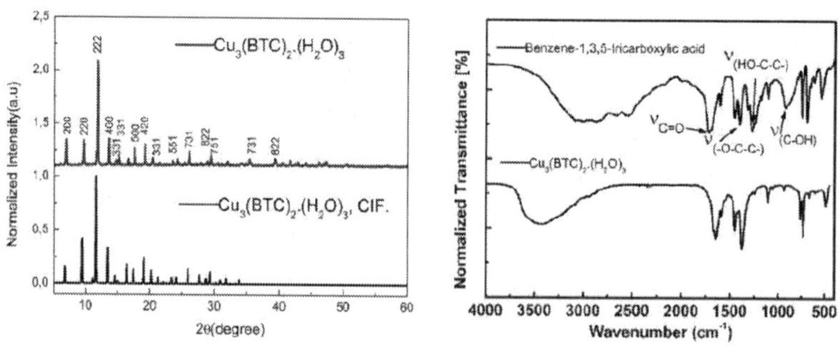

Figure 11. X-ray diffraction patterns (left) and FT–IR analysis (right) of $[Cu_3(BTC)_2]$-MOF prepared by laser ablation techniques at room temperature.

http://dx.doi.org/10.5772/66245

Figures 11 and **12** show the results of X-ray diffraction, infrared and scanning electron microscopy, which confirm that $Cu_2(BTC)_2(H_2O)_3$ MOFs has been successively prepared [20].

For the $Eu(BTC)(H_2O)_6$ preparation, it is noticed that the ablation of europium oxide (Eu_2O_3) make available Eu^{3+} ions in solution, which reacts with BTC molecules to form a luminescent network as shown in **Figure 13. Figure 14** shows the structural and spectroscopic characterization of the obtained compound, thus from the X-ray diffraction results we can conclude that the obtained LMOFs presents highly crystallinity with diffraction peaks isomorphic with $Eu(BTC)(H_2O)_6$. The FTIR analysis shows the disappearance of the carboxyl group in 1721 cm^{-1} and the appearance of the modes in 1404 and 1275 cm^{-1}, strongly suggesting that the Eu^{3+} ion is coordinated with the carbonyl group. Finally, the emission analysis shows that the compound exhibits an intense luminescence in the red region of the visible spectra λ = 611 nm, characteristic $^5D_0$-$^7F_2$ transition of Eu^{3+} ions when irradiated with UV light at λ = 254 nm, with a quantum efficiency of 10.8% and life time of 0.26 ms.

Figure 12. Morphologic analysis of $Cu_3(BTC)_2(H_2O)$ crystals prepared by the laser ablation technique as a function of the ablation time for: (a) 60 min, (b) 90 min.

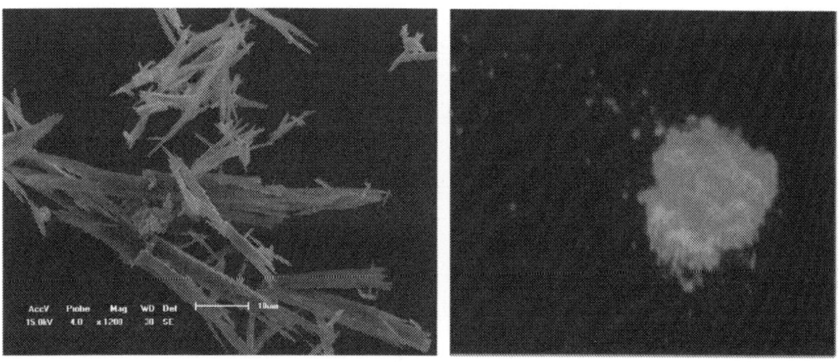

Figure 13. SEM image of $Eu(TMA)(H_2O)_4$ and a photograph under UV excitation λ = 254 nm.

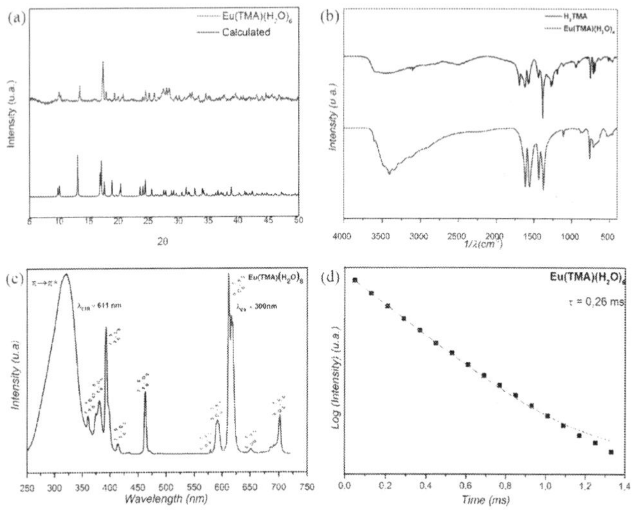

Figure 14. Structural and spectroscopic characterization: (a) XRD patterns of $Eu(BTC)(H_2O)_6$; (b) FT-IR spectrum of H_3TMA and $Eu(TMA)(H_2O)_6$; (c) photoluminescence and excitation spectra of $Eu(BTC)(H_2O)_6$ at room temperature; (d) luminescence decay curves at room temperature of the emitting $^5D_0$ level of the Eu^{3+} in $Eu(BTC)(H_2O)_6$.

From the results above, we conclude that we have succeeded in synthesizing a three-dimensional networks $Eu(TMA)(H_2O)_6$ and $Cu_3(BTC)_2.(H_2O)_2$ producing, for the first time, a metal-organic framework by laser ablation in liquid, wherein both obtained materials were crystalline and with well-defined morphology. Comparing with the conventional method, it is observed that this new synthesis route has the advantage of being a one-step synthesis, which allows obtaining submicromaterial with a high yield [21].

5. Conclusions

We have successfully developed a new technique for the synthesis of micro- and nanostructured materials using laser ablation in liquid environment, which consists of the combination of the target ablation in a liquid environment together with the photo physical or chemical decomposition of the solvent. With these method, several compounds, such as $Cu_2(BTC)_2(H_2O)_3$, $Eu(TMA)(H_2O)_4$, CdS quantum dots and magnetic materials such as Fe, FeO, $NaFeO_3$, Na_2FeO_4 have been prepared.

For each material prepared above, we observed that there was a different set of synthesis parameter to be adjusted in order to obtain the desired material. However, when compared with the conventional methods of synthesis, we can see that laser ablation in liquid is a straightforward technique with few adjustable parameter that allowed materials to be prepared with a high yield. As a matter of fact, with these examples we feel confident to propose this technique as the next-generation tool for the materials preparation.

Acknowledgements

This work was supported by Conselho Nacional de Desenvolvimento Cientifico e Tecnologico (CNPq) contract number 308008/2013-8, Coordenacao de Aperfeicoamento de Pessoal de Nivel Superior (CAPES), Fundacao de Amparo a Ciencia e Tecnologia do Estado de Pernambuco (FACEPE) and Programa de Recursos Humanos da ANP para o Setor Petróleo e Gás – PRH-ANP/MCTI. We also would like to thank Núcleo de Dispositivos Ópticos, Eletrônicos e Magnéticos (NUDOM), from NICIT-CAA and to Ms. Moralinda for grammar review.

Author details

Walter Mendes de Azevedo[1*], Sérgio de Lemos Campello[2*], Diego Leite da Cunha[3], Leonardo Tadeu Boaes de Mendonça[1] and Ohanna Maria Menezes Madeiro da Costa[4]

*Address all correspondence to: wma@ufpe.br and slcampello@yahoo.com.br

1 Department of Fundamental Chemistry, Federal University of Pernambuco, Recife, PE, Brazil

2 Interdisciplinary Nucleus of Exact Science and Technological Innovation, Federal University of Pernambuco, Recife, PE, Brazil

3 Faculdade Boa Viagem/DeVry Brasil, Recife, PE, Brazil

4 Postgraduate Program in Materials Science, Federal University of Pernambuco, Recife, PE, Brazil

References

[1] Srinivasan, R. and B. Braren, *Ultraviolet laser ablation of organic polymers*. Chemical Reviews, 1989. 89(6): pp. 1303–1316.

[2] Lazare, S. and V. Granier, *Ultraviolet laser photoablation of polymers: a review and recent results*. Laser Chemistry, 1989. 10(1): pp. 25–40.

[3] Bäuerle, D.W., *Laser processing and chemistry*. 2011: Springer-Verlag Berlin Heidelberg.

[4] Leon J. Radziemski, D.A.C., *Laser-induced plasma and applications*. 1989: CRC Press, New York.

[5] John C. Miller, R.F.H.J., *Laser ablation mechanisms and applications*. 1991: Springer New York, New York.

[6] Fogarassy, E. and S. Lazare, *Laser ablation of electronic materials: basic mechanisms and applications*. European Materials Research Society monographs. 1992, Amsterdam; New York: North-Holland. xi, 394 p.

[7] Chrisey D.B., Hubler G.K., Pulsed laser deposition of thin solid films. 1994: Wiley-Interscience, New York. 648.

[8] Yang, G.W., *Laser ablation in liquids: Applications in the synthesis of nanocrystals*. Progress in Materials Science, 2007. 52(4): pp. 648–698.

[9] Zeng, H., et al., *Nanomaterials via Laser Ablation/Irradiation in Liquid: A Review*. Advanced Functional Materials, 2012. 22(7): pp. 1333–1353.

[10] Furusawa, H., T. Sakka and Y.H. Ogata, *Characterization of ablated species in laser-induced plasma plume*. Journal of Applied Physics, 2004. 96(2): pp. 975–982.

[11] Mafuné, F., et al., *Full physical preparation of size-selected gold nanoparticles in solution: laser ablation and laser-induced size control*. The Journal of Physical Chemistry B, 2002. 106(31): pp. 7575–7577.

[12] Mafuné, F., et al., *Formation of stable platinum nanoparticles by laser ablation in water*. The Journal of Physical Chemistry B, 2003. 107(18): pp. 4218–4223.

[13] Mafuné, F., et al., *Structure and stability of silver nanoparticles in aqueous solution produced by laser ablation*. The Journal of Physical Chemistry B, 2000. 104(35): pp. 8333–8337.

[14] Mafuné, F., et al., *Formation and size control of silver nanoparticles by laser ablation in aqueous solution*. The Journal of Physical Chemistry B, 2000. 104(39): pp. 9111–9117.

[15] Pyatenko, A., et al., *Synthesis of silver nanoparticles by laser ablation in pure water*. Applied Physics A, 2004. 79(4-6): pp. 803–806.

[16] Takeuchi, Y., T. Ida and K. Kimura, *Colloidal stability of gold nanoparticles in 2-propanol under laser irradiation*. The Journal of Physical Chemistry B, 1997. 101(8): pp. 1322–1327.

[17] Sylvestre, J.-P., et al., *Stabilization and size control of gold nanoparticles during laser ablation in aqueous cyclodextrins*. Journal of the American Chemical Society, 2004. 126(23): pp. 7176–7177.

[18] da Cunha, D.L., et al., *Nanostructured hydrocerussite compound (Pb3(CO3)2(OH)2) prepared by laser ablation technique in liquid environment*. Materials Research Bulletin, 49(0): pp. 172–175.

[19] Mendonça, L.T.B. and W.M. de Azevedo, *A fast bottom-up route for preparing CdS quantum dots using laser ablation in a liquid environment*. Journal of Luminescence, 2016. 171: pp. 79–84.

[20] Campello, S.L., et al., *Laser ablation: A new technique for the preparation of metal-organic frameworks Cu3(BTC)2(H2O)3*. Materials Letters, 2015. 148: pp. 200–203.

[21] da Costa, O.M.M.M. and W.M. de Azevedo, *Highly luminescent metal organic framework $Eu(TMA)(H_2O)_4$ materials prepared by laser ablation technique in liquid*. Journal of Luminescence, 2016. 170, Part 2: pp. 648–653.

[22] Ma, R., et al., *Synthesis and fluorescence properties of ten lanthanide benzene-1,3,5-tricarboxylate complexes*. Spectrochimica Acta Part A: Molecular and Biomolecular Spectroscopy, 2010. 77(2): pp. 419–423.

[23] Souza, E.R., *Síntese de complexos benzenotricarboxilatos de terras raras e estudo de suas propriedades fotoluminescentes*, in *Instituto de Química*. 2008, Universidade de São Paulo: São Paulo.

[24] Martinetto, P., et al., *Synthetic hydrocerussite, 2PbCO3.Pb(OH)2, by X-ray powder diffraction*. Acta Crystallographica Section C, 2002. 58(6): pp. i82–i84.

[25] Welcomme, E., et al., *Investigation of white pigments used as make-up during the Greco-Roman period*. Applied Physics A, 2006. 83(4): pp. 551–556.

[26] Walter, P., et al., *Making make-up in Ancient Egypt*. Nature, 1999. 397(6719): pp. 483–484.

[27] Rachel Billinge, L.C., J. Dunkerton, S. Foister, J. Kirby, J. Pilc, A. Roy, M. Spring and R. White, Methods and materials of northern european painting in the national gallery, 1400–1550. National Gallery Technical Bulletin, 1997. 18, 6–55.

[28] Krivovichev, S.V. and P.C. Burns, *Crystal chemistry of basic lead carbonates. I. Crystal structure of synthetic shannonite, $Pb_2O(CO_3)$*. Mineralogical Magazine, 2000. 64(6): pp. 1063–1068.

[29] Frost, R.L., et al., *Raman spectroscopy of selected lead minerals of environmental significance*. Spectrochimica Acta Part A: Molecular and Biomolecular Spectroscopy, 2003. 59(12): pp. 2705–2711.

[30] Michell, E.W.J. and K.Y. Ng, *X-Ray powder diffraction studies of hydrocerussite (basic lead carbonate) as a stabiliser in plasticised polyvinyl chloride*. British Polymer Journal, 1980. 12(3): pp. 114–120.

[31] Steele, I.M.P., J. Joseph and A. Livingstone, *Crystal structure of susannite, Pb4SO4(CO3)2(OH)2: a trimorph with macphersonite and leadhillite*. European Journal of Mineralogy, 1999. 11(3): pp. 493–499.

[32] Porter, R.P. and W.A. Noyes, *Photochemical Studies. LIV. Methanol Vapor1*. Journal of the American Chemical Society, 1959. 81(10): pp. 2307–2311.

[33] Sakong, S. and A. Gross, *Total oxidation of methanol on Cu(110): a density functional theory study*. The Journal of Physical Chemistry A, 2007. 111(36): pp. 8814–8822.

[34] Walling, C. and R.W.R. Humphreys, *Photooxidation of alcohols by ferric ion*. The Journal of Organic Chemistry, 1981. 46(7): pp. 1260–1263.

[35] Fievet, F., et al., *Homogeneous and heterogeneous nucleations in the polyol process for the preparation of micron and submicron size metal particles*. Solid State Ionics, 1989. 32: pp. 198–205.

[36] Ducamp-Sanguesa, C., R. Herrera-Urbina and M. Figlarz, *Synthesis and characterization of fine and monodisperse silver particles of uniform shape*. Journal of Solid State Chemistry, 1992. 100(2): pp. 272–280.

[37] Fievet, F., et al., *Controlled nucleation and growth of micrometre-size copper particles prepared by the polyol process*. Journal of Materials Chemistry, 1993. 3(6): pp. 627–632.

[38] Ducamp-Sanguesa, C., R. Herrera-Urbina and M. Figlarz, *Fine palladium powders of uniform particle size and shape produced in ethylene glycol*. Solid State Ionics, 1993. 63: pp. 25–30.

[39] de Azevedo, W.M. and F.D. Menezes, *A new and straightforward synthesis route for preparing Cds quantum Dots*. Journal of Luminescence, 2012. 132(7): pp. 1740–1743.

[40] Bucca, M., et al., *Nucleation and crystallization of otavite, witherite, calcite, strontianite, hydrozincite and hydrocerussite by CO_2 membrane diffusion technique*. Chemical Geology, 2009. 266(3–4): pp. 143–156.

[41] Steigerwald, M.L. and L.E. Brus, *Semiconductor crystallites: a class of large molecules*. Accounts of Chemical Research, 1990. 23(6): pp. 183–188.

[42] Bawendi, M.G., M.L. Steigerwald and L.E. Brus, *The quantum mechanics of larger semiconductor clusters ("Quantum Dots")*. Annual Review of Physical Chemistry, 1990. 41(1): pp. 477–496.

[43] Alivisatos, A.P., *Semiconductor Clusters, Nanocrystals and Quantum Dots*. Science, 1996. 271(5251): pp. 933–937.

[44] Colvin, V.L., M.C. Schlamp and A.P. Alivisatos, *Light-emitting diodes made from cadmium selenide nanocrystals and a semiconducting polymer*. Nature, 1994. 370(6488): pp. 354–357.

[45] Klein, D.L., et al., *A single-electron transistor made from a cadmium selenide nanocrystal*. Nature, 1997. 389(6652): pp. 699–701.

[46] Bruchez, M., et al., *Semiconductor nanocrystals as fluorescent biological labels*. Science, 1998. 281(5385): pp. 2013–2016.

[47] Mandal, P., et al., *Orange-red luminescence from Cu doped CdS nanophosphor prepared using mixed Langmuir–Blodgett multilayers*. The Journal of Chemical Physics, 2008. 128(11): p. 114703.

[48] Unni, C., D. Philip and K.G. Gopchandran, *Studies on optical absorption and photoluminescence of thioglycerol-stabilized CdS quantum dots*. Spectrochimica Acta Part A: Molecular and Biomolecular Spectroscopy, 2008. 71(4): pp. 1402–1407.

[49] Unni, C., et al., *Aqueous synthesis and characterization of CdS, CdS:Zn2+ and CdS:Cu2+ quantum dots*. Spectrochimica Acta Part A: Molecular and Biomolecular Spectroscopy, 2009. 72(4): pp. 827–832.

[50] Brus, L., *Electronic wave functions in semiconductor clusters: experiment and theory*. The Journal of Physical Chemistry, 1986. 90(12): pp. 2555–2560.

[51] Patil, P., et al., *Pulsed-laser-induced reactive quenching at liquid-solid interface: Aqueous oxidation of iron*. Physical review letters, 1987. 58(3): pp. 238–241.

[52] Yang, G.-W., J.-B. Wang and Q.-X. Liu, *Preparation of nano-crystalline diamonds using pulsed laser induced reactive quenching*. Journal of Physics: Condensed Matter, 1998. 10(35): p. 7923.

[53] Alqudami, A. and S. Annapoorni, *Fluorescence from metallic silver and iron nanoparticles prepared by exploding wire technique*. Plasmonics, 2007. 2(1): pp. 5–13.

[54] Sharma, V.K., et al., *Ferrates (iron(VI) and iron(V)): environmentally friendly oxidants and disinfectants*. Journal of Water and Health, 2005. 3(1): pp. 45–58.

[55] Koltypin, M., et al., *The study of K2FeO4 (Fe6+-super iron compound) as a cathode material for rechargeable lithium batteries*. Journal of Power Sources, 2005. 146(1–2): pp. 723–726.

[56] He, W., et al., *Electrochemical preparation, characterization and discharge performance of solid K[3]Na(FeO[4])[2]*. Vol. 9. 2007, Amsterdam, PAYS-BAS: Elsevier. 4.

[57] Ni, Y., et al., *Synthesis and characterization of α-FeO(OH) nano-rods in situ via a solution-oxidation*. Materials Letters, 2001. 49(3–4): pp. 185–188.

[58] Ferey, G., *Hybrid porous solids: past, present, future*. Chemical Society Reviews, 2008. 37(1): pp. 191–214.

[59] Britt, D., et al., *Highly efficient separation of carbon dioxide by a metal-organic framework replete with open metal sites*. Proceedings of the National Academy of Sciences, 2009. 106(49): pp. 20637–20640.

[60] Kitaura, R., et al., *A pillared-layer coordination polymer network displaying hysteretic sorption: [Cu2(pzdc)2(dpyg)]n (pzdc= Pyrazine-2,3-dicarboxylate; dpyg=1,2-Di(4-pyridyl)glycol)*. Angewandte Chemie International Edition, 2002. 41(1): pp. 133–135.

[61] Min Wang, Q., et al., *Metallo-organic molecular sieve for gas separation and purification*. Microporous and Mesoporous Materials, 2002. 55(2): pp. 217–230.

[62] Alaerts, L., et al., *Probing the lewis acidity and catalytic activity of the metal–organic framework [Cu3(btc)2] (BTC=Benzene-1,3,5-tricarboxylate)*. Chemistry – A European Journal, 2006. 12(28): pp. 7353–7363.

[63] Liu, L., et al., *Ionothermal synthesis of the Metal-Organic Framework compound Cu3(BTC)2*, in *Studies in Surface Science and Catalysis*. 2008, Elsevier. pp. 459–462.

[64] Himeur, F., et al., *The ionothermal synthesis of metal organic frameworks, Ln(C9O6H3) ((CH3NH)2CO)2, using deep eutectic solvents.* Solid State Sciences. 12(4): pp. 418–421.

[65] Liao, J.-H., P.-C. Wu and Y.-H. Bai, *Eutectic mixture of choline chloride/urea as a green solvent in synthesis of a coordination polymer: [Zn(O3PCH2CO2)] Â · NH4.* Inorganic Chemistry Communications, 2005. 8(4): pp. 390–392.

[66] Zhang, J., et al., *Versatile structure-directing roles of deep-eutectic solvents and their implication in the generation of porosity and open metal sites for gas storage.* Angewandte Chemie International Edition, 2009. 48(19): pp. 3486–3490.

[67] Parnham, E.R., et al., *Ionothermal materials synthesis using unstable deep-eutectic solvents as template-delivery agents.* Angewandte Chemie, 2006. 118(30): pp. 5084–5088.

[68] Kuang, D., T. Brezesinski and B. Smarsly, *Hierarchical porous silica materials with a trimodal pore system using surfactant templates.* Journal of the American Chemical Society, 2004. 126(34): pp. 10534–10535.

[69] Rodrigues, M.O., et al., *Tb3+→Eu3+ energy transfer in mixed-lanthanide-organic frameworks.* The Journal of Physical Chemistry C, 2012. 116(37): pp. 19951–19957.

Chapter 10

Pulsed Laser Ablation in High-Pressure Gases, Pressurized Liquids and Supercritical Fluids: Generation, Fundamental Characteristics and Applications

Sven Stauss, Keiichiro Urabe, Hitoshi Muneoka and
Kazuo Terashima

Additional information is available at the end of the chapter

http://dx.doi.org/10.5772/65455

Abstract

Pulsed laser ablation (PLA) in high-density media—high-pressure gases, liquids, and supercritical fluids—has shown to be promising for nanomaterials fabrication and as an analysis technique in extreme environments, for example, the exploration of deep ocean levels and planetary atmospheres and surfaces. Despite the high potential of this technique, it is still not very widely used. The objective of the present chapter is to present the reader with an overview of recent advances in the use of pulsed laser ablation in pressurized media, the fundamental characteristics, especially the dynamics of cavitation bubbles and the optical emission, and the applications to the fabrication of metallic and semiconductor nanoparticles, and diamond molecules, the so-called diamondoids. Finally, a short overview of the use of pulsed laser ablation in pressurized media as a promising tool for the analysis of extreme environments is presented.

Keywords: pulsed laser ablation, high-density media, supercritical fluids, nanomaterials fabrication, diamondoids, cavitation bubble, shadowgraph imaging, optical emission spectroscopy

1. Introduction

Laser-based methods have opened new possibilities for materials treatment and processing, chemical synthesis, and analysis. Pulsed laser deposition (PLD) has been widely used for the fabrication of thin films (metallic, oxide, and complex heterostructures) and has become a mainstay for realizing high-quality thin films of different compositions [1].

Pulsed laser ablation (PLA) in liquids, which consists in the irradiation of a solid target placed inside liquids, has enabled the functionalization and fabrication of a wide range of nanomaterials. The main advantage of PLA is that it enables the fabrication of nanoparticles that cannot be obtained easily by other methods. Moreover, PLA realized in high-density media—high-pressure gases, liquids, and supercritical fluids—has opened up new possibilities for controlling nanomaterials fabrication by PLA and has also lead to the discovery of new phenomena related to plasma-fluid interactions.

In the context of this chapter, we define high-density media as gases, liquids, or supercritical fluids at and above atmospheric pressures, that is, at pressures $p \geq 0.1$ MPa and densities $\rho \gtrsim 1$ kg m^{-3}. While there have already been extensive reviews on PLA in liquids and their applications for nanomaterials processing [2, 3] including high-density media [4], the main aim of the present chapter is to present the main differences between PLA in atmospheric pressure liquids and PLA in high-density media—high-pressure gases, pressurized liquids, and supercritical fluids—and to point out possible advantages of using PLA in such media, both for fundamental research and applications in materials science and processing.

The chapter is structured as follows: The first part (section 2) gives a brief overview on the phenomena related to the formation of PLA plasmas in dense media. We introduce the techniques used for generating plasmas by PLA in high-density media, and the fundamental phenomena encountered under such conditions. This part is also devoted to the description of the experimental instrumentation and analysis techniques. In the third part (section 3) gives an overview of the use of pulsed laser ablation to nanomaterials' synthesis and the characterization of nanoparticle growth processes using X-ray scattering techniques. Finally, in the last two parts (sections 4 and 5), we present briefly other possible applications of PLA in high-density media and close the chapter with a short conclusion.

As mentioned above, while PLA in liquids or in atmospheric pressure gases has been investigated extensively, PLA in high-pressure/high-density media has not yet received as much attention, but we hope that the present review can help researchers gain an overview on this field.

2. Pulsed laser ablation plasmas in high-density media

This section gives a brief overview on the formation mechanism of PLA plasmas in high-density media, and the main experimental equipment required for carrying out such experiments.

The following section is devoted to introducing techniques for characterizing PLA plasmas in high-density fluids—direct imaging, shadowgraph and Schlieren imaging, laser scattering, optical emission spectroscopy (OES) [5], and Raman scattering [6].

http://dx.doi.org/10.5772/65455

2.1. Plasma generation in high-density media

To realize the pulsed laser plasma, different types of lasers can be used. The most common are second harmonic (532 nm) Nd:YAG pulsed laser sources (e.g., Ref. [7]) that are operated with a repetition rate of a few Hz and pulse widths of a few nanoseconds (~ 5 to 10 ns) and fluencies of a few tens of milli-Joules per pulse.

To realize the high-pressure conditions of the fluid, it is necessary to employ high-pressure pumps, for example, those used for high-performance liquid chromatography (HPLC) [8], or another method that allows compressing the gas or the liquid to be used. One simple approach is liquefaction that consists in cooling the source gas liquid nitrogen and then introducing it into the high-pressure cell, as this was used for PLA in supercritical xenon [7]. To keep the conditions in a well-defined state, the temperature and pressure in the cell are monitored by temperature and pressure sensors. Adjusting of the temperature can be achieved by using a heating/cooling circuit connected to the high-pressure cell.

One advantage of carrying out PLA in high-density and pressurized media is that varying the pressure and/or temperature of the medium allows further adjustment of the nanoparticle size and chemical composition (in addition to the laser fluence and pulse width).

For characterizing the plasma formation and the fluid evolution including the dynamics of the cavitation bubble, different fast imaging techniques are employed, the most common being shadowgraph and schlieren imaging. Finally, direct imaging can be used to generate the plasma formed by the laser irradiation. In shadowgraph imaging (cf. setup of **Figure 1**), the

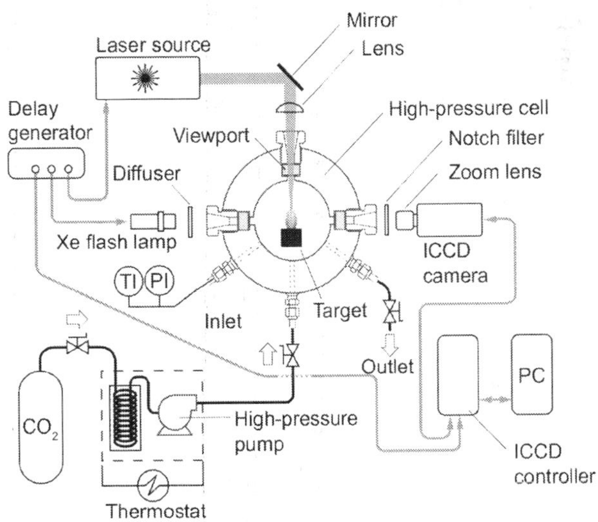

Figure 1. Schematic of experimental setup for direct and shadowgraph imaging in high-pressure conditions. Temperature and pressure indicators (TI, PI) are used to monitor the fluid conditions inside the high-pressure chamber. Figure adapted with permission from Ref. [10].

zone to be irradiated by the laser is illuminated from behind using a bright light source (a flash lamp or a laser source). The change in the fluid density leads to refraction of the light beams from the light source on the detector, which results in the formation of brighter and darker zones on the detector and correlates with the fluid density gradient.

PLA leads to the formation of shockwaves inside the target and the fluid. For 2D shocks produced in liquid water, pressures of up to 30 GPa and velocities up to Mach 6 have been reported [9].

Figure 1 illustrates the main components for conducting PLA in high-density media. The equipment consists of a reaction vessel capable of withholding pressures up to several megapascals. Usually, the reactor vessels are made of stainless steel (typically SUS316) or, in case of highly corrosive fluids such as supercritical water, other highly corrosion-resistant materials, mainly Ni-based alloys, for example, Hastelloy™, are used. As viewports, usually sapphire is used because of its superior hardness, high thermal conductivity, and chemical resistance. Another advantage is the large domain of optical transmission, from about 150 to 5000 nm.

To characterize the evolution of the plasma and the cavitation bubble, different types of imaging methods are used: The simplest is direct imaging, which is used for temporal and spatial evolution of the plasma and, when using bandpass filters of specific wavelengths, the spatial distribution of emissions corresponding to certain species can be monitored.

Finally, shadowgraph and Schlieren imaging allow the observation of changes in the fluid density, and optical emission spectroscopy can be used to characterize the plasma. Examples of these techniques employed for the characterization of PLA in high-density media will be presented in sections 2.2.2 and 2.2.3.

2.2. Characteristics of pulsed laser ablation plasmas in high-density media.

2.2.1. Plasma formation and characteristics

The formation of a plasma through laser irradiation is a highly complex process that involves several stages, which are illustrated in the case of a liquid in **Figure 2**. The laser irradiation and absorption of the laser energy by the target leads to the explosive vaporization of the target material and the formation of a cavitation bubble and subsequent nucleation and growth of nanoparticles that are ejected into the fluid.

While the detailed mechanisms are still not understood yet, depending on the type of laser, continuous wave (CW), nanosecond or pico-, respectively, and femtosecond, the mechanisms leading to the removal of material and plasma formation are different. In the case of CW lasers, material is removed primarily by melting, which creates a large heat-affected zone (HAZ), and material ejection is mainly dominated by thermal processes [11]. In nanosecond lasers, there are three main stages that lead to the formation of a plasma. In the first, laser photons couple both with electrons and phonons of the target material. The photon–electron coupling then results in an immediate rise of the electron temperature, leading to vaporization of the target. Compared to CW lasers, the HAZ created by nanosecond pulsed lasers is smaller.

http://dx.doi.org/10.5772/65455

With ultrafast pico- and femtosecond pulses, the laser pulse duration is much shorter than the timescale for energy transfer between free electrons and the material lattice, and electrons are excited to only a few or few tens of electron volts. Consequently, the lattice temperature of the target remains unchanged, and the main amount of the laser pulse energy is primarily absorbed in a thin layer of only a few microns close to the surface, where extremely high pressures and temperatures can be attained. The absorbed energy heats the material very quickly past the melting point, directly to the vapor phase with high kinetic energy, and the material is removed by vaporization. Consequently, in the case of pico- and femtosecond pulsed lasers, mainly the photon absorption depth governs the heated volume, the influence of thermal diffusion depth being smaller.

With nano-, pico-, and femtosecond pulsed lasers becoming more and more available, PLA has opened a wide range of new possibilities for materials processing: deposition of thin solid films, nanocrystal growth, surface cleaning, and the fabrication of microelectronic devices.

The evolution of the cavitation bubble is described in more detail in the next section.

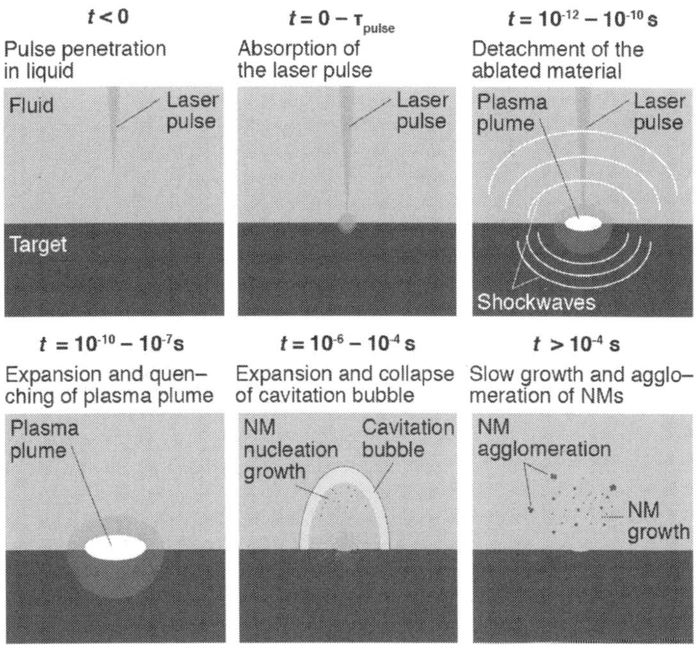

Figure 2. Schematic of the different events occurring after pulsed laser irradiation of a target inside a fluid. The rapid heating of the target and subsequent plasma formation leads to vaporization of the fluid and the formation of a cavitation bubble. Nanomaterial (NM) nucleation and growth occurs in the later stages, at $t \sim 10^{-6} - 10^{-4}$ s. Figure adapted with permission from Ref. [12].

2.2.2. *Evolution of cavitation bubbles*

One characteristic that distinguishes PLA in liquids or in pressurized media from PLD or PLA in vacuum is the confinement of the plasma plume by the surrounding fluid. As a consequence of this confinement and the large temperature rise of the target and the medium in the vicinity of the plasma plume, PLA in dense media is accompanied by the formation of a cavitation bubble. It is a region whose internal conditions—pressure and temperature—are different from those of the surrounding medium. Depending on the fluid conditions, expansion and compression of the cavitation bubble occur in several steps. It has been suggested that the bubble formed during pulsed laser ablation plays an important role in nanoparticle formation, as it confines the primary particles and redeposits them to the substrate.

In the first step, the cavitation bubble grows, until its internal pressure p_b becomes equal to the external pressure, p_0, after which the cavitation starts shrinking. Depending on the conditions of the medium, the expansion and shrinking can occur over several cycles.

To study the evolution of the cavitation bubble as a function of pressure, experiments in pressurized distilled water up to 3.5 × 10^7^ Pa on Ti targets at pulse widths of 10 ns and a laser fluence of 22 mJ pulse^{-1} were conducted [13]. **Figure 3** shows the variation of the size of the

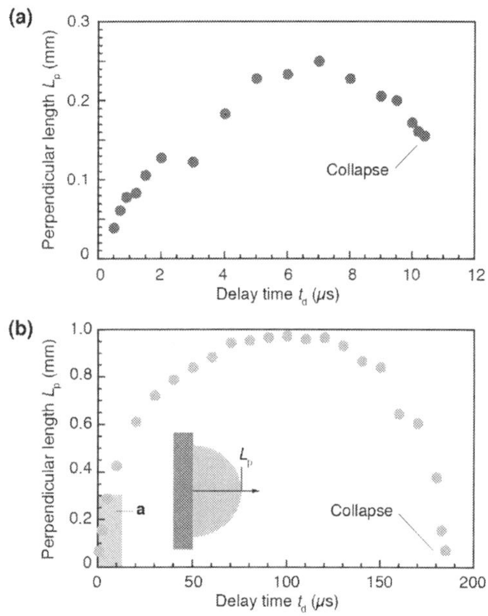

Figure 3. Variation of the length of the first cavitation bubble in the direction perpendicular to the target surface, L_p, as a function of the delay time t_d. **(a)** Water ambient pressure $p = 3 \times 10^6$ Pa. **(b)** Water ambient pressure $p = 1 \times 10^5$ Pa. The inset indicates the geometry of the cavitation bubble and the measure of L_p, and the shaded area labelled "a" indicates the x- and y-axes scales. Data adapted with permission from Ref. [13].

cavitation bubble in a direction perpendicular to the target surface as a function of time and for two different water pressures.

PLA in liquids has enabled the formation of a large variety of materials. A large variety of metallic nanoparticles [3], diamonds [14], and other carbon nanostructures [15].

The basis for calculating the variation of p and T inside a cavitation bubble can be estimated by using the the Rayleigh–Plesset equation:

$$\frac{p_b(t) - p(t)}{\rho} = r_b(t)\frac{d^2 r_b(t)}{dt^2} + \frac{3}{2}\left(\frac{dr_b(t)}{dt}\right)^2 + \frac{4v}{r_b(t)}\frac{dr_b(t)}{dt} + \frac{2S}{\rho_L r_b(t)} \tag{1}$$

where $r_b(t)$ is the radius of the cavitation bubble at time t, $p_b(t)$ the pressure inside the bubble, and $p(t)$ the pressure at a distance far from the bubble, ρ the density, v the kinematic viscosity, and S the surface tension of the surrounding fluid. The value of p_b at time t is given by

$$p_b(t) = p_V(T_b) + p_{G0}\left(\frac{T_b}{T_0}\right)\left(\frac{r_0}{r_b(t)}\right)^{3\gamma} \tag{2}$$

Here $p_v(T_b)$ is the vapor pressure at T_b, and T_0 is the temperature at a large distance from the cavitation bubble, $\gamma = C_p / C_v$ is the isentropic expansion factor, and p_{G0} is the pressure of a bubble present in the fluid before laser irradiation and r_0 the corresponding radius, related by $p_{G0} = p_0 - p_v(T_0) + 2S/ r_0$. The temperature in the bubble is given by

$$T_b(r) = \frac{T_0 r_0^{3(\gamma-1)}}{(r^3 - a^3)^{\gamma-1}} \tag{3}$$

with $a = r_0/8.86$ the hard core radius of the bubble. To improve the original Rayleigh–Plesset model, the authors took into account the hemispherical nature of the cavitation bubble and the effect of the contact angle between the bubble, the target, and the water [16]. **Figure 4** shows the variations of the pressure and temperature at the time of collapse inside the first cavitation bubble as a function of the external water pressure, p_0. As can be seen from the graph, the values of p_b reach values up to several TPa, with $p_b > 10$ at the time of collapse for $p_0 > 10$ MPa. On the other hand, the temperature at the collapse decreases only weakly with p_0, with values varying between $T \simeq 7000 - 13000$ K.

The variation of the volume as a function of water pressure $p = 0.1$, 10, 20 and 30 MPa is illustrated in **Figure 5**, which displays the variation of the cavity volumes as a function of time up to 2 μs following the laser irradiation. The volumes V were estimated by measuring the extension of the cavitation bubbles from the shadowgraph images, assuming half of an oblate spheroid ($V = \frac{4}{2 \cdot 3}\pi a^2 b$, where a is the semi-major axis and b the semi-minor axis). As can be

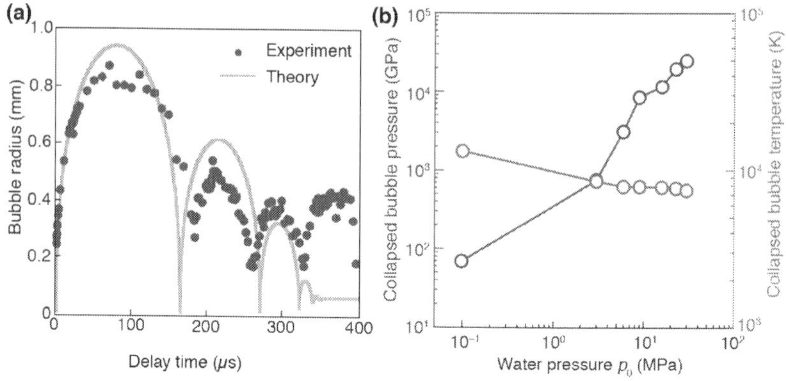

Figure 4. Dynamics of cavitation bubble and variation of pressure and temperature inside the bubble as a function of water pressure. **(a)** Experimental values of the evolution of the cavitation bubble radius as a function of time in comparison with theoretical model. **(b)** Variation of the bubble pressures and bubble temperatures at the time of collapse of the first cavitation bubble as a function of the water pressure p_0, using a modified Rayleigh–Plesset model. The dashed lines connecting the data points act as guides to the eye. Data adapted with permission from Ref. [16].

seen in **Figure 5** up to $t = 600$ ns, the volumes of the cavities are practically independent of the hydrostatic pressure. This is an indication that during the initial stages after laser irradiation, other mechanisms dominate the transient pressure profile surrounding the laser-ablated region [17].

The evolution of the cavitation bubble has been found to play an important role in the formation of nanoparticles (cf. Section 3.2). This control of the cavitation bubble dynamics can be achieved by several methods. One is by changing the laser fluence [18], or the viscosity of the medium [19]. Finally, another possibility is to pressurize the solution used for the PLA. By tuning these different parameters, the size, the chemical composition, and the type and concentration of defects in nanoparticles can be modified.

By changing the pressure of the surrounding medium, the plasma becomes confined and is restricted by the surrounding fluid.

Figure 6 shows a series of shadowgraph images obtained for a cavitation bubble generated in supercritical CO_2 (cf. **Figure 1** for a possible experimental setup for realizing shadowgraph images of PLA in high-density media). Near the critical point and depending on the fluid conditions, not only a single cavitation bubble but also a structure resembling a double-bubble can be observed [10, 20].

http://dx.doi.org/10.5772/65455

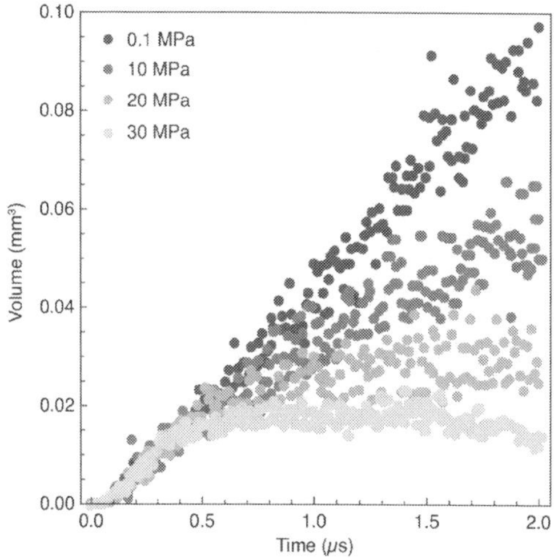

Figure 5. Evolution of the volume of the cavitation bubbles generated by single pulse irradiation on a brass target immersed in water pressurized at 0.1, 10, 20, and 30 MPa. Data adapted with permission from Ref. [17].

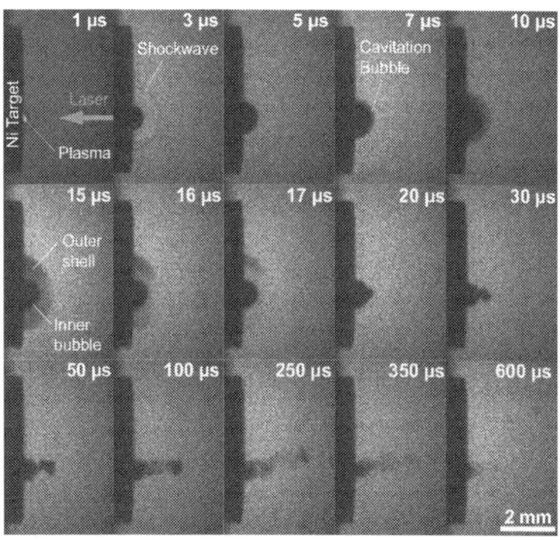

Figure 6. Series of shadowgraph images acquired for PLA on a Ni target in high-pressure liquid CO_2. The snapshots show different instants from t = 1 μs to 600 μs near the critical point (T = 302.0 K, p = 7.30 MPa) of CO_2. After the laser pulse, one can see the formation of a shockwave that emanates from the target (t = 3–7 μs). Then, from about 7 μs until 17 μs, the cavitation bubble exhibits the formation of a particular structure that consists of an inner bubble and an outer, partly transparent shell. Figure adapted with permission from Ref. [10].

The light source (e.g., a flash lamp or a light-emitting diode) is synchronized with a fast detector, for example, an intensified charge-coupled device (ICCD) or a streak camera. Flash lamps or diodes allow illumination times of typically $\Delta t \sim 10 - 100$ μs while Nd:YAG lasers allow higher fluencies and permit illuminations at pulse durations of $\Delta t \sim 3 - 10$ ns.

2.2.3. *Optical emission characteristics*

Figure 7(a) shows plasma emission images of PLA realized on a Ti target immersed in water at 0.1 and 30 MPa, respectively [21]. A crosssection of the emission intensity in a direction normal to target, in the middle of the plasma, is presented in **Figure 7(b)**. In both cases, the maximum emission is at a small distance from the target, and as the pressure is increased, the extension of the plasma becomes squeezed along the normal to the target.

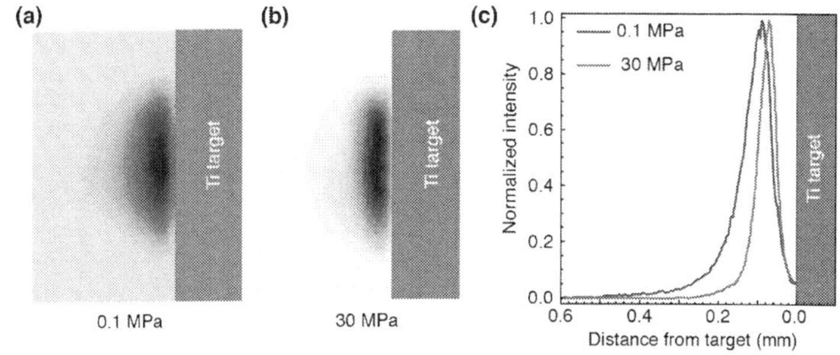

Figure 7. Optical emission intensities for pulsed laser ablation plasma in ambient and pressurized water. **(a)** Optical emission image observed at 0.1 MPa. **(b)** Optical emission image observed at 30 MPa. **(c)** Cross-sections of optical emission images in and along a direction normal to the target surface. Data adapted with permission from Ref. [21].

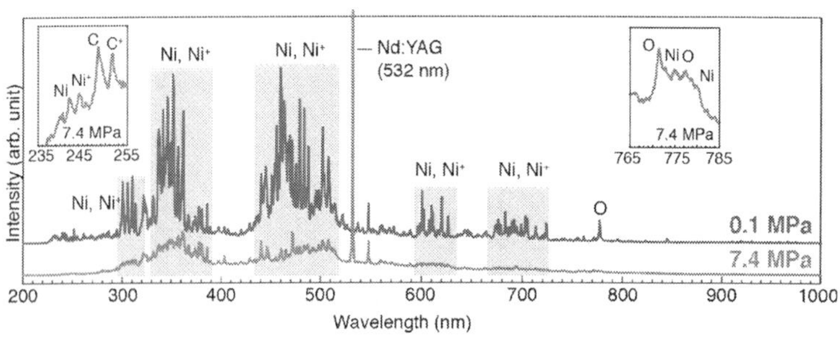

Figure 8. Optical emission spectra measured in CO_2 at $p = 0.1$ and 7.4 MPa for PLA on a Ni target. The dashed boxes indicate the domains of the spectrum where Ni lines are dominant. The inset on the left shows a close-up in the wavelength range between 235 and 255 nm, where peaks that can be attributed to atomic and ionized C can be found. The inset on the right shows the detailed spectrum in the region around 777 nm containing lines of atomic O. Data adapted with permission from Ref. [5].

As the pressure of the medium is increased, discrete peaks of emitting species are broadening.

Figure 8 shows two examples of OES data series recorded for PLA in atmospheric pressure and supercritical CO_2 (p = 7.4 MPa) [5]. In this work, the authors also found that the total emitted intensity reached a maximum near the critical point of CO_2, which was attributed to the maximum of the density fluctuation.

An additional spectroscopic technique that could be used for gaining information about the plasma characteristics is Raman spectroscopy. So far it has been used for characterizing laser-induced breakdown in water [6].

3. Nanomaterials synthesis and characterization

In this section, we give an overview of the application of PLA in high-density media for the fabrication of nanomaterials—mainly not only metallic and semiconductor nanoparticles but also organic molecules. We focus on the effect of the pressurized fluid on the particle characteristics, and advantages compared with PLA in atmospheric pressure liquids or gases.

While PLA in high-density media offers many advantages, the nucleation and growth mechanisms are still not understood very well. Methods that enable in situ characterization of nanoparticle formation, namely, fast X-ray Radiography (XRR) and Small-angle X-ray Scattering (SAXS) will be discussed in Section 3.2.

3.1. Synthesis of metallic, semiconductor, oxide and inorganic nanomaterials

3.1.1. Metallic nanoparticles

Metallic nanoparticles play an increasingly important role in many different fields. These include sensing, catalysis, electronics, and plasmonics, and especially the plasmonic properties of noble metallic nanoparticles has opened new possibilities in biotechnology and medicine, including DNA and protein sensing or new approaches for cancer therapy [22].

PLA in pressurized media is a promising approach for fabricating metallic nanoparticles and tailoring their properties. For example, gold nanoparticles were obtained by ablation in supercritical CO_2 at pressures up to 20 MPa [8, 23]. In addition to the particle size, the authors also investigated the influence of the pressure on the ablation depth, which was found to correlate with the constant volume heat capacity (C_V), that is, the largest particle removal rates were found for conditions $p \simeq 10$ MPa.

The density has also been found to affect the morphology and the size distribution of particles. **Figure 9(a)–(d)** show the morphologies of nanostructures obtained by PLA of Au targets in supercritical CO_2 [24]. For lower pressures (p = 4.29 MPa), the structures obtained resemble chains (**Figure 9(a)** and **(c)**), while at higher pressures (p = 14.5 MPa), spherical Au-nanoparticles are obtained (**Figure 9(b)** and **(d)**).

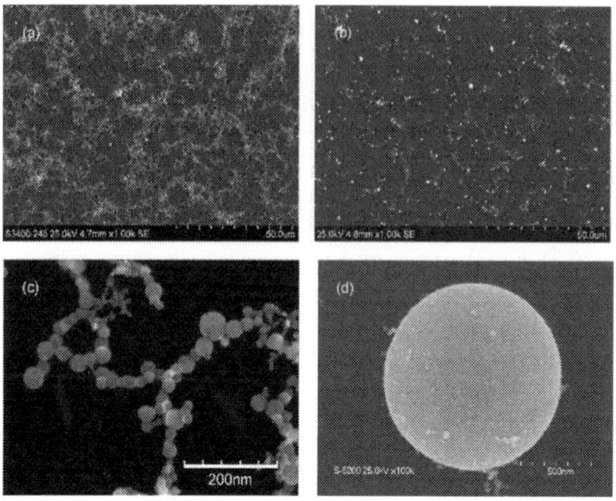

Figure 9. SEM images illustrating the influence of pressure on gold nanoparticle morphologies obtained by PLA in supercritical CO_2. **(a)** Nanoparticles generated by laser ablation at p = 4.29 MPa. **(b)** Nanoparticles generated by laser ablation at p = 14.5 MPa. **(c, d)** Enlarged images of (a) and (b). Reprinted (adapted) with permission from Ref. [24]. Copyright (2008) American Chemical Society.

In one report, the effect of laser fluence and fluidic pressure up to 200 MPa of PLA in water were investigated [25], while in a different study, PLA in pressurized CO_2 between 0.1 and 20 MPa was realized on gold and silver targets [26].

3.1.2. Semiconductor and oxide nanoparticles

Figure 10 shows a series of TEM images illustrating the change in morphology for ZnO nanoparticles formed by PLA in water at pressures of 0.1, 15, 22, and 31 MPa [27]. At a pressure of 0.1 MPa, the size distribution is quite large, with maximum particle sizes reaching values of 50 nm up to 100 nm. By increasing the water pressure, the authors found that the particle size

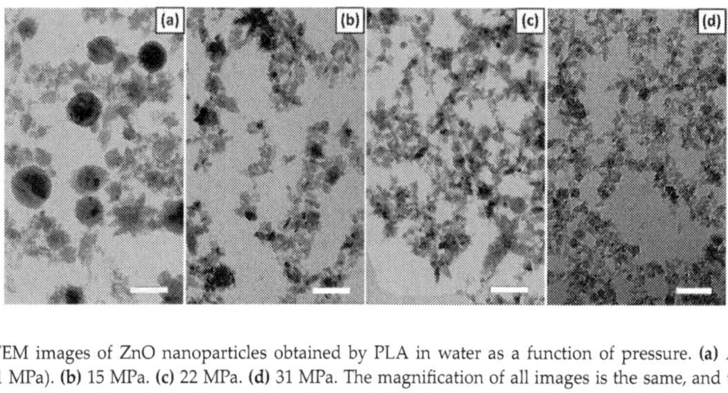

Figure 10. TEM images of ZnO nanoparticles obtained by PLA in water as a function of pressure. **(a)** Atmospheric pressure (0.1 MPa). **(b)** 15 MPa. **(c)** 22 MPa. **(d)** 31 MPa. The magnification of all images is the same, and the length of the scale bars is 100 nm. Reprinted with permission from Ref. [27]. Copyright (2013) American Institute of Physics.

http://dx.doi.org/10.5772/65455

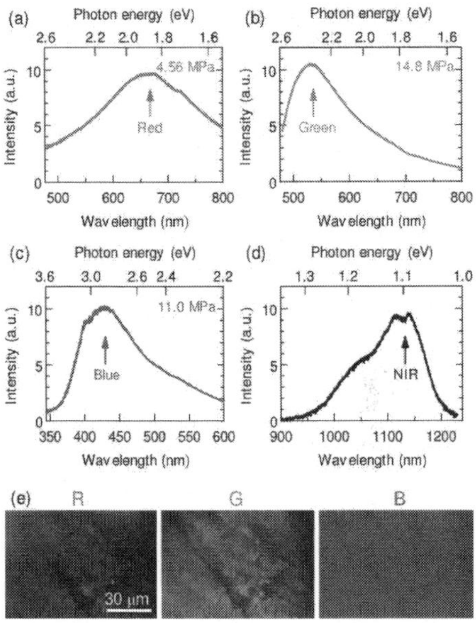

Figure 11. Photoluminescence spectra and luminescence images of silicon nanocrystals. **(a)** Red light-emitting silicon nanocrystals generated at $p = 4.56$ MPa in supercritical CO_2. **(b)** Green light-emitting silicon nanocrystals generated at $p = 14.8$ MPa in supercritical CO_2. **(c)** Blue light-emitting silicon nanocrystals generated at $p = 11$ MPa in supercritical CO_2. **(d)** Near-IR light-emitting bulk silicon measured by excitation at 632.8 nm of a He-Ne laser. **(e)** Photoluminescence images measured by a fluorescence microscope at an excitation wavelength of 375 nm. Reprinted with permission from Ref. [29]. Copyright (2009) American Chemical Society.

decreased and the difference between particle sizes also was reduced. Independent XRD measurements indicated particle sizes of from about 35 nm to ~ 15 – 20 nm.

Using ZnO nanoparticles with well-defined defects, mainly interstitial zinc atoms (Zn_i) and charged oxygen vacancies (V_O^+) showed promising properties as highly sensitive oxygen sensors [28].

PLA of silicon (Si) targets in $scCO_2$ permitted to control the size of Si-nanocrystals (Si-nc) [29].

The size distribution of Si-nc can be estimated from different characteristic cooling times τ, which can be expressed by

$$\tau = \frac{R_{Si}^2 \rho_{Si}^2 C_{Si}^2}{9 \rho_{CO_2} C_{CO_2} \lambda_{CO_2}} \tag{4}$$

with R_{Si} the radius of the Si-nc, ρ_{Si} its density, C_{Si} its specific heat capacity, and ρ_{CO_2} and C_{CO_2} the corresponding values of the supercritical CO_2, λ_{CO_2} being the thermal conductivity. By adjusting the pressure of the supercritical fluid, different cooling rates could be realized, and the obtained nanoparticles showed varying photoluminescence spectra. The different colors could be attributed to different types of defects caused by changes in the cooling rates (τ^{-1}). (**Figure 11**).

Using a similar approach of PLA in supercritical CO_2, white light-emitting Si nanoparticles could be obtained [30].

3.1.3. Carbon nanomaterials

In addition to metallic and semiconductor nanoparticles, PLA has also been used for the synthesis of carbon nanomaterials. One group used single and double pulse laser ablation of graphite targets placed in water that was pressurized at values ranging from $p = 1 - 146$ atm [31]. At $p = 146$ atm, the cooling rate was highest, resulting in the formation of carbon nanotubes.

Figure 12 shows the Raman spectra of the collected particles for water at 1 and 146 atm. At atmospheric pressure (**Figure 12 (a)**), the Raman spectrum of the ablated particles exhibit features that are characteristic of diamond-like carbon (DLC), that is, a mixture of carbon with different degrees of amorphousness [32]. As shown in **Figure 12(b)**, at a pressure of 146 atm, the Raman spectrum contains features that are characteristic of carbon nanotubes [33].

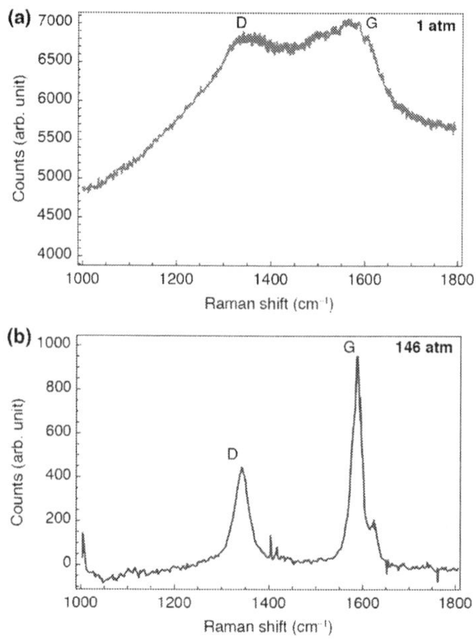

Figure 12. Raman spectra of particles produced by PLA of graphite target in H_2O. **(a)** Single pulse PLA at 1 atm liquid pressure. **(b)** Single pulse PLA at 146 atm. Data adapted with permission from Ref. [31].

PLA in supercritical xenon [7] and CO_2 [34] has also been employed for the synthesis of molecular diamond, so-called "diamondoids" [35]. Diamondoids are carbon nanomaterials, consisting of a $C(sp^3) – C(sp^3$ hybridized carbon cage structure in the form of adamantane units, which can be superimposed on a diamond lattice, and a H-terminated surface. Except for their isolation from crude oil and gas sources [36], diamondiods consisting of more than 4 units ($n \geq 4$) are very difficult or even impossible to synthesize. It was shown that by PLA in supercritical fluids diamondoids up to 10 units could be synthesized. **Figure 13** shows examples of mass spectra of diamondoids synthesized by PLA in supercritical xenon: diamantane ($n = 2$, molecular ion peak at mass-to-charge ratio m/z 188, **Figure 13(a)**); pentamantane ($n = 5$, molecular ion peak at m/z 330, **Figure 13(b)**), and octamantane ($n = 8$, molecular ion peak at m/z 472, **Figure 13(c)**).

From samples fabricated by PLA in supercritical CO_2, traces that could be attributed to "superadamantane," a highly symmetric diamondoid (point group T_d, $C_{35}H_{36}$) consisting of $n = 10$ units, were found [34].

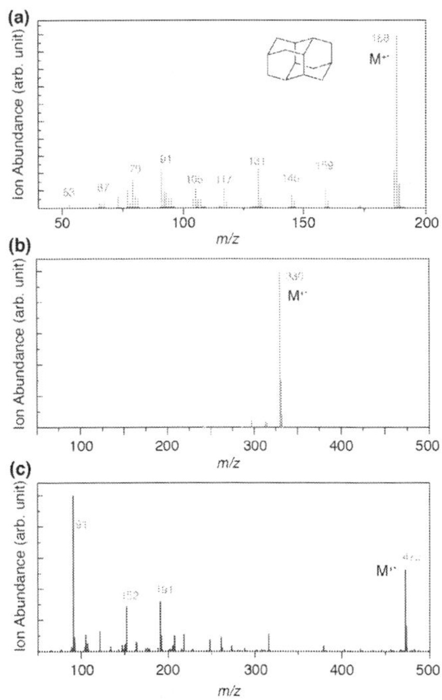

Figure 13. Mass spectra of products obtained by PLA in supercritical xenon. **(a)** Mass spectrum mad molecular structure of diamantane (molecular weight 188). **(b)** Mass spectrum with molecular ion peak ($M^{+•}$) at m/z 330 that may be attributed to pentamantane ($C_{25}H_{30}$). **(c)** Mass spectrum of species with molecular weight of 472 that could be assigned to octamantane $C_{36}H_{40}$). Data adapted with permission from Ref. [7].

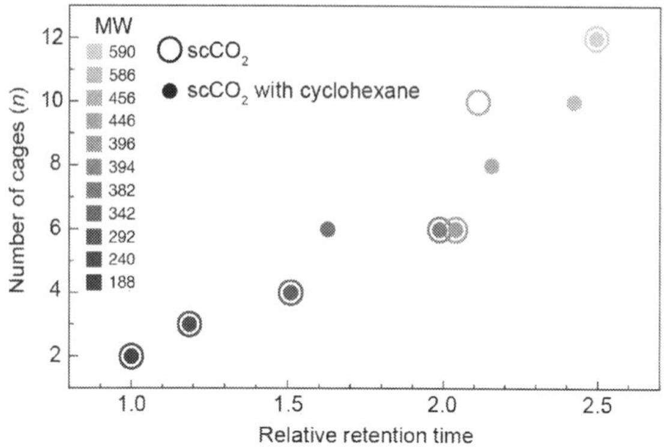

Figure 14. Variation of the number of diamondoid cages obtained by CO_2 with and without cyclohexane as a function of the relative retention time in the gas chromatography-mass spectrometry measurements. The retention time of diamantane is the reference retention time, and the color map indicates the molecular weight (MW) of the detected diamondoids. Diamondoids with higher MWs need increasingly longer elution times for being detected by mass spectrometry, the increase being almost linear. Data adapted with permission from Ref. [34].

Figure 14 shows the possible types of diamondoids (indicated by the number of cages n) that were obtained by PLA in CO_2, as a function of the relative retention time.

These examples illustrate that PLA in high-density media allows the synthesis of nanomaterials far from thermodynamic equilibrium that would be difficult or impossible to be achieved by other methods, the main reasons being the high pressures and temperatures that can be achieved by the PLA plasma, and the cooling rates that can be modified by changing the density of the fluid. At present, the main drawback of nanomaterials fabrication by PLA in high-density media is that the quantities of the formed nanomaterials cannot be increased easily to industrially relevant quantities.

3.2. Characterization of nanoparticle formation

Figure 15 illustrates a schematic of the experimental setup for carrying out in situ Small Angle X-ray Scattering (SAXS) measurements of nanoparticles during PLA. For continuous refreshing of the target surface, the target is in the form of a moving metallic ribbon (In this case silver (Ag), speed 10 cm s^{-1}). To avoid convolution of the SAXS measurements and overlapping of ablated material from previous laser shots, the fluid (H_2O) is also continuous (in this specific case, the authors used a flow rate of 25 l h^{-1}).

These time-resolved SAXS measurements of pulsed laser ablation in liquid water at atmospheric pressure revealed that after laser absorption by the target, a vapor-filled cavitation bubble is formed at the target surface which undergoes oscillation including a rebound and final collapses after 220 ms. Inside the cavitation bubble two types of particles can be identified, namely, compact primary particles of 8 – 10 nm size and bigger agglomerates of 40 – 60 nm size. While it cannot be ruled out, presently, SAXS experimental detection limits cannot prove or disprove the presence of very small particles or particle clusters with sizes < 2 nm.

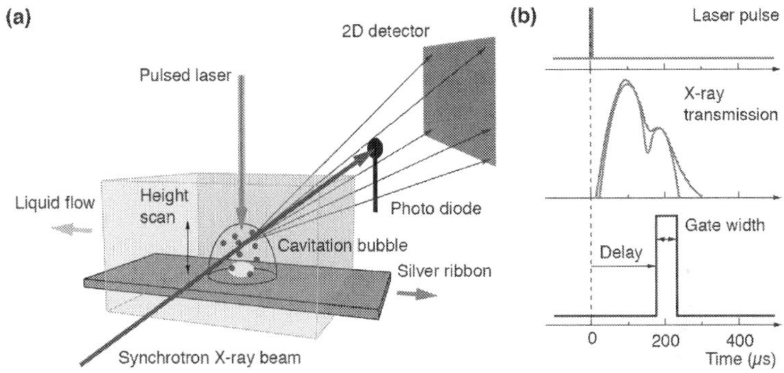

Figure 15. Schematic of PLA and synchrotron measurement during formation of nanoparticles. **(a)** Experimental setup of an X-ray scattering experiment, liquid flow conditions and moving target to provide reproducible experimental conditions for every laser pulse during time-gated data accumulation. **(b)** Schematic of the stroboscopic data acquisition with the detector being gated active for a fixed interval with delay with respect to the laser impact. The oscilloscope traces of the transmission change are recorded at the same time. Figure adapted with permission from Ref. [37].

Similar trends were observed using different water-ethanol mixtures from 0 to 100 % and pressures up to 13 MPa of the fluid, which enabled to control various types of defects in ZnO nanoparticles [38]. The higher pressures were associated with shorter lifetimes of the cavitation bubble, leading to higher quenching rates of the nanoparticles produced. The faster cooling rates also resulted in less agglomeration of the particles and higher surface-to-volume ratios. As a consequence, for higher pressure, the number of interstitial oxygen (O_i) defects was found to increase and at the same time, the contribution of near-band-edge emission, that is, emission in the UV (at $\lambda \leq 400$ nm), was found to decrease.

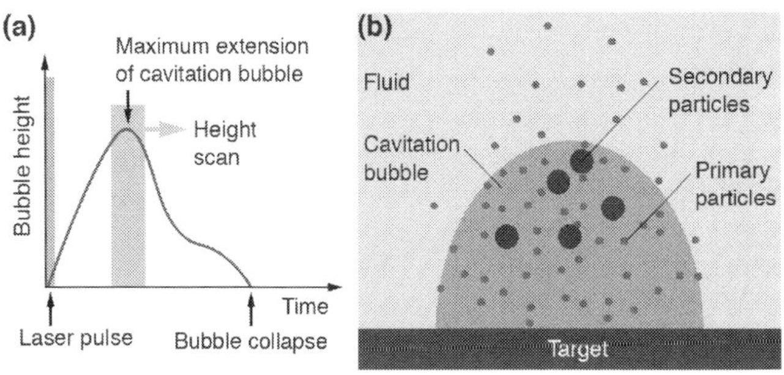

Figure 16. Schematic illustration of particle formation inside a laser-induced cavitation bubble and distribution at the time of its largest size for PLA inside atmospheric pressure water. **(a)** Time-line of laser irradiation and evolution of bubble height. **(b)** Evolution of cavitation bubble and evolution of primary and secondary particles. Figure adapted with permission from Ref. [40].

Figure 16 illustrates the particle formation inside a cavitation bubble generated by laser-irradiation of a target surface placed inside a liquid [39].

The time shown is at the instant of the largest extent of the bubble. After irradiation of the target by the laser, primary particles are formed. Over time, these primary particles coagulate to form larger, secondary particles. While the secondary particles are trapped inside the cavitation bubble, some of the primary particles can escape from the inside of the cavitation bubble to the surrounding fluid. In other words, the interface of the cavitation bubble is not an impenetrable wall or membrane, but instead can be crossed by particles.

Agglomeration occurs for the confined particles in the second cavitation bubble, which forms after the first bubble has collapsed. Additionally, upon the collapse of the second bubble, a jet of confined material is ejected perpendicularly to the target surface.

From this one can see that the lifetime of the cavitation bubble, the pressures and temperatures reached inside the cavitation bubble, and the conditions of the surrounding fluid all influence the nucleation and growth of nanoparticles.

In another study, to monitor the formation of nanoparticles in situ, a multipurpose time-resolved spectrometer was developed, that allows following the formation of nanoparticles over several timescales [41]. The spectrometer consists of three different parts, one that can be used for following the nanoparticle formation by time-resolved absorption spectroscopy in the wavelength range of 350 – 850 nm and on timescales of nanoseconds to milliseconds in the time following the laser pulse. The second part consists of an absorption spectrometer that allows following the nanoparticle formation on timescales of seconds to hours between 220 and 900 nm, and the third component, which dynamic light scattering for tracking nanoparticles with sizes ranging from 10 nm to 10, over timescales of seconds to hours.

3.2.1. Other characterization methods

In addition to the characterization methods mentioned above, it is worth noting that there are still other methods that can be used for investigating the effects of PLA. One is by numerical simulation. The mechanisms of PLA have been investigated for vacuum or ambient gas conditions [42].

4. Other applications

In this section, we give a brief overview of alternative applications of PLA for the characterization of exotic environments, for example, at deep sea levels, or the exploration of the surface of extraterrestrial bodies—planets and comets.

Spectroscopy techniques are increasingly becoming important in deep-sea exploration and geochemical assessment of solids and liquids, as they enable rapid, in-situ measurements. Laser-induced breakdown spectroscopy (LIBS) has also been used for the deep-sea exploration [17], that is, the examination of soil, or of water composition. The setup consists of a laser

http://dx.doi.org/10.5772/65455

source and an optical bench that is placed in a container that can sustain high hydrostatic pressures and has allowed real-time analysis of the ocean ground. It is expected that similar LIBS apparatus could be used in the future missions involving the exploration of planetary surfaces. In these cases, the laser source could be used to ionize materials of the surface to be probed, which are then introduced into a Time-of-Flight Mass Spectrometer (TOF-MS).

5. Conclusions

Pulsed laser ablation in high-density fluids is a versatile technique that enables the functionalization and fabrication of metallic, semiconductor, and organic nanomaterials, which cannot not be obtained easily by other methods. Moreover, pulsed laser ablation in high-density media can also be used as an analysis technique in extreme environments such as in deep sea environments, to gain information about elemental composition at such positions. There are still many phenomena that are not well-understood, especially with respect to the plasma formation, cavitation bubble dynamics, and the nanomaterials growth in the cavitation bubble.

As has been hopefully explained in the present review, the use of spectroscopy techniques—in particular SAXS and XRR—can aid in gaining a better understanding of growth processes of nanoparticles obtained by PAL of targets in liquids. To the best of our knowledge, so far these techniques have not been extended to the high pressure regimes yet, but we think that the same techniques could also be adapted to high pressure and SCF media, in order to further optimize the size, morphology, and composition of nanoparticles obtained by PLA.

We hope that the present review could give a short perspective on the current state in this field of research and that it can aid in contributing to the further expansion of this research.

Acknowledgements

Parts of the work presented in this chapter was carried out using laser facilities in the spectroscopy section of the Institute for Solid State Physics, the University of Tokyo, as a part of joint research. This work was financially supported in part by a Grant-in-Aid for Scientific Research on Innovative Areas, Frontier Science of Interactions between Plasmas and Nano-interfaces (No. 21110002), from the Ministry of Education, Culture, Sports, Science and Technology of Japan, and a Grant-in-Aid for JSPS Fellows (No. 12J03125) from the Japan Society for the Promotion of Science. The authors would like to thank previous members of the Terashima Laboratory for their contribution and experimental work to parts of the results presented: Mr. Sho Nakahara, Mr. Shoehei Himeno, and Mr. Yoshihiko Takizawa. Moreover, we would like to thank Prof. Emer. Tohru Suemoto (Institute of Solid State Physics, The University of Tokyo), and Prof. Motoyoshi Baba (Saitama Medical University), for their collaboration.

Author details

Sven Stauss[1*], Keiichiro Urabe[1,2,3], Hitoshi Muneoka[1,2,4] and Kazuo Terashima[1]

*Address all correspondence to: sven.stauss@plasma.k.u-tokyo.ac.jp

1 Department of Advanced Materials Science, Graduate School of Frontier Sciences, The University of Tokyo, Kashiwa, Japan

2 Japan Society for the Promotion of Science, Tokyo, Japan

3 K.K. Air Liquide Laboratories, Tsukuba, Japan

4 Nippon Steel & Sumitomo Metal, Kamisu, Japan

References

[1] P. R. Willmott and J. R. Huber. Pulsed laser vaporization and deposition. *Reviews of Modern Physics*, 72 (1): 315–328, 2000.

[2] G. W. Yang. Laser ablation in liquids: Applications in the synthesis of nanocrystals. *Progress in Materials Science*, 52 (4): 648–698, 2007. doi: 10.1016/j.pmatsci.2006.10.016.

[3] V. Amendola and M. Meneghetti. Laser ablation synthesis in solution and size manipulation of noble metal nanoparticles. *Physical Chemistry Chemical Physics*, 11 (20):3805–3821, 2009. doi: 10.1039/b900654k.

[4] K. Saitow. Nanoparticle Generation by Laser Ablation in Liquid and Supercritical Fluid. In: G. Yang, editor. Laser Ablation in Liquids: Principles and Applications in the Preparation of Nanomaterials. 1st ed. Singapore: Pan Stanford Publishing; 2012. p. 573–626.

[5] T. Kato, S. Stauss, S. Kato, K. Urabe, M. Baba, T. Suemoto, and K. Terashima. Pulsed laser ablation plasmas generated in CO_2 under high-pressure conditions up to supercritical fluid. *Applied Physics Letters*, 101(22):224103, 2012. doi: 10.1063/1.4767839.

[6] H. Yui, T. Tomai, M. Sawada, and K. Terashima. Generation of laser-induced plasma in supercritical water and vibrational spectroscopic study of accompanying stimulated raman scattering. *Applied Physics Letters*, 99(9):091504, 2011. doi: 10.1063/1.3627161.

[7] S. Nakahara, S. Stauss, H. Miyazoe, T. Shizuno, M. Suzuki, H. Kataoka, T. Sasaki, and K. Terashima. Pulsed laser ablation synthesis of diamond molecules

in supercritical fluids. *Applied Physics Express*, 3(9):096201, 2010. doi: 10.1143/APEX.3.096201.

[8] S. Machmudah, M. Goto, Wahyudiono, Y. Kuwahara, and M. Sasaki. Gold nanoparticles fabricated by pulsed laser ablation in supercritical CO_2. *Research on Chemical Intermediates*, 37(2–5): 515–522, 2011. doi: 10.1007/s11164-011-0279-x.

[9] T. Pezeril, G. Saini, D. Veysset, S. Kooi, P. Fidkowski, R. Radovitzky, and K. A. Nelson. Direct visualization of laser-driven focusing shock waves. *Physical Review Letters*, 106(21):214503, 2011. doi: 10.1103/PhysRevLett.106.214503.

[10] K. Urabe, T. Kato, S. Stauss, S. Himeno, S. Kato, H. Muneoka, M. Baba, T. Suemoto, and K. Terashima. Dynamics of pulsed laser ablation in high-density carbon dioxide including supercritical fluid state. *Journal of Applied Physics*, 114(14):143303, 2013. doi: 10.1063/1.4824538.

[11] M. N. R. Ashfold, F. Claeyssens, G. M. Fuge, and S. J. Henley. Pulsed laser ablation and deposition of thin films. *Chemical Society Reviews*, 33(1):23–31, 2004. doi: 10.1039/b207644f.

[12] S. Stauss and K. Terashima. *Diamondoids – Structure, Properties and Possible Applications, Isolation and Synthesis*. 1st ed. Singapore: Pan Stanford Publishing; 2016.

[13] K. Sasaki, T. Nakano, W. Soliman, and N. Takada. Effect of pressurization on the dynamics of a cavitation bubble induced by liquid-phase laser ablation. *Applied Physics Express*, 2(4):046501, 2009. doi: 10.1143/APEX.2.046501.

[14] D. Amans, A. C. Chenus, G. Ledoux, C. Dujardin, C. Reynaud, O. Sublemontier, K. Masenelli-Varlot, and O. Guillois. Nanodiamond synthesis by pulsed laser ablation in liquids. *Diamond and Related Materials*, 18(2–3):177–180, 2009. doi: 10.1016/j.diamond.2008.10.035.

[15] A. Santagata, A. De Bonis, A. De Giacomo, M. Dell'Aglio, A. Laurita, G. S. Senesi, R. Gaudiuso, S. Orlando, R. Teghil, and G. P. Parisi. Carbon-based nanostructures obtained in water by ultrashort laser pulses. *Journal of Physical Chemistry C*, 115(12):5160–5164, 2011. doi: 10.1021/jp1094239.

[16] W. Soliman, T. Nakano, N. Takada, and K. Sasaki. Modification of Rayleigh-Plesset theory for reproducing dynamics of cavitation bubbles in liquid-phase laser ablation. *Japanese Journal of Applied Physics*, 49(11):116202, 2010. doi: 10.1143/JJAP.49.116202.

[17] B. Thornton, T. Takahashi, T. Ura, and T. Sakka. Cavity formation and material ablation for single-pulse laser-ablated solids immersed in water at high pressure. *Applied Physics Express*, 5(10):102402, 2012. doi: 10.1143/APEX.5.102402.

[18] T. Tsuji, Y. Tsuboi, N. Kitamura, and M. Tsuji. Microsecond-resolved imaging of laser ablation at solid–liquid interface: investigation of formation process of nano-size metal colloids. *Applied Surface Science*, 229(1–4):365–371, 2004. doi: 10.1016/j.apsusc.2004.02.013.

[19] T. Tsuji, D. H. Thang, Y. Okazaki, M. Nakanishi, Y. Tsuboi, and M. Tsuji. Preparation of silver nanoparticles by laser ablation in polyvinylpyrrolidone solutions. *Applied Surface Science*, 254(16):5224–5230, 2008. doi: 10.1016/j.apsusc.2008.02.048.

[20] S. Himeno, T. Kato, K. Urabe, S. Stauss, S. Kato, H. Muneoka, M. Baba, T. Suemoto, and K. Terashima. Anomalous behavior of cavitation bubbles observed in pulsed laser ablation of Ni in liquid CO_2 near the critical point. *IEEE Transactions on Plasma Science*, 42(10):2630–2631, 2014. doi: 10.1109/TPS.2014.2327233.

[21] N. Takada, T. Nakano, and K. Sasaki. Influence of additional external pressure on optical emission intensity in liquid-phase laser ablation. *Applied Surface Science*, 255(24):9572–9575, 2009. doi: 10.1016/j.apsusc.2009.04.087.

[22] V. Amendola and M. Meneghetti. What controls the composition and the structure of nanomaterials generated by laser ablation in liquid solution? *Physical Chemistry Chemical Physics*, 15(9):3027–3046, 2013. doi: 10.1039/c2cp42895d.

[23] S. Machmudah, Wahyudiono, Y. Kuwahara, M. Sasaki, and M. Goto. Nano-structured particles production using pulsed laser ablation of gold plate in supercritical CO_2. *Journal of Supercritical Fluids*, 60:63–68, 2011. doi: 10.1016/j.supflu.2011.04.008.

[24] K. Saitow, T. Yamamura, and T. Minami. Gold nanospheres and nanonecklaces generated by laser ablation in supercritical fluid. *Journal of Physical Chemistry C*, 112(47): 8340–18349, 2008. doi: 10.1021/jp805978g.

[25] D. Werner, T. Ueki, and S. Hashimoto. Methodological improvement in pulsed laser-induced size reduction of aqueous colloidal gold nanoparticles by applying high pressure. *Journal of Physical Chemistry C*, 116(9):5482–5491, 2012. doi: 10.1021/jp300690z.

[26] S. Machmudah, N. T. Wahyudiono, H. Kanda, K. Sasaki, and M. Goto. Fabrication of gold and silver nanoparticles with pulsed laser ablation under pressurized CO_2. *Advances in Natural Sciences-Nanoscience and Nanotechnology*, 4(4):045011, 2013. doi: 10.1088/2043-6262/4/4/045011.

[27] S. A. Kulinich, T. Kondo, Y. Shimizu, and T. Ito. Pressure effect on zno nanoparticles prepared via laser ablation in water. *Journal of Applied Physics*, 113(3):033509, 2013. doi: 10.1063/1.4775733.

[28] T. Goto, Y. Shimizu, H. Yasuda, and T. Ito. Photoexcited zno nanoparticles with controlled defects as a highly sensitive oxygen sensor. *Applied Physics Letters*, 109(2): 023104, 2016.

[29] K. Saitow and T. Yamamura. Effective cooling generates efficient emission: blue, green, and red light-emitting si nanocrystals. *Journal of Physical Chemistry C*, 113(19):8465–8470, 2009. doi: 10.1021/jp900067s.

[30] S. Y. Wei, T. Yamamura, D. Kajiya, and K. Saitow. White-light-emitting silicon nano-crystal generated by pulsed laser ablation in supercritical fluid: Investigation of spectral

components as a function of excitation wavelengths and aging time. *Journal of Physical Chemistry C*, 116(6):3928–3934, 2012. doi: 10.1021/jp10080k.

[31] A. De Giacomo, A. De Bonis, M. Dell'Aglio, O. De Pascale, R. Gaudiuso, S. Orlando, A. Santagata, G. S. Senesi, F. Taccogna, and R. Teghil. Laser ablation of graphite in water in a range of pressure from 1 to 146 atm using single and double pulse techniques for the production of carbon nanostructures. *Journal of Physical Chemistry C*, 115(12):5123–5130, 2011. doi: 10.1021/jp109389c.

[32] A. C. Ferrari and J. Robertson. Interpretation of raman spectra of disordered and amorphous carbon. *Physical Review B*, 61(20):14095–14107, 2000.

[33] H. M. Heise, R. Kuckuk, A. K. Ojha, A. Srivastava, V. Srivastava, and B. P. Asthana. Characterisation of carbonaceous materials using raman spectroscopy: a comparison of carbon nanotube filters, single- and multi-walled nanotubes, graphitised porous carbon and graphite. *Journal of Raman Spectroscopy*, 40(3):344–353, 2009. doi: 10.1002/jrs.2120.

[34] S. Nakahara, S. Stauss, T. Kato, T. Sasaki, and K. TerashimaSynthesis of higher diamondoids by pulsed laser ablation plasmas in supercritical CO_2. *Journal of Applied Physics*, 109(12):123304, 2011. doi: 10.1063/1.3599887.

[35] H. Schwertfeger, A. A. Fokin, and P. R. Schreiner. Diamonds are a chemist's best friend: Diamondoid chemistry beyond adamantane. *Angewandte Chemie-International Edition*, 47(6):1022–1036, 2008. doi: 10.1002/anie.200701684.

[36] J. E. Dahl, S. G. Liu, and R. M. K. Carlson. Isolation and structure of higher diamondoids, nanometer-sized diamond molecules. *Science*, 299(5603):96–99, 2003.

[37] P. Wagener, S. Ibrahimkutty, A. Menzel, A. Plech, and S. Barcikowski. Dynamics of silver nanoparticle formation and agglomeration inside the cavitation bubble after pulsed laser ablation in liquid. *Physical Chemistry Chemical Physics*, 15(9):3068–3074, 2013. doi: 10.1039/c2cp42592k.

[38] T. Goto, M. Honda, S. A. Kulinich, Y. Shimizu, and T. Ito. Defects in zno nanoparticles laser-ablated in water-ethanol mixtures at different pressures. *Japanese Journal of Applied Physics*, 54(7):070305, 2015. doi: 10.7567/JJAP.54.070305.

[39] S. Ibrahimkutty, P. Wagener, A. Menzel, A. Plech, and S. Barcikowski. Nanoparticle formation in a cavitation bubble after pulsed laser ablation in liquid studied with high time resolution small angle X-ray scattering. *Applied Physics Letters*, 101(10):103104, 2012. doi: 10.1063/1.4750250.

[40] S. Ibrahimkutty, P. Wagener, T. D. Rolo, D. Karpov, A. Menzel, T. Baumbach, S. Barcikowski, and A. Plech. A hierarchical view on material formation during pulsed-laser synthesis of nanoparticles in liquid. *Scientific Reports*, 5:016313, 2015. doi: 10.1038/srep16313.

[41] S. Y. Wei and K. Saitow. In situ multipurpose time-resolved spectrometer for monitoring nanoparticle generation in a high-pressure fluid. *Review of Scientific Instruments*, 83(7): 073110, 2012. doi: 10.1063/1.4737886.

[42] L. V. Zhigilei, P. B. S. Kodali, and B. J. Garrison. A microscopic view of laser ablation. *Journal of Physical Chemistry B*, 102(16):2845–2853, 1998. doi: 10.1021/jp9733781.

Chapter 11

Metal Oxide Nanoparticle Preparation by Pulsed Laser Ablation of Metallic Targets in Liquid

Valery A. Svetlichnyi, Anastasiia V. Shabalina,
Ivan N. Lapin and Daria A. Goncharova

Additional information is available at the end of the chapter

http://dx.doi.org/10.5772/65430

Abstract

The basic mechanisms of pulsed laser ablation in liquids (PLAL) as a method for the synthesis of nanoparticles (NPs) were considered. Physical and chemical processes occurring during the PLAL that determine the formation, composition and structure of the nanoparticles obtained are described. The influence of the composition and properties of the target material, the solvent and the characteristics of the laser irradiation on the efficiency of the synthesis of nanoparticles is discussed. Separately, an influence of the absorption and scattering (including nonlinear) of laser radiation in the dispersion of nanoparticles on the primary synthetic processes and secondary transformations inside the colloidal solution is examined. The specificity of the characterization of the colloidal solutions of oxide particles produced by PLAL is highlighted. The most promising practical applications of nanomaterials obtained are identified and the examples of their successful use in catalytic research and biomedicine are provided.

Keywords: pulsed laser ablation, nanocolloids, chemical reactions, composition and structure of nanoparticles, metal oxide nanoparticles

1. Introduction

Interest in the study of fundamental processes of interaction of powerful optical radiation with matter and their practical applications arose simultaneously with the invention of lasers. Depending on the characteristics of the laser radiation (intensity, wavelength, pulse duration, polarization and coherence), the processes of interaction and the results of interaction may be different. At low intensities and high duration of the interaction of the radiation with the solid absorbing target, only heating of the target occurs. Increasing the intensity of the radiation can lead to the reaching of the melting temperature and to the further evaporation of the substance of the target. Under powerful pulse laser radiation, the evaporation of the substance occurs with the formation of plasma cloud that contains ions, clusters and small particles. For such an "explosive" evaporation, the term "laser ablation" (from Lat. "ablatio"—taking, removal) is used. Sometimes in the literature, the term "ablation" is interpreted more widely, covering the removal of substances from the surface as a result of any physical and chemical processes that occur under high-energy impact on the object. In the present chapter, the term "ablation" will be used only for the threshold process of explosive vaporization of material from a surface of solid targets with the formation of gas (vapor) plasma cloud via the rapid absorption of energy of high-power laser pulses in a limited volume.

First of all, the process of pulsed laser ablation (PLA), when micro-, nano- and later femtosecond pulses are exposed to the target, had found a successful application in laser materials processing for punching, marking and precise removal of layers of material in electronics [1]. Next, the PLA in gas phase and vacuum became successfully used for atomization of targets in mass spectrometry, obtaining thin films and ultrafine powders [2–4].

Using PLA of bulk targets in liquids to obtain nanoparticles (NPs) was initiated by the rapid development of nanotechnology in the 1990s of the twentieth century. The first deliberate use of this technique for the synthesis of nanocolloids took place in 1993 when the dispersions of Ag, Au, Pt, Pd and Cu nanoparticles in water and organic solvents for the surface-enhanced Raman scattering (SERS) spectroscopy were obtained [5].

Over the past two decades, pulsed laser ablation in liquids (PLAL) has become an effective and popular tool for obtaining nanosized materials. The absence of a mechanical interaction in the synthesis process and the ability to prepare "pure" nanoparticles without additional chemicals in the pure solvents straightaway in the form of stable colloids make this method very attractive for biological and medical applications. Catalysis, electronics and nonlinear optics are also the areas where the nanocolloids synthesized by the PLAL method and nanocrystalline powder obtained via further drying of the dispersions are used. The relatively simple experimental technique and the possibility of obtaining different types of nanoparticles (from metals to ceramics and polymers) on the same installation make this method a suitable tool for obtaining nanomaterials for the study of fundamental properties of substances in the nanostate. An opportunity to change the parameters of laser pulses and use various solvents with additives of precursors provides additional options to vary the composition, structure and dimensional characteristics of the nanoparticles obtained.

As of now, thousands of original research works, reviews and monographs on various aspects of the PLAL for the synthesis of nanostructures have been published [6–8]. They consider common mechanisms of PLAL and obtaining and characterization of specific nanomaterials as well. However, the interest in such research work continues unabated. On the one hand, this is due to the demand for nanomaterials with specific functional properties for various applications. On the other hand, there are three main aims that have not been achieved yet: effective control of structure and dimensional characteristics of nanoparticles obtained;

http://dx.doi.org/10.5772/65430

obtaining of multicomponent particles and particles with complex structure, such as core shell; and initiation of the required chemical reactions in the synthesis process.

Another important point for PLAL as a synthesis method of nanoparticles is its low productivity caused by the physical mechanisms of the ablation process itself. It hinders the broad practical use of this technology. That is why the optimization of the experimental conditions to achieve maximum yield of nanoparticles in liquid is important. Individual factors that influence the productivity of particle synthesis via PLAL are, for example, the thermophysical characteristics of the target and its optical properties and secondary interaction of radiation with a plasma torch and with particles in a colloid. Developed theoretical models of the ablation process, to some extent, take into account the parameters of the medium and the characteristics of the laser irradiation. So far, however, focused systematic experimental investigation of the influence of the factors on the productivity of synthesis has not been carried out. In addition to solving a practical task of increasing the productivity of the method, these studies are important for obtaining new fundamental knowledge about the processes taking place in a reactor under the repetitively pulsed irradiation of the target in the presence of the concentrated colloids of particles with high chemical activity stimulated by laser radiation. Moreover, in addition to the changing of the effectiveness of the PLAL process, finite properties of nanomaterials may vary significantly.

In this section, the basic experimental aspects influencing the process of oxide nanoparticles obtaining under nanosecond-pulsed laser ablation of some reactive metals in liquids are discussed and the characteristics and some applications of such prepared nanoparticles are considered.

2. Preparation

2.1. Basics of laser ablation in liquids

In this section, the general processes occurring during PLAL of solid bulk targets are briefly described. A more detailed description can be found, for example, in Ref. [6].

Initially, laser ablation models that describe the process of absorption of the radiation by the target, its explosive vaporization with the formation and diffusion dynamics of plasma and the radiation interaction with plasma have been developed for the case of ablation in the gas phase. Our primary interest is in the description of the ablation process under the action of nanosecond laser pulses. In contrast to the very short (fs) pulses where the processes of excitation (isochoric heating) and further thermodynamic behavior of the system can be divided, more long (ns) pulses lead to the simultaneous course of these processes, making it difficult to describe them. In this case, the ablation is described well enough by the modern thermal model [9], which takes into account the absorption of radiation by plasma and uses a movable boundary of evaporation and evaporation obeying the Hertz-Knudsen law. This model works well not only near the ablation threshold but also when there is a large pumping over the threshold where a dense plasma plume appears.

The main differences between ablation in liquids and ablation in gas phase/vacuum are as follows. Firstly, ablation occurs at the boundary between the solid and liquid phase. Secondly, the formation and distribution of the plasma plume occurs in the liquid environment. Because the liquid has a much higher density and thermal conductivity than gas, this environment significantly affects the thermodynamic characteristics of the process. Thirdly, a large number of chemical transformations are possible in liquid. In characterization of the ablation in liquids, an approach described in Ref. [10] seems to be effective. It describes the process mechanisms, aggregate states and phase transitions depending on the pulse duration in temperature-density coordinates.

From the practical point of view, ablation in liquids leads to the following consequences. First of all, the threshold for ablation in liquids is higher than in gas and the material of the target removal rate, i.e., the productivity of obtaining nanoparticles, is much lower. Secondly, a vapor plasma plume that is much denser (up to 1023 cm^{-3}) is formed. At flame temperatures of 4000–5000 K, its pressure reaches 10 GPa and higher. This leads to efficient chemical transformations, as well as to the possible formation of metastable phases. Moreover, because the plasma is limited and is held by the dense liquid, its expansion occurs adiabatically at supersonic speed with the formation of a powerful shock wave.

In terms of practical use of PLAL in NPs synthesis, the use of ns excitation sources is the most promising. The greatest efficiency in the synthesis of large quantities of nanoparticles is observed exactly when nanosecond laser pulses are used. When short (femto- and picosecond) pulses are used, the productivity at the beginning of the process can be higher than for nanosecond pulses, but with the increasing of the particles' concentration in the solution, the efficiency of the process decreases sharply. This happens, for example, due to the secondary interaction of laser radiation with particles in solution that leads to its strong attenuation and scattering. In addition, the cost of a sufficiently powerful picosecond laser suitable for PLAL is several times higher and femtosecond (oscillator-amplifier) more than 10 times exceeds the cost of the usual nanosecond Q-switch Nd:YAG laser. In conjunction with the complexity of the operation and relatively low resources of such short-pulse lasers, the cost of synthesis of meaningful quantities of nanoparticles (except for scientific research and a number of high-tech biomedical applications) will not be justified.

The use of longer μs pulses significantly reduces the efficiency of the ablation process. This is connected with the higher threshold for ablation in liquids compared to gas/vacuum systems, as well as with the strong screening effect of the plasma for microsecond pulses. Besides, a large energy contribution to the media leads to the heating of the liquid that limits the power of lasers used. In addition, when long pulses (from several tens of ns to hundreds of μs) are used, other ablation mechanisms take place [11], which will not be covered in this chapter.

The scheme of the ablation process is presented in **Figure 1**. **Figure 1a** shows the formation of a plasma cloud, **Figure 1b** shows interaction of plasma with radiation, **Figure 1c** shows chemical processes in plasma and **Figure 1d** shows the result: nanoparticles in liquid and surface erosion.

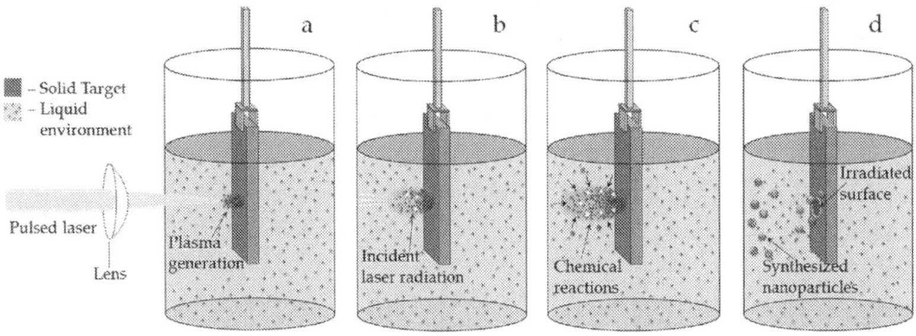

Figure 1. Nanosecond-pulsed laser ablation in liquids: (a) the formation of a plasma plume resulting from the absorption of laser pulse, (b) the evolution of the laser plasma in liquid during the nanosecond pulse and its interaction with radiation, (c) the further evolution of the plasma after the pulse and the chemical reactions inside the plume and at the border and (d) clusters (nanoparticles) in liquid and a crater on the surface of the target after the process.

There are three significant moments that should be noted:

— Interaction of a plasma cloud with the liquid that limits it and with laser radiation leads to the formation of powerful shock waves in the environment.

— Heated dense plasma cloud stimulates several types of chemical reactions inside the plasma and on the interface between three components: the target material, clusters (particles) that are formed and the solvent molecules.

— Generated particles remain in the liquid in the path of laser radiation and participate in a secondary interaction with the laser beam at long pulse-periodic irradiation of the target influencing the processes of plasma formation, diffusion dynamics and chemical reactions.

2.2. General experimental technique of PLAL

Experimental technique for nanosecond PLAL is simple; that makes this method attractive for laboratory applications. To start the process of ablation, it is enough to focus the radiation of a nanosecond laser source on the target surface submerged in the liquid. At the same time, to obtain high-quality nanocolloids with reproducible properties, a variety of experimental procedures are required.

There are two variants of introducing radiation into the reactor: from above, across the air/liquid border and sideways through the transparent window/side wall reactor. From the point of reproducibility of the experiment and control of the focus conditions, the second option is preferred. When the radiation is introduced to the reactor from above, there are difficulties related to the control of the liquid layer over the target and the formation of bubbles during ablation that leads to scattering and bad focusing. In addition, when the thickness of the layer of liquid above the target is small, the spraying of the liquid takes place and contamination of focusing elements and protective windows may occur. When the radiation is introduced through the side wall of the reactor, these obstacles do not appear. But in this case, the main point is the radiation resistance of a window or transparent reactor wall that is affected by the laser beam. To minimize the power density of radiation in the window or wall of the reactor, one has to use short-focus lenses. In practice, the focal length of the lenses in such experiments is $F = 25–50$ mm.

As a result of the PLAL process, a thin layer of the material is removed from the surface of the target. After repeated multiple irradiation of the same place on the surface of the target, a crater appears and then a through-hole is formed. For uniform irradiation and to prevent crater formation, scanning of a laser beam on the surface of the target is used. To do this, one can either redirect the laser beam or move the target itself. There are two common methods for moving the target. For cylindrical targets, the rotation around their axis with a slow linear displacement is preferable. Most often, however, the target is a parallelepiped and in this case, scanning in the XY plane orthogonal laser beam is used.

The removal of the nanoparticles obtained by PLAL from the optical path of the radiation occurs either naturally through the convection or by the agitation of the liquid in a reactor, for example, using a magnetic stirrer. To produce large amounts of colloidal solutions with a concentration of nanoparticles of 10–50 mg/l, it is advisable to use flow reactors with the control of concentration in the synthesis process. Possible ways of technical implementation of such methods are considered, for example, in our work [12].

To obtain concentrated dispersions for further use and preparation of the nanocrystalline powders, it is better to carry out the optimization of synthesis conditions (considered below) instead of using flow reactors.

2.3. Effect of thermophysical parameters of the target on the PLAL process targets

One of the fundamental factors that determine the efficiency of ablation is a combination of thermophysical characteristics of the target material: melting point and evaporation temperature, heat capacity, heat of fusion and boiling and thermal conductivity. The dependence of ablation threshold and the thickness of the removed layer from the thermal characteristics of the target can be easily seen on the example of chemically inert materials, such as noble metals. For targets made of chemically active metals, this dependence may be more complicated. This is connected with the formation of oxides and other compounds on the surface of the target and these compounds' thermophysical characteristics differ from those of the target material. On the other hand, during the frequency pulse-periodic irradiation, the thickness of the modified layer on the surface is very thin and the thermodynamic characteristics of the PLAL are mainly determined by the properties of the target material. An additional factor, particularly for relatively refractory metals (Ti, Fe, etc.), is the decomposition of organic solvents on the surface of the target and formation of carbides, nitrides and other compounds.

Table 1 presents the basic thermophysical characteristics of the materials that were used as targets for nanosecond PLAL: T_{melt}, melting point; T_b, boiling point; H_{fus}, heat of fusion; H_{vap}, heat of vaporization; k, thermal conductivity; $E°$, standard electrode potential; W, pulsed power density of radiation on the surface of target; and υ – rate of the target weight loss. Analyzing the table, one can establish the relationship between the thermophysical characteristics of the target, the power density that is used in the PLAL and the rate of nanoparticle production, which is estimated from the target weight loss per unit of time.

For example, the maximum productivity is observed for PLAL of cerium targets, which has a relatively low melting point and heat of fusion, as well as low thermal conductivity. This allows reaching high local temperatures easily and overcoming a high boiling point. It is interesting to note relatively low productivity for aluminum nanoparticles by PLAL. This is connected not only with the low weight of this metal but also with its high thermal conductivity that is ~15 times greater than for cerium. Low ablation productivity was found for magnetic 3d metals Fe, Ni and Co, as well as for Cu that has a very high thermal conductivity at a much lower boiling point. Despite the relatively low thermal conductivity, the rate for Ti nanoparticle production is not high; that is associated with large energy consumption due to high melting point and heat of fusion.

Target/ solution		T_{melt}, K	H_{fus}, kJ/mol	T_b, K	H_{vap}, kJ/mol	k, W/(m·K)	E°, V	W, MW/cm²	υ, mg/h	Consist
Mg	DW	923	9.2	1363	131.8	171	−2.372	150	40	$Mg(OH)_2$
	Alc									$MgO@Mg(OH)_2$
Al	DW	933.5	10.75	2792	284.1	238	−1.700	200	25	$Al(OH)_3$
	Alc									Not determined
Ce	DW	1072	5.2	3699	398	10.9	−2.336	200	50	CeO_2
	Alc									CeO_2
Ti	DW	1935	18.8	3560	422.6	15.5	−1.63	800	10	TiO_2
	Alc						−1.19			$Ti@TiO_2$
Zn	DW	693	7.28	1179	114.8	113	−0.763	300	20	$ZnO/Zn(OH)_2$
	Alc									ZnO
Fe	DW	1812	13.8	3134	340	72.4	−0.441	600	12	Fe_3O_4
	Alc									$Fe@Fe_3O_4$
Co	DW	1768	15.48	3143	389.1	69	−0.28	600	12	$Co(OH)_2$
	Alc									Co_2O_3/Co_3O_4
Ni	DW	1726	17.61	3005	378.6	60.5	−0.234	600	12	Ni@NiO
	Alc									Ni
Sn	DW	505	7.07	3146	296	63	−0.14	300	30	Sn@SnO
	Alc									Sn
Bi	DW	545	11.3	1837	172	53.3	+0.317	250	40	Bi_2O
	Alc									$Bi@Bi_2O$
Cu	DW	1356	13.01	2840	304.6	398	+0.33	600	10	Cu_2O
	Alc						+0.522			$Cu@Cu_2O$

Q-switch Nd:YAG laser, λ = 1064 nm, t_{puls} = 7 ns, E_{puls} = 120 mJ, f = 20 Hz.

Table 1. Thermophysical and electrochemical characteristics of the targets, characteristics of the synthesis and composition of NPs obtained by PLAL.

2.4. Chemical transformations during PLAL

High temperature and pressure in plasma plume in PLAL stimulate various chemical reactions. Commonly, there are four types of chemical reactions [13] that are shown in **Figure 2**. The first two types of reactions occur within the plasma cloud. The first type is the interaction of ionized clusters of the target material with each other inside the plume. As a result of these interactions, metastable phases are often formed. The second type of chemical reactions occurs between the ionized molecules of the liquid and the target material. This is possible when the liquid is heated on the plasma/liquid boundary to the temperature of the plasma. Then so-called liquid plasma is formed and mixed with the main plasma cloud consisting of the target material. The third type of chemical reaction takes place on the plasma/liquid boundary at high pressure and temperature between the particles of the target material and molecules of the liquid, including the processes under the additional stimulation by the laser radiation. The fourth type of chemical reaction occurs after the collapse of the plasma plume directly in the liquid when the particles formed continue interacting with the molecules of the liquid environment.

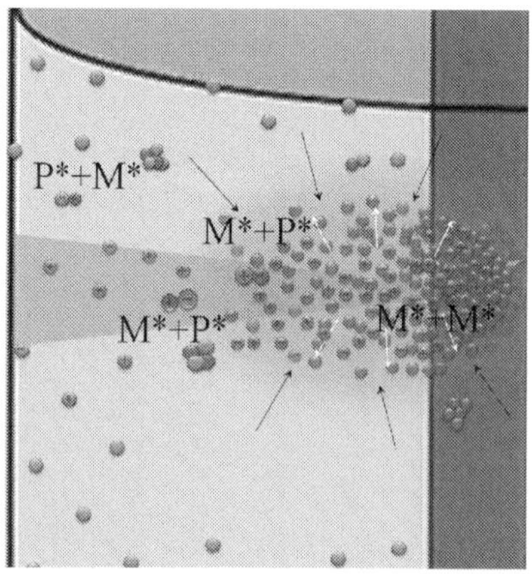

Figure 2. Types of chemical reactions that occur within PLAL.

Because the particles remain in solution, they can interact with the radiation when they appear on its path. This can lead to a significant change of particles' size distribution as a result of their melting and defragmenting under the influence of the laser beam. This also can effectively stimulate further chemical transformations of particles with different compositions and structures and molecules of the solvent. Secondary interaction is particularly strong when the high-energy photons of the visible and ultraviolet (UV) range of the spectrum are used. Moreover, these photons are better absorbed by the particles thanks to plasmon or exciton absorption in this wavelength range. Thus, the PLAL can be used to synthesize a large variety of compounds.

In addition to the parameters of the plasma plume, the chemical interactions, obviously, depend on the chemical properties of the target material and the solvent. The most common chemical reactions are associated with the interaction with oxygen both in a free form and as part of the solvent molecules. **Table 1** shows the data on the chemical composition of the NPs obtained via ablation of different metals in distilled water and ethanol. Depending on the reactivity of the metal target that is characterized by a parameter *E*° (standard electrode potential) and the type of the solvent, one can control the composition of the particles in the dispersion. The majority of metals react chemically during the PLAL in the water with the formation of oxides or hydroxides. Chemically active metals such as magnesium and aluminum hydrolyze during ablation in water (**Table 1**), while, for example, the PLAL of more inert Ni and Sn produces core-shell-type particles with the metal core and the oxide layer on the surface.

The use of different precursors provides additional opportunities to control the structure of the product particles. For example, adding different sulfur-containing precursors—thioacetamide (TAM) and hydrogen sulfide (H_2S)—to water during the ablation of metallic cadmium, CdS nanoparticles can be obtained [14].

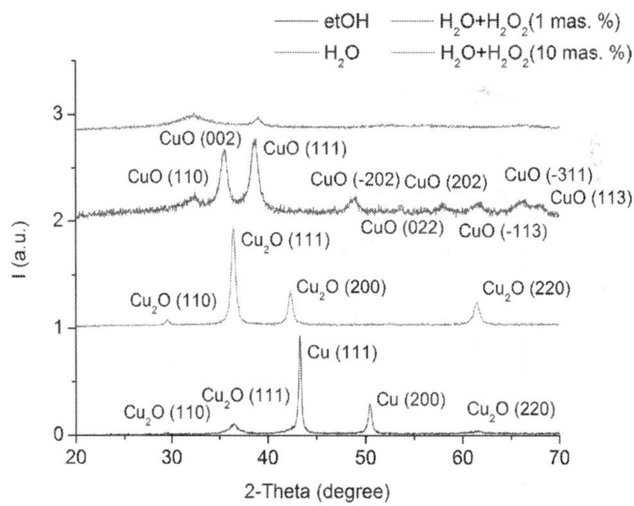

Figure 3. X-ray diffraction patterns of powders of NPs obtained via PLAL of metallic copper target in various solvents.

Figure 3 shows the X-ray diffraction (XRD) data for powders obtained by ablation of the copper metallic target. Using alcohol, water and additional oxidizer—hydrogen peroxide of various concentrations—as solvents, crystalline NPs of mostly metallic copper (I) oxide (Cu_2O), copper (II) oxide (CuO), as well as X-ray amorphous copper (I) hydroxide (CuOH) were obtained, respectively.

2.5. Influence of the secondary interaction of radiation with the NPs in solution on the productivity of the synthesis of nanoparticles and their composition

Because the ablated particles remain in the liquid after PLAL, secondary interaction of laser radiation with the synthesized particles has a significant impact, especially during the long process of synthesis of highly concentrated colloidal solution. Interaction of laser radiation with colloid NPs occurs in the liquid in front of the target. As a result of the absorption and scattering of the radiation, its intensity on the surface of the target can significantly decrease. To minimize these energy losses, the optimal layer thickness and focus radiation conditions have to be found.

Normally, the linear absorption and scattering of the dispersions can be measured and considered easily while selecting the conditions of PLAL. In our experiments, IR radiation of the fundamental harmonic of Nd:YAG laser with a wavelength of 1064 nm was used. It is known that metal targets have a large reflection in this range of the spectrum, compared to the radiation in the visible range and especially the UV range, which leads to the large energy losses. However, under the long pulse-periodic radiation, the reflectance of the material decreases up to several times [1]; that increases the efficiency of pumping. Additionally, due to minimizing of other losses, the use of IR radiation, especially for concentrated dispersions, can be evaluated. This is connected with the fact that almost all the metals do not have plasmon absorption in the near-IR range, plus there is no or minimal absorption of oxide particles and the contribution of linear scattering (compared to visible and UV ranges) is low.

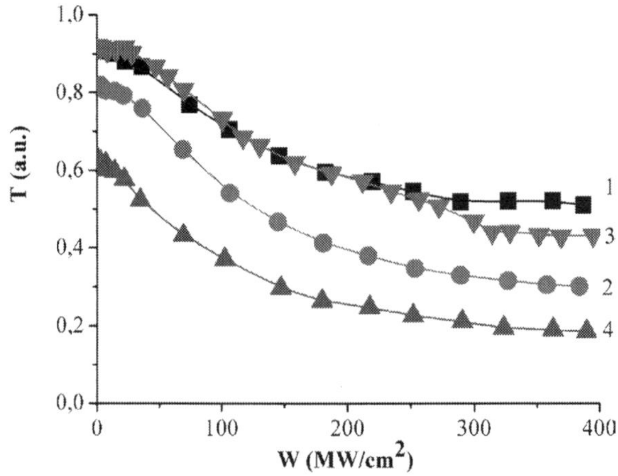

Figure 4. The dependence of transmittance of nanoparticle dispersions obtained by PLAL of zinc and copper targets in various solvents from the power density of ns laser radiation with the wavelength of 1064 nm.

In addition to the linear absorption, nonlinear processes of the interaction of radiation with the NPs have much greater influence on the productivity in the conditions of the powerful pulse excitation in colloidal solutions. These nonlinear processes show complicated dependence on/from the nature and concentration of the particles as well as the parameters of the laser pulses [15]. **Figure 4** shows the dependence of the deletion from the power density for fresh dispersions of nanoparticles obtained by PLAL of zinc targets (curves 1, 2) and copper (curves 3, 4) in water (curves 1, 3) and alcohol (curves 2, 4). Mass concentration of particles in solutions was about ~50 mg/l. From these figures, it is clearly seen that the nonlinear radiation limiting by the interaction with the NPs in the solution can significantly reduce the power density on the surface of the target and, therefore, the effectiveness of the PLAL even at the low concentrations and short optical length of the interaction.

Another important consequence of the interaction of colloidal NPs with pulsed laser radiation, as has been already noted in 2.4, is the initiation of chemical transformations and change of the dimensional characteristics of the particles. For inert metal particles, e.g., gold and silver, the processes of sintering, defragmentation and alloy particle formation are characteristic. For chemically active metals, chemical reactions are stimulated. For example, let us consider the changing of a spectrum of 2 ml of dispersion obtained by PLAL of metallic zinc target in alcohol ($C = 0.1$ g/l) after exposure to radiation of the fundamental harmonic of Nd:YAG laser (1.064 nm, 7 ns, 20 Hz) with the power density of 100 MW/cm^2 within 5 min. The formation of a characteristic band in the region of 300–350 nm for the irradiated solution points on the formation of ZnO oxide, whereas in the initial non-radiated solution, there is a significant portion of non-oxidized zinc metal phase (**Figure 5a**). Without irradiation, similar changes in the spectrum of the initial solution that is associated with the oxidation of metallic Zn particles in alcohol occur within a few hours.

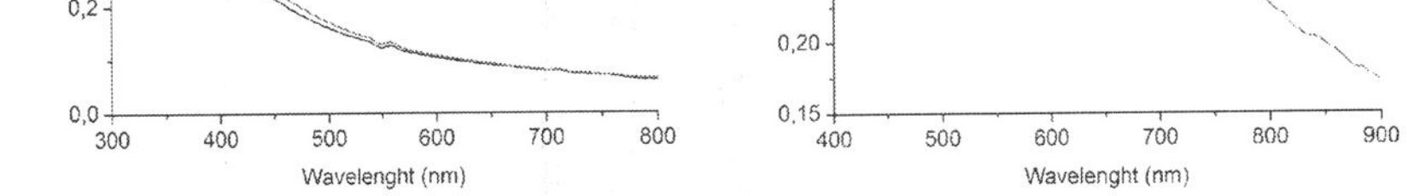

Figure 5. Absorption spectra of the dispersions of nanoparticles obtained by PLAL of zinc (a) and copper (b) targets in ethanol.

Figure 5b shows the spectra of colloidal solutions obtained by PLAL of copper in ethanol. The irradiation in the similar conditions as for zinc in **Figure 5a** does not lead to oxidation of more inert copper (a peak of a surface plasmon resonance of copper at 590 nm does not change), but scattering decreases and a wide band in the wavelength longer than 700 nm that belongs to the absorption of agglomerates disappears.

The examples show that in the process of colloidal solutions obtaining, especially under long-term irradiation and large concentrations, complex chemical processes in the dispersions occur. These processes are stimulated by the secondary interactions of NPs with the laser radiation. Thus, they can not only affect the composition and structure of the particles obtained, but significantly affect the optical properties of the environment and hence influence the PLAL process.

3. Characterization

When nanoparticles are characterized, it is often advisable to divide their properties into two classes. The first class consists of general physical-chemical properties that describe dimension parameters, composition and structure. The second class is presented by the functional properties that are closely linked/strongly connected with physical-chemical characteristics and define specific areas of NP practical application. For example, their fundamental characteristics—particle size, surface composition and charge parameters—determine their ability to penetrate into specific cells, i.e., their functional properties for therapy or diagnostics. Another example: The absorption spectrum connected with the nature of the particles and their size defines the functional ability to block UV radiation in the composition of protective sunscreen materials.

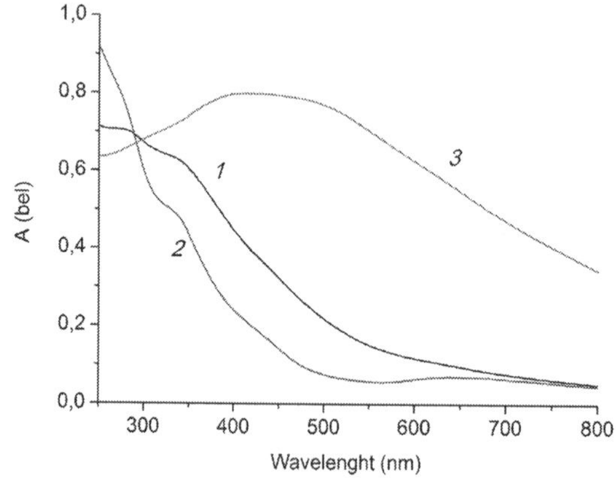

Figure 6. The dynamics of the absorption spectra of colloidal nanoparticles obtained by PLAL of metallic copper target in the water after synthesis (1), after 5 hours (2) and after 10 days (3).

The characterization of colloidal particles obtained by PLAL has numerous features related exactly to their condition. The state of the nanoparticles produced may be quite specific as a result of multiple chemical reactions that are initiated (even for relatively chemically inert materials) during high-power synthesis in the presence of a dense liquid phase. That is why the particles obtained have complicated structure with anisotropic composition, containing metastable phases and high defectiveness. Moreover, a unique solvate shell is formed around the particles and it can determine their properties in many respects. Colloid solution is often the dynamic non-stable system and continues to undergo significant changes at storage. Chemical reactions between active ablated particles, dissolved components of the target (ions) and gasses and molecules of solvent and impurities (or precursors) go on in the colloid. As an example, **Figure 6** shows the changes in the spectrum of colloidal nanoparticles' obtained PLAL of metal copper target in water during its storage. These changes are connected to the processes of oxidation and hydrolysis that occur in the colloid.

Drying of the dispersion is required to make tests of structure and composition of the nanoparticles by microscopic or X-ray methods. But even drying in the soft conditions without high temperature and oxygen can significantly transform the initial structure and composition of the nanomaterials, so it is important to be able to study the properties of NPs in their dispersions.

An effective tool to study the characteristics of NPs directly in colloids is modern optical methods (UV-Vis absorption and fluorescent spectroscopy, Raman spectroscopy, confocal fluorescent microscopy, photon correlation spectroscopy) that allow evaluation of the size and shape, as well as the composition and structure, of the nano-objects.

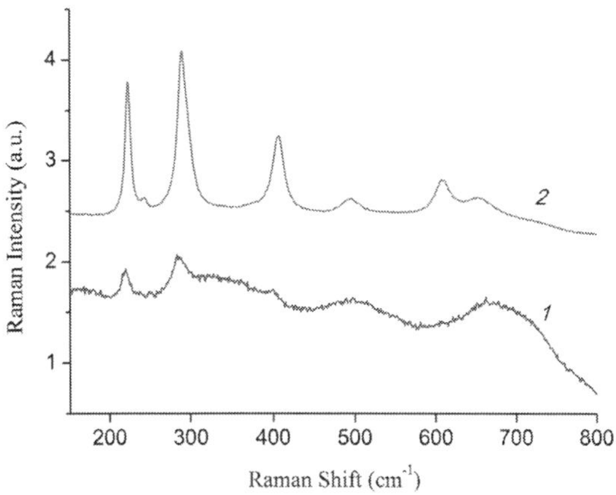

Figure 7. Raman spectra of iron oxide nanoparticles obtained by PLAL of metallic iron target in water: (1) powder after preparation and drying and (2) powder after laser irradiation.

So to determine the particles' size, apart from the widely used method of photon correlation spectroscopy based on the dynamic light scattering by the nanoparticles, there are empirical dependences between the diameter of NPs and the band gap in a certain range of sizes for a number of quantum dots including oxides [16, 17]. These dependences allow assessment of the particles' size from the absorption spectra. UV-Vis absorption spectra are a simple, express method to study the formation of oxides in PLAL of a large number of metals (Ce, Ti, Cu, Zn, Mg, etc.), because a characteristic peak of exciton absorption appears in the spectrum.

Raman spectroscopy allows evaluation of the particles' structure both in powders and in dispersions. For example, this method makes it possible to distinguish between three iron oxide forms, magnetite, maghemite and hematite [18], that are difficult to recognize by other methods, including X-ray diffraction. **Figure 7** shows the Raman spectra of iron oxide nanoparticles obtained by PLAL of a metallic iron target in water. Freshly obtained magnetite with superparamagnetic properties (curve 1) during heating or irradiation by laser easily transforms into hematite (curve 2). Also, shifts and widening of bands in Raman spectra allow one to reveal the structure, including amorphous materials [19], defects and NP size [20, 21].

Fluorescence spectroscopy is used for the study of defective states of NPs of many oxides in dispersions and powders. For example, spectra and kinetics of fluorescence give information on different oxygen defects in zinc oxides [22], titanium oxide [23], tin oxide [24], etc.

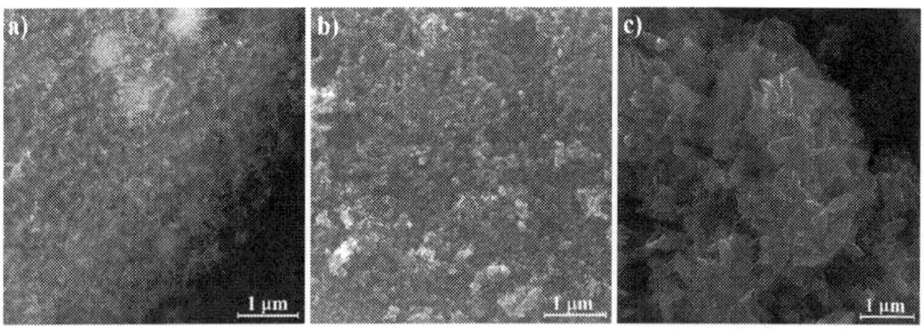

Figure 8. SEM images of CuO nanoparticles obtained at various conditions: drying of water dispersion with the subsequent annealing at 500°C (a), vacuum drying of water dispersion with addition of 1% (weight) hydrogen peroxide (b) and drying of water dispersion with addition of 0.01 M of nitric acid followed by annealing at 300°C (c).

Obviously, classic methods for determining particles' dimensional characteristics are important and are also used for the study of the NPs obtained by the PLAL method. A typical example of such a study is presented in **Figure 8**, which shows microphotographs of copper (II) oxide crystal powders. An important feature of using different solvents and precursors in PLAL is the ability to control not only the composition and structure of the particles but also their size and morphology. Additional treatments of the particles in the dispersions obtained by PLAL and different methods of nanopowder preparation from them (deposition, thermal and vacuum drying, subsequent annealing) allow, for example, synthesizing nanomaterials with different characteristics but the same chemical composition. Copper (II) oxide CuO powders in **Figure 8** are obtained in different experimental conditions of PLAL and further treatments: drying of water dispersion with the subsequent annealing at 500°C (a), vacuum drying of water dispersion with addition of 1% (weight) hydrogen peroxide (b) and drying of water dispersion with addition of 0.01 M of nitric acid followed by annealing at 300°C (c). In the first case, the oxidation of Cu_2O occurs; in the second case, CuO was obtained directly in the dispersion during the PLAL; in the last version, NPs of $Cu_2NO_3(OH)_3$ were a result of ablation and after annealing, it transformed into copper (II) oxide. NPs of copper (II) oxide obtained by PLAL in water with additive of 1% (weight) of hydrogen peroxide exhibit needle-like shape. The average size of crystals is 10 nm and specific surface is sufficiently large—66 m^2/g (**Figure 8a**). The particles obtained by the annealing of copper (I) oxide are larger and faceted (**Figure 8b**) and the specific surface area is of 12 m^2/g (prior to annealing, BET surface area or specific surface area measured via BET method (S_{BET}) was 30 m^2/g). At the same time, the PLAL with the use of aqueous solutions of HNO_3 and subsequent annealing resulted in the formation of well-crystallized large particles (**Figure 8c**) with the specific surface area of ~2 m^2/g.

4. Application

The scope of application of metal oxide NPs is extremely wide and varied. However, for the oxide nanoparticles obtained by PLAL, there are two most promising applications: scientific research and biomedicine. In both cases, there is no demand for large quantities of nanomaterials and the disadvantage of the PLAL connected with the relatively low productivity is immaterial. Instead, the important advantages of PLAL are express preparation of a wide range of nanomaterials in laboratory conditions for scientific research and "pure" nanoparticles directly in the form of colloidal solutions for biomedicine.

Thus, in our laboratory, oxide nanoparticles obtained by PLAL are used in the study of catalytic processes for searching for the structure and composition of the most effective catalysts for various chemical and photochemical processes. In Ref. [25], the results of the study of catalytic CO oxidation reaction for CeO_2-Pd composites obtained by PLAL of Ce and Pd targets in various solvents are presented. The change of the synthesis parameters of PLAL allows control of the size, chemical composition and superficial properties of obtained particles and hence modification of the catalytic properties. In addition to the cerium dioxide, we used Al_2O_3, SnO, SnO_2, CuO and CuO obtained by PLAL as carriers and catalysts. Composites based on zinc oxide, titanium dioxide and copper oxide nanoparticles were used for the study of photocatalytic processes.

Nanocolloids obtained by PLAL are suitable material for the study of the influence of nanomaterials on the environment. On the one hand, inert enough particles of noble metals can be used to define their migration paths in different ecosystems and, on the other, to investigate the toxicity. Authors of Ref. [26] used several types of nanoparticles including particles obtained by PLAL to study their influence on different aquatic organisms.

Search for tools and methods of protection from antibiotic-resistant pathogenic bacteria strains is an actual problem of modern medicine. The use of nanoparticles as antibiotics that does not cause an adaptation effect among bacteria is one of the promising approaches. In this regard, a new area of research in biomedicine has appeared and the term "nano-antibiotics" [27] has been invented. Effects of zinc oxide nanoparticles obtained by PLAL on pathogenic bacteria were examined in Ref. [28] compared to silver nanoparticles and classic antibiotics. Then, we continued research using NPs of Cu_2O that are more movable and have the best antibacterial properties compared to NPs of ZnO. **Figure 9** shows a picture of cotton fabric samples with

ZnO and Cu_2O particles deposited from water dispersions obtained by PLAL; the composites obtained were covered by *Escherichia coli* bacteria. The figure illustrates the suppression of bacteria growth on fabrics with nanoparticles, while bacteria multiply on the test sample.

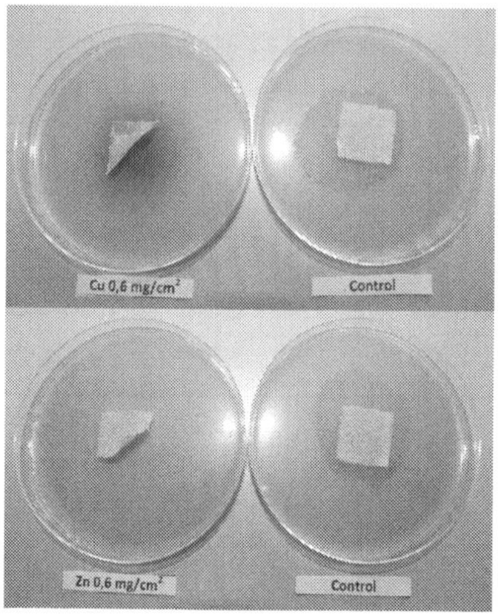

Figure 9. *E. coli* growth inhibition by zinc oxide and copper oxide nanoparticles.

Among other applications in biomedicine, the prospects of using CeO_2, TiO_2 and ZnO obtained by PLAL to protect the skin from UV radiation (sunscreen) are worth noting. Lately, CeO_2 has attracted more and more attention in this field [29] because its toxicity is much lower than standard TiO_2 and its photocatalytic effects are lower as well.

The new direction that has developed effectively in nanomedicine in the last decade is the use of magnetic nanoparticles for modern theranostics—to deliver drugs, for contrasting of pathological object and in magnetic therapy. The use of PLAL for obtaining nanoparticles with special magnetic properties based on oxides of 3d metals directly in the biologically compatible liquids, their functionalization by gold particles and biopolymers have great prospects for the applications mentioned above.

5. Conclusions

Pulsed laser ablation in liquids is a unique physical-chemical method of obtaining nanostructures in colloidal solutions. At the beginning of the process, physical high-energy pulsed laser irradiation stimulates the chain of both physical and chemical processes that lead to the formation of nanoparticles. Careful control of the process allows a large variety of nanostructures to be obtained. This method cannot be considered as the classic top-down approach of nanomaterial synthesis, despite the fact that initially bulk targets are used. As a result of the PLAL process, a plasma cloud of ions and small clusters appears and nanoparticles are formed from it.

The potential use of PLAL for the synthesis of pure nanocolloids for high-tech applications in science, biomedicine and a number of other areas overlap with disadvantages associated with its relatively small productivity. The specific characteristics of nanosecond laser synthesis, including the factors influencing the effectiveness of PLAL; chemical transformations in the liquids; the specificity of nanodispersion characterization, including optical methods; and future areas of such prepared materials application discussed in the chapter using concrete examples are designed to help scientists and professionals who deal with synthesis, characterization and applications of the similar nanocolloids and powders in their studies.

Acknowledgements

This work was carried out under the State Assignment No. 2014/223 of the Ministry of Education and Science of the Russian Federation (Project No. 1347).

Author details

Valery A. Svetlichnyi*, Anastasiia V. Shabalina, Ivan N. Lapin and Daria A. Goncharova

*Address all correspondence to: v_svetlichnyi@bk.ru

Siberian Physical-Technical Institute of Tomsk State University, Tomsk, Russia

References

[1] Ready J.F. Industrial Applications of Lasers. 2nd ed. San Diego: Academic Press; 1997. 600 p.

[2] Miller J.C. editor. Laser Ablation: Principles and Applications. Berlin: Springer-Verlag; 1994. 197 p. DOI: 10.1007/978-3-642-78720-1

[3] Eason R. editor. Pulsed Laser Deposition of Thin Films: Applications-Led Growth of Functional Materials. Hoboken: John Wiley & Sons, Inc.; 2007. 707 p. DOI: 10.1002/0470052120

[4] Ullmann M., Friedlander S.K., Schmidt-Ott A. Nanoparticle formation by laser ablation. J. Nanopart. Res. 2002; 4: 499-509. DOI: 10.1023/A:1022840924336

[5] Neddersen J., Chumanov G., Cotton T.M. Laser Ablation of Metals: A New Method for Preparing SERS Active Colloids. Appl. Spectrosc. 1993; 47: 1959-1964. DOI: 10.1366/0003702934066460

[6] Yang G. editor. Laser Ablation in Liquids: Principles and Applications in the Preparation of Nanomaterials. Singapore: Pan Stanford Publishing Pt. Ltd.; 2012. 1164 p. DOI: 10.4032/9789814241526

[7] Zeng H., Du X.-W., Singh S.C., Kulinich S.A., Yang S., He J., Cai W. Nanomaterials via laser ablation/irradiation in liquid: A review. Adv. Funct. Mater. 2012; 22: 1333-1353. DOI: 10.1002/adfm.201102295

[8] Tan D., Zhou S., Qiu J., Khusro N. Preparation of functional nanomaterials with femtosecond laser ablation in solution. J. Photochem. Photobiol. C. 2013; 17: 50-68. DOI: 10.1016/j.jphotochemrev.2013.08.002

[9] Bulgakova N.M., Bulgakov A.V. Pulsed laser ablation of solids: Transition from normal vaporization to phase explosion. Appl. Phys. A. 2001; 73: 199-208. DOI: 10.1007/s003390000686

[10] Lorazo P., Lewis L.J., Meunier M. Short-pulse laser ablation of solids: From phase explosion to fragmentation. Phys. Rev. Lett. 2003; 91: 225502-1-225502-4. DOI: 10.1103/PhysRevLett.91.225502

[11] Leitz K.-H., Redlingshöfer B., Reg Y. Otto A., Schmidt M. Metal ablation with short and ultrashort laser pulses. Phys. Procedia. 2011; 12: 230-238. DOI: 10.1016/j.phpro.2011.03.128

[12] Lapin I.N., Svetlichnyi V.A. Features of the synthesis of nanocolloid oxides by laser ablation of bulk metal targets in solutions. Proc. of SPIE. 2015; 9810: 98100T1-7. DOI: 10.1117/12.2224699

[13] Yang G.W. Laser ablation in liquids: Applications in the synthesis of nanocrystals. Prog. Mater. Sci. 2007; 52: 648-698. DOI: 10.1016/j.pmatsci.2006.10.016

[14] Goncharova D., Lapin I., Svetlichnyi V. Synthesis of CdS nanoparticles by laser ablation of metallic cadmium target in presence different precursors. Adv. Mater. Res. 2015; 1085: 182-186. DOI: 10.4028/www.scientific.net/AMR.1085.182

[15] Sun Y.-P., Riggs J.E., Henbest K.B., Martin R.B. Nanomaterials as optical limiters. J. Nonlinear Opt. Phys. Mater. 2000; 9: 481-503. DOI: 10.1142/S0218863500000315

[16] Kippeny T., Swafford L.A., Rosenthal S.J. Semiconductor nanocrystals: A powerful visual aid for introducing the particle in a box. J. Chem. Educ. 2002; 79: 1094-1100. DOI: 10.1021/ed079p1094

[17] Zhang F., Jin Q., Chan S.-W.Ceria nanoparticles: Size, size distribution and shape. J. Appl. Phys. 2004; 95: 4319-4326. DOI: 10.1063/1.1667251

[18] Shebanova O.N., Lazor P. Raman study of magnetite (Fe_3O_4): Laser-induced thermal effects and oxidation. J. Raman Spectrosc. 2003; 34: 845-852. DOI: 10.1002/jrs.1056

[19] Yoshitake H., Abe D. Raman spectroscopic study of the framework structure of amorphous mesoporous titania. Micropor. Mesopor. Mater. 2009; 119: 267-275. DOI: 10.1016/j.micromeso.2008.10.027

[20] Spanier J.E., Robinson R.D., Zhang F., Chan S.-W., Herman I.P. Size-dependent properties of CeO_{2-y} nanoparticles as studied by Raman scattering. Physical Review B. 2001; 64: 245407-1-245407-8. DOI: 10.1103/PhysRevB.64.245407

[21] Choi H.C., Jung Y.M., Kim S.B. Size effects in the Raman spectra of TiO_2 nanoparticles. Vib. Spectrosco. 2005; 37: 33-38. DOI: 10.1016/j.vibspec.2004.05.006

[22] Svetlichnyi V.A., Lapin I.N. Structure and properties of nanoparticles fabricated by laser ablation of Zn metal targets in water and ethanol. Russ. Phys. J. 2013; 56: 581-587. DOI: 10.1007/s11182-013-0071-z

[23] Mathew S., Prasad A. K., Benoy T., Rakesh P.P., Hari M., Libish T.M., Radhakrishnan P., Nampoori V.P.N., Vallabhan C.P.G. UV-visible photoluminescence of TiO_2 nanoparticles prepared by hydrothermal method. J. Fluoresc. 2012; 22: 1563-1569. DOI: 10.1007/s10895-012-1096-3

[24] Gondal M.A., Drmosh Q.A., Saleh T.A. Preparation and characterization of SnO_2 nanoparticles using high power pulsed laser. Appl. Surf. Sci. 2010; 256: 7067-7070. DOI: 10.1016/j.apsusc.2010.05.027

[25] Slavinskaya E.M., Kardash T.Y., Gulyaev R.V., Stonkus O.A., Lapin I.N., Svetlichnyi V.A., Boronin A.I. Metal-support interaction in Pd/CeO_2 model catalysts for CO oxidation: From pulsed laser-ablated nanoparticles to highly-active state of the catalyst. Catal. Sci. Technol. 2016; 6 (17): 6650–6666. DOI: 10.1039/C6CY00319B

[26] Morgalev S., Morgaleva T., Gosteva I., Morgalev Y. Evaluation of the toxicity of superfine materials to change the physiological functions of aquatic organisms of different trophic levels. IOP Conf. Series. Mater. Sci. Eng. 2015; 98: 012006-1-012006-8. DOI: 10.1088/1757-899X/98/1/012006.

[27] Huh A.J., Kwon Y.J. "Nanoantibiotics": A new paradigm for treating infectious diseases using nanomaterials in the antibiotics resistant era. J. Control. Release. 2011; 156: 128-145. DOI: 10.1016/j.jconrel.2011.07.002

[28] Svetlichnyi V., Shabalina A., Lapin I., Goncharova D., Nemoykina A. ZnO nanoparticles obtained by pulsed laser ablation and their composite with cotton fabric: Preparation and study of antibacterial activity. Appl. Surf. Sci. 2016; 372: 20-29. DOI: 10.1016/j.apsusc.2016.03.043

[29] Yabe S., Sato T. Cerium oxide for sunscreen cosmetics. J. Solid State Chem. 2013; 171: 7-11. DOI: 10.1016/S0022-4596(02)00139-1

Chapter 12

Application of Liquid Laser Ablation: Organic Nanoparticle Formation and Hydrogen Gas Generation

Ikuko Akimoto and Nobuhiko Ozaki

Additional information is available at the end of the chapter

http://dx.doi.org/10.5772/64939

Abstract

Laser ablation is induced by a heating process of materials through the absorption of laser light and results in an explosive expansion of materials. For materials located in liquid, in contrast to those in vacuum, laser ablation proceeds under rather mild conditions via a cycle of heating and cooling by mediated solvent; therefore, it is applicable for organic solids to fragment into nanoparticles. Alternatively, for effective light absorbers, the irradiated site becomes the reaction centre of a photochemical reaction even in liquids, resulting in hydrogen gas generation. In this chapter, two topics of laser ablation in the liquid phase are presented: nanoparticle formation of organic materials and hydrogen gas generation from solid carbon in water. Thereby, the extended abilities of liquid laser ablation to transform ordinary materials into functional ones are introduced.

Keywords: organic nanoparticles, colloidal solution, hydrogen generation, carbon-based nanoparticles, spectroscopy, nanosecond laser pulses

1. Introduction

Laser ablation in the liquid phase, which is a breakdown method leading to nanoparticle formations developed in the study of laser-processing techniques using short and ultrashort laser pulses [1, 2], was applied to organic materials in the middle of the 1990s by Masuhara and coworkers [3–14]. In the liquid phase, laser ablation proceeds under rather mild conditions via a cycle of heating by light absorption and cooling by mediated solvent molecules under ambient atmospheric pressure and temperature, in contrast to the laser ablation in the gas phase such as laser deposition of metals under vacuum [2]. Therefore, it is applicable for an organic material that is more intolerant of optical radiation than a metal or an inorganic semiconductor.

Laser ablation of an organic material in the liquid phase results in impressive phenomena. Microcrystals of an organic material suspended in a poor solvent, typically water, are irradiated by laser pulses; thereby, the cloudy suspension is transformed into transparent-colored water. The resultant is a colloidal solution containing organic fragments of a few tens to hundreds nanometers in diameter, that is, nanoparticles. The colloidal solution is stable for a long time without any surfactants, typically for weeks or more, because of persistent Brownian motion of the nanoparticles in the solvent [12]. Therefore, laser ablation in the liquid phase produces a uniform solution of small organic materials. The size of the nanoparticles can be controlled by the ablation conditions such as the irradiated laser fluence [15], laser pulse duration [16], and wavelength [10]. The easy and high collectability of the created nanoparticles by treating the solution is a useful characteristic of laser ablation in the liquid phase.

Such a colloidal solution is applicable in an ink print, a drag delivery, and cosmetic applications. In particular, it is expected to provide a low-cost wet process in the device fabrication field. In recent years, organic materials are increasingly utilized in optical and electronic devices such as organic light-emitting diodes (OLEDs) [17] and organic field-effect transistors (OFETs) [18]. However, high-cost vacuum sublimation processes are required for the fabrication of such devices, because small organic molecules are difficult to dissolve in solvents without the addition of hydrophilic moieties on the molecule or any surfactants. The attachment of additional moieties generally does not guarantee the original functions of molecules, and surfactants may affect the functions of systems. Therefore, the formation of a stable solution of the target organic materials by laser ablation is a promising technique for the device fabrication. The colloidal solution gives a significant advantage to the industrial field with regard to low-cost and eco-friendly products, by applying a print technique as introduced in Ref. [19].

Alternatively, an irradiated site in laser ablation can be a reaction centre of photochemistry for energy conversion from light to a fuel gas. A novel hydrogen generation method was discovered during laser ablation of solid carbon in water [20, 21]. Hydrogen gas is a clean gaseous fuel, and hydrogen generation methods have been intensively studied. Classical coal gasification [22] and optically induced water-splitting in the presence of a photocatalyst titanium oxide [23] are representative phenomena in hydrogen generation methods. In the photocatalytic water-splitting reaction, the addition of solid carbon assisted the reaction by avoiding the reverse reaction through oxidization [24, 25]. Thus, solid carbon has been utilized for hydrogen gas production. However, the optical activity of solid carbon itself was not known before the discovery.

As by-products of this laser-induced hydrogen gas generation reaction, nanoparticles of carbon-based materials were also found in the post-irradiated solution. Thus far, novel carbon-based materials, such as graphene oxide (GO) [26–28], nano-diamond [29, 30], and diamond-like carbon (DLC) [29, 31], have been studied as photocatalysts in the water splitting reaction [26, 30] and reduction of carbon oxide [27, 28]. These carbon-based materials can be produced by laser ablation of solid carbon as demonstrated in Refs. [29, 31]. Therefore, laser ablation of solid carbon in an aqueous solution is attractive for at least two different interests: hydrogen generation and carbon-based nanoparticle production.

In this chapter, two topics of laser ablation in the liquid phase using nanosecond laser pulses are presented. The first is nanoparticle formation of organic materials, particularly from the points of particle size control by irradiated laser fluence and absorption spectrum properties depending on particle size. The second is hydrogen gas generation from solid carbon in water accompanied with nanoparticle generation. Through these topics, the extended abilities of liquid laser ablation to transform ordinary materials into functional ones are introduced.

2. Laser ablation techniques

2.1. Procedure of laser ablation in the liquid phase

Laser ablation in liquid phase is performed for a solid material suspended in poor solvent, in which the material is difficult to dissolve, by irradiation of laser pulses through the side wall of a transparent cuvette/bottle or the top surface of the solution. The suspension was stirred by a magnetic stirrer during the irradiation. The experiments presented in this chapter were performed with laser pulses of 5 ns pulse duration from a tunable optical parametric oscillator excited by a Q-switched YAG laser operated at a 10 Hz repetition rate (Spectra Physics, MOPO) or from a second harmonics of a Q-switched YAG laser (Continuum Surelite) operated at a 10 Hz repetition rate. The incident laser power was controlled using a Glan-Laser calcite polarizer. By maintaining the laser beam unfocused or loosely focused, the energy density was maintained below a few hundred milli-joule per square centimeter in order to not achieve a plasma state [29, 32, 33]. The experimental details are described in the following sections. All laser ablation experiments were performed at room temperature (24°C).

2.2. Nanoparticle formation for organic molecules in water

Laser ablation of organic molecules was performed with dispersed mixture of organic molecules in the distilled water in quartz cuvette of $1 \times 1 \times 5$ cm^3 at a concentration in the range from 2×10^{-2} to 5×10^{-5} mol/l. The microcrystals in suspension were irradiated by laser pulses for a few minutes. The wavelengths of the irradiated laser pulse were selected corresponding to absorption band of molecules. Here, results for two materials are presented; a yellow pigment quinacridone quinone (QQ, Aldrich) was irradiated at the wavelength 430 nm, and rubrene (Rb, Aldrich, sublimated grade) was irradiated at 520 nm. More details of preparation procedure were described in Refs. [34, 35].

Post-irradiated solution was investigated by UV-VIS absorption measurement with a conventional system (JASCO, V-560) and a dynamic light scattering (DLS) measurement (HORIBA Scientific, nanopartica, or Otsuka Electronics, Photal). The mode diameter, which indicates the most frequent diameter of nanoparticles in the ensemble, was employed to estimate the size of particles. Dried nanoparticles in deposited films were visualized by an atomic force microscopy (AFM) (SII, SPA400) and the transmission electron microscopy (TEM) (JEOL,

2000EX). Surface electric potential on the nanoparticles was obtained by a ζ-potential measurement (HORIBA Scientific, nanopartica, or Otsuka Electronics, Photal). A film of the QQ nanoparticles was prepared on a glass substrate covered by an indium-tin-oxide (ITO) transparent electrode by the electrophoretic deposition (EPD) method, and its UV-VIS absorption spectrum was compared with that of a vapor-deposited film of QQ with a thickness of 46.5 nm and that of solutions.

Time-resolved EPR measurement was carried out in QQ 2-methyltetrahydrofuran solution at 90 K and C_{60} toluene solutions at 100 K by the excitation at 430 and 532 nm, respectively, with an instrument (Bruker, ELEXSYS E580).

2.3. Gas generation via laser ablation of carbon materials in aqueous solution

In the laser ablation of carbon in aqueous solution, binchotan charcoal powder of a mean diameter 5 μm (A, Latest Coop., Wakayama, Japan), high-grade carbon powder of a mean diameter 5 μm (B, SEC, SCN-5, 99.5%), and graphite powder of a mean diameter less than 45 μm (C, Wako, 072-03845) were used. Surface area of powder was measured by the BET method developed by Brunauer, Emmett, and Teller. The mixture of the powder and distilled water was irradiated by an unfocused beam (6.2 mm in diameter) of laser pulses for 30 or 60 min. The wavelengths of the laser pulses were selected in the VIS-near-infrared (NIR) region. The generated gas was collected by the water displacement method. The collected gas volume was measured with a scale on a tube at a resolution of 0.05 mL.

Gas components were analyzed by quadrupole mass spectrometry (Nuclear Engineering Co., Ltd., Ibaraki, Japan) for gases generated under argon atmosphere. The gas components were compared with two gas samples generated from binchotan powder in 100% distilled water and in 50% ethanol aqueous solution. The portion of molecules N_2, CO, and C_2H_4 of the same molecular mass at 28 was determined by filtered partial presser measurements and mass fragments at N and C in quadrupole mass spectrometer. More details of preparation procedure were described in Refs. [20, 21].

During the hydrogen generation, optical radiation from the irradiated site was observed for a commercial binchotan charcoal block and a carbon electrode block (99.9%) in distilled water (H_2O) or in 50% ethanol aqueous solution (EtOH/H_2O). The block was irradiated by loosely focused nanosecond laser pulses (5 ns, 10 Hz, 532 nm) with a laser beam size of 0.50 × 0.25 cm^2. The emission spectrum was detected using an intensified charge-coupled device (ICCD) (Roper Scientific, PI-MAX) attached to a monochromator (Acton, 300i) with 4 nm spectral resolution. Strong light scattering was blocked by a super notch filter designed for 532 nm incident light. More details of preparation procedure were described in Ref. [36].

Post-irradiated solutions and ablation products were investigated by UV-VIS, DLS, and TEM methods, similar to the organic nanoparticles as mentioned in Section 2.2. In addition, FT-IR spectrum of dried carbon-based nanoparticles deposited on a pure silicon substrate was observed.

http://dx.doi.org/10.5772/64939

3. Organic nanoparticle formation

The validity of the laser ablation in the liquid phase for nanoparticle generation has been demonstrated by several groups for organic systems including poly-diacetylene (poly-DCHD) [4], metallo-phthalocyanines [6–9], dendronized perylenediimide (DPDI) [13, 14], perylene [12], pentacene [37], and a series of pigments including quinacridone (QA) [10–12] and its derivatives [34, 35]. Fullerene C_{60} was also fragmented into nanoparticles by laser ablation in water [15], although it is an inorganic molecule. The number of molecules is rather limited. The diversity of applicable organic molecules for fragmentation by laser ablation is limited by the photodegradation of a molecule [34] even under the mild conditions in the liquid phase. As a result, the optical properties of the colloidal solutions need case by case interpretations. Further investigation is required for nanoparticle formation of various organic molecules by laser ablation.

The most successful organic system for laser ablation is a class of pigments, QA, and quinacridone-quinone (QQ), whose molecular structures are shown in **Figure 1**. They exhibit an excellent tolerance for photolysis. A yellow pigment QQ has a simple monomorphism crystal phase [38], which made spectral analysis easy. On the other hand, a red pigment QA has polymorphism in α, β, or γ forms, and significant spectral change was observed depending on the crystal form [10]. Therefore, QQ is more suitable to investigate the optical properties of colloidal solutions. Here, the size-dependent optical properties of the QQ colloidal solutions prepared by laser ablation in the liquid phase are presented.

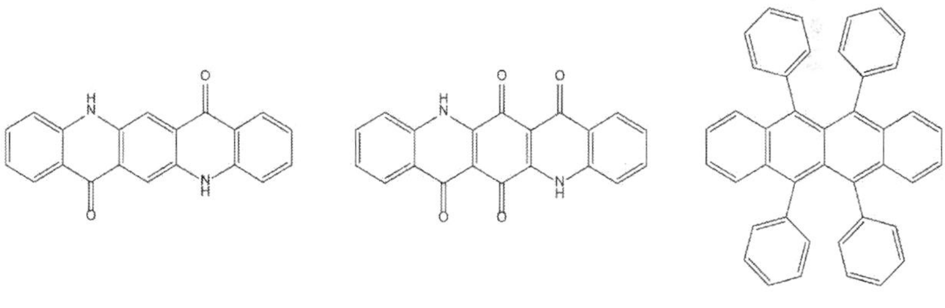

Figure 1. Molecular structures of QQ, QA, and rubrene (Rb).

3.1. Nanoparticle generation: in the case of quinacridone quinone

Figure 2a shows the absorption spectra of the supernatants before and after laser irradiation at the irradiation fluence of 11 mJ/cm^2 and the wavelength 430 nm [35]. Light extinction in the suspension is due to the absorption and scattering of light. For prolonged laser irradiation, a characteristic absorption peak at 2.88 eV increased and a scattering tails at 2.24 eV decreased. Such changes were accompanied with a visible disappearance of precipitants and appearance of a transparent yellow solution. This visible change is a characteristic of laser ablation in the liquid phase, caused by converting precipitants to QQ nanoparticles, as described in the following section. The transparency is maintained for months, imparting interests for various applications [19].

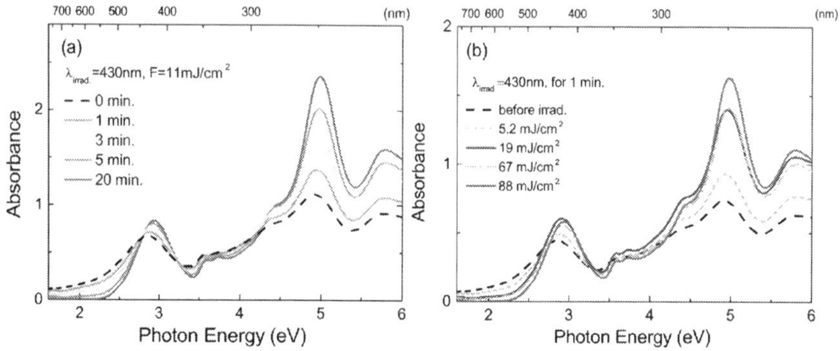

Figure 2. (a) Absorption spectra before and after various irradiation times at a laser fluence of 11 mJ/cm². (b) Absorption spectra before and after 1 min irradiation at various laser fluences of 5.2, 19, 67, and 88 mJ/cm². Adapted with permission from Ref. [35].

A similar change was also observed for increase in the irradiated laser fluence. **Figure 2b** shows the absorption spectra of the QQ supernatants before and after laser irradiation for 1 min at laser fluences of 5.2, 19, 67, and 88 mJ/cm^2. As the laser fluence increased, the absorbance increased. In addition, the peak energy clearly shifted to the higher energy side (blueshift), and each full width at half maximum (FWHM) became narrower (see **Figure 4a**). Similar spectral changes were observed above a threshold fluence, which depends on the specimens for QA in water [10], QQ in chloroform, and others [7, 14], whereas the threshold was smeared for QQ in water due to the relatively higher solubility in water [35]. Furthermore, in the case of QQ, nanoparticle generation by laser ablation was confirmed even in pH-controlled water from 2.5 to 10 by ion-exchange resin.

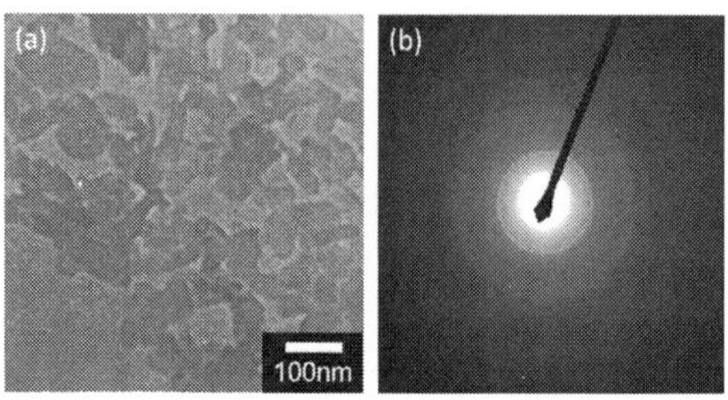

Figure 3. (a) Typical TEM image and (b) electron diffraction pattern of QQ nanoparticles prepared with the laser fluence 19.2 mJ/cm² for 1 min. Adapted with permission from Ref. [35].

The formation of nanoparticles was confirmed by AFM and TEM images of dried specimens as well as by DLS measurements of the solution. A typical TEM image of the QQ nanoparticles, which was prepared at a fluence of 19.2 mJ/cm² for 1 min, indicates the shape of distorted ellipsoid dispersed uniformly, as shown in **Figure 3a**. The image of the nanoparticles was observed typically in size around 90 nm, which is not far from the mode diameter observed by the DLS (78 nm). An electron diffraction pattern of the nanoparticle ensemble shows multiple Debye-Scherrer rings (**Figure 3b**), which means that the nanoparticles consist of a crystalline structure. The lattice spacings were coincident with those obtained by X-ray diffraction measurements for the QQ powder within analytical accuracy. Therefore, the crystalline structure of the QQ powder is maintained after the laser ablation in water.

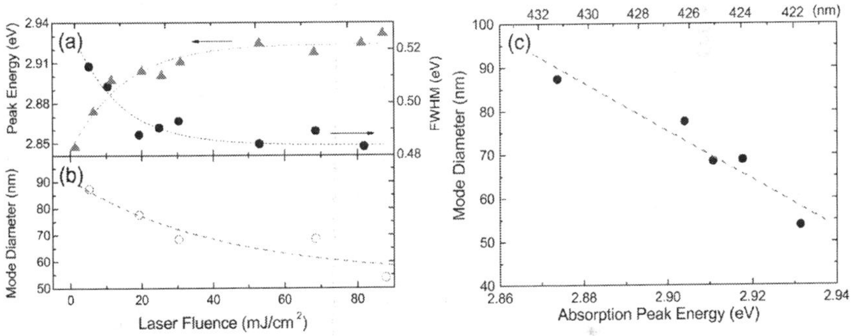

Figure 4. (a) Fluence dependence of the lowest absorption peak energy (solid triangles) and its FWHM (solid circles). (b) Fluence dependence of the mode diameter estimated by DLS (open circles). Dotted curves are guides for eyes. (c) Correlation of the mode diameter to the lowest peak energy obtained at various fluences (solid circles). A broken line is a fitting curve obtained by the least squares method. Adapted with permission from Ref. [35].

Note the relationship between the blueshift of the lowest absorption peak energy and the mode diameters of the nanoparticles contained in the solutions estimated by DLS. As shown in **Figure 4b**, the observed mode diameter was smaller for the irradiation at higher laser fluences. A linear correlation between the mode diameters and the lowest absorption peak energies was found as shown in **Figure 4c**. This relationship provides us the possibility of simple estimation of the most frequent nanoparticle diameter in an ensemble by observation of the absorption peak energy, at least, within the range from 55 to 90 nm in QQ. Such a size dependence was hidden by spectral variation due to polymorphism in QA [10] and by superposition of light scattering in DPDI [14].

The size dependence of the energy shift in the absorption spectrum can be considered in relation to the surface states. The ratio of a surface area (S) to a particle volume (V) is larger for smaller particles with the dependence of S/V=6/D, where D is the diameter. Thus, the larger shifts with the smaller particles imply influences from surface states. Indeed, the QQ nanoparticles were under a negative surface potential (ζ-potential) of −69 to −44 mV [34] in water, indicating the creation of charge or polarization on the surface by laser irradiation.

Furthermore, by utilizing the negative ζ-potential on the nanoparticles, a film was fabricated on an ITO glass electrode from the colloidal solution by electrophoretic deposition (EPD). As shown in **Figure 5**, the absorption spectrum of the nanoparticle film (d) showed that the lowest energy peak was apparent at the same position as that in the colloidal solution (c) but was shifted from those of a vapor-deposited film (b) and solution before irradiation (a) [34]. The characteristic of the colloidal solution formed by liquid laser ablation was maintained in the EPD film. A preliminary device with nanoparticles by the EPD method was demonstrated for QA [11]. Such negative surface potential response to the electric field is also applicable for the roll-to-roll fabrication method [39].

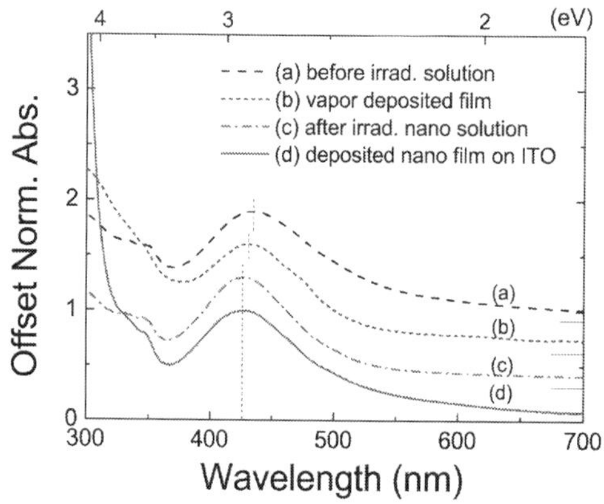

Figure 5. Comparison of normalized absorption spectra of (a) a starting aqueous QQ solution before irradiation, (b) a vapor-deposited film, (c) a colloidal solution before preparing an EPD deposit film, and (d) an EPD deposit film on an ITO electrode. Adapted with permission from Ref. [34].

3.2. Suitable organic materials for nanoparticle generation by laser ablation

No signal of photodegradation appeared in the yellow pigment QQ under the aforementioned irradiation conditions. The rigid molecular structure and stacking of flat molecules in QQ and QA systems enhance the tolerance for laser irradiation [38]. In contrast, the laser ablation of fragile and luminous molecules, such as rubrene (Rb) whose molecular structure is shown **Figure 1**, failed because Rb underwent photodissociation upon irradiation by the laser pulses [34].

Besides a molecular structure, it is worth taking into account the relaxation processes after optical excitation in order to understand the necessary condition for laser fragmentation. Because the fragmentation of organic powders into nanoparticles proceeds by rapid photothermal conversion on the surface layers of a solid [10], non-radiative relaxation processes, such as intersystem crossing and/or internal conversions, are potential thermal sources in the molecules. Population into an excited triplet state is a plausible entrance of the de-excitation path into thermal energy generation.

The population of an excited triplet state was observed for QQ and C_{60} by transient EPR measurements as shown in **Figure 6**; however, it was unobservable for Rb. Transient microwave signals of emission and absorption decayed with a lifetime of 57 μs in the QQ solution at 90 K and 3–6 μs in the C_{60} solution at 100 K, respectively. These signals arise from transitions between sublevels of an excited triplet state which were populated via efficient intersystem crossing from an excited singlet state after optical excitation. Although the observed lifetime in the C_{60} solution was much faster than the value in literature [40] due to oxidation and a high concentration of the solution, a large intersystem crossing and triplet population were obvious. Therefore, a sufficient population of an excited triplet state is one possible necessary condition for laser fragmentation via photothermal conversion in organic materials. In addition, photoluminescence (PL) from QQ was hardly observed, similar to C_{60}, in which PL was somewhat observed with a radiative quantum yield of 10^{-4} [41]. The absent of PL may also be a good signal for proceeding with the photothermal conversion.

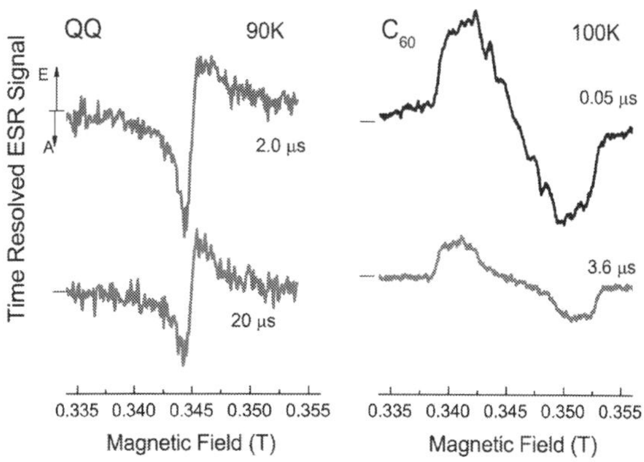

Figure 6. Time-resolved EPR signals for a QQ solution at 90 K and electron-spin-echo-detected time-resolved EPR signals for the C_{60} solution at 100 K.

4. Hydrogen generation

From another viewpoint, the irradiated site in the laser ablation can be a reaction centre of energy conversion from light to a fuel gas. In this section, our recent discovery of novel hydrogen gas generation during laser ablation of carbon in an aqueous solution [20, 21] is presented. This reaction proceeded via a photochemical reaction that carried the temperature elevation of the irradiated sites in water [36]. Electrodes and any other photocatalysts were not necessary for such reactions. Furthermore, the hydrogen generation was accompanied by simultaneous carbon-based nanoparticle production. Therefore, the laser ablation of carbon in water demonstrated two different interests: the hydrogen generation and the carbon-based nanoparticle production.

This hydrogen generation reaction occurred under a lower irradiation energy than that required for plasma-state generation. It has been known that the plasma state is induced when the laser pulse energy is focused on materials with an energy density over a few joule per square centimeter [29, 32]. Such an exploded plasma gas is the result of material dissociation and has been investigated, for example, in laser-induced breakdown spectroscopy (LIBS) by measuring the luminescence from the plasma state [29, 32]. In contrast, in the present reaction, no evidence of a plasma state was observed, but temperature elevation at the irradiated site was confirmed by spectroscopy [36] as described in Section 4.5.

In this reaction, a high-grade Japanese charcoal, known as binchotan in Japan, is adopted as the carbon source because of its high carbonization over 93%. Among various carbon materials, charcoal is a sustainable carbon source, because it is made of wood and intermediates Earth's carbon cycle. Laser ablation effects are compared to other carbon materials.

4.1. Hydrogen generation from carbon in water

During a trial of nanoparticle generation by laser ablation in the liquid phase, we found that bubbles rose from the irradiated site inside a bottle which contained powder of binchotan charcoal and water. After preliminary discovery of explosive combustion of the generated gas by ignition, the collected gas was analyzed to find that hydrogen gas was contained. **Table 1** shows the ratio of the generated gas components, where argon portion from the argon-purged water was excluded. In the collected gas, roughly 50% of hydrogen and 20% of carbon monoxide were contained, whereas the amount of oxygen was very low [20, 21]. No gas was generated from pure water itself under the same irradiation conditions. From these facts, it was concluded that the reaction is due to photochemical reaction of carbon with water, instead of ideal water-splitting. Alcohol additive enhanced the generated gas volume with 56% of hydrogen concentration [21] as shown in the third row in **Table 1**. Details are described in Section 4.3.

	H_2 (%)	O_2 (%)	CO (%)	CO_2 (%)	N_2 (%)	CH (%)
100% H_2O	48.7	1.3	20.5	0.5	5.1	23.1
50% EtOH/H_2O	56.2	2.7	25.2	0.0	5.5	8.5

In both rows, values show a ratio excluding the argon. Adapted with permission from Refs. [20, 21].

Table 1. Ratio of gas component obtained from binchotan suspension by laser irradiation under argon atmosphere with argon-purged water (100% H_2O) and 50 wt% ethanol/argon-purged water (50% EtOH/H_2O).

From the ratio of the generated gas components, the reaction resembles to that of coal gasification, which is a classical technique of syngas production by steaming of coal under high pressure (a few MPa) and high temperature (>800°C) (HPHT) [42] via the following:

$$C(s) + H_2O(g) \rightarrow H_2(g) + CO(g) \tag{1}$$

$$C(s) + 2H_2O(g) \rightarrow 2H_2(g) + CO_2(g) \tag{2}$$

$$3\,C(s) + O_2(g) + H_2O(g) \rightarrow H_2(g) + 3CO(g) \tag{3}$$

$$2\,C(s) + O_2(g) + 2H_2O(g) \rightarrow 2H_2(g) + 2CO_2(g) \tag{4}$$

$$2\,C(s) + 2H_2O(g) \rightarrow CH_4(g) + CO_2(g) \tag{5}$$

In the present laser-induced reaction, the water temperature rose from 22 to 29°C during the 30-min irradiation at 182 mJ/cm^2 for 9.5 mL volume of water. Further evidence of temperature elevation in the laser pulse duration was witnessed by optical emission spectroscopy as discussed later in Section 4.5.

4.2. Laser fluence dependence

The hydrogen-included gas was generated above a threshold fluence of nanosecond laser pulse irradiation. **Figure 7a** shows the laser fluence dependence of a generated gas volume after 30 min of irradiation at a laser wavelength of 532 nm for three kinds of carbon powders. Macroscopic gas volumes of more than 0.05 mL were detectable only above a laser fluence of ca. 50 mJ/cm^2. The gas volume generated with binchotan charcoal powder of 5 μm in diameter (A, red solid circles) was almost twice that of the high-grade carbon powder of 5 μm (B, black open circles) and graphite powder of less than 45 μm (C, blue solid triangles) under the same irradiation conditions, whereas the threshold laser fluences were nearly coincident. The generated gas volume increased by irradiation time within one hour, but the further prolonged

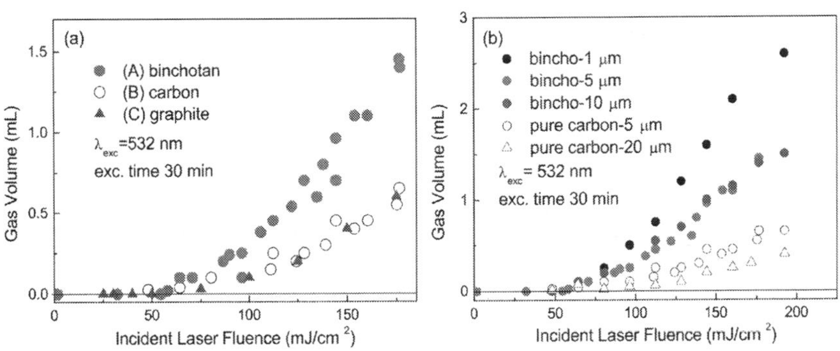

Figure 7. (a) Laser fluence dependence of generated gas volume with irradiation at 532 nm for 30 min, 26 mg of binchotan charcoal powder (A, red solid circles), high-grade carbon powder (B, black open circles), and graphite powder (C, blue solid triangles). The results for sample A and B were adapted with permission from Ref. [20]. (b) Laser fluence dependence of generated gas volume for powders of different sizes in diameter.

irradiation at 144 mJ/cm^2 made irreducible change, and the generated gas volume decreased [20]. This hydrogen generation reaction did not proceed by the irradiation of 30 fs laser pulses which affords a laser fluence of 80 mJ/cm^2 [20], implying the present reaction is classified into the thermal ablation process [1].

The BET surface area of the binchotan of 5 μm (A) was 22 ± 3 m^2/g, which was almost twice that of the high-grade carbon powder of 5 μm (B) 13 ± 3 m^2/g [20] or the graphite powder of less than 45 μm (C) 9 ± 3 m^2/g. Therefore, the higher gas generation is mainly attributed to its larger surface area. However, when the gas volume was compared to binchotan powders of different sizes of 1, 5, and 10 μm in mean diameter, whose BET surface areas were 120, 22, and 8 m^2/g, the generated volume did not depend on the surface area ratio linearly as shown in **Figure 7b**. This fact implies that there are other factors affecting the reaction efficiency.

The gas generation was observed from VIS to near-infrared (NIR) irradiation for both cases with binchotan (A) and pure carbon (B). **Figure 8** shows the generated gas volume versus the irradiated wavelength, obtained for 30 min irradiation at a laser fluence of 112 mJ/cm^2. The gas volume generated with binchotan (A, red solid circles) was more than twice that of the pure carbon (B, black open circles) under the same irradiation conditions in the VIS-NIR range. A tendency for a reduction in gas yield at longer wavelengths was anticorrelated to the optical reflectivity [20].

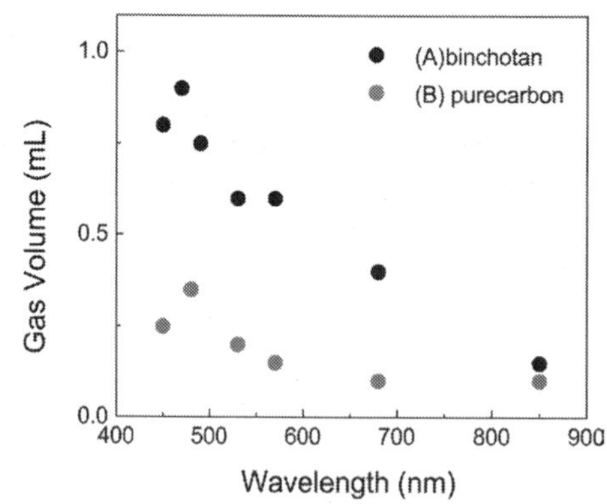

Figure 8. Irradiated wavelength dependence of the generated gas volume for binchotan powder (A, solid circles) and pure carbon powder (B, red circles) obtained by pulse irradiation of 112 mJ/cm^2 for 30 min. The result for sample A was adapted with permission from Ref. [20].

4.3. Alcohol additive effect

An alcohol additive in the binchotan water suspension enhanced the hydrogen generation efficiency for the laser fluences above the threshold of ca. 50 mJ/cm^2 [21]. Among methanol, ethanol, and isopropanol, ethanol was the most efficient additive and raised twice the generated volume. The generated volume increased according to the increase of ethanol additive and saturated at 40–50% of ethanol in water. The generated gas contained 56% of hydrogen as shown in **Table 1**.

Alcohol is known to act as an oxygen scavenger preventing the reverse reaction into water [43, 44] and/or as a current doubler [45] in photocatalytic water splitting. As the present photochemical reaction is different from the photocatalytic water-splitting reaction, the enhancement of the reaction is partially due to a photochemical reaction of ethanol itself. Endothermic reactions of a steam reformation of ethanol, $C_2H_5OH + 3H_2O \rightarrow 2CO_2 + 6H_2$ and/or $C_2H_5OH + H_2O \rightarrow 2CO + 4H_2$, which usually progress under HPHT [40], might occur by the laser irradiation, in addition to the oxidation reactions of solid carbon.

The generated gas volume of 7.3 mL, which was obtained after 1 h of irradiation with a 209 mJ/pulse at 532 nm by the 50% ethanol additive [21], was quite small. According to the hydrogen ratio of 56%, the hydrogen amount included in the volume was calculated as 0.17 mmol. That is, one hydrogen molecule per 126 photons was generated, assuming that 64% of the irradiated laser power was used in the reaction [20]. Although this gas volume was comparable to the production by a photocatalytic water reduction with hydrogen-terminated nano-diamonds [30], it was much less than the carbon-assisted electrochemical hydrogen generation by electric power [46].

4.4. Nanoparticle generation

A post-irradiated solution was investigated using UV-VIS absorption and DLS methods. As shown in **Figure 9**, light extinction by UV absorption and light scattering appeared in the centrifuged solution following the irradiation. This change indicates the production of new materials by laser ablation as by-products of the hydrogen generation. DLS measurements showed that this solution contained nanoparticles of 125 nm in mean diameter (see the inset). Such nanoparticles could be measured only for the irradiation above the threshold fluence that is the same as that for gas generation. The mean diameter of the nanoparticles was independent

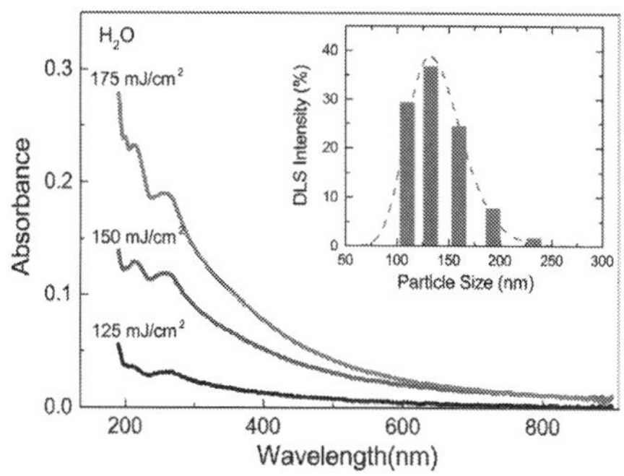

Figure 9. UV-Vis absorption spectra of centrifuged solutions after laser irradiation at 125 (black line), 150 (blue line), and 175 (red line) mJ/cm² for binchotan in water. Inset: size distribution of nanoparticles (bars) and logarithmic normal distribution function (broken line) in the centrifuged solution created by laser irradiation at 175 mJ/cm², measured using the DLS method. Adapted with permission from Ref. [36].

of the laser fluence, whereas light extinction was enhanced at higher irradiation fluences corresponding to an increase in the nanoparticle number. Furthermore, the generated nanoparticles are slightly dressed by a negative ζ-potential (−15 MeV) in water. Therefore, the nanoparticles suspended in water are expected to be stable for a long time as in the case of the organic QQ nanoparticles mentioned in Section 3.

A TEM image of the nanoparticles is shown in **Figure 10a**. Nanoparticles with sizes of around 100 nm were typically observed, as indicated by a yellow circle for a typical one. The sizes of the nanoparticles are consistent with the mean diameter observed by the DLS. A selected area electron diffraction (SAED) pattern of the nanoparticles (**Figure 10b**) shows clear diffraction spots in addition to diffused halo rings, whereas the Debye-Scherrer rings from the carbon structure were observed in a SAED pattern from the nonirradiated particle ensemble (**Figure 10c**). Some of spots in **Figure 10b** was located on the rings derived from the lattice spacings of diamond, and other parts of spots were on those of the C8 and n-diamond that were produced by laser ablation of a graphite target covered by water [29]. There were still other diffraction spots that could not be assigned to diffraction patterns of known structures. These results indicate that various crystalline/amorphous carbon structures including nanocrystalline carbon/diamond were created by laser ablation of binchotan charcoal in the liquid phase.

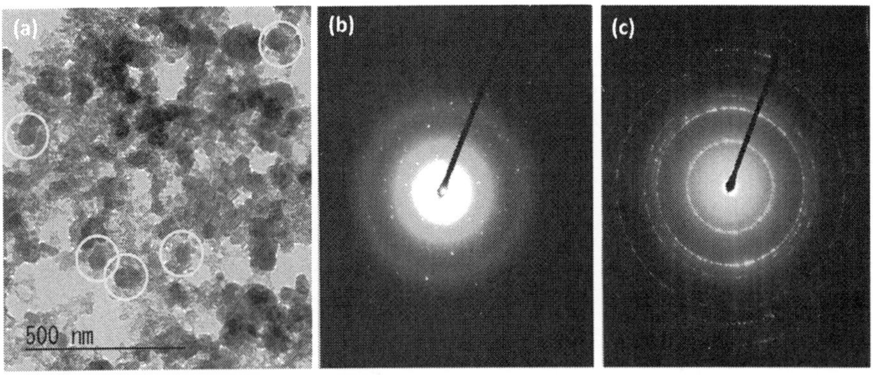

Figure 10. (a) A TEM image of binchotan nanoparticles produced by laser ablation. (b) SAED pattern obtained from the nanoparticles. (c) SAED pattern obtained from binchotan powder before irradiation.

Furthermore, the dried nanoparticles on a silicon substrate showed new IR peaks at 797, 873, 1019, 1261, 1425, 2906, and 2963 cm^{-1} as shown in **Figure 11**. Generally, vibrations of aromatic molecules are observed in the fingerprint range of 500–1500 cm^{-1}, and C—H stretch modes are in 2800–3000 cm^{-1} by an FT-IR measurement. The Raman peaks of O—H bonding were also observed during the reaction as described in Section 4.5. Therefore, the appearance of the peaks in these ranges indicated the creation of small carbon networks including the bonding of C—H and O—H groups.

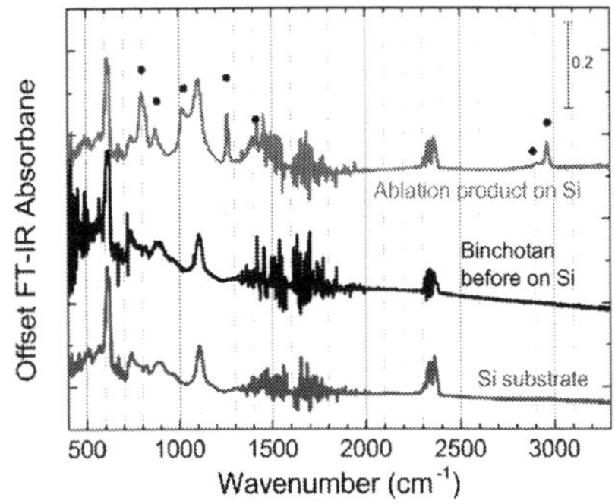

Figure 11. FT-IR spectra of dried post-irradiated nanoparticles (a red line) and nonirradiated binchotan powder (a black line) on a silicon substrate at room temperature. For comparison, the spectrum of a silicon substrate (a blue line) is also shown.

It is known that charcoal constitutes a form of amorphous carbons consisting of sp^2 and sp^3 bonding [47]. For the creation of new networks, bond breaking and reconstruction occur during the laser ablation by nanosecond pulses. Light energy at the threshold is necessary for such reactions. Surprisingly, in graphite powder, no nanoparticle was measured, and no additional IR peaks were observed.

4.5. Mechanism of the hydrogen generation

A clue to understand the mechanism behind the hydrogen generation via intense light irradiation is to clarify the nonequilibrium conditions at the irradiated site within a nanosecond time period. Investigation by time-integrated/-resolved spectroscopy during the hydrogen generation provided us crucial information regarding on-site nonequilibrium conditions including temperature increases [36].

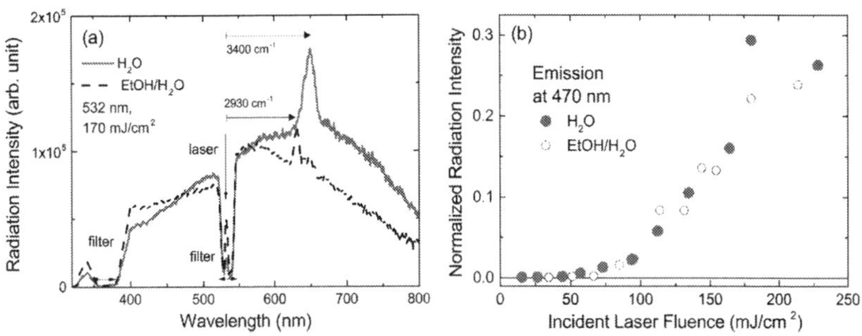

Figure 12. (a) Optical emission spectra from binchotan block in water (red solid line) and in 50% ethanol aqueous solution (broken black line) excited by laser pulses with 170 mJ/cm² energy density and 532 nm wavelength. (b) Incident laser fluence dependence of emission intensity at 470 nm for binchotan block in water (red solid circles) and in 50% ethanol aqueous solution (blue open circles). Adapted with permission from Ref. [36].

White-light emission was observed during the reaction from a binchotan block in water. As shown in **Figure 12a**, a broad spectrum over the visible range is apparent on both sides of the 532 nm excitation wavelength, across the penetration gap of the super notch filter, in water (solid red line), or in 50% ethanol aqueous solution (broken black line). No emission was observed from the water itself. The relatively narrow peaks at 650 and 630 nm are attributed to the Raman scattering lines at 3400 and 2930 cm^{-1}, because the peak positions changed following excitation wavelengths. There was no indication of the plasma emission from neutral/ionized atoms typically observed in LIBS. The Raman scattering lines are assigned to vibration of the O—H stretch mode under a hydrogen bond and Raman-active C—H vibrational modes of ethanol [48].

The white-light emission appeared only above a threshold excitation energy density. As shown in **Figure 12b**, the emission intensity at 470 nm increased nonlinearly in accordance with variations in the incident laser fluence. The threshold at 50 mJ/cm^2 was identical for both specimens in the water (red solid circles) and 50% ethanol aqueous solution (blue open circles). Note that the threshold for the appearance of the white light is coincident with the threshold for hydrogen generation (**Figure 7**). Therefore, it is reasonable to consider that the white-light emission is a simultaneous product of the hydrogen generation reaction. With a carbon electrode (99.9%), one fifth of emission intensity was observed above similar threshold excitation energy, and the generated gas volume was also small.

Spectral shape at shorter than 650 nm is well reproduced by Planck's law at a temperature 3860 K. Furthermore, time-resolved spectrum revealed a repetitive spectral change due to the temperature variation in the duration of laser pulse [36]. From these experimental facts, it was confirmed that the laser pulse supplies heat energy through optical absorption, and the white-light emission can reasonably be attributed to blackbody radiation from the irradiated site. It implies that hydrogen generation induced by laser irradiation proceeds similarly to classical coal gasification, which features reactions at HPHT. Finally, it was concluded that the hydrogen generation induced by the laser pulse irradiation occurs under high-pressure and high-temperature conditions.

5. Conclusion

The extended abilities of laser ablation in liquid phase were presented through two topics. The first is nanoparticle formation of an organic material, which produced a colloidal solution of a small organic material. In a yellow pigment QQ, a systematic blueshift of the absorption peak corresponding to the decrease of particle size in colloidal solutions was discoverd. This dependence provides an easy estimation method of the averaged diameter of the ensemble that will be applied to organic devices by a wet process. Furthermore, the population of an excited triplet state through optical excitation might be one guideline to select and synthesize materials for laser fragmentation.

The second is hydrogen gas generation from solid carbon in water by a photochemical reaction. Even under a lower energy irradiation that achieves no plasma state, the irradiated site can be a reaction centre of energy conversion from light to a fuel gas, although the gas generation efficiency is very low. Simultaneously produced carbon-based nanoparticles possess a characteristic structure including those of diamond, nano-diamond, and diamond-like carbon, supplying new carbon materials from binchotan charcoal. A rise of temperature during the irradiated pulse duration was witnessed by the observation of white-light emission from the site ascribed to the blackbody radiation. Therefore, we conclude that the hydrogen generation induced by the laser pulse irradiation occurs under high-pressure and high-temperature conditions.

Laser ablation in the liquid phase is a useful technique to transform ordinary materials into functional ones under relatively mild condition. Although to date this technique has been applied to rather limited number of materials, further studies from various interests will extend the application field.

Acknowledgements

This work was supported by the Original Research Support Project of Wakayama University, 2011–2012. All experiments presented here were done with students of master's and bachelor's courses in Wakayama University from 2011 to 2016. TEM observations were performed under the Inter-university Cooperative Research Program of the Institute for Materials Research, Tohoku University. This publishing project was supported by the Kansai Research Foundation for technology promotion (2016P001).

Author details

Ikuko Akimoto* and Nobuhiko Ozaki

*Address all correspondence to: akimoto@sys.wakayama-u.ac.jp

Wakayama University, Sakaedani, Wakayama, Japan

References

[1] B. C. Stuart, M. D. Feit, S. Herman, A. M. Rubenchik, B. W. Shore, and M. D. Perry, Optical ablation by high-power short-pulse lasers, J. Opt. Soc. Am. B 1996, 13, 459-468.

[2] H. Zeng, X.-W. Du, S. C. Singh, S. A. Kulinich, S. Yang, J. He, and W. Cai, Nanomaterials via laser ablation/irradiation in liquid: A review, Adv. Funct. Mater. 2012, 22, 1333-1353.

[3] H. Fukuhara and H. Masuhara, The mechanism of dopant-induced laser ablation: Possibility of cyclic multiphotonic absorption in excited states, Chem. Phys. Lett. 1994, 221, 373-378.

[4] H. Fujiwara, H. Fukuhara, and H. Masuhara, Laser ablation of a pyrene-doped poly(methyl methacrylate) film: Dynamics of pyrene transient species by spectroscopic measurements, J. Phys. Chem. 1995, 99, 11844-11853.

[5] V. V. Volkov, T. Asahi, H. Masuhara, A. Masuhara, H. Kasai, H. Oikawa, and H. Nakanishi, Size-dependent optical properties of polydiacetylene nanocrystal, J. Phys. Chem. B 2004, 108, 7674-7680.

[6] Y. Tamaki, T. Asahi, and H. Masuhara, Tailoring nanoparticles of aromatic and dye molecules by excimer laser irradiation, Appl. Surf. Sci. 2000, 168, 85-88.

[7] Y. Tamaki, T. Asahi, and H. Masuhara, Nanoparticle formation of vanadyl phthalocyanine by laser ablation of its crystalline powder in a poor solvent, J. Phys. Chem. A 2002, 106, 2135-2139.

[8] Y. Tamaki, T. Asahi, and H. Masuhara, Solvent-dependent size and phase of vanadyl phthalocyanine nanoparticles formed by laser ablation of VOPc crystal-dispersed solution, Jpn. J. Appl. Phys. 2003, 42, 2725-2729.

[9] Y. Hosokawa, M. Yashiro, T. Asahi, and H. Masuhara, Photothermal conversion dynamics in femtosecond and picosecond discrete laser etching of Cu-phthalocyanine amorphous film analysed by ultrafast UV–VIS absorption spectroscopy, J. Photochem. Photobio. A: Chem. 2001, 142, 197-207.

[10] T. Sugiyama, T. Asahi, T. Takeuchi, and H. Masuhara, Size and phase control in quinacridone nanoparticle formation by laser ablation in water, Jpn. J. Appl. Phys. 2006, 45, 384-388.

[11] H.-G. Jeon, T. Sugiyama, H. Masuhara, and T. Asahi, Preparation and photoconductive property of electrophoretically deposited film of quinacridone nanoparticles prepared by laser ablation in water, Jpn. J. Appl. Phys. 2007, 46, L733-L735.

[12] T. Asahi, T. Sugiyama, and H. Masuhara, Laser fabrication and spectroscopy of organic nanoparticles, Acc. Chem. Resea. 2008, 41, 1790-1798.

[13] R. Yasukuni, T. Asahi, T. Sugiyama, H. Masuhara, M. Sliwa, J. Hofkens, F.C. De Schryver, M. van der Auweraer, A. Herrmann, and K. Muellen, Fabrication of fluorescent nanoparticles of dendronized perylenediimide by laser ablation in water, Appl. Phys. A 2008, 93, 5-9.

[14] R. Yasukuni, M. Sliwa, J. Hofkens, F.C. De Schryver, A. Herrmann, K. Muellen, and T. Asahi, Size-dependent optical properties of dendronized perylenediimide nanoparticle prepared by laser ablation in water, Jpn. J. of Appl. Phys 2009, 48, 065002 (6 pp).

[15] H. Tabata, M. Akamatsu, M. Fujii, and S. Hayashi, Formation of C_{60} colloidal particles suspended in poor solvent by pulsed laser irradiation, Jpn. J. Appl. Phys. 2007, 46, 4338-4343.

[16] L. V. Zhigilei and B. J. Garrison, Microscopic mechanisms of laser ablation of organic solids in the thermal and stress confinement irradiation regimes, J. Appl. Phys. 2000, 88, 1281-1297.

[17] C. W. Tang and S. A. VanSlyke, Organic electroluminescent diodes, Appl. Phys. Lett. 1987, 51, 913-915.

[18] H. Klauk, D. J. Gundlach, J. A. Nichols, and T. N. Jackson, Pentacene organic thin-film transistors for circuit and display applications, IEEE Trans. Electr. Dev. 1999, 46, 1258-1263.

[19] D. R. Gamota, P. Brazis, K. Kalyanasundaram, J. Zhang, Printed Organic and Molecular Electronics, Springer Science & Business Media, 2013, Chapter 1 and 3.

[20] I. Akimoto, K. Maeda, and N. Ozaki, Hydrogen generation by laser irradiation of carbon powder in water, J. Phys. Chem. C 2013, 117, 18281-18285.

[21] K. Maeda, N. Ozaki, and I. Akimoto, Alcohol additive effect in hydrogen generation from water with carbon by photochemical reaction, Jpn. J. Appl. Phys. 2014, 53, 05FZ03 (3 pp).

[22] A. M. Squires, Clean fuels from coal gasification, Science 1974, 184, 340-346.

[23] A. Fujishima and K. Honda, Electrochemical photolysis of water at a semiconductor electrode, Nature 1972, 238, 37-38.

[24] T. Kawai and T. Sakata, Hydrogen evolution from water using solid carbon and light energy, Nature 1979, 282, 283-284.

[25] S. Sato and J. M. White, Photocatalytic reaction of water with carbon over platinized titania, J. Phys. Chem. 1981, 85, 336-341.

[26] T.-F. Yeh, J.-M. Syu, C. Cheng, T.-H. Chang, and H. Teng, Graphite oxide as a photocatalyst for hydrogen production from water, Adv. Funct. Mater. 2010, 20, 2255-2262.

[27] H.-C. Hsu, I. Shown, H.-Y. Wei, Y.-C. Chang, H.-Y. Du, Y.-G. Lin, C.-A. Tseng, C.-H. Wang, L.-C. Chen, Y.-C. Lind, and K.-H. Chen, Graphene oxide as a promising photocatalyst for CO_2 to methanol conversion, Nanoscale, 2013, 5, 262-268.

[28] I. Shown, H.-C. Hsu, Y.-C. Chang, C.-H. Lin, P. K. Roy, A. Ganguly, C.-H. Wang, J.-K. Chang, C.-I. Wu, L.-C. Chen, and K.-H. Chen, Highly efficient visible light photocatalytic reduction of CO_2 to hydrocarbon fuels by Cu-nanoparticle decorated graphene oxide, Nano Lett. 2014, 14, 6097-6103.

[29] S. Z. Mortazavi, P. Parvin, A. Reyhani, S. Mirershadi, and R. Sadighi-Bonabi, Generation of various carbon nanostructures in water using IR/UV laser ablation, J. Phys. D: Appl. Phys. 2013, 46, 165303 (9 pp).

[30] D. M. Jang, Y. Myung, H. S. Im, Y. S. Seo, Y. J. Cho, C. W. Lee, J. Park, A.-Y. Jee, and M. Lee, Nanodiamonds as photocatalysts for reduction of water and graphene oxide, Chem. Comm. 2012, 48, 696-698.

[31] A. A. Voevodin and M. S. Donley, Preparation of amorphous diamond-like carbon by pulsed laser deposition: a critical review, Surf. Coat. Tech. 1996, 82, 199-213.

[32] M. Capitelli, A. Casavola, G. Colonna, and A. De Giacomo, Laser-induced plasma expansion: theoretical and experimental aspects, Spectrochimica Acta Part B 2004, 59, 271-289.

[33] L. J. Radziemski, T. R. Loree, D. A. Cremers, and N. M. Hoffman, Time-resolved laser-induced breakdown spectrometry of aerosols, Anal. Chem. 1983, 55, 1246-1252.

[34] I. Akimoto, M. Ohata, N. Ozaki, and G. Ping, Formation of nanoparticles of organic molecules by liquid laser ablation, MRS Proceedings. 2012, 1455: mrss12-1455-ii08-04 (6 pp), (DOI: 10.1557/opl.2012.1346).

[35] I. Akimoto, M. Ohata, N. Ozaki, and P. Gu, Size dependent optical properties of quinacridonequinone nanoparticles prepared by liquid laser ablation in water, Chem. Phys. Lett. 2012, 552, 102-107.

[36] I. Akimoto, S. Yamamoto, and K. Maeda, White-light emission from solid carbon in aqueous solution during hydrogen generation induced by nanosecond laser pulse irradiation, Appl. Phys. A 2016, 122, 675 (6 pp).

[37] S. Kita, S. Masuo, S. Machida, and A. Itaya, Nanoparticle formation of pentacene by laser irradiation in ethanol solution, Jpn. J. Appl. Phys. 2006, 45, 6501-6507.

[38] H. Nishi, Chemistry of quinacridone, Seisan-kenkyu 1965, 17, 159-169 (in Japanese).

[39] M. Vilkman, T. Hassinen, M. Keranen, R. Pretot, P. van der Schaaf, T. Ruotsalainen, H. G. O. Sandberg, Fully roll-to-roll processed organic top gate transistors using a printable etchant for bottom electrode patterning, Org. Electron. 2015, 20, 8-14.

[40] M. S. Dresselhaus, G. Dresselhaus, and P. C. Eklund, Science of Fullerenes and Carbon Nanotubes, Academic Press, 1996, Chapter 13.

[41] I. Akimoto and K. Kan'no, Photoluminescence and near-edge optical absorption in the low-temperature phase of pristine C_{60} single crystals, J. Phys. Soc. Jpn. 2002, 71, 630-643.

[42] L. M. Gandia, G. Arzamendi, and P. M. Dieguez, editors, Renewable Hydrogen Technologies: Production, Purification, Storage, Applications and Safety, Elsevier, Amsterdam. 2013, Chaps. 6 and 7.

[43] T. Kawai and T. Sakata, Photocatalytic hydrogen production from liquid methanol and water, J. Chem. Soc.: Chem. Commun. 1980, 694-695, (DOI: 10.1039/C39800000694).

[44] Michael Bowker, Sustainable hydrogen production by the application of ambient temperature photocatalysis, Green Chem. 2011, 13, 2235-2246.

[45] S. R. Morrison and T. Freund, Chemical role of holes and electrons in ZnO photocatalysis, J. Chem. Phys. 1967, 47, 1543-1551.

[46] M. S. Seehra, S. Bollineni, Nanocarbon boosts energy-efficient hydrogen production in carbon-assisted water electrolysis, Int. J. Hydrogen Energy, 2009, 34, 6078-6084.

[47] K. Ishimaru, T. Hata, P. Bronsveld, T. Nishizawa, and Y. Imamura, Characterization of sp^2- and sp^3-bonded carbon in wood charcoal, J. Wood Sci. 2007, 53, 442-448.

[48] A. D'Aprano, A. Lizzio, V. T. Liveri, F. Aliotta, C. Vasi, and P. Migliardo, Aggregation states of water in reversed AOT micelles: Raman evidence, J. Phys. Chem. 1988, 92, 4436-4439.

Chapter 13

High-Quality Carbon Nanomaterials Synthesized by Excimer Laser Ablation

Calin-Constantin Moise and Marius Enachescu

Additional information is available at the end of the chapter

http://dx.doi.org/10.5772/65309

Abstract

Due to their special physical and chemical properties and potential applications from hydrogen storage to medical implantation, the carbon-based nanomaterials are in the frame of attention for many research groups all over the world. As synthesis techniques, we highlight arc discharge, chemical vapor deposition (CVD) and laser ablation. Even an expensive technique, laser ablation is suitable for single-wall carbon nanotubes (SWCNTs) synthesis, providing the highest yield of over 70%, while arc discharge yield is about 30% and CVD is about 42%. The most common investigation methods for carbon nanomaterials are micro-Raman spectroscopy, thermo-gravimetric analysis (TGA) and morphological and topographic studies done by atomic force microscopy (AFM), scanning electron microscopy (SEM), transmission electron microscopy (TEM) and high-resolution transmission electron microscopy (HR-TEM). We also emphasize in this work that by involving a home-designed reactor, we successfully synthesized SWCNTs, carbon nano-onions (CNOs) as well as graphene in the same reactor. Tuning the experimental parameters, we switch the end type of nanomaterials. We have done comprehensive studies regarding the carbon nanomaterials synthesis.

Keywords: laser ablation, SWCNTs, carbon nano-onions (CNOs), graphene, AFM, SEM, TEM, STEM, HR-TEM, HR-STEM, EDX, micro-Raman spectroscopy

1. Introduction

1.1. Short history of carbon nanomaterials

During the last three decades multiple types of carbon nanomaterials were discovered: fullerenes in 1985 (Nobel Prize 1996) [1], carbon nanotubes (CNTs) in 1991 [2], followed by single-wall carbon nanotubes (SWCNTs) in 1993 [3] and graphene in 2007 (Nobel Prize 2010) [4]. Due to their physical and chemical properties, carbon nanomaterial applications cover a wide range, from hydrogen storage to medical implantation.

1.2. Physical properties of carbon nanotubes

Figure 1 illustrates the comparison of SWCNTs properties against all known materials. For instance, the electrical conductivity is 1000 times higher than cooper, Young modulus almost same as diamond, tensile strength up to 14 times higher than Kevlar and thermal conductivity almost same as aluminum.

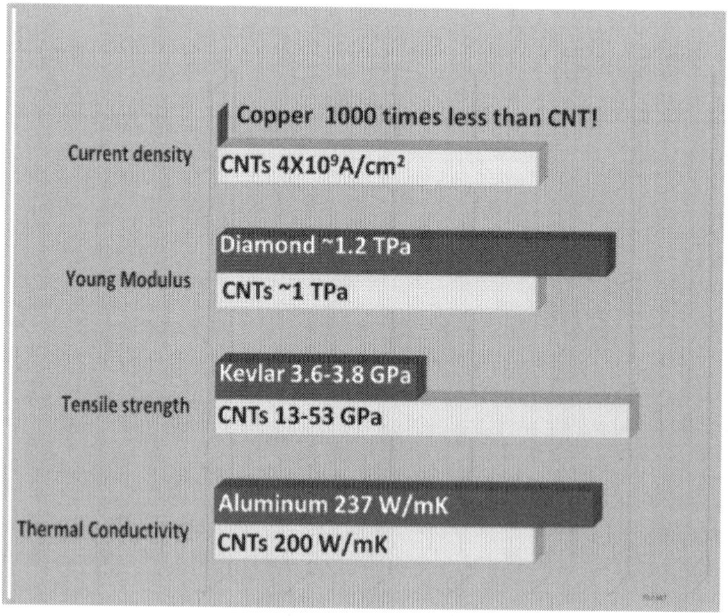

Figure 1. Properties of SWCNTs versus known materials.

1.3. Potential applications of carbon nanomaterials

In **Figure 2**, we select some of the potential applications for SWCNTs. One of the interesting application of SWCNTs is hydrogen storage. SWCNTs have high surface area; curvature creates cavities of molecular dimensions and superpositions of potential fields in cavities lead to large attractive force on hydrogen molecules [5]. SWCNTs are capable of adsorbing hydrogen quickly, to high density, at ambient temperatures and pressures [6–8]. The SWCNTs have been used in different types of applications such as biosensors [9], chemical sensors [10], scanning microscope tips [11], nanoelectronics [12], and third generation solar cell [13].

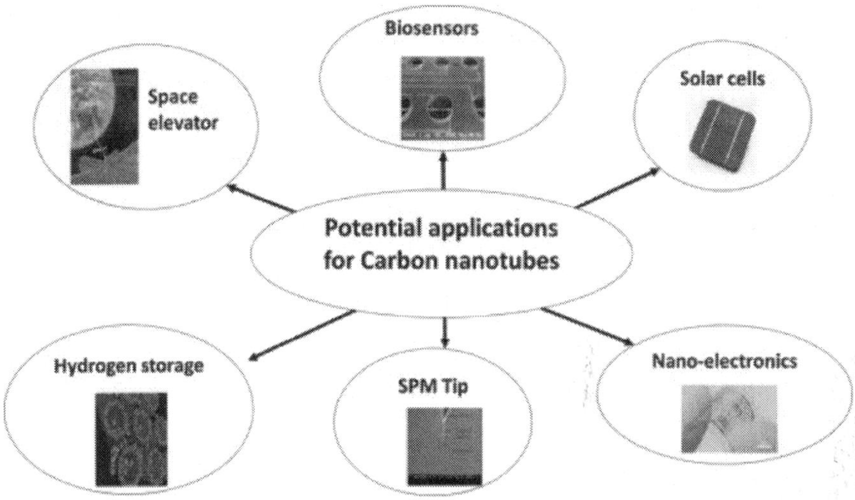

Figure 2. Potential applications of SWCNTs.

1.4. The routes for synthesis of carbon nanomaterials

There are several methods employed to obtain carbon nanomaterials. They are described very well by Tomanek in [14].

The nanotubes synthesis involve high energetic or temperature processes such as: arc evaporation (AE), chemical vapor deposition (CVD), conversion of carbon monoxide at high pressure in HiPCO process, ball milling (BM), as well as, the pulsed laser vaporization (PLV).

Arc discharge or arc evaporation (AE) can be used successfully for fullerenes, multi-wall carbon nanotubes and single-wall carbon nanotubes synthesis. Direct current of 10-20V is supplied to electrodes in a cooling inert gas environment. Nanotubes are formed on cathode during electrode evaporation in the direction of electric field lines. On the other hand fullerenes are founded in the soot on chamber walls. For syntheses of single-wall nanotubes the presence of catalyst into electrodes is required [2, 15].

The chemical vapor deposition (CVD) is a very common process for covert hydrocarbons gases into a diamond-like carbon (DLC) coating or carbon nanotubes. Conversion takes place at high temperature in thermal CVD, or can be assisted by plasma in plasma-enhanced CVD (PE-CVD). In the presence of water and by using a substrate (usually silicon) vertically aligned nanotube arrays (VANTAs) can be obtained by CVD. This process is called "Supergrowth".

A low-cost synthesis technique for nanotubes and related nanostructures from layered substances, such as graphite is the ball milling (BM). The BM is performed at room temperature, but the impact of a macroscopic ball on a powdered substance leads locally to a temperature increase that enables rearrangement of chemical bonds. Changes are induced in chemical bonding and structure by pressure, often introduced by hitting the material with a foreign object.

Pulsed laser vaporization (PLV) or laser ablation is a technique to produce high-quality carbon nanostructures. A laser beam is focused on a heated carbon target. The resulting plume and vapors are carried by an inert gas to a cooled cooper collector called cold finger (CF), where the raw material condenses. A sea urchin, also known as a dahlia structure, is a colloquial designation for the compact agglomeration of carbon nano-horns formed during the laser ablation process.

PLV or PLD (pulsed laser deposition) is the technique that will be in great detail described below along the results of the synthesis obtained using this method.

2. Laser ablation

We will focused in this section on synthesis of SWCNTs and other carbon nanomaterials via laser ablation since this is the technique used in ours laboratories. The pulsed laser ablation process for production of SWCNTs has first been reported by Guo *et al.* [16]. Nanotubes produced by laser ablation are up to 90% pure [17]. The target consists of mainly graphite to which small amounts of metal catalysts are added.

The laser ablation takes place inside a quartz tube heated to 800–1200°C [16, 18, 19] through which an ablation gas is used to control the dynamics of the plume and carry the ablation products to a cooled collector. Among the gases investigated Argon has been the most studied [16, 18, 19], but we and others proved that other gases like nitrogen [20, 21], krypton [20, 22], neon [20, 23], and helium [24] are also suitable for SWCNTs production by laser ablation.

2.1. Different types of reactors for laser ablation experiments

Previous laser ablation experiments have been focused on a chamber design commonly referred to as "front pumped counter flow pulse laser vaporization (PLV)". The design, presented in **Figure 3(a)** consists of an electric oven having a quartz tube inserted into it. The carbon target doped with small amount of metal catalyst (1) is located in the mid-position inside the quartz tube. The laser beam passes a quartz window (3) and is focused on the target. The ablation gas is fed from the front of reactor and a vacuum pump is located downstream of the cold finger (2) generating an even flow and pressure inside of the tube. The plume resulting from ablation of the target and evaporated species are transported by the inert gas toward the cold finger where they condense. Guo *et al.* reported for the first time SWCNTs production by laser ablation using a typically front pumped counter flow PLV [16].

Some other experiments have been carried out using a similar chamber design [18, 19].

Another chamber design includes a smaller quartz tube through which the carrier gas can be fed close to the target in order to confine the plume to a smaller region [17, 25].

In **Figure 3(b)**, the schematically design is represented. These simple designs proved to be effective.

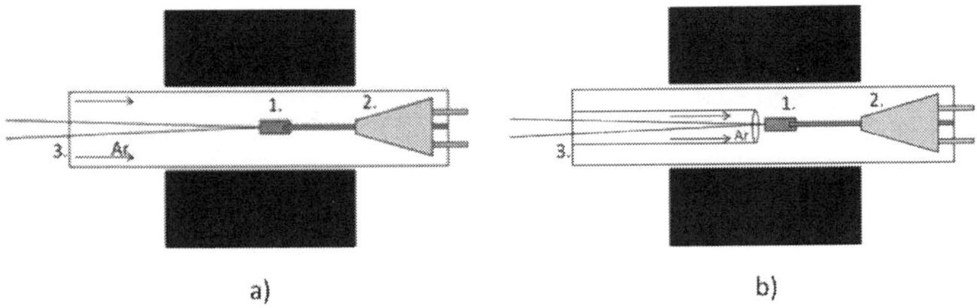

Figure 3. (a) Design of front pumped counter flow PLV system [16, 18, 19]; (b) design of front pumped counter flow PLV system with inner tube [17, 25].

The short length of the oven is the main disadvantage of these designs. From here is resulting a small zone were the ablation medium could be maintained at a certain needed temperature. Short oven is equivalent with a high temperature gradient into the ablation chamber. In our new design this disadvantage is avoided.

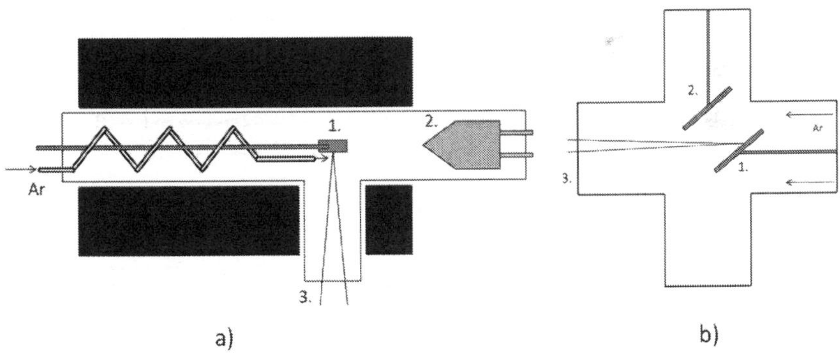

Figure 4. (a) Design of "T"-shaped PLV chamber with laser entrance from the side [26]; (b) "X"-shaped PLV chamber [27].

For increasing the ablated surface Holloway *et al.* proposed A "T" shaped quartz tube where the laser enters the chamber from the side [26]. The design is schematically represented in **Figure 4(a)**. The design includes feeding the ablation gas through a narrow tube coiled around the transfer rod in order to obtain a higher surface, thus increasing the heat transfer. The narrow tube takes the hot gas close to the target so that the products are carried away towards the cold finger. By this method can be ablated a bigger surface of the target. Complexity of the system which implies a custom made quartz tube with a side viewport for laser is the main disadvantage of this design. Geometry of the oven providing heating and temperature into the chamber is changed by this design.

The geometry becoming cumbersome and harder to control, resulting in high temperature gradients and a turbulent flow of the gas.

A chamber in "X" shape similar to those used for thin film depositions was proposed by Yahya *et al.* [27]. In **Figure 4(b)** is shown this design. The target (1) is positioned at an angle (usually 45°) parallel with water cooled substrate (2) at different distances. A quartz tube is not needed for this design, only the laser entrance window resulting in a more robust and easy to fabricate system. The system is usefully since it can be used for other purposes such as film deposition. With this type of reactor only multi walled carbon nanotubes (MWCNTs) were obtained. The system is not heated into an oven making impossible for the plume to reach eutectic point. May be this is one of the reasons that SWCNTs have not been obtained using such a chamber design. Other disadvantage of such reactor is the inert gas is not directed to carry the products from target to substrate, but only provides an inert atmosphere inside.

2.2. Our new chamber design

Figure 5 presents a cut-away side view of the novel experimental set-up developed and used in our research for the laser ablation of the targets [28]. The laser ablation chamber consists of a quartz tube, 50 mm in diameter and 1260 mm long, inserted into an oven. The quartz tube is O-ring sealed and operates from 10^{-3} Torr up to atmospheric pressure and temperature is controlled from 30 to 1200°C.

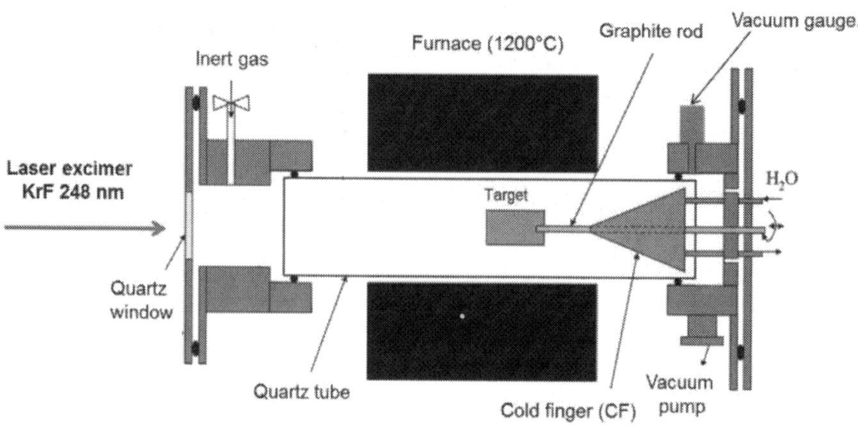

Figure 5. New reactor – Patent pending.

Laser ablation starts by passing the laser beam through a UV transparent quartz window and enters into the quartz tube hitting the target, when the target material begins to be ablated. The target having 20 mm in diameter is mounted on a graphite rod. Target was rotated during ablation with constant speed to get uniform ablation. Then the inert gas, which enters from the left-up side of the reaction chamber, moves through the quartz tube to the heated area, e.g., at 1100°C, where the reaction takes place, transporting the ablation product toward a copper condenser called cold finger (CF), where it will be deposited as black soot. The cold finger was cooled to 12°C using water supplied by a chiller. The inert atmosphere and the transportation of the ablated material to the CF were maintained by using the carrier gas at a certain flow rate, e.g., 70 L/h. The pressure is maintained by pumping through a needle valve and measured by a vacuum gauge sitting just outside the chamber.

The KrF excimer laser used for the ablation is a Coherent COMPex Pro 205 equipment, having the wavelength of 248 nm and the pulse duration of 20 ns.

Our system has some special technical features:

(i) The length of the oven was doubled compared with previous designs [16, 29], to ensure a more uniform temperature in the ablation reactor, allowing the product to travel longer time in the constant heated zone favoring the growth of the SWCNTs.

Figure 6(a) shows a schematic comparison between size accurate representations of our chamber design and other chamber designs [16, 18, 19]. Ratio of the length of the new chamber versus older designs is over 2:1. The temperature distribution expected as a result of increased oven length is represented by the red semicircles. Kataura *et al.* have shown that increasing the furnace temperature and the time spent by the ablation products at this elevated temperature increases the SWCNTs production yield [30].

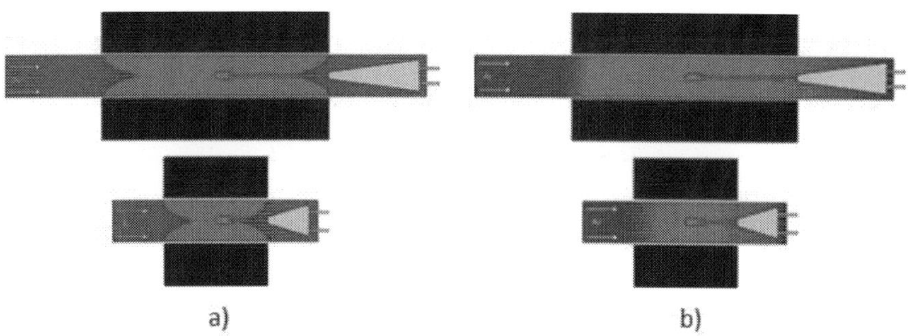

Figure 6. Our design (above) [28] and previously used chamber designs (below) [16, 18, 19]: (a) schematic representation of the heated area inside the ablation chamber; (b). Schematic representation of temperature gradient given by the water cooled cold finger. Ratio of the length of the new cold finger versus older design is 4.5:1.

(ii) The cold finger is longer (260 mm) with a larger surface than the previous one [28, 31], increasing the temperature gradient over its length and improving the capture of the product. A comparison of them is displayed in **Figure 7**, from where we can see that ratio of the length of the new cold finger versus older designs is over 4.5:1.

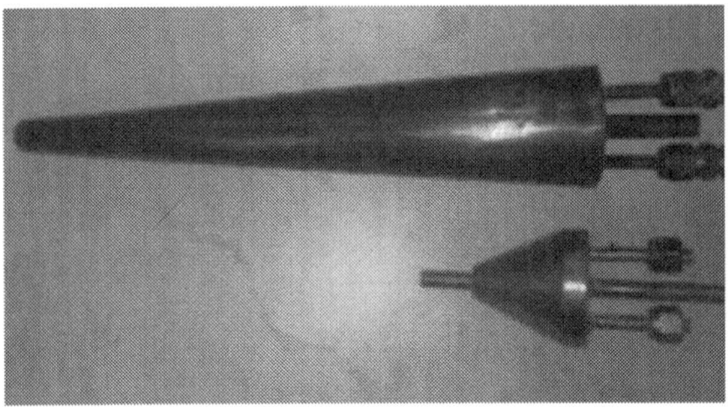

Figure 7. Photograph of the new and previous designed cold fingers.

The longer cold finger provides a larger surface resulting in a better percentage collection of the ablation products. The temperature gradient induced by the collector outside the oven area is distributed on a larger length of the cold finger, as presented in **Figure 6(b)**. This provides the means to study the influence the cold finger temperature gradient has on the condensation of the ablation products, i.e., the final products.

(iii) The attached Alicat flow meter controller allows controlling accurately the flow rate of the carrier gas over a wide range, 0–10 normal liters per minute (NLPM), with a resolution of 0.01 NLPM.

2.3. Original target preparation method

The target was prepared by following a recipe developed by Enachescu's group [28, 31, 32]. This recipe involves the mixing of graphite cement (GC 8010-B from Metal Forming Lubricants) with metal catalysts powder (Sigma Aldrich). We have studied several target compositions using uni- as well as bi-components metal catalysts. From all the combinations the best composition for SWCNTs synthesis was found to be Co 0.6%, Ni 0.6%, and C 98.8% (atomic percentage).

The mixture was transferred into a Teflon mold (20 mm diameter) and then cured 4 h at 130°C in air to increase the mechanical strength. Further another heat treatment of the target was applied for 1 h at 800°C in inert atmosphere to remove all the remaining organic compounds.

Material type	Pressure (Torr)	*T* (°C)	*E* (mJ)	RR (Hz)	Flow L/h	Catalyst
SWCNTs	500	1100	650	30	70	yes
CNOs	7	900	700	10	300	no
Graphene	100	1100	700	30	300	no

Table 1. The parameters used for carbon nanomaterial synthesis.

2.4. Recipes for carbon nanomaterial synthesis

We successfully synthesized SWCNTs, carbon nano-onions (CNOs) as well as graphene in the same reactor. Tuning the experimental parameters we switch the end type of nanomaterial. **Table 1** shows the parameters used for carbon nanomaterial synthesis. For SWCNTs synthesis, target must contain metallic catalyst grains. The parameters in the table are pressure in reaction chamber during ablation, oven temperature, laser pulse energy, repetition rate of laser pulses and the flow of inert gas.

3. Characterization experiments for SWCNTs

In previous work [28, 32], we found the suitable parameters for SWCNTs production:

- target composition Co 0.6%, Ni 0.6%, and C 98.8% (atomic percentage);

- oven temperature 1100°C;

- repetition rate of laser pulses 30 Hz;

- inert carrier gas pressure 500 Torr.

In the next subchapters we present the characterisation of SWCNTs obtained under above described conditions.

3.1. Micro-Raman spectroscopy experiments

Micro-Raman spectroscopy is a great investigation tool for SWCNTs characterization. In **Figure 8**, is a comparison of the typical micro-Raman spectra for SWCNTs obtained in our laboratories (blue) and high-quality commercially one (black), provided by Sigma Aldrich. As can be observed the spectra are quite similar leading to the conclusion that quality of SWCNTs obtained in our laboratory is comparable to commercial one.

The excitation laser was green (532 nm). In the zone of radial breathing mode (RBM) in **Figure 8** left top we found peaks characteristic only to SWCNTs and whose frequencies are strongly SWCNTs diameter dependent. The diameters were calculated in accordance with Eq. (1):

ω – frequency for vibrations in the radial direction [cm^{-1}];

$c1$, $c2$ – constants [cm^{-1}]; $c1$ = 215 [cm^{-1}]; $c2$ = 18 [cm^{-1}];

d – diameter of the nanotube [nm].

$$d = \frac{c1}{\omega - c2} \qquad (1)$$

The diameters of our SWCNTs are in the range of d = 1.1–1.6 nm with average d_A = 1.35 nm.

The G band in the Raman spectra clearly demonstrates, depending on the ratio of I_{G+}/I_{G-} that semiconducting SWCNTs were obtained, **Figure 8** (top right). However by using red excitation laser we obtain evidence of metallic SWCNTs existence (not shown here). So we conclude that our material is a mixture of semiconducting and metallic nanotubes.

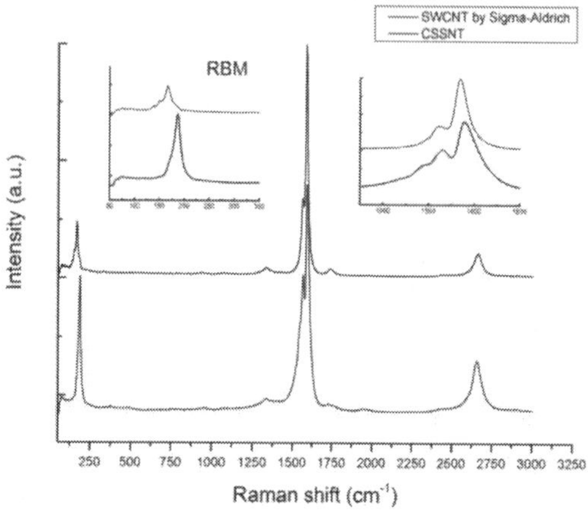

Figure 8. Micro-Raman spectra of SWCNTs – blue: our product, black: commercial product.

3.2. AFM experiments

Atomic force microscopy (AFM) was involved for the synthesized SWCNTs characterization as well as for SWCNTs mixed with poly(3-octylthiophene) (P3OT) polymer, for solar cell applications.

In **Figure 9** we present AFM images for a synthesized SWCNTs bundle on Si substrate: (a) 2D and; (b) 3D. In (c) is presented the topography of P3OT polymer on glass and (d) is 12% SWCNTs composite in P3OT polymer on glass and we can observe the SWCNTs bundles. We investigated also the adhesions forces on both surface and they increase with the increase of SWCNTs content into polymer (not shown here).

3.3. SEM experiments

In **Figure 10** is a collection of images obtained by a field emission scanning electron microscopy (SEM) for synthesized SWCNTs. The high resolution of SEM system of 0.6 nm, is making possible direct visualization and direct measuring of the SWCNTs bundles into SEM images. Usually such measurements are performed in high resolution transmission electron microscopy (HR-TEM), where the sample preparation is laboriously comparing with SEM sample investigations. We can identify well defined nanotubes shapes with μm length and nm diameters and observe their abundance. It is very interesting to observe their parallel arrangement.

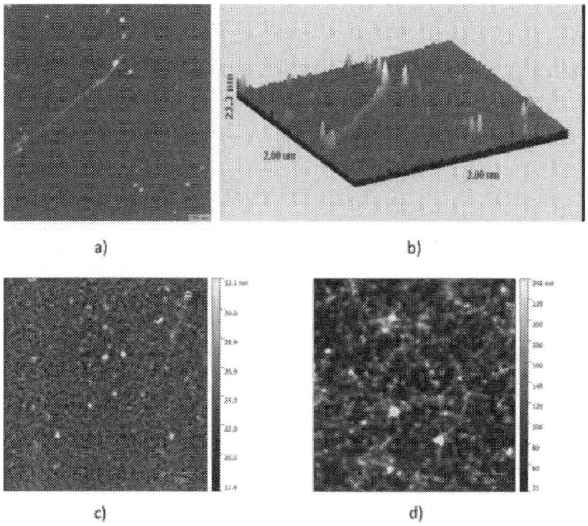

Figure 9. AFM images (a) 2D; (b) 3D; (c) 100% polymer; (d) 12% SWCNTs in polymer.

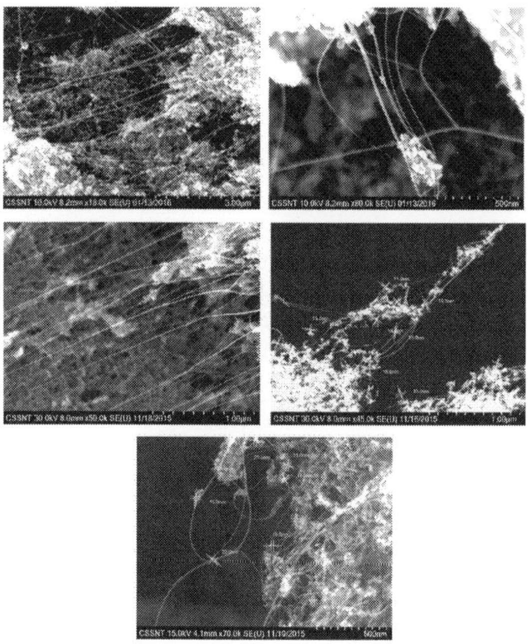

Figure 10. SEM images of synthesized SWCNTs.

3.4. SEM-Z contrast-STEM-EDX combined experiments

An advanced surface science system, capable of SEM and scanning transmission electron microscopy (STEM) investigations, performed at the same location of a specimen, was involved for the synthesized SWCNTs.

The special features of the synthesized SWCNTs are presented in **Figure 11** based on the capabilities of the special equipment presented above: *SWCNTs images in SEM, Z contrast (atomic mass contrast), STEM and energy dispersive X-ray spectroscopy (EDX) modes at the same sample location*. The images allow to a clear identification of catalysts grains and amorphous carbon (EDX mapping sustained also by Z contrast image) and in **Figure 11(d)** we observe that carbon is wide-spread. In SEM and STEM mode we can see the nanotubes shape and observe the opposite contrast of the images (a) and (c).

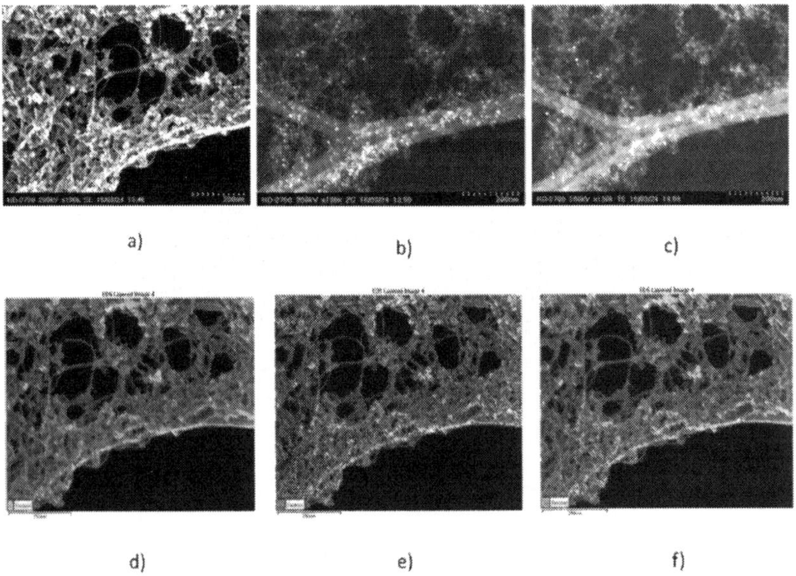

Figure 11. SWCNTs images taken at the same sample location in (a) SEM mode; (b) Z-contrast mode; (c) STEM mode; (d–f) EDX mapping mode for C, Co, Ni, respectively.

3.5. HR-STEM experiments

In order to get direct information on the morphology of the synthesized SWCNTs, we performed HR-STEM investigations.

We measured the diameters of several SWCNTs using HR-STEM images as in **Figure 12**. The measured diameters distribution and its histogram are plotted in **Figure 13(a)** and **(b)**, respectively. The histogram of diameters distribution was in the range of 1.0–1.7 nm, and the center between 1.25 and 1.35 nm [23]. This result is in great agreement with the result obtained from micro-Raman measurements (average d_A = 1.35 nm).

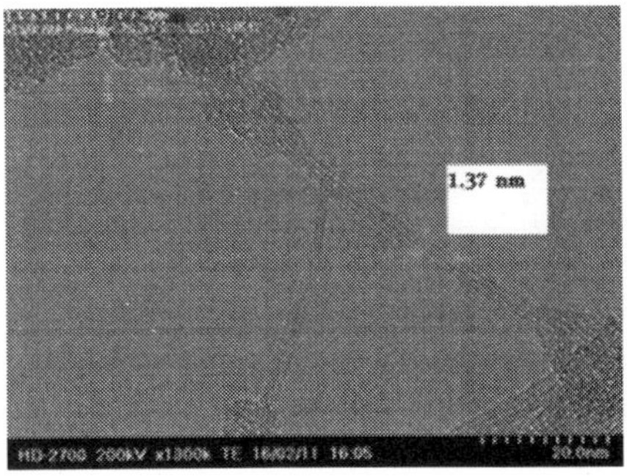

Figure 12. The measurement of SWCNTs diameter by HR-STEM.

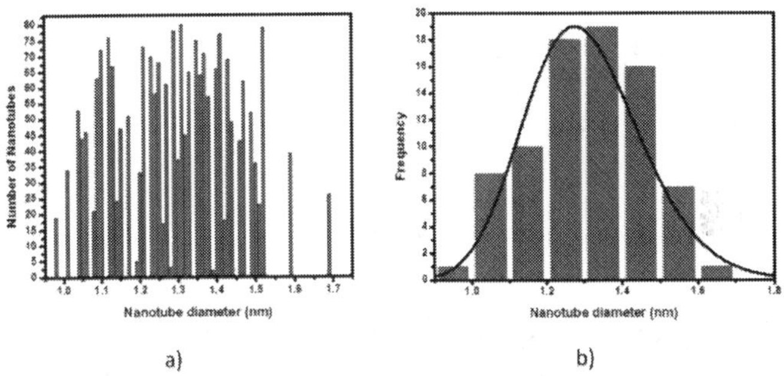

Figure 13. (a) Measured diameters distribution of SWCNTs. (b) Histograms of diameter distribution of the SWCNTs.

3.6. TGA experiments

Thermo-gravimetric analysis (TGA) was involved for the synthesized SWCNTs characterization as well as for the target composition used for obtaining the SWCNTs.

We analyzed by TGA, graphite powder, and materials used for our target. They burn in the same zone denoting that our target is graphitized. On the other hand collected material from cold finger (SWCNTs, green curve) burn in totally different place confirming it is a different material. What we also depicted in **Figure 14** is that ablation take place with respecting the target stoichiometry since the target materials and SWCNTs contain same percentage of remaining catalyst metals. From TGA curves we estimate a 70% yield of SWCNTs in raw product.

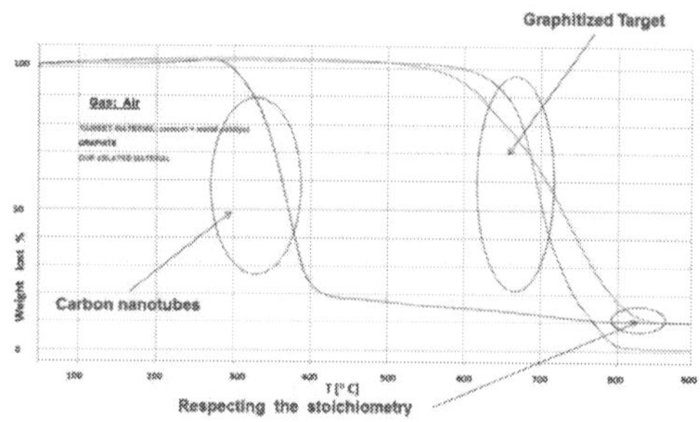

Figure 14. TGA curves for graphite powder (black), target material (red), SWCNTs (green).

3.7. Influence of carrier gas over SWCNTs synthesis

3.7.1. Experimental description

In an effort to compare our work with the work of Nishide *et al.* [20] which used Nd-YAG (neodymium-doped yttrium aluminium garnet; Nd:Y3Al5O12) laser, we study the SWCNTs synthesis versus different inert carrier gas used during the ablation with KrF excimer laser. For this study we varied the ablation gas types while the other parameters: laser type, laser energy, pulse repetition rate, oven temperature, gas pressure, gas flow rate, and ablation time remained constant (**Table 2**). The laser ablation experiments were carried out in four different inert gases: argon, nitrogen, neon, and helium.

Parameter	Value
Laser type	Excimer KrF, 248 nm
Laser energy	600 mJ
Repetition rate	30 Hz
Pulse period	20 ns
Oven temperature	1100°C
Gas pressure	500 Torr
Ablation time	60 min

Table 2. Parameters used for SWCNTs synthesis in different gas carrier environment.

3.7.2. Ablated mass and collected mass

We measured the mass of the target before and after ablation and also the raw product collected mass, i.e., SWCNTs. We observed linear decrease of ablated mass with the increase of carrier gas molecular mass. Also the collected mass follows the same trend, so we can increase and optimize the SWCNTs production by changing carrier gas. Fitting lines are shown in **Figure 15**.

http://dx.doi.org/10.5772/65309

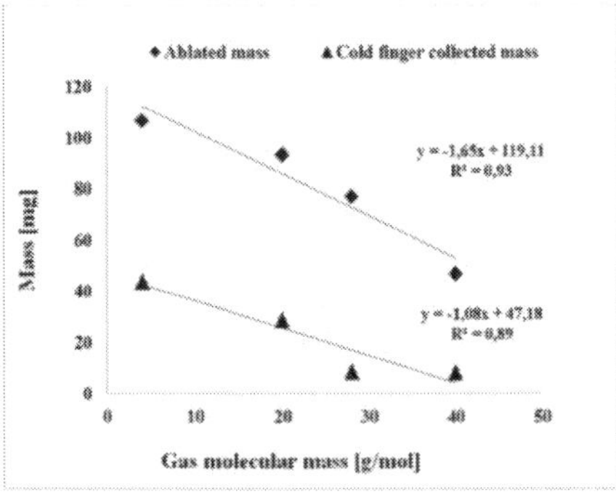

Figure 15. Ablated mass and collected mass versus gas molecular mass.

3.7.3. *Micro-Raman experiments*

Typical micro-Raman spectra for SWCNTs obtained in different inert gases are shown in **Figure 16(a)**. The excitation laser was green, 532nm.

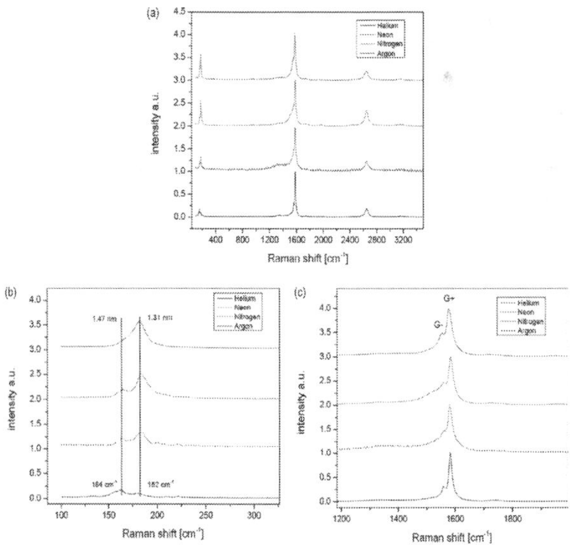

Figure 16. (a) Micro-Raman spectra for SWCNTs obtained using different ablation gases; (b) Radial breathing mode (RBM) part of the spectra; (c) G band part of the spectra.

In the zone of radial breathing mode (RBM) **Figure 16(b)** we found two peaks whose frequencies are strongly SWCNTs diameter dependent. The diameters were calculated in accordance with Eq. (1). For all gases two diameters distribution of 1.31 and 1.47 nm were calculated see **Figure 16(b)** (black lines) in great agreement with the result obtained from statistically HR-STEM measurements 1.25–1.35 nm (**Figure 13**).

3.7.4. TGA experiments

As can be seen in **Figure 17** we found different yields of SWCNTs versus the gas carrier. In **Table 3** using the yields and collected mass we calculate the mass of SWCNTs contained in the soot and the highest value was obtained in helium, almost 10 times more than in argon.

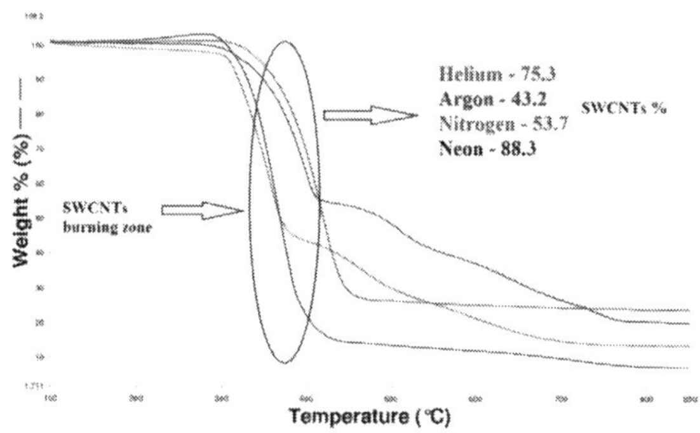

Figure 17. TGA curves of the ablation product, i.e., SWCNTs, obtained in different inert gases (Gas: air, T = 100–850°C).

Carrier gas	Mass of deposition onto cold finger [mg]	Percentage of SWCNTs in the deposition from TGA curves	Mass of SWCNTs produced [mg]
He	44	75.3	33.1
Ne	28.7	88.3	25.3
N_2	8.6	53.7	4.6
Ar	8.4	43.2	3.6

Table 3. SWCNTs mass calculated from the TGA curves.

4. Characterization experiments for CNOs

Under condition described in **Table 1** we obtain in our laboratories high quality CNOs [33].

4.1. Micro-Raman experiments

In **Figure 18** the induced defects and disorder are related to D band (1350 cm^{-1}), G band (1550–1620 cm^{-1}) represents vibrations in graphene plane and D' band (2500–3000 cm^{-1}) correspond to secondary order Raman scattering (second order of D and G bands).

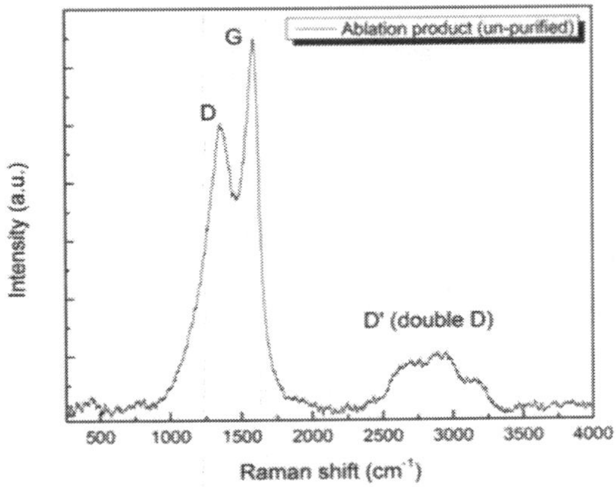

Figure 18. Raman band characteristics to CNOs.

The ratio of the D and G band intensity (I_D/I_G), which is related with the crystalline perfection was calculated and is 0.63. This value indicates that the CNOs synthesized by laser ablation shown a high crystallinity. Pimenta proposed an empirical formula (2) for determine in-plane crystallite size L_a in nm [34].

$$L_a = \frac{560}{E^4}\frac{I_G}{I_D} \tag{2}$$

where E is the excitation energy in eV (2.33 eV). Thus, resulting L_a value based on this formula is about 30 nm.

4.2. SEM experiments

As can be observed in **Figure 19** by ablating the target under conditions for CNOs we obtain totally different material comparing with SWCNTs images from **Figure 10**. The raw material looks as well defined spherical shapes.

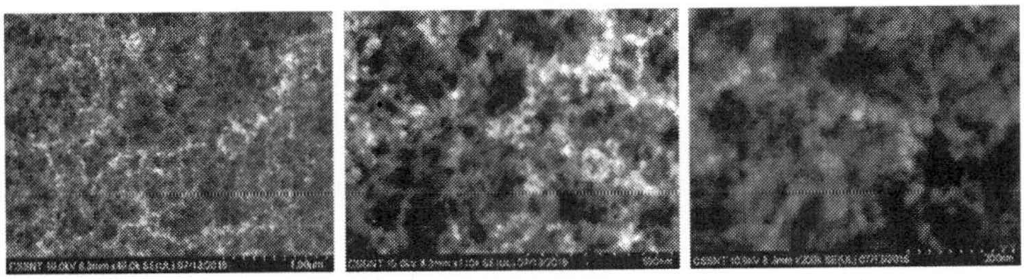

Figure 19. SEM images of CNOs.

4.3. HR-TEM experiments

The products obtained by ablating the pure graphite target using conditions for CNOs synthesis in **Table 1**, look like well-defined nano-onions, both individual and clustered through a matrix of amorphous carbon. As can be seen in **Figure 20(a)** the diameter is between 10 and 25 nm which is in good agreement with the dimensions obtained from micro-Raman spectroscopy using Pimenta formula (2).

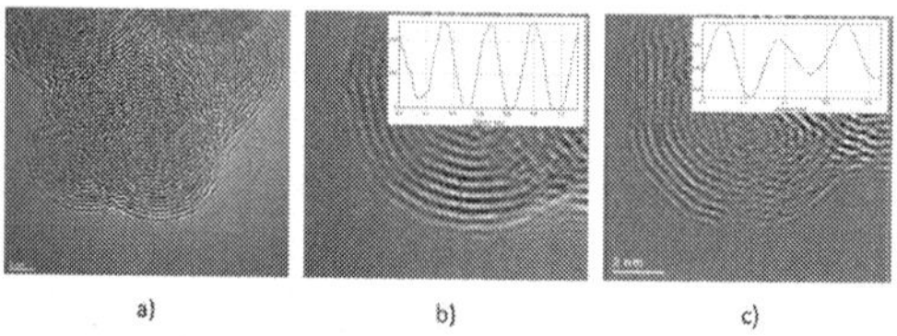

Figure 20. (a) CNOs clustered; (b) CNOs profile perpendicular to shells (c) CNOs profile along of the shell.

If we measure the graphitic interlayer distance of the CNOs from perpendicular direction to shells profile we found 0.35 nm in great agreement with the graphene monolayer thickness (**Figure 20(b)**). On the other hand if we analyze the profile in along of shell we found the distance between two consecutive atoms to be 0.24 nm, in good agreement with the atomic lattice on graphene **Figure 20(c)**.

4.4. TGA experiments

In order to measure the purity of obtained CNOs we perform thermo-gravimetric analysis (TGA) as can be seen in **Figure 21**. The percentage of weight lost has been measured using a constant heating rate of 5°C/min under a 20 mL/min Ar flux that limited the oxygen content in the furnace.

The mass loss between 350 and 600°C was assigned to the oxidation of the CNOs and is about 94%.

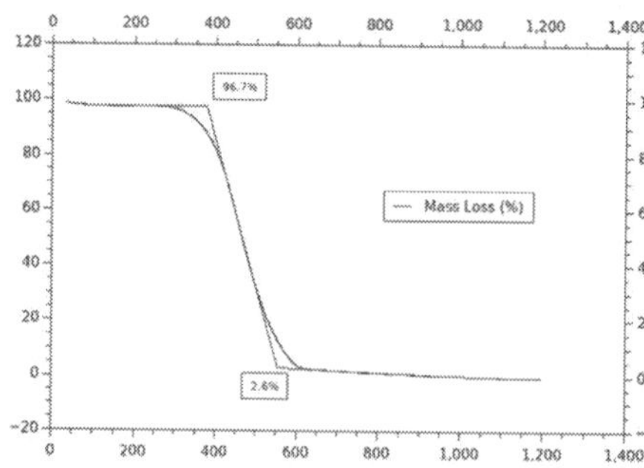

Figure 21. TGA curves of the CNOs (Gas: argon, T = 50–1200°C).

5. Characterization experiments for graphene

The products obtained by ablating the pure graphite target under condition for graphene in **Table 1**, look like well-defined graphene sheets.

5.1. SEM experiments

High quality graphene sheets obtained by laser ablation in our reactor are shown in **Figure 22**. As can be seen the amorphous carbon content is small. We observe in **Figure 22** that by ablating the target under conditions for graphene product we obtain totally different material comparing with SWCNTs images from **Figure 10** and CNOs images from **Figure 19**.

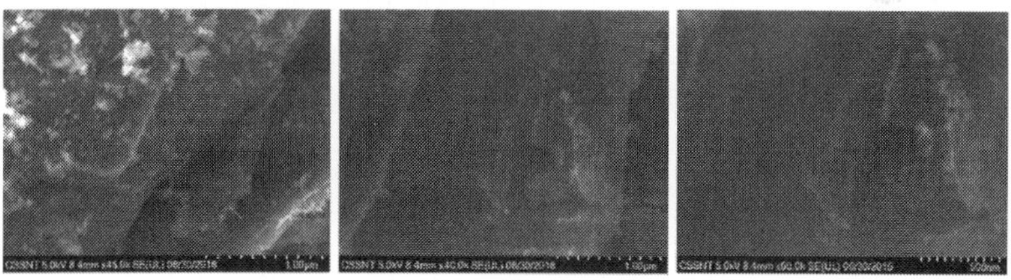

Figure 22. SEM images of graphene sheets.

5.2. TEM and electron diffraction experiments

In **Figure 23** we can observe nice transparent graphene flakes obtained in our reactor. The electron diffraction pattern is confirming the presence of graphene (nicely hexagonally concentric shapes).

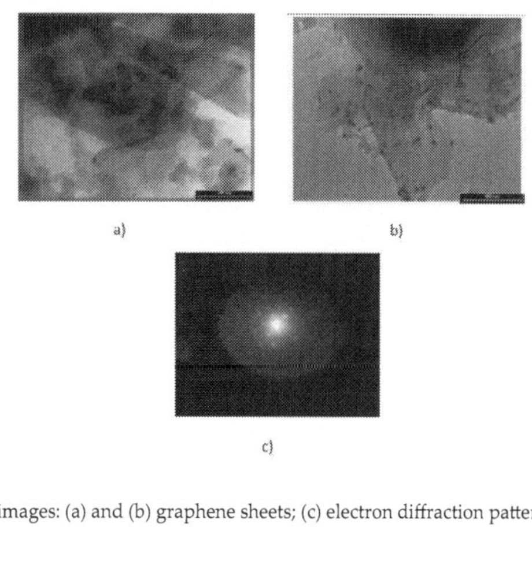

a) b) c)

Figure 23. Graphene TEM images: (a) and (b) graphene sheets; (c) electron diffraction pattern.

5.3. SEM-Z contrast-STEM combined experiments

The special features of the synthesized graphene are presented in **Figure 24** based on the capabilities of the special equipment presented above: *graphene images in SEM mode, Z contrast and STEM modes at the same sample location*. In these amazing images the SEM mode image suggest two graphene sheets one top of the other. This is confirmed by STEM image, since the top part is black denoting thicker structure. On the other hand the Z contrast image is showing only the presence of carbon in the sample. *This is one of the great advantages to have images in different modes on the same sample location.*

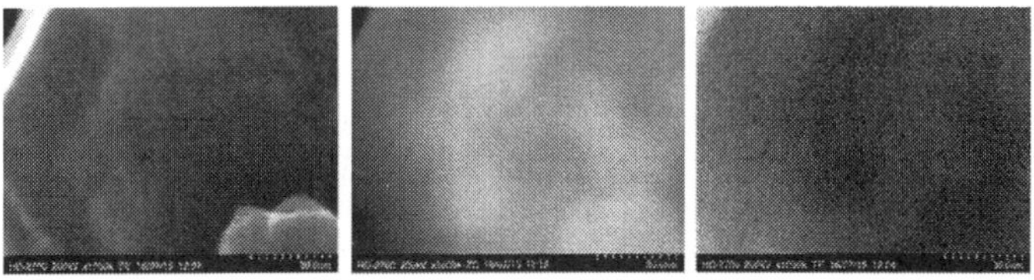

a) b) c)

Figure 24. Graphene images in (a) SEM mode; (b) Z-contrast mode; (c) STEM mode.

5.4. HR-STEM experiments

Figure 25(a) is a very good quality HR-STEM image of graphene sheet. We can observe the carbon atoms and their periodically disposal into lattice. Also we performed fast Fourier transform (FFT) and as can be seen in **Figure 25(b)** we identified two directions with crystalline planes.

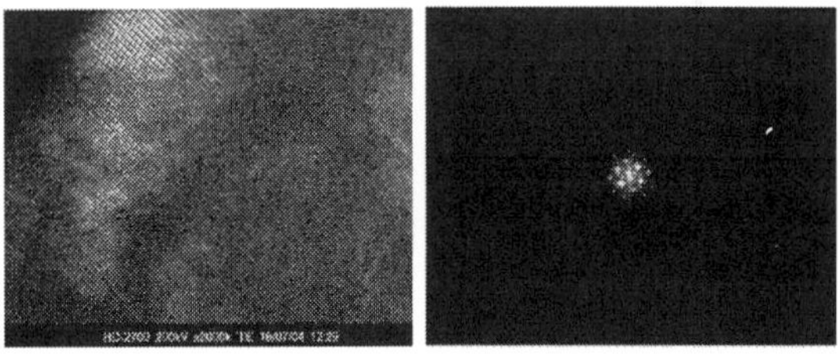

a) b)

Figure 25. Graphene: (a) HR-STEM image (b) FFT image.

5.5. Nano-diffraction experiments

To confirm the presence of graphene in our product we performed electron nano-diffraction measurement. We can see in **Figure 26** that the diffraction pattern is confirming the presence of graphene (nicely hexagonally shape).

Figure 26. Nano-diffraction experiment on graphene.

6. Conclusion

The present work highlighted the important specific synthesis way of carbon based nanomaterials, especially SWCNTs, and described the most used production routes using laser ablation technique.

We proved by micro-Raman spectroscopy, AFM, SEM, TEM, HR-TEM, STEM, HR-STEM, SEM-STEM, and TGA that high quality of carbon nanomaterials have been synthesized via laser ablation technique in ours laboratories.

Tuning experimental parameters we demonstrate how to switch to different type of synthesized carbon nanomaterials, i.e., SWCNTs, CNOs, and graphene.

We show for the first time images of *SWCNTs in SEM, Z contrast, TEM, and EDX modes at the same sample location* leading to the identification of SWCNTs, catalyst grains and amorphous carbon.

Acknowledgements

This work was supported by Romanian Ministry of Education and by Executive Agency for Higher Education, Research, Development and Innovation Funding, under projects PCCA 2-nr. 166/2012 and ENIAC 04/2014.

Author details

Calin-Constantin Moise and Marius Enachescu*

*Address all correspondence to: marius.enachescu@upb.ro

Center for Surface Science and Nanotechnology, University Politehnica of Bucharest, Bucharest, Romania

References

[1] Kroto H.W., Heath J. R., O'Brien S. C., Curl R. F., Smalley R. E. (1985). C60: Buckminsterfullerene, *Nature*, 318, 162-163.

[2] Iijima S. (1991). Helical microtubules of graphitic carbon, *Nature*, 354, 56-58.

[3] Iijima S., Ichihashi T. (1993). Single-shell carbon nanotubes of 1-nm diameter, *Nature*, vol. 363, no. 6430, pp. 603–605.

[4] Geim A. K., Novoselov K. S. (2007). The rise of graphene, *Nature Materials*, 6, 183–191.

[5] Barbaro P., Bianchini C. (2009). Catalysis for Sustainable Energy Production, WILEY-VCH Verlag GmbH & Co. KGaA, pp. 123.

[6] Li J., Furuta T., Goto H., Ohashi T., Yip S. (2003). Theoretical evaluation of hydrogen storage capacity in pure carbon nanostructures, *Journal of Chemical Physics*, 119, 2376.

[7] Dillon A. C., Heben M. J.(2001). Hydrogen storage using carbon adsorbents: past, present and future, *Appl. Phys. A*, 72, 133–142.

[8] Liu C., Chen Y., Wu C.-Z., Xu S.-T., Cheng H.-M. (2010).Hydrogen storage in carbon nanotubes revisited, *Carbon* 48, 452–455.

[9] Besteman K., Lee J.-O., Wiertz F. G. M., Heering H. A., Dekker C. (2003). Enzyme-coated carbon nanotubes as single-molecule biosensors, *Nano Letters*, vol. 3, no. 6, pp. 727–730.

[10] Li J., Lu Y., Ye Q., Cinke M., Han J., Meyyappan M., (2003). Carbon nanotube sensors for gas and organic vapor detection, *Nano Letters*, vol. 3, no. 7, pp. 929–933.

[11] Uchihashi T., Choi N., Ashino M., Sugawara Y., Nishijima H., Akita S. (2000). Carbon-nanotubes tip for highly-reproducible imaging of deoxyribonucleic acid helical turns by noncontact atomic force microscopy, *Japanese Journal of Applied physics*, vol. 39, no. 8, pp. 667–889.

[12] Li J., Papadopoulos C., Xu J. (1999). Growing Y-junction carbon nanotubes, *Nature*, vol. 402, no. 6759, pp. 253–254.

[13] Dorobantu D., Bota P., Badea M., Boerasu I., Bojin D., Enachescu M., (June 2013). High quality and reliable carbon nano-structures used for 3rd generation of solar cells, *in Proceedings of the 37th Annual Congress of the American Romanian Academy of Arts and Sciences (ARA '37), Chisinau, Moldova.*

[14] Tomanek D. (2014) Guide Through the Nanocarbon Jungle Buckyballs, Nanotubes, Graphene, and Beyond, Michigan State University, Morgan & Claypool Publishers

[15] Krätschmer W., Lamb L. D., Fostiropoulos K., Huffman D. R. (1990). Solid C60: a new form of carbon, *Nature*, 347, 354.

[16] Guo T., Nikolaev P., Thess A., Colbert D. T., Smalley R. E. (1995). Catalytic growth of single-walled nanotubes by laser vaporization, *Chem. Phys. Lett.*, 243, 49–54.

[17] Scott C. D., Arepalli S., Nikolaev P., Smalley R. E. (2001). Growth mechanisms for single-wall carbon nanotubesin a laser-ablation process, *Appl. Phys. A*, 72, 573–580.

[18] Puretzky A. A., Geohegan D. B., Fan X., Pennycook S. J. (2000). Dynamics of single-wall carbon nanotube synthesis by laser vaporization, *Appl. Phys. A*, 70, 153–160.

[19] Puretzky A. A., Schittenhelm H., Fan X., Lance M. J., Allard L. F., Geohegan D. B. (2002). Investigations of single-wall carbon nanotube growth by time-restricted laser vaporization, *Phys. Rev. B*, 65, 245425.

[20] Nishide D., Kataura H., Suzuki S., Tsukagoshi K., Aoyagi Y., Achiba Y. (2003). High-yield production of single-wall carbon nanotubes in nitrogen gas, *Chemical Physics Letters*, vol. 372, no. 1–2, pp. 45–50.

[21] Al-Zanganawee J., Moise C., Katona A., Bojin D., Enachescu M. (2015). Characterization of single wall carbon nanotubes synthesized by KrF excimer laser ablation in nitrogen atmosphere, *Digest Journal of Nanomaterials and Biostructures*, vol. 10, no. 4, October–December, pp. 1119–1127.

[22] Al-zanganawee J., Katona A., Moise C., Bojin D., Enachescu M.(2015). Krypton gas for high quality single wall carbon nanotubes synthesis by KrF excimer laser ablation, *Journal of Nanomaterials,* Volume 2015, Article ID 909072, 7 pages http://dx.doi.org/10.1155/2015/909072 .

[23] Al-Zanganawee J., Mubarak T., Katona A., Moise C., Balan D., Dorobantu D., Bojin D., Enachescu M. (2016). Raman spectroscopy and morphology characterizations of SWCNTs synthesized by KrF excimer laser ablation under neon gas atmosphere, *Digest Journal of Nanomaterials and Biostructures*, vol. 11, no. 2, April–June, pp. 525–536.

[24] Moise C., Katona A., Dinescu D., Al-zanganawee J., Bojin D., Enachescu M.(2015). KrF excimer laser ablation in helium yields the highest amount of SWCNTs over other inert gases, *published in the Proceeding of The 39th American Romanian Academy of Arts and Sciences Congress National Institute of Nuclear Physics, Frascati, Rome, July 28-31, 2015.*

[25] Arepalli S., Nikolaev P., Holmes W., Bradley S. (2001). Production and measurements of individual single-wall nanotubes and small ropes of carbon, *Appl. Phys. Lett.*, 78, 1610.

[26] Holloway B. C., Eklund P. C., Smith M. W., Jordan K. C., Shinn M. (2010). April 6, *US Patent No. 7692116 B1.*

[27] Yahya N., Guan B. H., Hashim M. (2005). Development of pulsed laser deposition system for the formation of web-liked carbon nanotubes, *American Journal of Applied Sciences*, 11, 1509.

[28] Bota P. M., Dorobantu D., Boerasu I., Bojin D., Enachescu M. (2015). New laser ablation chamber for producing carbon nanomaterials using excimer laser, *Materials Research Innovations*, vol. 50, no. 1, pp. 33–39.

[29] Braidy N., El Khakani M. A., Botton G. A. (2002). Single-wall carbon nanotubes synthesis by means of UV laser vaporization, *Chemical Physics Letters*, vol. 354, no. 1-2, pp. 88–92.

[30] Kataura H., Kuzmazawa Y., Maniwa Y., Ohtuska Y., Sen R., Suzuki S., Achiba Y. (2000). Diameter control of single-walled carbon nanotubes, *Carbon*, 38, 1691-1697.

[31] Bota P. M., Dorobantu D., Boerasu I., Bojin D., Enachescu M. (2014). Synthesis of single-wall carbon nanotubes by excimerlaser ablation, *Surface Engineering and Applied Electrochemistry*, vol. 50, no. 4, pp. 11–15.

[32] Boerasu I., Enachescu M. (2013). Pulsed Laser Ablation Synthesis of Carbon Nano-Structures: Effect of Target Composition and Laser Ablation Condition on Their Yield and Morphology, Micro and Nanoengineering 2013 (http://www.nano-link.net/mne/volume22/5.pdf)

[33] Dorobantu D., Bota P. M., Boerasu I., Bojin D., Enachescu M. (2014). Pulse laser ablation system for carbon nano-onions fabrication, *Surface Engineering and Applied Electrochemistry*, 50 (5), 19–23.

[34] Pimenta M. A., Dresselhaus G., Dresselhaus M. S., Cancado L. G., Jorio A., Saito R. (2007). Studying disorder in graphite-based systems by Raman spectroscopy. *Physical Chemistry Chemical Physics*, 9, 1276–1290.

Section 3

Surface Modification

Chapter 14

Femtosecond Laser-Induced Surface Modification and its Application

Kenji Goya, Masahiko Shiraishi, Yusuke Fuchiwaki,
Kazuhiro Watanabe and Toshihiko Ooie

Additional information is available at the end of the chapter

http://dx.doi.org/10.5772/65428

Abstract

In this chapter, we present femtosecond laser micromachining to fiber optics, focusing on surface qualities. Some techniques applied in the field are introduced to date and a review of some of the current applications for this type of technology. Section 2 describes laser-induced periodic surface structures (LIPSSs), which are induced in low- and high-fluence regime. Section 3 describes the influences of laser-induced structures for the fabrication of fiber-optic sensors, with experimental techniques and results in our research group. These sections explore ultrashort laser pulses applications, roughly going from lower to higher energy (power, intensity) ones.

Keywords: LIPSS, formation mechanism, femtosecond laser microfabrication, high-fluence regime, fiber-optic sensor

1. Introduction

Ultrashort pulse laser interaction has been attractively applied to the modification of material properties [1–8] in various materials. The ultrashort pulse laser applications to micromachining [9–16] have received much attention. This is mainly because of two reasons that are extremely high-peak power and ultrashort pulse less than material thermal relaxation. In the high-peak laser regime, multiphoton absorption can be easily induced even in transparent materials. Furthermore, the effects of thermal diffusion during material processing can be extremely minimized with ultrashort pulses that produce a thermal non-equilibrium state between electrons and lattice. Owing to such non-thermal and spatially localized effects can facilitate to locate a specified volume ablation without collateral thermal damage, and microdevices can be precisely machined using femtosecond processing to produce high functionality even in a microscale platform such as thin optical fibers [17–25] or analytical microchip for biochemical application. For this reason, such processing for functional microstructures has been previously reported in the past decade.

Using femtosecond laser micromachining for such optical devices, surface characteristics should be carefully checked because the optical/mechanical properties on the laser-induced surface structures provide completely different response with dependence on micro/nanoscale conditions which is varied with laser irradiation parameters. Such surface responses can be used for controlling phenomena like wettability, reflectivity, and friction as well as biomimetics [26]. This is the reason why nanotechnology has attracted wide interest in many fields.

In this chapter, we present femtosecond laser micromachining to microdevice such as fiber optics with a focus on surface qualities. Section 2 is dedicated to laser-induced periodic surface structures (LIPSSs), which are induced in low- and high-fluence regime. The influences of laser-induced structures for the fabrication of fiber-optic sensors are described in Section 3, with experimental techniques and results in our research group. Conclusions are presented in Section 4.

2. Femtosecond laser-induced surface structures

Laser-induced surface structures have been widely investigated in various materials such as metals, ceramics, glass, and polymers. A periodicity of about the wavelength λ of the laser radiation is commonly observed, which are usually called as ripples or laser-induced periodic surface structures (LIPSS). It should be noted that such LIPSS show water-repellent properties [12, 27–29], diffusive reflection [30–34], and friction change [35–38]. The spatial periodicity of low-spatial frequency LIPSS (LSFL) is known to be approximately laser wavelength dependent, and the LSFL is mainly orientated perpendicular to the direction of the laser polarization. LSFL and high-spatial frequency LIPSS (HSFL) generated by femtosecond laser pulses have been recently observed. Using high-energy laser pulses, micrometer-size-rugged surfaces are formed inside the channel structures ablated by femtosecond laser pulses [39–41]. It is also important to understand the formation mechanism of micrometer-scale surfaces for femtosecond laser direct writing technologies in which relatively higher laser energies are used to fabricate a 3D structure such as fluidic channels [3, 42, 43], sampling cells [19, 21, 22], and interferometers [44–46]. Different generating processes will induce different mechanical responses, which will be detailed in the following.

2.1. Laser-induced periodic surface structures (LIPSS)

In 1965, LIPSS formation induced by exposing semiconductors to a ruby laser was first observed by [47]. Although LIPSS had been unfavorable as irregular structures inevitably accompanied by laser processing, the development of ultrashort pulse laser allows us to achieve not only stable periodic structures, but also non-conventional advantages based on nanotechnologies mentioned above. Applications of LIPSS for mechanical engineering such as water-repellent property, strain relief [48], and friction control have been reported as well as for biomedical applications such as cell adhesion and its growth [49, 50]. For the case of optics, LIPSS are applied to structural coloring [51], surface-enhanced Raman spectroscopy (SERS) [52], and anti-reflection surface. Mechanism of LIPSS formation has been discussed in many literatures. Sipe et al. established a first principle for LIPSS formation process, by considering a surface-scattered wave from the surface roughness and modeling the effect of the roughness on the electromagnetic field [53–55]. **Figure 1** shows a simple model for the formation of LIPSS, showing geometry of a laser beam idealized as a finite plane wave onto a rough surface. While LSFL can be obtained by either a CW (continuous wave) laser or a pulsed laser irradiation, HSFL can only be observed for the case of using pulse laser in the duration of picosecond or femtosecond range. For linearly polarized light at a normal direction to the plane, an LIPSS periodicity becomes much smaller than the laser wavelength and their direction can be parallel [56] or orthogonal [57] to the polarization, depending on the material properties and the irradiation parameters. The nature of HSFL is still under controversially discussion, and many theories have been proposed to explain the formation mechanism: surface-plasmon polaritons [58–60], second harmonic generation [57, 61], self-organization [62–67], waveguides modes [68], parametric process [69, 70], or interference along with a modification of the optical properties during the pulse [56].

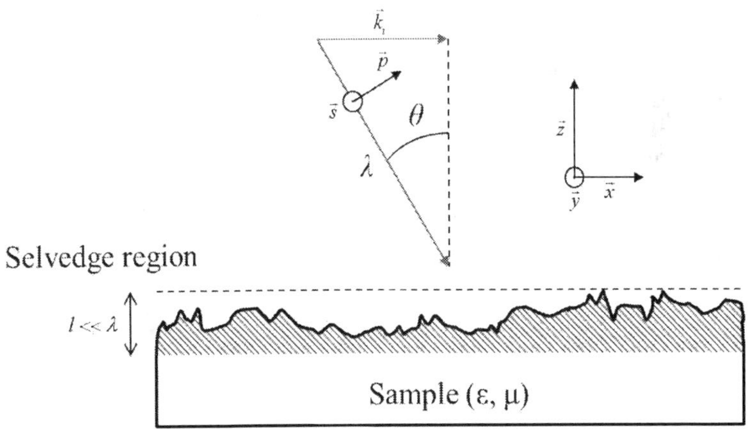

Figure 1. Geometry of a one-dimensional rough surface of length L. The medium below the rough interface is electrically defined by the permittivity ε and permeability μ. The surface roughness is assumed to be confined in a "selvedge region" in a range between $z = 0$ and $z = 1$, where $1 \ll \lambda$.

Discussion of a contribution of surface plasmon polaritons has been presented by Bonse et al. [71, 72] and Miyaji et al. [58, 59], where the formation of LSFL and HSFL was experimentally observed with the change of optical properties induced by an electron excitation under laser irradiation. Huang et al. showed that self-organized LSFL on dielectric materials was formed via a metallic-like behavior by an interference of initial surface plasmon and laser pulses followed by the grating assisted surface plasmon-laser coupling [67]. They also obtained a good agreement with their experimental results, by taking account into the effect of surface plasmon. Reif et al. reported that HSFL is due to self-organization structure formation during the relaxation of highly non-equilibrium condition after explosive ion emission [63]. Straub et al. demonstrated that generation of a dense electron-hole plasma at a carrier concentration slightly beyond the critical plasma density allows for the excitation of surface plasma waves at the high-spatial frequencies that are required for HSFL formation [60]. Bonse et al. explained HSFL formation based on the interference between the surface-scattered wave of the laser and its second harmonic [57, 71, 73]. Okumuro et al. found that the periodicity of HSFL can be explained by induction of a surface-plasma wave through the parametric process of laser light [69]. They also showed that laser fluence dependence of periodic grating structures was formed on metal surfaces under femtosecond laser pulse irradiation. The results indicate that the formation of periodic grating structures depends not only on target properties but also on the electron density of plasma produced in the femtosecond excitation regime [70].

2.2. Surface microstructures in higher energy regime

Femtosecond laser machining with high fluence can selectively remove large volumes of material only in the focal area. Thermal effect could be observed in higher fluence more than the threshold value of materials even though heat diffusion outside the focal area is comparatively minimized in contrast to longer pulse lasers. Several literatures have provided evidence for the existence of a molten region surrounding ablation area, even for the ultrashort pulse processing [39–41, 74]. Thermocapillary force and hydrodynamic force will be dominated in the high-fluence regime. Theoretical model for the regime has also been fabricated and seems to be in good agreement with experimental results [41]. Yaker et al. showed reliable evidence regarding the formation mechanism of surface microstructure induced by high-fluence femtosecond pulses. They tried to fabricate a microflow channel for microfluidic devices by scanning femtosecond laser beam on a borosilicate glass. The experiment was performed by mutually overlapping thin rims surrounding the laser-induced smooth crater to produce surface microstructures as shown in **Figure 2**.

The crater surrounded by thin rim is created by resolidification of a molten layer generated during the ablation process. They found that a thin rim is formed around the craters and is created on the undamaged surface. To investigate the mechanism of granulated rim formation by successive laser pulses, they also demonstrated theoretical analysis of the thermal and fluid processes taking into account ablation process in the femtosecond regime. The series of the experiment indicate that a thin molten layer below the surface and the rim is generated by the high pressure resulted from a dense plasma producing a pressure-driven fluid motion of the molten material outwards from the center of laser focal point as shown in **Figures 2** and **3**. The second laser pulse is diffracted by the rim created by the first laser pulse, resulting in diffraction pattern on the crater area overlapped. The distance between the diffraction rims seems to be approximately equal to the wavelength of the laser light, which means that diffraction by the rim plays an important role for the formation of granulated structures. Flow forces applied to

http://dx.doi.org/10.5772/65428

a molten layer are generated from expanding plasma which can be happened by thermocapillary forces and forces driven by plasma expansion above the irradiated surface. There are a number of processes in the regime of femtosecond pulse-induced phenomena, including nonlinear absorption, plasmas, shock propagation, melt propagation, and resolidification. These processes can roughly be broken down into three different time domains as illustrated in **Figure 4**.

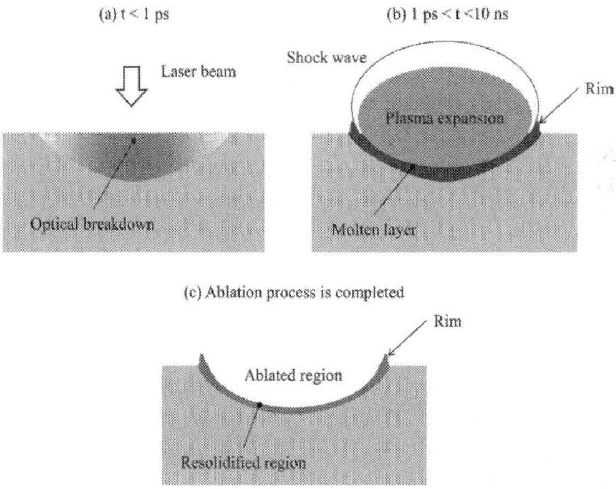

Figure 2. A schematic of ablation process with a femtosecond laser pulse in the characteristic time (a)–(c) of excitation and relaxation process.

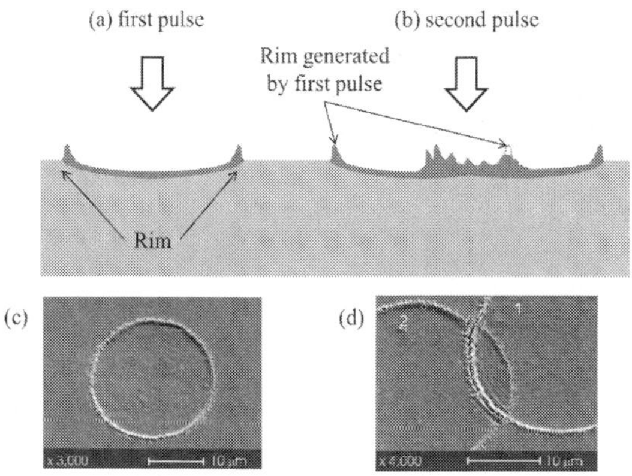

Figure 3. Femtosecond laser-induced crater rim and its diffraction pattern generated by overlapping two pulses. (a), (c) One laser pulse. (b), (d) Two overlapping pulses. (SEM pictures adapted with permission from Ref. [41], © IOP Publishing. Reproduced by permission of IOP Publishing. All rights reserved).

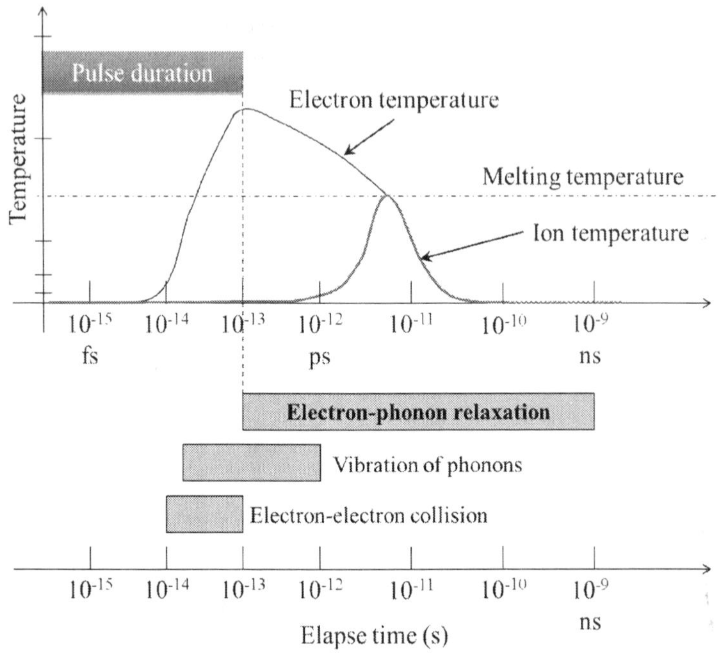

Figure 4. Characteristic time of excitation and relaxation process.

The physics and discussions mentioned in this section are quite important for the case of using femtosecond laser microstructuring into microdevices such as a thin optical fiber to be highly functionalized. The following sections show the functionalization of optical fibers using femtosecond laser drilling, the operation of which requires a large number of pulses, and high fluence much larger than a material threshold to fabricate 3D microstructures. The surface-granulated structures found in the above approach are also discussed with recent experimental results in our research groups.

3. Femtosecond laser hole drilling to optical fibers

In the past three decades, fiber-optic sensor technology has been developed for different applications. The progresses in research have developed the fiber-optic sensors for practical applications. For instance, distributed fiber-optic sensors have been installed in dams and bridges to monitor the performance of these facilities. With the rapid growth of optical networks, the cost of fiber-optic sensors has significantly decreased as key components in fiber-optic sensors (such as optical fiber, light sources, and photo detectors) are commercially available in optical communications industries. Therefore, it can be anticipated that fiber-optic sensors will be used in more extensive applications in sensing technology. Recent interest in the technologies has not only been focused on the development of novel ideas but also on technologies for integrating high-performance devices of various functionalities onto a single optical fiber, which are sometimes referred to as "lab-on-fiber" technologies. The authors have previously reported a new approach to the development of a fiber-optic spectrometer by embedding a microhole to be a spectroscopic sample cell in multimode optical fiber, using a near-ultraviolet (NUV) femtosecond laser [22].

Optical fiber sensors machined by femtosecond laser drilling are summarized in **Table 1**. The fiber-optic sensor with a microhole working as a sensing area in single-mode optical fibers [19, 20] was demonstrated to monitor the refractive index (RI) of liquid, in which the microhole structures were fabricated by femtosecond laser irradiation of near infrared (NIR) at 800 nm with a 120-fs duration pulse at 1 kHz repetition rate during 15 s and pulse energy of 11 μJ. The sensor performances seem to be limited for introducing liquid into the microhole when using high viscous liquids because the shapes of microholes are sharply tapered over the entire hole due to objective lens used to focus laser beam and its fabrication procedure of one side laser irradiation. The other experiment [21] was conducted by using 1030-nm NIR femtosecond laser with a 280-fs pulse of pulse energy of 1–5 μJ and 100-kHz high repetition rate during 5 min, and immersion oil is also required to reduce the laser-induced thermal energy. An optical fiber was immersed in water to avoid the thermal influence, where helical drilling method was also employed in the fabrication procedure in order to reduce tapering the microhole. However, the tapering in the lateral direction of optical axis still remains at the opposite side of cladding because laser beam is focused by passing through the cylindrical body of the optical fiber processed. For the case of femtosecond laser deep drilling to dielectrics, laser energy can be nonlinearly transferred to electrons via multiphoton absorption, where the order of N-photon ionization process is determined by photon energy. This fact indicates that ablation threshold becomes lower with decreasing wavelength [75, 76] because efficient energy transfer to electrons can be easily achieved via a shorter-wavelength femtosecond laser via efficient multiphoton absorption with reducing extra energy diffusion by collision cascade [77, 78]. Consequently, it can be supposed that a shorter-wavelength laser with a higher photon energy is comparatively favorable for microstructuring in microscale platforms such as thin optical fibers because the shorter-wavelength beam can reduce thermal debris inevitably generated through resolidification of a molten material. Such debris disturbs growth of microhole by absorbing/scattering subsequent laser pulses and redeposition of debris on the hole surface, resulting in poor surface quality of the inner wall. Actually, the experiment conducted on the fabrication of microhole with 1030-nm femtosecond laser in air atmosphere was tested experimentally and failed [21]. Moreover, it was found that the surface microstructures on the inner wall are remarkably different between using NIR and NUV femtosecond laser drilling. Further advantage of use of 400-nm laser deep drilling is found in the fact that Rayleigh range can be longer. In contrast to the hole drilling using NIR wavelength (e.g. 800 nm [19, 20], 1030 nm [21]) which were inevitably accompanied with debris and longer processing period more than 10 s or 5 min, efficient deep hole drilling in glass optical fiber was successfully demonstrated with reducing debris in a very short irradiation time of approximately 1.0 s to fabricate a fiber-optic inline spectrometers, using a 400-nm, 1-kHz, and 350-fs laser irradiation.

Laser and its irradiation condition							Cell					
References	Pulse width (fs)	Wave length (nm)	Repetition (kHz)	Pulse energy	Irradiation time	Fabrication process	Lens NA	Number of hole	Shape (taper angle)	Hole diameter (μm)	Roughness	Volume (single cell) (pL)
[19]	120	800	1	11 μJ	5 s	Single direction	0.25	1	Dead-end (7°–10°)	6	–	1.2
					10 s					8		6.3
					15 s					11		12
[20]				–	–			1	Through hole (4°–18°)	15	>3 μm	60–80
[21]	280	1030	100	1–5 μJ	5 min	Helical drilling	0.68	SM: 1 MM: 3	Through hole (1°–4°)	20	300 nm (no evidence)	40
[22]	350	400	1	15 μJ	1.2 s	Two direction	0.40	MM: 1, 10	Through hole (3.7°)	10	<500 nm	20

Table 1. Optical fiber sensors with a microhole produced by femtosecond laser drilling.

3.1. Fabrication of microholes into optical fibers

The following sections show results obtained in our recent study regarding the fabrication of optical fiber sensors using femtosecond laser with high-fluence regimes. As shown in **Figures 5** and **6**, the microholes are intentionally designed and machined to easily guide a liquid sample into microhole working as a sensing area without immersion liquids during fabrication process. The sensor part has microhole array penetrating through the whole fiber core by drilling from both sides of the fiber cladding in order to avoid tapering the opening apertures inlet/outlet of through hole.

The microholes were fabricated with optical train, as illustrated in **Figure 5**. A multimode optical fiber (MMF, core diameter 62.5 μm) was irradiated with femtosecond pulses from two directions (not simultaneously) to avoid tapering of the inlet/outlet of the cell. The fundamental of a 1 kHz Ti-sapphire laser (IFRIT Cyber Laser Inc., 1 mJ/pulse, λ = 800 nm, 210 fs pulse duration) was converted to the second harmonic (240 μJ/pulse, 350 fs pulse duration) in a wavelength converter [second harmonic generation (SHG) unit, manufactured by Cyber Laser Inc.]. The second harmonic pulses were introduced to an objective lens through an optical train to guide and focus the beam at the target optical fiber. The collimating optics reduce the beam diameter from 6.0 to 2.8 mm so that a longer Rayleigh distance compared with the initial beam can be obtained using a negative and positive lens array. This prevents plasma generation in the air between the lens combination, which is important because the laser beam could be defocused and the energy could be lost by plasma creation during collimation and thereby affected by diffraction before the final focusing. An optical fiber irradiated was mounted on a three-dimensional motor-controlled translation stage equipped with a rotary mechanism for rotation of the fiber. The laser focal point was carefully adjusted by moving the translation stage in the X, Y, Z, and θx directions. The focal point, defined as the center of the Rayleigh distance, was set to be 5 μm beyond the cladding surface of the inlet.

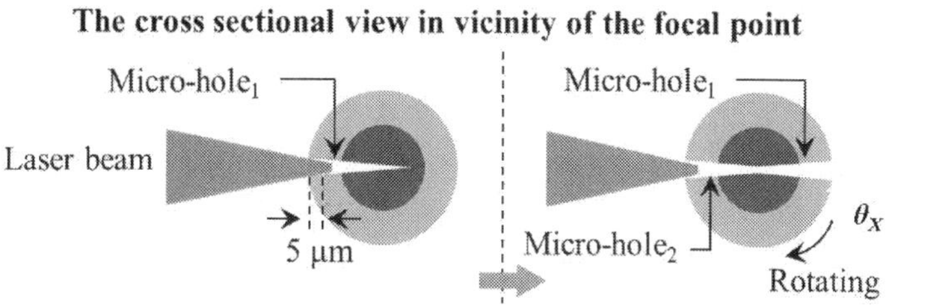

	Objective lens	NA	Focal length (mm)	Effective diameter (mm)	Spot size (μm)	Rayleigh distance (μm)
(a)	M Plan NUV 20x (0.40)	0.4	10	8	1.82	6.5
(b)	M Plan SL 20x (0.28)	0.28	20	5.6	3.64	26

Figure 5. The cross-sectional view in vicinity of the focal point and two different types of collimations.

3.2. Surface roughness on inner wall

Figure 6(a) and **(b)** indicates the schematic images of side and cross-sectional views of the through holes embedded in optical fiber lines, respectively. Based on the microscopic image showing the fabricated through hole (c), the hole diameters were found to be approximately 18 and 10 μm at the opening aperture and waist, respectively, the volume of which was calculated to 19.8 pL by assuming that its shape could be approximated by two truncated cones with taking into account the taper angle of the microhole. Optical micrographs of the through hole of the side and top view are presented in **Figure 6(c)** and **(d)**, respectively, showing that the microholes are connected together so as to produce a through hole. Taking a look at the inside of the through hole by using scanning electron microscope (SEM) as shown in **Figure 6(e)**, a morphology modification was formed to be granulated structure, which have been reported so far in the previous works as mentioned above in the femtosecond regime [41]. The sample of **Figure 6(e)** was prepared in such a way that the inner surface of a part of hole was exposed by intentionally cleaving the fabricated fiber along an off-axis plane of the through hole to clearly see the rugged surface with SEM. The rough surface could be generated with folded debris or microrims surrounding laser-induced craters by successive laser pulses. Importantly, the size of the rugged particle is found to be approximately a few hundred nanometers. Such rough surface gives more diffusive reflection on the hole boundary, hence higher optical scattering, especially for shorter-wavelength light wave. Additionally, it should also be noted that such particle-like structure will be more water repellent to a liquid by which water inside becomes easier to flow. As shown in **Table 1**, the surface roughness should be depended on the fabrication method. The shorter wavelength laser irradiation can reduce the increasing thermal debris and its redeposition on the inner surface, the fact of which can be seen from the comparison between Refs. [20] and [22].

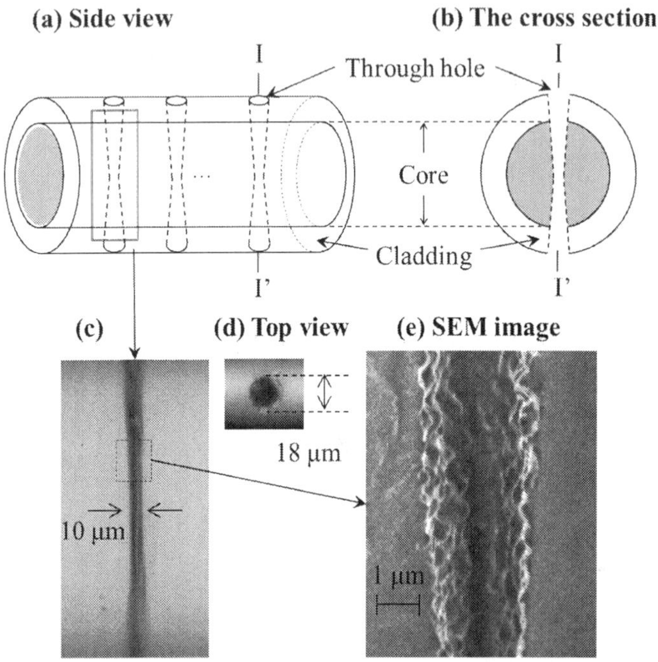

Figure 6. The schematic drawings of microholes in a fiber-optic line monitored from side view (a) and the cross section (b). The photographs of (c) and (d) give examples which show the shape of the through hole and the hole opening, respectively. (e) SEM picture of a part of hole. Adapted with permission from Ref. [22].

3.3. Influences of surface roughness for sensing

To evaluate the intrusion/discharge velocity of liquids sucked into/drained out from a through hole, the real-time response of optical intensity change was measured in such a way that the sensing part with a single through hole is alternatively immersed in water and ethanol as indicated in **Figure 7**. Sample liquids were immediately sucked into microholes by capillary driving force and as was firstly confirmed by monitoring the optical intensity of transmitting light and staining the microhole with a color dye simultaneously. As soon as the sensing part was lifted from a liquid pool, the liquid held in the microhole seems to be immediately drained out in a few seconds because of the hydrophobic-repellent structure.

Figure 8 shows transmitting spectra to investigate sensing response for RI changes in a wavelength between 400 and 1000 nm. In this measurement, two types of sensor samples which had (a) a single cell and (b) 10 cells were prepared to compare the spectroscopic measurement between them. Taking a look at **Figure 8(b)**, remarkable decreases in transmission can be confirmed in a shorter wavelength range ($\lambda < 600$ nm), because light transmitted through the fiber core was scattered at the hole surfaces by the fact that the wavelength of transmitted light becomes approximately equal to the scale of surface roughness. By reducing or controlling the surface roughness on the inner wall, sensing performances of fiber optics with an inline sample cell in terms of sensitivity, reproducibility, response time, and recovery time as well as versatility for sensing target could be improved for practical use. High-quality surface finish therefore is required for the improvement of the sensing performances even in high-fluence regimes. The roughness with a periodicity of <100 nm could be useful for reducing the scattering in a shorter wavelength range, by roughly estimating the reflectivity on the inner wall based on Mie-scattering effect.

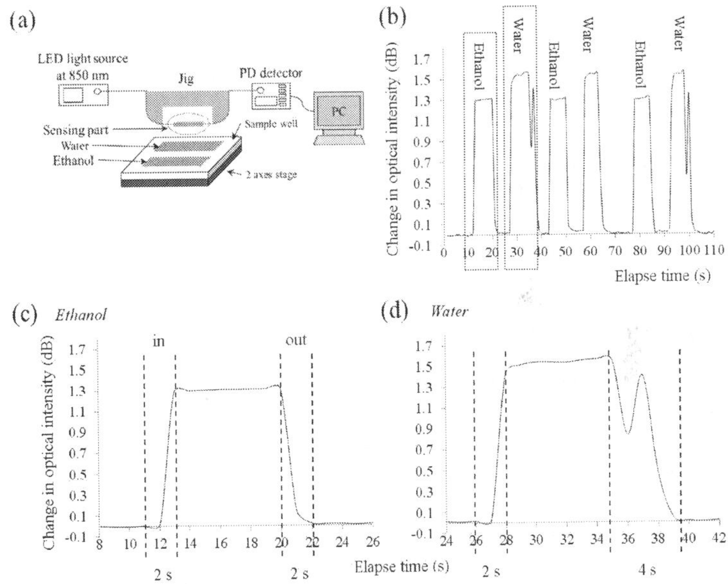

Figure 7. (a) Experimental setup to measure intrusion/discharge velocity of liquids. (b) The real-time response of optical intensity change measured by alternatively immersing in water and ethanol. (c) and (d) show one of a response waveform using ethanol and water, respectively.

According to the experimental data [19–22] (summarized in **Table 1**), the roughness might be reduced by immersing an optical fiber in water or immersion oils even though the fabrication period becomes unfavorably much longer. Instead of using liquid immersion processing, we have been trying to reduce the roughness using a NUV 400-nm femtosecond laser so as to achieve a high-quality inner surface of microhole in a high-throughput fashion. **Figure 9** shows SEM pictures of the inner surface of a dead-end microhole fabricated in a flat silica glass plate. The experiment was performed to make a smooth surface on the inner wall of the hole with varying laser irradiation parameters. As can be seen from **Figure 9(a)**, wave-like nanostructures with a spatial periodicity of 100–200 nm are observed on the bottom of the hole. **Figure 9(b)** and **(c)**, respectively, shows the inner surface of the hole at different depths near and far from the hole inlet. Surface roughness is still observed at the periphery of the inlet of **Figure 9(b)** because the surface roughness could be worse as the number of irradiated pulses becomes greater. On the other hand, the case of the inner surface **Figure 8(c)** seems to be much smoother than the surface (b). To figure out the formation mechanism of the granulated surface in a high-energy regime, further experiment was carried out on this point.

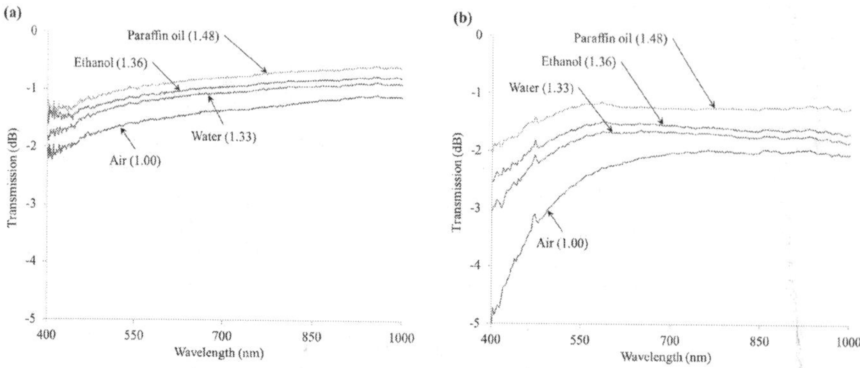

Figure 8. Transmission spectra for the case of sensor samples (a) single cell and (b) 10 cells when immersion in three types of liquids. Adapted with permission from Ref. [22].

The surface morphology of shallow craters was generated by single and multiple pulses with an emphasis on rims surrounding the craters. In this experiment, shallow craters were formed on a fused silica plate by overlapping focused laser pulses on the same spot as shown in **Figure 10**. The shape of the crater created in this experiment is depended on incident laser beam, the pattern of which is deformed through a second harmonic generator (SHG unit manufactured by Cyber Laser Inc.). A crater surrounded by a single rim was observed by SEM analysis of **Figure 11(b)**. The average thickness of the first rim is approximately 500 nm. **Figure 11(c)** shows an overlapped crater created by two pulses. Molten materials are driven away from the crater to almost 10 μm distance. The second image (**Figure 11(d)**) shows that a new rim is formed inside the first one by overlapping a second pulse. The distance between the two rims is approximately equal to the wavelength of the irradiated laser beam. We also found a belt-shape structure on a part of the rim. The composition and formation mechanism still need to be explored. For the case of **Figure 11(f)**: overlapping three pulses, it was interestingly found that the height of third rim is comparatively low simultaneously accompanied with nanofibers [79–81] with a diameter ranging from a few to few tens of nanometers. It should be noted that, instead of the growth of the rim, nanofibers were grown by only a few pulses. In order to obtain optimum surface conditions for the fiber-optic inline spectrometer, further investigations are in progress to improve the surface finish.

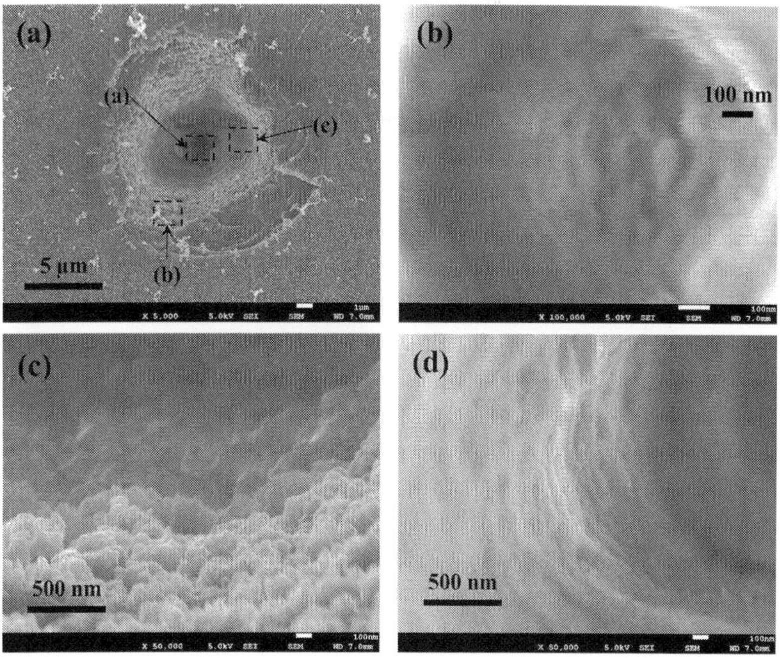

Figure 9. SEM images of a microhole generated on a flat silica glass plate, by successive laser pulses of the energy of 20 µJ and using focusing optics (inserted table (b) in **Figure 5**). (a) The whole fabricated microhole at 5000× magnification and higher resolution SEM pictures at hole bottom (b), the side wall near the hole inlet (c), and the side wall near the bottom (d).

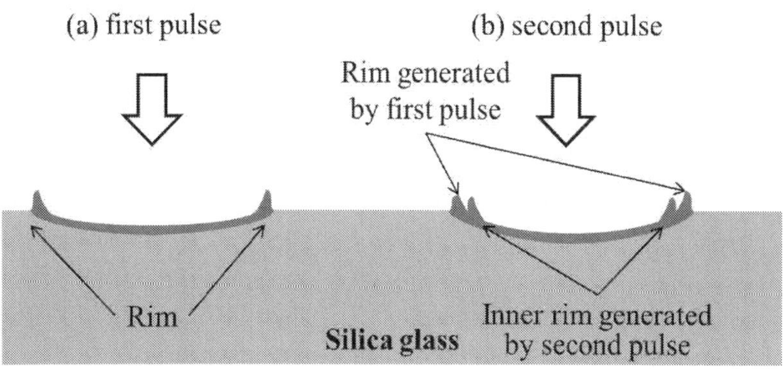

Figure 10. The schematic image of crater rims produced by overlapping focused laser pulses on the same spot. (a) One laser pulse. (b) Two overlapping pulses.

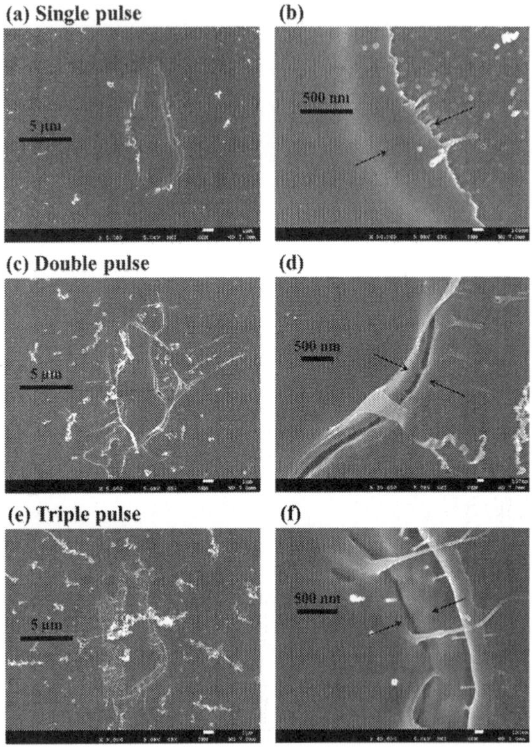

Figure 11. SEM images of shallow craters with thin rim generated on a flat silica glass plate with varying the number of pulses. The whole crater at 5000× magnification (a, c, e) and higher resolution (b, d, f).

4. Conclusion

In this chapter, the authors presented an overview of laser-induced periodic surface structures (LIPSS) produced by ultrashort laser pulses. A part of its historical background was provided; the formation mechanism of LIPSS in femtosecond pulse regimes was described. Experimental results obtained by our studies were also presented as a practical use of femtosecond micro-machining even though a part of the result is still under investigation.

Section 4 shows the functionalization of optical fibers using femtosecond laser drilling. High-fluence femtosecond laser microstructuring was described with recent experimental data in our research group. The investigation not only to explain the formation mechanism of LIPSS but also to have a microscale perspective should be important for material removal because a high-throughput fabrication of 3D microstructures to microdevices requires a large number of pulses and high fluence. We therefore expect that the obtained results can contribute to practical aspects.

Acknowledgements

This work has been partially supported by JSPS KAKENHI Grant Number 15K04735.

Author details

Kenji Goya[1*], Masahiko Shiraishi[2], Yusuke Fuchiwaki[1], Kazuhiro Watanabe[2] and Toshihiko Ooie[1]

*Address all correspondence to: k-gouya@aist.go.jp

1 National Institute of Advanced Industrial Science and Technology (AIST), Hayashi-cho, Takamatsu, Kagawa, Japan

2 Soka University, Tangi-machi, Hachioji, Tokyo, Japan

References

[1] S.S. Mao, F. Qu′er′, S. Guizard, X. Mao, R.E. Russo, G. Petite, P. Martin, Dynamics of femtosecond laser interactions with dielectrics, Appl. Phys. A Mater. Sci. Process. 79 (2004) 1695–1709. doi:10.1007/s00339-004-2684-0.

[2] S. Kanehira, K. Miura, K. Hirao, Ion exchange in glass using femtosecond laser irradiation, Appl. Phys. Lett. 93 (2008) 2–4. doi:10.1063/1.2959820.

[3] Y. Liao, Y. Shen, L. Qiao, D. Chen, Y. Cheng, K. Sugioka, K. Midorikawa, Femtosecond laser nanostructuring in porous glass with sub-50 nm feature sizes., Opt. Lett. 38 (2013) 187–189. doi:10.1364/OL.38.000187.

[4] B.N. Chicbkov, C. Momma, S. Nolte, F. von Alvensleben, A. Tuinnermann, Femtosecond, picosecond and nanosecond laser ablation of solids, Appl. Phys. A Mater. Sci. Process. 115 (1996) 109–115.

[5] D. Vipparty, B. Tan, K. Venkatakrishnan, Nanostructures synthesis by femtosecond laser ablation of glasses, J. Appl. Phys. 112 (2012) 073109:1–7. doi:10.1063/1.4754864.

[6] G. Dumitru, V. Romano, H.P. Weber, M. Sentis, W. Marine, Femtosecond ablation of ultrahard materials, Appl. Phys. A Mater. Sci. Process. 74 (2002) 729–739. doi:10.1007/s003390101183.

[7] K. Furusawa, K. Takahashi, H. Kumagai, K. Midorikawa, M. Obara, Ablation characteristics of Au, Ag, and Cu metals using a femtosecond Ti: sapphire laser, Appl. Phys. A Mater. Sci. Process. 69 (1999) S359–S366. doi:10.1007/s003399900393.

[8] S. Ahsan, Y. Gyu, M. Seop, Formation mechanism of nanostructures in soda – lime glass using femtosecond laser, J. Non. Cryst. Solids. 357 (2011) 851–857. doi:10.1016/j.jnoncrysol.2010.11.006.

[9] Y. Guan, C. Li, X. Cheng, B. Wang, R. Sun, X. Liang, J. Zhao, H. Chen, G. Lu, Highly sensitive mixed-potential-type NO2 sensor with YSZ processed using femtosecond laser direct writing technology, Sensors Actuators, B Chem. 198 (2014) 110–113. doi: 10.1016/j.snb.2014.02.098.

[10] X. Zhao, Y.C. Shin, Femtosecond laser drilling of high-aspect ratio microchannels in glass, Appl. Phys. A Mater. Sci. Process. 104 (2011) 713–719. doi:10.1007/s00339-011-6326-z.

[11] Y. Shimotuma, H. Kazuyuki, P.G. Kazansky, J. Qiu, Invited Review Paper Three-Dimensional Micro- and Nano-Fabrication in Transparent Materials by Femtosecond Laser, 44 (2005) 4735–4748. doi:10.1143/JJAP.44.4735.

[12] M. Barberoglou, V. Zorba, E. Stratakis, E. Spanakis, P. Tzanetakis, S.H. Anastasiadis, C. Fotakis, Bio-inspired water repellent surfaces produced by ultrafast laser structuring of silicon, Appl. Surf. Sci. 255 (2009) 5425–5429. doi:10.1016/j.apsusc.2008.07.130.

[13] C. Hnatovsky, R.S. Taylor, E. Simova, P.P. Rajeev, D.M. Rayner, V.R. Bhardwaj, P.B. Corkum, Fabrication of microchannels in glass using focused femtosecond laser radiation and selective chemical etching, Appl. Phys. A. 84 (2006) 47–61. doi:10.1007/s00339-006-3590-4.

[14] M. Duocastella, J.M. Fernández-Pradas, J.L. Morenza, D. Zafra, P. Serra, Novel laser printing technique for miniaturized biosensors preparation, Sensors Actuators, B Chem. 145 (2010) 596–600. doi:10.1016/j.snb.2009.11.055.

[15] C. Cugnet, O. Zaouak, A. René, C. Pécheyran, M. Potin-Gautier, L. Authier, A novel microelectrode array combining screen-printing and femtosecond laser ablation technologies: development, characterization and application to cadmium detection, Sens. Act. B Chem. 143 (2009) 158–163. doi:10.1016/j.snb.2009.07.059.

[16] M. Yamaji, H. Kawashima, J. Suzuki, S. Tanaka, Three dimensional micromachining inside a transparent material by single pulse femtosecond laser through a hologram, Appl. Phys. Lett. 93 (2008) 041116:1–3.

[17] I. Sohn, Y. Kim, Y. Noh, I.W. Lee, J.K. Kim, H. Lee, Femtosecond laser and arc discharge induced microstructuring on optical fiber tip for the multidirectional firing, Opt. Express. 18 (2010) 19755–19760. doi:10.1364/OE.18.019755.

[18] C.H. Chen, T.C. Tsao, W.Y. Li, W.C. Shen, C.W. Cheng, J.L. Tang, C.P. Jen, L.K. Chau, W. Te Wu, Novel U-shape gold nanoparticles-modified optical fiber for localized plasmon

resonance chemical sensing, Microsyst. Technol., 2010: pp. 1207–1214. doi:10.1007/s00542-009-0945-8.

[19] Y. Wang, D.N. Wang, M. Yang, W. Hong, P. Lu, Refractive index sensor based on a microhole in single-mode fiber created by the use of femtosecond laser micromachining, Opt. Lett. 34 (2009) 3328–3330. http://www.ncbi.nlm.nih.gov/pubmed/19881583.

[20] Y. Wang, M. Yang, D.N. Wang, S. Liu, P. Lu, Fiber in-line Mach–Zehnder interferometer fabricated by femtosecond laser micromachining for refractive index measurement with high sensitivity, J. Opt. Soc. Am. B. 27 (2010) 370–374. doi:10.1364/JOSAB.27.000370.

[21] R. Buividas, M. Mikutis, G. Gervinskas, D. Day, G. Slekys, S. Juodkazis, Femtosecond laser drilling of optical fibers for sensing in microfluidic applications, In: Proc. SPIE – Int. Soc. Opt. Eng., 2012: p. 84630T. doi:10.1117/12.929607.

[22] K. Goya, T. Itoh, A. Seki, K. Watanabe, Efficient deep-hole drilling by a femtosecond, 400 nm second harmonic Ti: Sapphire laser for a fiber optic in-line/pico-liter spectrometer, Sensors Actuators, B Chem. 210 (2015) 685–691.

[23] Z. Tian, S.S. Yam, H. Loock, Refractive index sensor based on an abrupt taper Michelson interferometer in a single-mode fiber, 33 (2008) 1105–1107.

[24] F.C. Favero, L. Araujo, G. Bouwmans, V. Finazzi, J. Villatoro, V. Pruneri, Spheroidal Fabry-Perot microcavities in optical fibers for high-sensitivity sensing, Opt. Express. 20 (2012) 7112. doi:10.1364/OE.20.007112.

[25] Y.-J. Rao, M. Deng, D.-W. Duan, X.-C. Yang, T. Zhu, G.-H. Cheng, Micro Fabry-Perot interferometers in silica fibers machined by femtosecond laser, Opt. Express. 15 (2007) 14123–14128. doi:10.1364/OE.15.014123.

[26] F. Müller, C. Kunz, S. Gräf, Bio-inspired functional surfaces based on laser-induced periodic surface structures, Materials (Basel). 9 (2016) 476. doi:10.3390/ma9060476.

[27] M. Yamaguchi, Y. Kaneko, M. Sasaki, Fabrication of nano-periodic structure on plastic film for water repellent using femtosecond laser, Transaction on Control and Mechanical Systems 1 (2012) 306–311.

[28] M.R. Cardoso, V. Tribuzi, D.T. Balogh, L. Misoguti, C.R. Mendona, Laser microstructuring for fabricating superhydrophobic polymeric surfaces, Appl. Surf. Sci. 257 (2011) 3281–3284. doi:10.1016/j.apsusc.2010.10.156.

[29] A.Y. Vorobyev, C. Guo, Multifunctional surfaces produced by femtosecond laser pulses, J. Appl. Phys. 117 (2015) 033103:1–5. doi:10.1063/1.4905616.

[30] J. Hiller, J.D. Mendelsohn, M.F. Rubner, Reversibly erasable nanoporous anti-reflection coatings from polyelectrolyte multilayers, Nat. Mater. 1 (2002) 59–63. doi:10.1038/nmat719.

[31] S.-H. Hong, B.-J. Bae, K.-S. Han, E.-J. Hong, H. Lee, K.-W. Choi, Imprinted moth-eye antireflection patterns on glass substrate, Electron. Mater. Lett. 5 (2009) 39–42. doi: 10.3365/eml.2009.03.039.

[32] K. Han, J. Shin, H. Lee, Enhanced transmittance of glass plates for solar cells using nano-imprint lithography, Sol. Energy Mater. Sol. Cells. 94 (2010) 583–587. doi:10.1016/j.solmat.2009.12.001.

[33] S. Liu, C. Liao, Fast fabrication of nano-structured anti- reflection layers for enhancement of solar cells performance using plasma sputtering and infrared assisted roller embossing techniques, 20 (2012) 2908–2912.

[34] A.A. Ionin, Y.M. Klimachev, A.Y. Kozlov, S.I. Kudryashov, A.E. Ligachev, S. V. Makarov, L. V. Seleznev, D. V. Sinitsyn, A.A. Rudenko, R.A. Khmelnitsky, Direct femtosecond laser fabrication of antireflective layer on GaAs surface, Appl. Phys. B Lasers Opt. 111 (2013) 419–423. doi:10.1007/s00340-013-5350-4.

[35] N. Tagawa, M. Takada, A. Mori, H. Sawada, K. Kawahara, Development of contact sliders with nanotextures by femtosecond laser processing, Tribol. Lett. 24 (2006) 143–149. doi:10.1007/s11249-006-9142-4.

[36] A.A. Voevodin, J.S. Zabinski, Laser surface texturing for adaptive solid lubrication, Wear. 261 (2006) 1285–1292. doi:10.1016/j.wear.2006.03.013.

[37] K.J. Kubiak, M.C.T. Wilson, T.G. Mathia, P. Carval, Wettability versus roughness of engineering surfaces, Wear. 271 (2011) 523–528. doi:10.1016/j.wear.2010.03.029.

[38] J. Eichstädt, G.R.B.E. Römer, A.J. Huis, Towards friction control using laser-induced periodic surface structures, Phys. Procedia. 12 (2011) 7–15. doi:10.1016/j.phpro.2011.03.099.

[39] A. Ben-Yakar, R.L. Byer, A. Harkin, J. Ashmore, H.A. Stone, M. Shen, E. Mazur, Morphology of femtosecond-laser-ablated borosilicate glass surfaces, Appl. Phys. Lett. 83 (2003) 3030–3032. doi:10.1063/1.1619560.

[40] A. Ben-Yakar, R.L. Byer, Femtosecond laser ablation properties of borosilicate glass, J. Appl. Phys. 96 (2004) 5316–5323. doi:10.1063/1.1787145.

[41] A. Ben-Yakar, A. Harkin, J. Ashmore, R.L. Byer, H.A. Stone, Thermal and fluid processes of a thin melt zone during femtosecond laser ablation of glass: The formation of rims by single laser pulses, J. Phys. D. Appl. Phys. 40 (2007) 1447–1459. doi: 10.1088/0022-3727/40/5/021.

[42] K. Sugioka, Y. Cheng, Ultrafast lasers — reliable tools for advanced materials processing, Light: Science & Applications 3 (2014), e149:1-12. doi:10.1038/lsa.2014.30.

[43] K. Sugioka, D. Wu, K. Midorikawa, Ship-in-a-bottle biomicrochips fabricated by hybrid femtosecond laser processing, MATEC Web of Conferences, 8 (2013) 05005:1-2.

[44] S. Her, C. Huang, Effect of coating on the strain transfer of optical fiber sensors, Sensors 11 Sensors 11 (2011) 6926–6941. doi:10.3390/s110706926.

[45] B. Li, L. Jiang, S. Wang, Q.C. Mengmeng Wang, J. Yang, A new Mach-Zehnder interferometer in a thinned-cladding fiber fabricated by electric arc for high sensitivity refractive index sensing, Opt. Lasers Eng. 50 (2012) 829–832. doi:10.1016/j.optlaseng.2012.01.024.

[46] Z. Feng, J. Li, X. Qiao, L. Li, H. Yang, M. Hu, A thermally annealed Mach-Zehnder interferometer for high temperature measurement, Sensors (Basel). 14 (2014) 14210–14221. doi:10.3390/s140814210.

[47] M. Birnbaum, Semiconductor surface damage produced by Ruby lasers, J. Appl. Phys. 36 (1965) 3688–3689. doi:10.1063/1.1703071.

[48] C. Teichert, Self-organization of nanostructures in semiconductor heteroepitaxy, Phys. Rep. 365 (2002) 335–432. doi:10.1016/S0370-1573(02)00009-1.

[49] X. Wang, C.A. Ohlin, Q. Lu, J. Hu, Cell directional migration and oriented division on three-dimensional laser-induced periodic surface structures on polystyrene, Biomaterials. 29 (2008) 2049–2059. doi:10.1016/j.biomaterials.2007.12.047.

[50] K. Oya, S. Aoki, K. Shimomura, N. Sugita, K. Suzuki, Morphological observations of mesenchymal stem cell adhesion to a nanoperiodic-structured titanium surface patterned using femtosecond laser processing, Jpn. J. Appl. Phys. 51 (2012) 125203. doi:10.1143/JJAP.51.125203.

[51] B. Dusser, Z. Sagan, H. Soder, N. Faure, J.P. Colombier, M. Jourlin, E. Audouard, Controlled nanostructures formation by ultra fast laser pulses for color marking., Opt. Express. 18 (2010) 2913–2924. doi:10.1364/OE.18.002913.

[52] B.-B. Xu, Z.-C. Ma, L. Wang, R. Zhang, L.-G. Niu, Z. Yang, Y.-L. Zhang, W.-H. Zheng, B. Zhao, Y. Xu, Q.-D. Chen, H. Xia, H.-B. Sun, Localized flexible integration of high-efficiency surface enhanced Raman scattering (SERS) monitors into microfluidic channels., Lab Chip. 11 (2011) 3347–3351. doi:10.1039/c1lc20397e.

[53] J.E. Sipe, J.F. Young, J.S. Preston, H.M. Van Driel, Laser-induced periodic surface structure. I. Theory, Phys. Rev. B. 27 (1983) 1141–1154. doi:10.1103/PhysRevB.27.1141.

[54] J.F. Young, J.S. Preston, H.M. Van Driel, J.E. Sipe, Laser-induced periodic surface structure. II. Experiments on Ge, Si, Al, and brass, Phys. Rev. B. 27 (1983) 1155–1172. doi:10.1103/PhysRevB.27.1155.

[55] J.F. Young, J.E. Sipe, H.M. Van Driel, Laser-induced periodic surface structure. III. Fluence regimes, the role of feedback, and details of the induced topography in germanium material, Phys. Rev. B. 30 (1984) 2001-2015. doi:10.1103/PhysRevB.30.2001.

[56] Q. Wu, Y. Ma, R. Fang, Y. Liao, Q. Yu, X. Chen, K. Wang, Femtosecond laser-induced periodic surface structure on diamond film, Appl. Phys. Lett. 82 (2003) 1703–1705. doi: 10.1063/1.1561581.

[57] A. Borowiec, H.K. Haugen, Subwavelength ripple formation on the surfaces of compound semiconductors irradiated with femtosecond laser pulses, Appl. Phys. Lett. 82 (2003) 4462–4464. doi:10.1063/1.1586457.

[58] G. Miyaji, K. Miyazaki, Nanoscale ablation on patterned diamondlike carbon film with femtosecond laser pulses, Appl. Phys. Lett. 91 (2007) 2005–2008. doi:10.1063/1.2784966.

[59] G. Miyaji, K. Miyazaki, Origin of periodicity in nanostructuring on thin film surfaces ablated with femtosecond laser pulses, Opt. Express. 16 (2008) 16265–16271. doi: 10.1364/OE.16.016265.

[60] M. Straub, M. Afshar, D. Feili, H. Seidel, K. König, Surface plasmon polariton model of high-spatial frequency laser-induced periodic surface structure generation in silicon, J. Appl. Phys. 111 (2012) 124315:1-6. doi:10.1063/1.4730381.

[61] R. Le Harzic, D. Dörr, D. Sauer, M. Neumeier, M. Epple, H. Zimmermann, F. Stracke, Large-area, uniform, high-spatial-frequency ripples generated on silicon using a nanojoule-femtosecond laser at high repetition rate, Opt. Lett. 36 (2011) 229–231. doi: 10.1364/OL.36.000229.

[62] M. Henyk, N. Vogel, D. Wolfframm, A. Tempel, J. Reif, Femtosecond laser ablation from dielectric materials: Comparison to arc discharge erosion, Appl. Phys. A Mater. Sci. Process. 69 (1999) 355–358. doi:10.1007/s003399900411.

[63] J. Reif, F. Costache, M. Henyk, S. V Pandelov, Ripples revisited: Non-classical morphology at the bottom of femtosecond laser ablation craters in transparent dielectrics, Appl. Surf. Sci. (2002) 891–895. doi:10.1016/S0169-4332(02)00450-6.

[64] F. Costache, S. Kouteve-arguirova, J. Reif, Sub – damage – threshold femtosecond laser ablation from crystalline Si◉: Surface nanostructures and phase transformation, Appl. Phys. A Mater. Sci. Process. 1432 (2004) 1429–1432. doi:10.1007/s00339-004-2803-y.

[65] J. Reif, O. Varlamova, F. Costache, Femtosecond laser induced nanostructure formation: Self-organization control parameters, Appl. Phys. A Mater. Sci. Process. 92 (2008) 1019–1024. doi:10.1007/s00339-008-4671-3.

[66] J. Reif, F. Costache, O. Varlamova, G. Jia, M. Ratzke, Self-organized regular surface patterning by pulsed laser ablation, Phys. Status Solidi Curr. Top. Solid State Phys. 6 (2009) 681–686. doi:10.1002/pssc.200880719.

[67] M. Huang, F. Zhao, Y. Cheng, N. Xu, Z. Xu, Origin of laser-induced near-subwavelength ripples: Interference between surface plasmons and incident laser, ACS Nano. 3 (2009) 4062–4070. doi:10.1021/nn900654v.

[68] G.A. Martsinovskii, G.D. Shandybina, D.S. Smirnov, S. V Zabotnov, L.A. Golovan, V.Y. Timoshenko, P.K. Kashkarov, Ultrashort excitations of surface polaritons and waveguide modes in semiconductors, Opt. Spectrosc. 105 (2008) 67–72. doi:10.1134/S0030400x08070114.

[69] S. Sakabe, M. Hashida, S. Tokita, S. Namba, K. Okamuro, Mechanism for self-formation of periodic grating structures on a metal surface by a femtosecond laser pulse, Phys. Rev. B – Condens. Matter Mater. Phys. 79 (2009) 1–4. doi:10.1103/PhysRevB.79.033409.

[70] K. Okamuro, M. Hashida, Y. Miyasaka, Y. Ikuta, S. Tokita, S. Sakabe, Laser fluence dependence of periodic grating structures formed on metal surfaces under femtosecond laser pulse irradiation, Phys. Rev. B - Condens. Matter Mater. Phys. 82 (2010) 1–5. doi:10.1103/PhysRevB.82.165417.

[71] J. Bonse, M. Munz, H. Sturm, Structure formation on the surface of indium phosphide irradiated by femtosecond laser pulses, J. Appl. Phys. 97 (2005) 013538:1-9. doi: 10.1063/1.1827919.

[72] J. Bonse, A. Rosenfeld, J. Kruger, On the role of surface plasmon polaritons in the formation of laser-induced periodic surface structures upon irradiation of silicon by femtosecond-laser pulses, J. Appl. Phys. 106 (2009) 104910:1-5. doi:10.1063/1.3261734.

[73] D. Dufft, A. Rosenfeld, S.K. Das, R. Grunwald, J. Bonse, Femtosecond laser-induced periodic surface structures revisited: A comparative study on ZnO, J. Appl. Phys. 105 (2009) 034908:1-9. doi:10.1063/1.3074106.

[74] F. Baset, K. Popov, A. Villafranca, A.M. Alshehri, J.M. Guay, L. Ramunno, V.R. Bhardwaj, Nanopillar formation from two-shot femtosecond laser ablation of poly-methyl methacrylate, Appl. Surf. Sci. 357 (2015) 273–281. doi:10.1016/j.apsusc.2015.09.007.

[75] E.G. Gamaly, A. V Rode, V.T. Tikhonchuk, E.G. Gamaly, A. V Rode, Ablation of solids by femtosecond lasers: Ablation mechanism and ablation thresholds for metals and dielectrics, Phys. Plasmas. 9 (2002) 949–957. doi:10.1063/1.1447555.

[76] A. Borowiec, H.F. Tiedje, H.K. Haugen, Wavelength dependence of the single pulse femtosecond laser ablation threshold of indium phosphide in the 400–2050 nm range, Appl. Surf. Sci. 243 (2005) 129–137. doi:10.1016/j.apsusc.2004.09.105.

[77] J. Noack, A. Vogel, Laser-induced plasma formation in water at nanosecond to femtosecond time scales: Calculation of thresholds, absorption coefficients, and energy density, IEEE J. Quant. Electr. 35 (1999) 1156-1167.

[78] A. Kaiser, Microscopic processes in dielectrics under irradiation by subpicosecond laser pulses, Phys. Rev. B 61 (2000) 11437–11450.

[79] H.M. Saulius, Juodkazis, Oleg A Louchev, Kenji Kitamura, Femtosecond laser ablation of chalcogenide glass: Explosive formation of nano-fibres against thermo-capillary growth of micro-spheres, Nanotechnology. 17 (2006) 4. doi: 10.1088/0957-4484/17/19/003.

[80] K. Venkatakrishnan, D. Vipparty, B. Tan, Nanofibre fabrication by femtosecond laser ablation of silica glass, Opt. Express. 19 (2011) 15770. doi:10.1364/OE.19.015770.

[81] A. V. Rode, E.G. Gamaly, B. Luther-Davies, Formation of cluster-assembled carbon nano-foam by high-repetition-rate laser ablation, Appl. Phys. A Mater. Sci. Process. 70 (2000) 135–144. doi:10.1007/s003390050025.

Chapter 15

Laser Ablation of Polymethylmethacrylate (PMMA) by Phase-Controlled Femtosecond Two-Color Synthesized Waveforms

Ci-Ling Pan, Chih-Hsuan Lin, Chan-Shan Yang and Alexey Zaytsev

Additional information is available at the end of the chapter

http://dx.doi.org/10.5772/65637

Abstract

Single-shot laser ablation of polymethylmethacrylate (PMMA) was studied using dual-color waveform synthesis of the fundamental (ω) and its second harmonic (2ω) of a femtosecond Ti: Sapphire laser. Changing the relative phase of the fundamental (ω) and second-harmonic (2ω) outputs of the exciting laser resulted in clear modulation of the ablated area. The modulation as well as the dependence of the ablation threshold on the relative phase between the ω and 2ω beams correlated closely with the theoretical model of laser breakdown (ablation) of transparent materials through photoionization in the intermediate regime (Keldysh parameter $\gamma \approx 1.5$). Our study illustrates the potential applications of using phase-controlled synthesized waveform for laser processing of materials.

Keywords: femtosecond pulses, laser ablation, coherent control, breakdown, polymethylmethacrylate (PMMA), polymer, Keldysh parameter

1. Introduction

Femtosecond (fs) laser micromachining has been studied intensively for the past two decades. One of the advantages of using ultrashort laser pulses rather than longer pulses for laser material processing pertains to the nonthermal ablation mechanism. By considerably reducing the area of heat-affected zones, precise laser micro- and nanomachining have become feasible for fs machining. To date, fs laser micromachining has been performed on a variety of wide-band-gap materials, such as polymers [1, 2], fused silica [3–6], and silicon [7–10]. However, almost all of these studies employed one-color laser pulses. More recently, coherent waveform-synthesized two-color laser pulses have been successfully used for increasing plasma generation [11], generating high harmonics [12], and producing broadband terahertz radiation [13]. By studying femtosecond laser ablation of polymethylmethacrylate (PMMA), our group demonstrated that the ablated hole areas exhibited clear modulation with a contrast of 22% by varying the relative phase between the ω and 2ω beams [14]. It was assumed that different peak intensity for the synthesized waveform was responsible for the observed phenomena. The physical mechanism was not clear.

In general, ultrafast laser ablation of dielectrics, such as PMMA, has been explained by the photochemical, photothermal, and photophysical models [15]. In the photochemical model, direct bond breaking in PMMA is achieved by exposing it to an ultrashort laser pulse for producing several reaction products, such as CO, CO_2, CH_4, CH_3OH, and $HCOOCH_3$. In the photothermal model, electronic excitation by picosecond laser pulses results in thermal bond breaking, leading to the formation of PMMA monomers. Among these models, the most interesting one is the photophysical one, in which both thermal and nonthermal bond breaking occur simultaneously. In thermal bond breaking, electronic excitation by ultrashort laser pulses results in ultrashort-laser-induced ionization in the picosecond (ps) and fs ranges. The three main processes of photophysical laser-induced breakdown are (i) excitation of conduction band electrons through ionization, (ii) heating of conduction band electrons through irradiation of the dielectric, and (iii) plasma energy transfer to the lattice, which causes bond breaking [16–19].

The Keldysh formalism, describing electron tunneling through a barrier created by the electric field of a laser, is often employed for modeling laser breakdown of materials by photoionization, including both multiphoton and tunneling cases. The Keldysh parameter can be expressed as the square root of the ratio between the ionization potential and twice the value of the ponderomotive potential of the laser pulse. Alternatively, it can be expressed as the ratio of tunneling frequency to the laser frequency. The tunneling time or the inverse of the tunneling frequency is given by the mean free time of an electron passing through a barrier width, $l_{tunneling} = I_p/eE(t, \varphi)$, where I_p is the ionization potential, e is the electron charge, and $E(t, \varphi)$ is the optical field.

Depending on the laser intensity used for above-threshold ionization [20–22], two regions of photoionization exist: the tunneling ionization region [20, 23, 24] and multiphoton ionization region [25–28]. In tunneling ionization, the electric field is extremely strong. The Coulomb well can be suppressed to cause the bound electron to tunnel through the barrier and be ionized. At lower laser intensities, the electron can absorb several photons simultaneously. The electron makes the transition from the valence band to the conduction band if the total energy of the absorbed photons is greater than or equal to the band gap of the material.

The boundary between tunneling ionization and multiphoton ionization is unclear. Schumacher et al. showed that there should be a so-called intermediate region that exhibit both tunneling and multiphoton characteristics. Mazur et al., following the Keldysh formalism, estimated that the intermediate region corresponded to a Keldysh parameter $\gamma \approx 1.5$ [29].

http://dx.doi.org/10.5772/65637

The tunneling ionization rate is a function of the electric field. It is well known that the multiphoton ionization rate can be expressed as $\varpi_{mpi} \propto \sigma_k I^k$, where I is the laser intensity and σ_k is the multiphoton absorption coefficient for k photons [30]. When the ionization occurs in the intermediate region, an electron can absorb several photons and be ionized by the tunneling effect. In this regime, the ionization rate can also vary with phase of the exciting electric field. The laser intensity required, however, is considerably lower than that in the pure tunneling ionization case.

In this chapter, we present the current progress on laser ablation of polymethylmethacrylate (PMMA) by phase-controlled femtosecond two-color synthesized waveforms. Significantly, laser breakdown (ablation) of transparent materials through photoionization in the intermediate regime (Keldysh parameter $\gamma \approx 1.5$) was demonstrated for the first time. The modulation of ablated hole area as well as the dependence of the ablation threshold on the relative phase between the ω and 2ω beams were observed. The data correlated closely with the theoretically predicted phase dependence of the photoionization rate using the Keldysh formalism.

2. The physical mechanisms of ultrafast laser ablation

2.1. Introduction

In this section, some key concepts of ultrafast laser ablation will be summarized. This includes light-matter interaction mechanisms such as photochemical, photothermal and photophysical. Dielectric breakdown due to ionization by tunneling, multiphoton and avalanche processes are described. Most relevant for this work, the so-called intermediate regime of photoionization, will be formulated by using the Keldysh equation, defining the Keldysh parameter used throughout this chapter.

2.2. Photoexcitation processes

Laser ablation is one of the manifestations of light-matter interactions. As expected, the ablation processes depend on characteristics of the irradiating laser, such as its intensity, wavelength, and polarization. When ultrafast laser are used, ablation mechanism become more complicated. For polymeric materials, not only photoionization but also the direct bond breaking will lead the ablation process. The main mechanisms are photochemical, photothermal, and photophysical. These three effects are located in different regions of laser pulse.

For ultrafast laser ablation of PMMA, there are two dominant mechanisms, i.e., photochemical and photothermal. In photochemical events, absorption of photons by the material being processed lead directly to covalent bond breaking [15]. The polymeric materials, such as PMMA, are generally made of a wide variety of chromophores, which may dissociate into reactive fragments by absorption of energetic UV photons. Absorption of less energetic photons, e.g., those in the visible or near infrared band, can also lead to the above photochemical processes [15]. Photothermal effect is another basic mechanism of laser ablation. Irradiated by ultrashort laser pulses, the irradiated material absorbs photons and transfer energy to electrons such that photoionization of the material can occur. In this case, excited electrons can heat up the lattice and induce bond breaking [15]. Depending on fluence of the irradiating laser, ablation could be originated through either tunneling ionization or above-threshold ionization (ATI). Multiphoton and avalanche ionization are two main mechanisms of ATI.

In a large band gap material, it is difficult to ionize the constituent atoms by absorbing only one photon from commonly available lasers. Theoretically, an atom might absorb two or more photons simultaneously, giving electrons sufficient energy to cross the band gap from the valence band to the conduction band. This is illustrated schematically in **Figure 1**. For multi-photon ionization to occur, the laser intensity needs to be in the range of 10^{12} –10^{16} W/cm^2. In contrast, the avalanche ionization mechanism, for which the laser intensity required is in the range of 10^9 –10^{12} W/cm^2, depicts the process whereas a small number of initial electrons of the materials are accelerated to a high value of kinetic energy. Afterwards, high-energy electrons will collide with another electron of lower energy, which is shown schematically in **Figure 1**. Afterwards, the two electrons are accelerated by the laser field collide with other electrons in an avalanche-like process, leading to large amount of electrons with high energies to form a plasma.

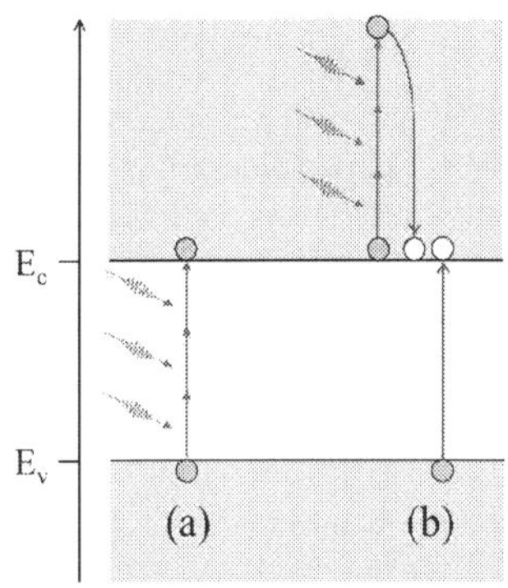

Figure 1. Schematic diagrams illustrating the processes of (a) multiphoton ionization and (b) avalanche ionization of materials.

Finally, in the photophysical mechanism, nonthermal, photochemical, thermal, and photo-thermal processes all play their respective roles. Two independent mechanisms of bond breaking could be present. Further, bond breaking energies for the ground state and the excited state chromophores are, in general, different. The photophysical mechanism of ablation usually applies for irradiating lasers with short laser pulses, of which the pulse duration is in the ps and fs range.

2.3. Basics of femtosecond ablation dynamics of PMMA in the intermediate regime

In a class paper, Keldysh showed that the total photoionization rate of a material upon irradiation by a laser can be written as [31, 32]:

$$\varpi = \frac{2\omega}{9\pi}\left(\frac{\sqrt{1+\gamma^2}}{\gamma}\frac{m\omega}{\hbar}\right)^{3/2} Q\left(\gamma,\frac{\tilde{\Delta}}{\hbar\omega}\right)\exp\left\{-\pi < \frac{\tilde{\Delta}}{\hbar\omega}+1 > \times\left[K\left(\frac{\gamma}{\sqrt{1+\gamma^2}}\right)-E\left(\frac{\gamma}{\sqrt{1+\gamma^2}}\right)\right]/E\left(\frac{1}{\sqrt{1+\gamma^2}}\right)\right\} \quad (1)$$

where $\gamma = \omega(mI_p)^{1/2}/eF$ is the so-called Keldysh parameter, ω is laser frequency, m is the electronic mass, e is the electronic charge, $\hbar$ is plank constant, and $\tilde{\Delta}$ is the effective ionization potential,

$$\tilde{\Delta} = \frac{2}{\pi}I_p\frac{\sqrt{1+\gamma^2}}{\gamma}E\left(\frac{1}{\sqrt{1+\gamma^2}}\right). \quad (2)$$

The symbol $<\tilde{\Delta}/\hbar\omega + 1>$ in Eq. (1) is the integer part of the number, $\tilde{\Delta}/\hbar\omega + 1$, while $Q(\gamma, \tilde{\Delta}/\hbar\omega)$ is defined by Eq. (3):

$$Q\left(\gamma,\frac{\tilde{\Delta}}{\hbar\omega}\right) = \sqrt{\frac{\pi}{2K\left(1/\sqrt{1+\gamma^2}\right)}} \times\sum_{n=0}^{\infty}\exp\left\{-\pi n\left(K\left(\gamma/\sqrt{1+\gamma^2}\right)-E\left(\gamma/\sqrt{1+\gamma^2}\right)\right)/E\left(1/\sqrt{1+\gamma^2}\right)\right\} \times\Phi\left\{\sqrt{\frac{\pi^2\left(2<\frac{\tilde{\Delta}}{\hbar\omega}+1>-2\frac{\tilde{\Delta}}{\hbar\omega}+n\right)}{2K\left(1/\sqrt{1+\gamma^2}\right)E\left(1/\sqrt{1+\gamma^2}\right)}}\right\}, \quad (3)$$

where K and E are first and second kind of the complete elliptic integrals and Φ is the Dawson integral.

In the presence of high-intensity or strong electric field of the laser, we are in the region of tunneling ionization or $\gamma \ll 1$. The rate of tunneling ionization is given by

$$\varpi_{\text{tunneling}} = \frac{2}{9\pi^2}\frac{I_p}{\hbar}\left(\frac{mI_p}{\hbar^2}\right)^{3/2}\left(\frac{e\hbar F}{m^{1/2}I_p^{3/2}}\right)^{5/2} \times\exp\left\{-\frac{\pi}{2}\frac{m^{1/2}I_p^{3/2}}{e\hbar F}\left(1-\frac{1}{8}\frac{m\omega^2 I_p}{e^2F^2}\right)\right\}. \quad (4)$$

On the other hand, if $\gamma >> 1$, the ionization is in the regime of multiphoton absorption. The probability of multiphoton absorption is given by

$$\varpi_{\text{multiphoton}} = \frac{2}{9\pi}\omega\left(\frac{m\omega}{\hbar}\right)^{3/2}\Phi\left[\sqrt{2<\frac{\tilde{\Delta}}{\hbar\omega}+1>-\frac{2\tilde{\Delta}}{\hbar\omega}}\right]$$
$$\times\exp\left\{2<\frac{\tilde{\Delta}}{\hbar\omega}+1>\left(1-\frac{e^2F^2}{4m\omega^2 I_p}\right)\right\}\left(\frac{e^2F^2}{16m\omega^2 I_p}\right)^{<\tilde{\Delta}/\hbar\omega+1>}. \tag{5}$$

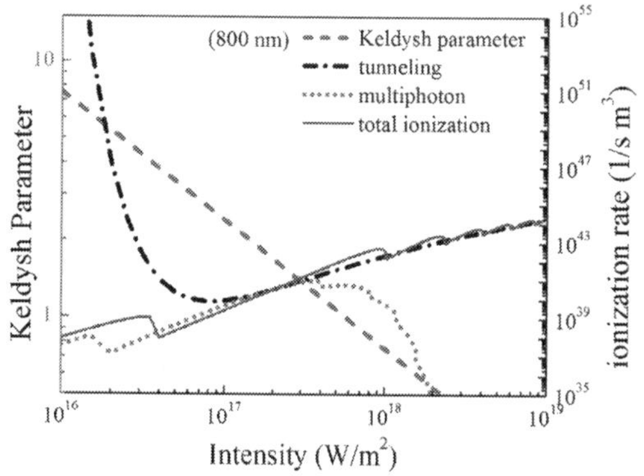

Figure 2. The photonionization rate and Keldysh parameter are plotted as a function of laser intensity (λ = 800 nm) in PMMA.

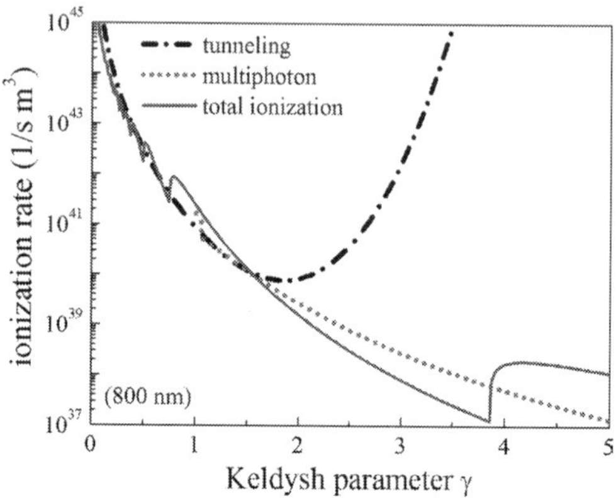

Figure 3. Photonionization rates are plotted as a function of the Keldysh parameter for NIR (λ = 800 nm) ultrafast laser ablation of PMMA.

Below, we have plotted the effective ionization potential $\tilde{\Delta} = \Delta + e^2F^2/4m\omega^2$ and the Keldysh parameter as a function of laser intensity in PMMA in **Figure 2**. The laser wavelength is assumed to be in the near infrared (λ = 800 nm). Besides the total ionization rate, contributions by tunneling and multiphoton ionization, respectively, are also shown. The dependence of the photoionization rate on the Keldysh parameter is also enlightening. This is illustrated in **Figure 3**. Similar curves for ultrafast laser ablation of PMMA using NUV (λ = 400 nm) are presented in **Figures 4** and **5**.

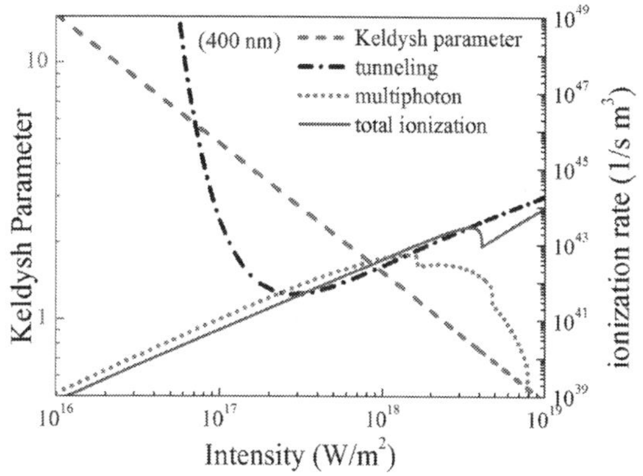

Figure 4. Photonionization rates and Keldysh parameter are plotted as a function of laser intensity for NUV (λ = 400 nm) laser ablation of PMMA.

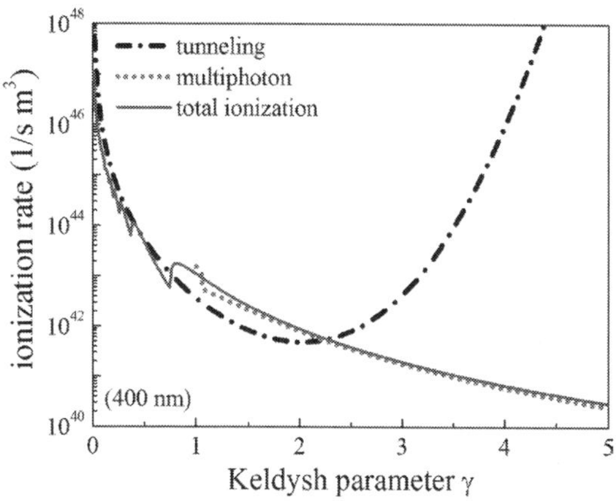

Figure 5. Photonionization rates are plotted as a function of the Keldysh parameter for NUV (λ =400 nm) ultrafast laser ablation of PMMA.

The solid blue lines in **Figures 2–5** correspond to the photoionization rate of PMMA calculated using Eq. (1). The full expression of Keldysh formula (Eq. 1) take into account both tunneling and multiphoton ionization processes. The dashed black line and dotted red line represent the tunneling ionization and multiphoton ionization rates determined from Eqs. (4) and (5), respectively. When the Keldysh parameter, $\gamma \approx 1.5$, tunneling and multiphoton ionization rates overlap each other for PMMA irradiated with either NIR (800 nm) or NUV (400 nm) beams. We defined this overlapping region as an intermediate regime. To predict the ionization rate in the intermediate regime, we assume the process resembles that of tunneling ionization. The ionization rate depends on the instantaneous field amplitude, $F_L(t)$, which is given by [22]:

$$\varpi(t) = 4I_p^{5/2}\frac{1}{F_L(t,\varphi)}\exp\left(-\frac{2}{3}I_p^{3/2}\frac{1}{F_L(t,\varphi)}\right) \tag{6}$$

In order to understand the phase dependence of observed dual-color laser ablation phenomena, we proceed as follows: assume that the irradiating laser consists of beams at two commensurate laser frequencies, i.e., the fundamental (ω) laser beam and its second harmonic (2ω). The dual-color laser field $F(t)$ can then be written as

$$F_L(t,\varphi) = F_\omega e^{-2\ln 2t^2/\tau^2}\cos\omega t + F_{2\omega}e^{-2\ln 2t^2/\tau^2}\cos(2\omega t+\varphi) \tag{7}$$

where F_ω and $F_{2\omega}$ are the envelope function of the fundamental and second-harmonic laser fields, respectively; φ is the relative phase of the second-harmonic (2ω) beam with respect to that of the fundamental (ω) beam.

The Keldysh model above can be used to describe the photoionization phenomenon due to either the multiphoton or tunneling route. Typically, the Keldysh parameter is defined by the square root of the ratio of ionization potential and twice the ponderomotive potential of the laser pulse. Some researchers also define it as the ratio of tunneling frequency to the laser frequency [33]. In the original derivation, the Keldysh parameter was used to describe the phenomenon of an electron tunneling through a barrier created by the optical field. The tunneling time or the inverse of the tunneling frequency is determined by the mean free time of the electron passing through a barrier width l.

$$l_{\text{tunneling}} = I_p / eF_L(t,\varphi) \tag{8}$$

where I_p is the ionization potential, e is the electron charge and $F_L(t,\varphi)$ is the electric field of the incident laser. The average velocity of an electron can be written as,

$$\langle v \rangle = \sqrt{\frac{2I_p}{m_e}}, \tag{9}$$

where m_e is the mass of an electron. By combining Eqs. (8) and (9), the tunneling time is given by

$$t_{\text{tunneling}}(t,\varphi) = \frac{l}{<v>} = \frac{\sqrt{I_p m_e}}{\sqrt{2} e F_L(t,\varphi)} = \frac{1}{v_{\text{tunneling}}(t,\varphi)} \tag{10}$$

Tunneling can occur if the mean tunneling time, which is given by Eq. (10), is less than half the period of the laser. Taking this into account, we modify the Keldysh parameter,γ as appropriate for this study as

$$\gamma = \frac{2t_{\text{tunneling}}}{t_{\text{laser}}} = \frac{l_{\text{tunneling}}}{l_{\text{laser}}} = v_{\text{laser}} \frac{\sqrt{2I_p m_e}}{eF_L(t,\varphi)} \tag{11}$$

where t_{laser} is the period of laser, l_{laser} is the mean distance that an electron moves during half of period t_{laser} at a mean velocity of $<v>$, and v_{laser} is the laser frequency. When the Keldysh parameter γ has relative phase dependence at dual-color synthesized waveform condition, the ionization rate can be calculated after we determine the effective frequency of the dual-color laser pulse.

For the purpose of defining the envelope equation for single-cycle pulse of the synthesized waveform, we express the complex electric field as [34]

$$E(t) = \tilde{E}_a(t) e^{-i\omega_0' t + i\psi} + c.c. = \tilde{E}(t) + c.c. = \frac{1}{\sqrt{2\pi}} \int_0^{\infty} \tilde{E}(\omega) e^{-i\omega t} d\omega + c.c., \tag{12}$$

where ω_0 is the effective carrier frequency denoted as

$$\omega_0 = \frac{\int_0^{\infty} \omega |E(\omega)|^2 d\omega}{\int_0^{\infty} |E(\omega)|^2 d\omega} \tag{13}$$

In Eqs. (12) and (13), $E(\omega)$ is the Fourier transform of $E(t)$ and ψ is the imaginary part of the complex envelope. Substituting the effective carrier frequency into Eq. (1), we can plot the photoionization rate as a function of laser intensity and the relative phase between the fundamental (ω) and second-harmonic (2ω) beams in our experiment, which is shown in **Figure 6**.

According to **Figure 6**, the ionization rate is predicted to be dependent on the relative phase of the fundamental (800 nm) and second-harmonic (400 nm) beams. Further, the modulation of ionization rate is more pronounced at higher laser intensities.

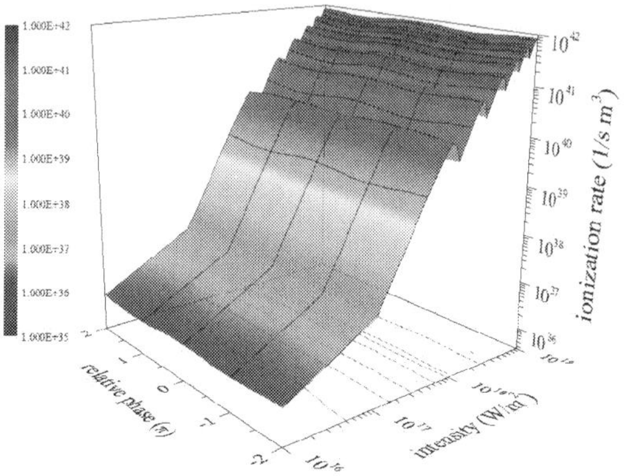

Figure 6. The total ionization rate versus laser intensity and relative phase of fundamental (ω) and second-harmonic (2ω) laser beams.

Recall that ablation by a laser with low and high intensities would fall into the regimes governed by multiphoton and tunneling ionization mechanisms, respectively. Conventionally, tunneling ionization corresponds to a regime in which the Keldysh parameter $\gamma \ll 1$. In this limit, the strength of the field is more than the value necessary to overcome the barrier. For a weaker field such that the Keldysh parameter $\gamma \gg 1$, the main mechanism for ionization is due to the multiphoton ionization. In this regime, the electric filed strength is below the value that required for overcoming the barrier. In order to calculate the Keldysh parameter for the dual-color case, we need to define the period of the laser t_{laser}. It can be easily shown that the period of dual-color synthesized waveform by NIR (800 nm) and NUV (400 nm) beams is essentially that of the period of fundamental (ω) beam. Therefore, Eqs. (1) and (11) can be combined to determine the ionization rate as shown in **Figure 6**.

In the intermediate ionization regime, which is defined by $\gamma \approx 1.5$, the ionization rate has a strong phase dependence. Likewise, the electron tunnel time now depends on the phase difference between the two colors. Note that the electric field is actually lower than the value required for electrons to overcome the barrier.

Ablation threshold is an important parameter for laser material processing. It is a function of the laser pulse duration, wavelength, and intensity. According to the simplified Fokker-Planck equation [35]:

$$\frac{\partial n}{\partial t} = \beta(I)n + P(I), \tag{14}$$

where n is the free electron number, β is the avalanche ionization factor, assuming that the photogenerated electron distribution grows in magnitude without changing its shape. $P(I)$ is the multiphoton ionization rate. It can be approximated by the tunneling ionization rate by using Keldysh equation.

First of all, we need to calculate the number of free electrons generated by the laser pulse, assumed to be Gaussian in shape, $I(t) = I_0\exp(-4\ln2t^2/\tau^2)$, where τ is the pulse duration. By solving the equation above, the free electron number can be obtained as

$$n = n_0 \exp\left(\int_0^\infty \beta dt\right) = n_0 \exp\left(\alpha \int_0^\infty I dt\right) = n_0 \exp\left(\alpha I_0 \int_0^\infty e^{-4\ln 2t^2/\tau^2} dt\right) = n_0 \exp\left(\frac{\alpha I_0 \tau}{4}\sqrt{\frac{\pi}{\ln 2}}\right) \quad (15)$$

where α is the absorption coefficient of the material and n_0 is the total number of free electrons which is generated through multiphoton or tunneling ionization mechanism,

$$n_0 = \int_{-\infty}^{\infty} P(I)dt \quad (16)$$

The ablation threshold, F_{th}, can then be written as

$$F_{th} = \frac{2}{\alpha}\ln\left(\frac{n_{cr}}{n_0}\right) \quad (17)$$

where the density of free electrons, n_{cr} correspond to the threshold fluence, F_{th}. For ablation with ultrafast laser pulses, contribution by the avalanche ionization mechanism is not significant. Dielectric breakdown or ablation is through the processes of photoionization by tunneling and multiphoton ionization mechanisms. When the free electron number increases, the ablation threshold decreases.

3. Experimental methods

The experimental setup for laser ablation by dual-color femtosecond synthesized waveform [14] is shown schematically in **Figure 7**. The laser source was an amplified Ti: Sapphire laser system (Spitfire, Spectra Physics), which generates 70 fs laser pulses at a central wavelength of 800 nm (λ_1) with an energy up to 1.5 mJ at 1 kHz.

As shown in **Figure 7**, we adopt an inline arrangement for phase control of the fundamental and second harmonic of the laser output. The 800-nm fs pulses were focused onto the sample surface by a single convex lens with a focal length of 300 mm. Meanwhile, the fundamental beam frequency was doubled in a 100-μm-thick type-I Beta Barium Borate (β-BBO) crystal in

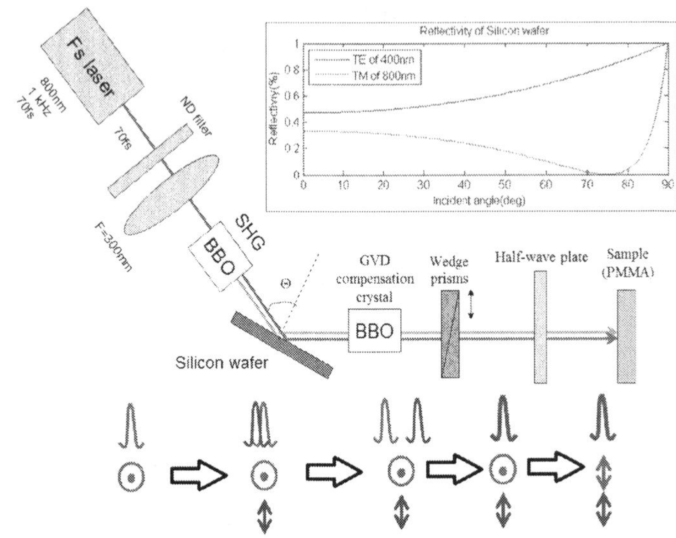

Figure 7. Experimental setup for laser ablation of PMMA by femtosecond dual-colour synthesized waveforms. The polarizations of fundamental and second harmonic pulses were controlled by the half-wave plate. ND: neutral density filter; BBO: Barium borate; GVD: group velocity dispersion. The inset shows the reflectivity of the silicon wafer as a function of the incident angle for both polarizations.

the same beam path to generate 2ω pulses at 400 nm (λ_2). Both beams were reflected from the silicon wafer at some incident angle, taking advantage of the fact that reflectivity of silicon varies with wavelengths and polarizations of the fundamental and second-harmonic beams (see the inset in **Figure 7**). We can control the intensity ratio $P_{2\omega}/P_{\omega}$ of the two (collinear) beams by adjusting the incident angle. A pair of wedge prisms with controllable optical path difference was used to precisely adjust the relative time delay between the ω and 2ω pulses. We also employed a 5-mm-thick β-BBO to compensate the group velocity mismatch (GVM) of the two colors in the beam path. Besides, the ω and 2ω fields with original polarizations perpendicular to each other, were passed through a dual-color zero-order wave plate serving as a half-wave plate for 800 nm to make polarizations of the two colors parallel. Finally, the ω and 2ω pulses, overlapped in time with the same linear polarizations were focused on the sample. The spatial and temporal overlap and adjustment of the phase difference between the ω and 2ω fields were conducted using a procedure described previously [14].

4. Single-color femtosecond laser ablation of PMMA

4.1. Introduction

Ultrafast laser-induced ablation or breakdown of wide band gap materials, such as polymers [1, 2, 36–41], fused silica [6], and silicon [7, 42] have already been intensively studied. Among them, various kinds of polymers, such as polymethylmethacrylate (PMMA) [2, 36, 38–41], polyimide (PI) [1], polyethylene (PE) [37], polypropylene (PP), and polycarbonate (PC) [2], have drawn a lot of attention due to their potential industrial applications. Compared to ns-laser ablation, the energy ablation threshold fluence of fs-laser at approximately the same incident wavelength is known to be reduced [43]. This can be attributed to the fact that the breakdown intensities in the fs regime approach that of the threshold of multiphoton ionization of which the electron densities is high enough to cause damage [35]. On the other hand, because the induced energy absorbed by electrons is much faster than that transferred to a lattice [35]; therefore, the nonthermal ablation nature of such behavior achieved by applying fs-lasers could lead to a significant reduction of heat-affected zones. Also of interest is the possibility of decreasing the threshold for ablation. For example, Stuart et al. observed a continuously decreasing threshold with a gradual transition from the thermal regime where the longer pulses (>100 ps) dominated the ablation compared with the shorter pulses (<10 ps), which is caused by multiphoton ionization and plasma formation [17].

To date, studies of single-color femtosecond laser ablation of PMMA were overwhelmingly conducted using the Ti: sapphire laser system of which the central wavelength is around 800 nm [44–46]. On the other hand, photoablation of materials with ultraviolet (UV) lasers has also gained in popularity [36, 47–49]. The mechanism for ablation of materials by UV light is mainly through the photochemical process by one-photon absorption. Most of the dielectrics, such as glass and polymer, have relatively high absorption coefficient in the UV region. This is in contrast to the commonly accepted mechanism for ablation of PMMA using 800 nm laser pulses, such as photothermal, photophysical, or multiphoton ionization and tunneling ionization, as mentioned previously. Therefore, it is of interest to conduct a comparative study of single-color femtosecond laser ablation of PMMA using the Ti: sapphire laser and its second harmonic.

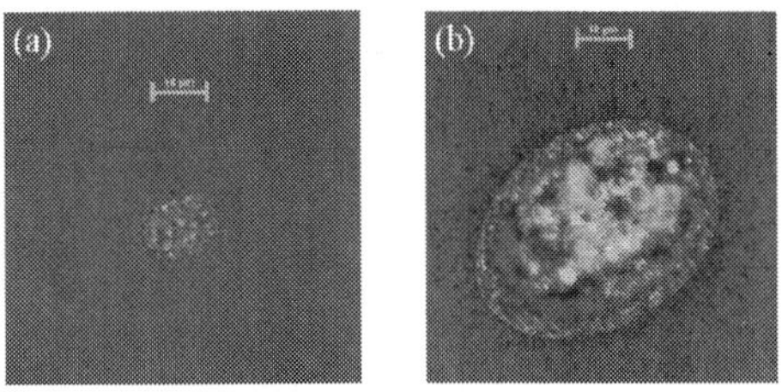

Figure 8. Images of single-color (800 nm) and single-shot ablated holes. The input laser fluence are equal to (a) 2.63 J/cm^2 and (b) 5.90 J/cm^2, respectively.

4.2. Single-shot single-color (800 nm) femtosecond laser ablation of PMMA

In **Figure 8**, we show images of single-shot ablated holes in PMMA irradiated with femtosecond pulses at the wavelength of 800 nm. By changing the input laser fluence from 2.63 to 5.90 J/cm^2, areas of the holes are found to be equal to 155.25 and 1359.50 μm^2, respectively.

The photon energy for 800 nm is equal to 1.55 eV and the material band gap of PMMA is 4.58 eV. Therefore, more than three incident photons are needed for photoabsorption, leading to ablation. For such studies, one of the key parameter for studying the mechanism of ablation is its threshold. The method we used to define the ablation threshold value is measuring the ablated hole areas by using an optical microscope. In **Figure 9**, we have plotted hole-area of the ablated holes as a function of the irradiating laser fluence.

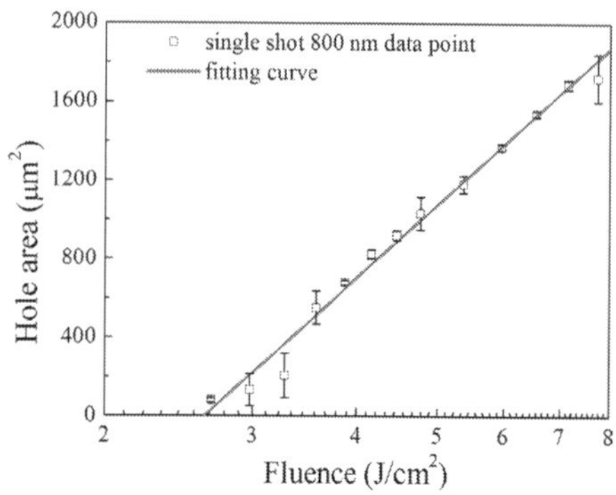

Figure 9. Hole-areas of the single-shot, single-color (800 nm) femtosecond laser ablated holes are plotted as a function of the irradiating laser fluence. Error bars are indicated.

Assuming the irradiating beam has a Gaussian spatial profile, the generally accepted scaling law for ablated holes for incident laser fluence is given by

$$D^2 = 2w^2 \ln\left(\frac{F}{F_{th}}\right) \tag{18}$$

where D is the diameter of the ablated region, w is effective laser beam width, F is the incident laser fluence and F_{th} denotes the ablation threshold (unit here is J/cm^2). Following Eq. (18), the ablation threshold F_{th} can be determined by fitting the experimental data to be 2.63 J/cm^2.

4.3. Single-shot single-color (400 nm) femtosecond laser ablation of PMMA

To compare, we conducted similar ablation studies with exciting wavelength at 400 nm. Recall that the material band gap of PMMA is 4.58 eV, which means the dominated mechanism for photoablation on PMMA at 400 nm is also multiphoton absorption. The photon energy for 400 nm is equal to 3.1 eV. Therefore, more than two incident photons are needed for photoabsorption, leading to ablation. In **Figure 10**, we show images of single-shot ablated holes in PMMA irradiated with femtosecond pulses at the wavelength of 400 nm. By changing the input laser fluence from 1.78 to 3.92 J/cm^2, the hole areas are found to increase from 155.25 to 1359.50 μm^2, respectively.

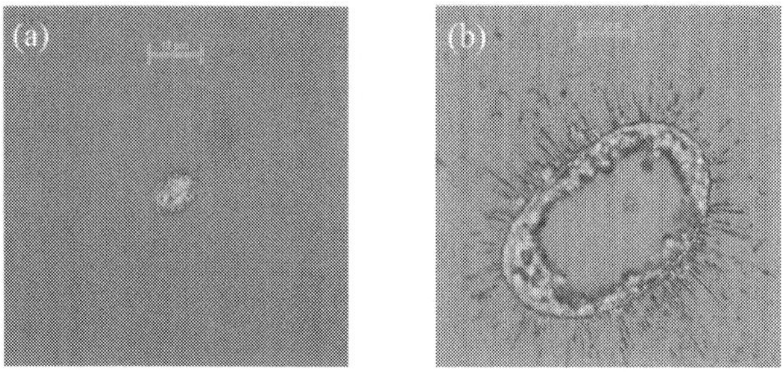

Figure 10. Images of single-color (400 nm) and single-shot ablated holes. The input laser fluence are equal to (a) 1.78 J/cm^2 and (b) 3.92 J/cm^2, respectively.

Figure 11 shows determination of the ablation threshold in the case of exciting wavelength at 400 nm. It can be seen that the same scaling behavior is observed in the case of ablation by the near IR beam. Our data show that the ablation threshold for PMMA irradiated by the near UV light of 400 nm is about 1.38 J/cm^2.

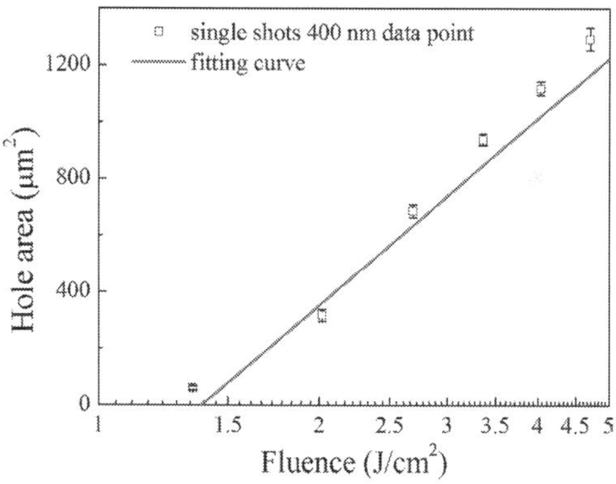

Figure 11. Single-colour ablation results for PMMA irradiated by femtosecond laser pulses with a central wavelength 400 nm. The fitted ablation threshold F_{th} is equal to 1. 38 J/cm^2.

5. Laser ablation of PMMA with femtosecond two-color synthesized waveforms

5.1. Introduction

As we noted earlier, laser ablation studies were conducted almost exclusively with single-color laser beams [6, 7, 37, 42]. There are a few studies that employed two-color lasers. These studies can be organized into two categories: incoherent combination and coherent superposition of the two-color laser beams. An example of ablation by incoherently combined two-color beams is the work of Théberge et al., in which the authors observed an increase in volume of the ejected material by applying the superposition of fs and ns pulses. This was attributed to the free electrons and defect sites induced by the fs pulses, which could be exploited by the ns pulses [6]. Besides, Okoshi et al. reported that dual-color fs pulses with a fluence ratio of (2ω:ω = 2:78 mJ/cm^2) could etch PE deeper and faster. It was proposed that an isolated carbon, in addition to C=O and C=C–H bonds, was formed on the ablated surface after treating PMMA with 2ω or dual-color pulses. The higher photon energy of 2ω pulses then cuts the chemical bonds of PE to form the modified layer on the ablated surface [37]. In related studies of fused silica, because of the creation of defect states or free electron plasma by dual-color fs pulses at zero delay, the enhancement of absorption/reflection was observed [6]. For silicon, upon using ns and picosecond (ps)-laser pulses, it was also shown that a weak 2ω beam can be beneficial in exciting electrons into conduction band to launch the ablation process of silicon [42]. In contrast, for fs pulses, where a sufficient population on the conduction band can be created by multiphoton absorption, this effect became insignificant [42].

All the above studies employ relatively long-time delays between the two colors, on the scale of the carrier lifetime (≈ picoseconds). If the relative delay is of the order of an oscillation period between dual-color fs pulses, interesting phenomena could unfold. In other fields, a dual-color coherently superposed beams achieved by relative-phase control of each color were applied to study the physical mechanism of intense-field photoionization, especially in the gas phase [11, 24]. Schumacher et al., for example, studied the electric-field phase-dependent photoelectrons created in a regime including the multiphoton and tunneling signatures simultaneously by changing the dual-color relative phase [24]. Later, Gao et al. claimed this phase-difference effect resembled the phenomenon of quantum interference (QI) between the different channels characterized by the number of photons. In other words, phase-dependent photoemission is not a classical-wave effect, but rather a quantum-mechanical one. Recently, in comparison with monochromatic excitation, the threshold of plasma formation has been demonstrated to be significantly improved with the superposition of an ns infrared laser pulses and its second-harmonic field [11]. The authors explained their measurements by the effect of a field-dependent ionization cross section [11]. In the following, we report results of our studies of the ablation of PMMA using dual-color waveform synthesis of ω and 2ω beams of an fs Ti: sapphire laser.

5.2. Single-shot dual-color ablation of PMMA

In **Figure 12**, we show images of single-shot ablated holes in PMMA irradiated with dual-color (ω and 2ω) femtosecond pulses. In this experiment, the average powers of ω and 2ω beams are 200 and 40 mW, respectively. The corresponding laser fluence for the fundamental (NIR) beam is equal to 7.55 J/cm^2. Phase dependence of ablated holes was observed. **Figure 12(a)**, for which the relative phase $\varphi = \pi$, the ablated hole area is equal to 844.95 μm^2. When the relative phase φ = 0, area of the ablated hole is 982.31 μm^2 (see **Figure 12b**).

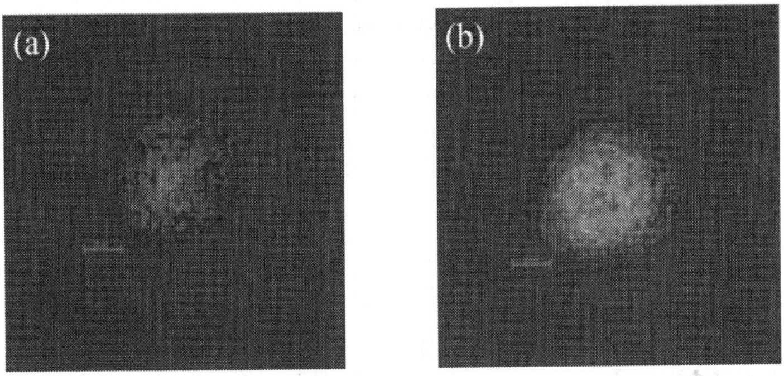

Figure 12. Images of single-shot ablated holes in PMMA irradiated by dual-colour (ω and 2ω) femtosecond lasers. The laser fluence for the fundamental beam is equal to 7.55 J/cm^2 and ratio of second-harmonic to the fundamental beams was 1:5. The hole areas of (a) 844.95 μm^2 (b) 982.31 μm^2 were observed when the relative phase φ were set to π and 0, respectively. The length of the red double-arrows in Fig. 12 (a) and (b) are both equal to 10 μm.

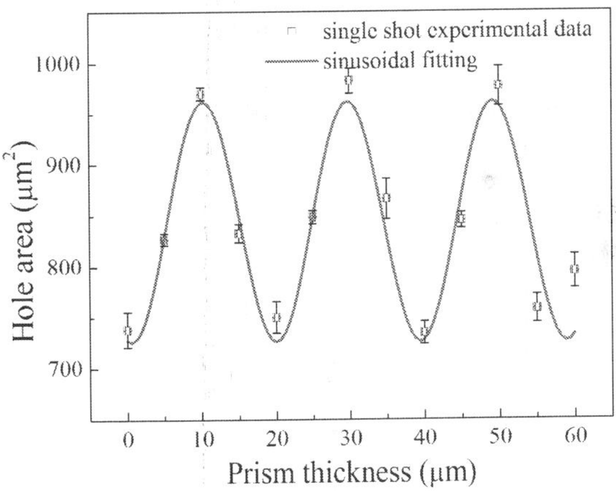

Figure 13. The ablated hole area versus relative prisms' thickness with sinusoidal fitting in the case of single shot. The period for the ablated hole areas' change is equal to 19.5 μm.

By varying the prism thickness traversed by the laser beams (see **Figure 7**), we observed that hole areas oscillated, as shown in **Figure 13**. Theoretically, we expect a sinusoidal variation with a period (relative phase change of 2π) of 20 μm. This is in good agreement with experimentally determined period of 19.5 μm in **Figure 7**.

According to our model, the two-color ionization rate would depend on the relative phase. In **Figure 14**, we have plotted the ionization rate according to Eq. (6) for synthesized dual-color instantaneous field from Eq. (7) as a function of the relative phase. The corresponding ablated hole areas are also plotted for comparison. The difference in period between the fitting curve in **Figure 13** and the simulation curve in **Figure 14** is only 1.3%.

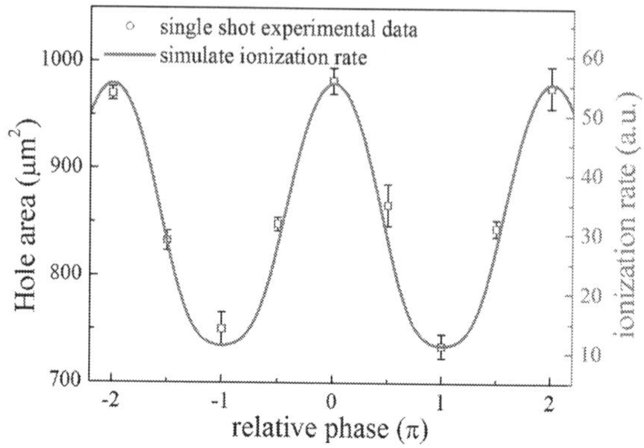

Figure 14. The ablated hole area and simulated dual-color ionization rate versus relative phase in the case of single shot. The observed modulation contrast in ablated area is ≈28% (peak to peak).

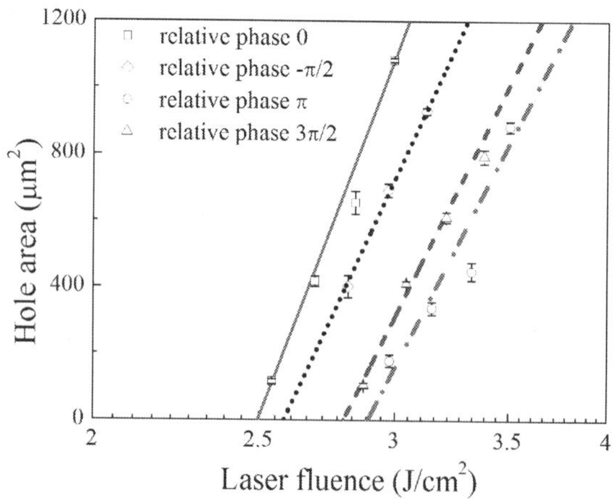

Figure 15. The ablation threshold measurement. The single-shot ablated hole areas in PMMA irradiated by femtosecond dual-color synthesized waveforms are plotted as a function of laser fluence. Four sets of data for different values of relative phases are shown.

In order to study how the relative phase affects the ablation threshold, we conducted a series of experiments in which the wedge prism's thickness was fixed at some value and the laser fluence varied. The family of experimentally measured ablated hole areas for three values of relative phase as a function or irradiating laser fluence are plotted in **Figure 15**. The same scaling law for the single-color case was used to fit the experimental data. In this manner, we were able to determine the ablation threshold for a given value of relative phase. The ablation thresholds are 2.49, 2.58, 2.89 and 2.80 J/cm^2, respectively, for the relative phase to be equal to 0, $-\pi/2$, π and $3\pi/2$.

Interestingly, the fitted ablation thresholds also exhibit apparent dependence on the relative phase between ω to 2ω beams for our dual-color pulses. This is shown in **Figure 16**. The period of the sinusoidal oscillation is $\approx 2.4\pi$.

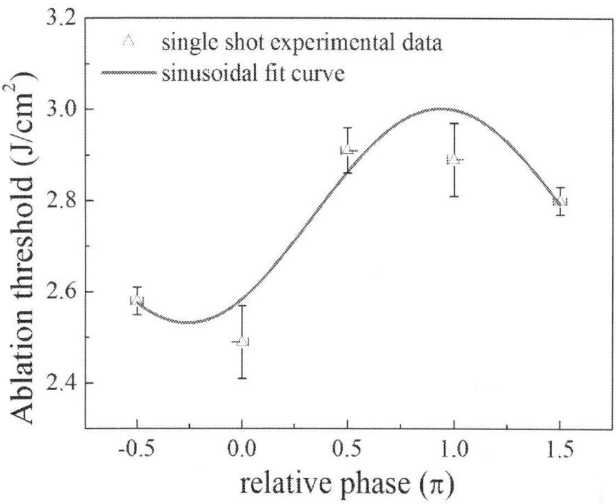

Figure 16. The ablation threshold in PMMA irradiated by femtosecond dual-color synthesized waveforms is plotted as a function of relative phase changes. The period for the ablation thresholds' change is $\approx 2.4\pi$.

Because the ablation threshold is dependent upon the number of free electrons created in the material [35, 50], we believe the observed periodicity in ablation threshold in **Figure 16** demonstrates how electric field of the synthesized waveform affects the variation of ablation threshold. In the above experiments, the beam waists for every condition deliberately kept to be approximately the same. These are equal to 49.42, 54.58, 46.46, 46.69 and 47.51 μm in the cases of relative phase set at $-\pi/2$, 0, $\pi/2$, π and $3\pi/2$. That is, variation in the beam spot size is small, ±3.90%.

6. Summary

In this work, we have investigated single-shot laser ablation of polymethylmethacrylate (PMMA) using dual-color waveform synthesis of the fundamental (ω) and its second harmonic (2ω) of a femtosecond Ti: Sapphire laser. For comparison, single-color studies were also conducted. The threshold fluence for single-color ablation of PMMA irradiated by fundamental (ω) (NIR) and its second-harmonic (2ω) (NUV) beams were found to be respectively, 2.63

and 1.38 J/cm^2. Changing the relative phase of the fundamental (ω) and second-harmonic (2ω) outputs of the exciting laser resulted in clear modulation of the ablated area. The modulation as well as the ablation threshold depends on the relative phase between the ω and 2ω beams. The ablation thresholds for ablation of PMMA irradiated by two-color femtosecond frequency-synthesized waves are 2.49, 2.58, 2.89 and 2.80 J/cm^2, respectively, for the relative phase to be equal to 0, $-\pi/2$, π and $3\pi/2$. The results correlated well with the theoretical model of laser breakdown (ablation) of transparent materials through photoionization in the intermediate regime (Keldysh parameter $\gamma \approx 1.5$). Our study clearly illustrates the potential applications of using phase-controlled synthesized waveform for laser processing of materials.

Acknowledgements

This work was funded by the grant of the National Science Council 102-2622-E-007-021-CC2, 101-2221-E-007-103-MY3, and 101-2112-M-007-019-MY3. The authors would like to thank Dr. Wei-Jen Chen for many useful discussions. They would also like to thank Prof. Ru-Pin Pan for the use of the microscope.

Author details

Ci-Ling Pan[1,2,3*], Chih-Hsuan Lin[2], Chan-Shan Yang[1,4] and Alexey Zaytsev[1]

*Address all correspondence to: clpan@phys.nthu.edu.tw

1 Department of Physics, National Tsing Hua University, Hsinchu, Taiwan

2 Institute of Photonics Technology, National Tsing Hua University, Hsinchu, Taiwan

3 Frontier Research Center on Fundamental and Applied Sciences of Matters, National Tsing Hua University, Hsinchu, Taiwan

4 Department of Physics, University of California, Berkeley, CA, USA

References

[1] S. Baudach, J. Bonse, and W. Krautek, "Ablation experiments on polyimide with femtosecond laser pulses," Applied Physics A-Materials Science & Processing, 69, S395–S398 (1999).

[2] S. Baudach, J. Bonse, J. Kruger, and W. Kautek, "Ultrashort pulse laser ablation of polycarbonate and polymethylmethacrylate," Applied Surface Science, 154–155, 555–560 (2000).

[3] P.P. Rajeev, M. Gertsvolf, E. Simova, C. Hnatovsky, R.S. Taylor, V.R. Bhardwaj, D.M. Rayner, and P.B. Corkum, "Memory in nonlinear ionization of transparent solids," Physical Review Letters, 97, 253001-1-5 (2006).

[4] N. Sanner, O. Uteza, B. Bussiere, G. Coustillier, A. Leray, T. Itina, and M. Sentis, "Measurement of femtosecond laser-induced damage and ablation thresholds in dielectrics," Applied Physics A-Materials Science & Processing, 94, 889–897(2009).

[5] Y.P. Deng, X.H. Xie, H. Xiong, Y.X. Leng, C.F. Cheng, H.H. Lu, R.X. Li, and Z.Z. Xu, "Optical breakdown for silica and silicon with double femtosecond laser pulses," Optics Express, 13(8), 3096–3103 (2005).

[6] I.H. Chowdhury, X. Xu, and A.M. Weiner, "Ultrafast two-color ablation of fused silica," Applied Physics A-Materials Science & Processing, 83, 49–52 (2006).

[7] S. Zoppel, R. Merz, J. Zehetner, and G.A. Reider, "Enhancement of laser ablation yield by two color excitation," Applied Physics A-Materials Science & Processing, 81, 847–850 (2005).

[8] T.J. Derrien, T. Sarnet, M. Sentis, and T.E. Itina, "Application of a two-temperature model for the investigation of the periodic structure formation on Si surface in femtosecond laser interactions," Journal of Optoelectronics and Advanced Materials, 12(3), 610–615 (2010).

[9] M.D. Perry, B.C. Stuart, P.S. Banks, M.D. Feit, V. Yanovsky, and A.M. Rubenchik, "Ultrashort-pulse laser machining of dielectric materials," Journal of Applied Physics, 85(9), 6803–6810 (1999).

[10] B. Chimier, O. Uteza, N. Sanner, M. Sentis, T. Itina, P. Lassonde, F. Legare, F. Vidal, and J.C. Kieffer, "Damage and ablation thresholds of fused-silica in femtosecond regime," Physical Review B, 84, 094104-1-10 (2011).

[11] E. Schwarz and G.A. Reider, "Laser-induced plasma by two-color excitation," Applied Physics B-Lasers and Optics, 107(1), 23–30 (2012).

[12] X. He, J.M. Dahlstrom, R. Rakowski, C.M. Heyl, A. Persson, J. Mauritsson, and A. L'Huillier, "Interference effects in two-color high-order harmonic generation," Physical Review A, 82, 033410 (2010).

[13] X. Xie, J. Dai, and X. C. Zhang, "Coherent control of THz wave generation in ambient air," Physical Review Letters, 96, 075005-1-4 (2006).

[14] C.-S. Yang, C.-H. Lin, A. Zaytsev, K.-C. Teng, T.-H. Her, and C.-L. Pan, "Femtosecond laser ablation polymethylmethacrylate via dual-color synthesized waveform," Applied Physics Letters, 106, 051902-1-5 (2015).

[15] S.R. Cain, "A photothermal model for polymer ablation: chemical modification," Journal of Physical Chemistry, 97, 051902-1-5(1993).

[16] R.R. Gattass and E. Mazur, "Femtosecond laser micromachining in transparent materials," Nature Photonics, 2, 219–225 (2008).

[17] B.C. Stuart, M.D. Feit, A.M. Rubenchik, B.W. Shore, and M.D. Perry, "Laser-induced damage in dielectrics with nanosecond to subpicosecond pulses," Physical Review Letters, 74(12), 2248–2251 (1995).

[18] M. Lenzner, J. Kruger, S. Sartania, Z. Cheng, C. Spielmann, G. Mourou, W. Kautek, and F. Krausz, "Femtosecond optical breakdown in dielectrics," Physical Review Letters, 80(18), 4076–4079 (1998).

[19] M. Li, S. Menon, J.P. Nibarger, and G.N. Gibson, "Ultrafast electron dynamics in femtosecond optical breakdown of dielectrics," Physical Review Letters, 82(11), 2394–2397 (1999).

[20] D.W. Schumacher, F. Weihe, H.G. Muller, and P.H. Bucksbaum, "Phase dependence of intense field ionization: a study using two colors," Physical Review Letters, 73(10), 1344–1347 (1994).

[21] R.M. Potvliege and P.H.G. Smith, "2-Color multiphoton ionization of hydrogen by an intense laser field and one of its harmonics," Journal of Physics B: Atomic and Molecular Physics, 25, 2501–2516 (1992).

[22] P.B. Corkum, N.H. Burnett, and F. Brunel, "Above-threshold ionization in the long-wavelength limit," Physical Review Letters, 62(11), 1259–1262 (1989).

[23] A.C. Tien, S. Backus, H. Kapteyn, M. Murnane, and G. Mourou, "Short-pulse laser damage in transparent materials as a function of pulse duration," Physical Review Letters, 82(19), 3883–3886 (1999).

[24] D.W. Schumacher and P.H. Bucksbaum, "Phase dependence of intense-field ionization," Physical Review A, 54(5), 4271–4278 (1996).

[25] R. Sauerbrey and G.H. Pettit, "Theory for the etching of organic materials by ultraviolet laser pulses," Applied Physics Letters, 55(5), 421–423 (1989).

[26] A.V. Sokolov, D.R. Walker, D.D. Yavuz, G.Y. Yin, and S.E. Harris, "Femtosecond light source for phase-controlled multiphoton ionization," Physical Review Letters, 87(3), 033402-1-4 (2001).

[27] H. Kumagai, K. Midorikawa, K. Toyoda, S. Nakamura, T. Okamoto, and M. Obara, "Ablation of polymer films by femtosecond high-peak-power Ti: sapphire laser at 798 nm," Applied Physics Letters, 65(14), 1850–1852 (1994).

[28] G.H. Pettit and R. Sauerbrey, "Pulsed ultraviolet laser ablation," Applied Physics A-Materials Science & Processing, 56, 51–63 (1993).

[29] C.B. Schaffer, A. Brodeur, and E. Mazur, "Laser-induced breakdown and damage in bulk transparent materials induced by tightly focused femtosecond laser pulses," Measurement Science & Technology, 12, 1784–1794 (2001).

[30] S.C. Jones, P. Braunlich, R.T. Casper, X.A. Shen, and P. Kelly, "Recent progress on laser-induced modifications and intrinsic bulk damage of wide-gap optical materials," Optical Engineering, 28, 1039–1068 (1989).

[31] L.V. Keldysh, "Ionization in field of a strong electromagnetic wave," Soviet Physics JETP-USSR, 20(5), 1307 (1965).

[32] P. Balling and J. Schou, "Femtosecond-laser ablation dynamics of dielectrics: basics and applications for thin films," Reports on Progress in Physics, 76(3), 39 (2013).

[33] M.J. DeWitt and R.J. Levis, "Calculating the Keldysh adiabaticity parameter for atomic, diatomic, and polyatomic molecules," The Journal of Chemical Physics, 108(18), 7739–7742 (1998).

[34] T. Brabec and F. Krausz, "Nonlinear optical pulse propagation in the single-cycle regime," Physical Review Letters, 78(17), 3282 (1997).

[35] B.C. Stuart, M.D. Feit, S. Herman, A.M. Rubenchik, B.W. Shore, and M.D. Perry, "Nanosecond-to-femtosecond laser-induced breakdown in dielectrics," Physical Review B, 53(4), 1749–1761 (1996).

[36] R. Srinivasan, "Ablation of polymers and biological tissue by ultraviolet-lasers," Science, 234(4776), 559–565 (1986).

[37] M. Okoshi and N. Inoue, "Laser ablation of polymers using 395 nm and 790 nm femtosecond lasers," Applied Physics A-Materials Science & Processing, 79(4–6), 841–844 (2004).

[38] J.R. Nam, K.S. Lim, and S.C. Jeoung, "Femtosecond laser ablation of polymethylmethacrylate doped with dye molecules and formation of a grating structure," Journal of the Korean Physical Society, 52(5), 1661–1664 (2008).

[39] D. Klinger, R. Sobierajski, R. Nietubyc, J. Krzywinski, J. Pelka, L. Juha, M. Jurek, D. Zymierska, S. Guizard, and H. Merdji, "Surface modification of polymethylmethacrylate irradiated with 60 fs single laser pulses," Radiation Physics and Chemistry, 78, S71–S74 (2009).

[40] J.M. Guay, A. Villafranca, F. Baset, K. Popov, L. Ramunno, and V.R. Bhardwaj, "Polarization-dependent femtosecond laser ablation of poly-methyl methacrylate," New Journal of Physics, 14, 085010 (2012).

[41] F. Baset, K. Popov, A. Villafranca, J.M. Guay, Z. Al-Rekabi, A.E. Pelling, L. Ramunno, and R. Bhardwaj, "Femtosecond laser induced surface swelling in poly-methyl methacrylate," Optics Express, 21(10), 12527–12538 (2013).

[42] S. Zoppel, J. Zehetner, and G.A. Reider, "Two color laser ablation: enhanced yield, improved machining," Applied Surface Science, 253(19), 7692–7695 (2007).

[43] S. Kuper and M. Stuke, "Femtosecond UV excimer laser ablation," Applied Physics B-Photophysics and Laser Chemistry, 44(4), 199–204 (1987).

[44] P.J.S. Alexandra Baum, W. Perrie, M. Sharp, K.G. Watkins, D. Jones, R. Issac, and D.A. Jaroszynski, "NUV and NIR femtosecond laser modification of PMMA," LPM20007 - The 8th International Symposium on Laser Precision Microfabrication, pp. 1–5 (2007).

[45] A.A. Serafetinides, M. Makropoulou, E. Fabrikesi, E. Spyratou, C. Bacharis, R.R. Thomson, and A.K. Kar, "Ultrashort laser ablation of PMMA and intraocular lenses," Applied Physics A-Materials Science & Processing, 93(1), 111–116 (2008).

[46] C. De Marco, S.M. Eaton, R. Suriano, S. Turri, M. Levi, R. Ramponi, G. Cerullo, and R. Osellame, "Surface properties of femtosecond laser ablated PMMA," ACS Applied Materials & Interfaces, 2(8), 2377–2384 (2010).

[47] H. Schmidt, J. Ihlemann, B. Wolff-Rottke, K. Luther, and J. Troe, "Ultraviolet laser ablation of polymers: spot size, pulse duration, and plume attenuation effects explained," Journal of Applied Physics, 83(10), 5458–5468 (1998).

[48] S. Lazare and V. Granier, "Ultraviolet-laser photoablation of polymers – a review and recent results," Laser Chemistry, 10(1), 25–40 (1989).

[49] R. Srinivasan, E. Sutcliffe, and B. Braren, "Ablation and etching of polymethylmethacrylate by very short (160 fs) ultraviolet (308 nm) laser-pulses," Applied Physics Letters, 51(16), 1285–1287 (1987).

[50] B.C. Stuart, M.D. Feit, S. Herman, A.M. Rubenchik, B.W. Shore, and M.D. Perry, "Optical ablation by high-power short-pulse lasers," Journal of the Optical Society of America B-Optical Physics, 13(2), 459–468 (1996).

Chapter 16

Surface Modification of Polymer Materials Induced by Laser Irradiation

Rozalina Zakaria

Additional information is available at the end of the chapter

http://dx.doi.org/10.5772/66377

Abstract

We report on the surface modification effects using allyl-diglycol CR39 polymer induced by laser irradiation at 157 nm F_2 laser (VUV) and 248 nm KrF laser. The motivation is to investigate the ablation effects on this polymer in optical waveguides application the ablation effects on this polymer in optical waveguides. Fabrication of waveguides has been observed using continuous wave (CW) at 244 nm argon ion laser. Ablation effects on the surface of this polymer have been characterized including ablation threshold at different wavelengths from the assorted depth of craters formed from UV pulsed laser. Application of this polymer in optical waveguide application is corroborated by the refractive index value on the CR39 channels that varied as fluences changed when using the continuous wave UV irradiation. A limit for upper fluence at the point where laser ablation originates on this polymer has also been determined.

Keywords: CR39 laser ablation, surface modification, refractive index modification, channel waveguide

1. Introduction

Study of interaction of lasers with polymer materials that induce surface modification and ablation has been an interesting topic for decades [1]. Polymeric materials have been used in various applications such as high-performance photonics devices and engineering applications such as micro/nanofluidics channel fabrication, micromachining/microdrilling [2, 3], splitters, waveguide gratings and filters, and also in optical waveguide fabrication [3–5]. One of the most significance in the research field of photonics is refractive index modification of germanium-doped silica glass using 244 nm UV-laser irradiation as well as studies on nonlinear refractive index change of glass by femtosecond laser irradiation [6]. It has been reported through time regarding the interaction of polymer materials with laser at different photoetching techniques by means of infrared nanosecond laser, femtosecond laser as well as excimer laser [7]. Nevertheless, several parameters in combination including, among others, the material as well as the laser pulse and energy affect the material removing processes [8].

In this work, we focus on using F_2 (157 nm laser) and KrF (248 nm laser) to obtain ablation threshold for these materials at these wavelengths. Refractive index of CR39 is changed by varying the laser fluences using continuous wave laser. To obtain higher fluence in respect of the changes to be made, the laser spot size is focused down to microns in diameter.

For pulse laser, an aperture size of 6 × 3 mm is aligned and positioned in front of the lens to ensure substantial edge of craters can be seen on the film. Subsequently, examination of the pulse crater is carried out using a microscope to conclude the depth of each crater. After a few pulses, the correlation between adjustments of fluence and etch depth is determined. The exact number of pulses depends upon the fluence, wavelength, and absorption of the polymer in which the system will settle down to a constant etch depth per pulse. On the other hand, when continuous wave laser is used, the refractive index calculated after laser induced depends on the numerical aperture (NA) measurement on the written waveguides. Positive refractive index is perceived, and consequently, the association between irradiation fluence and refractive index change can then be concluded. In cases where laser ablation is initiated above this perimeter, an upper fluence limit could also be attained.

2. Experimental procedure

2.1. VUV F_2 laser

A 157 nm VUV F_2 laser (Lambda Physik LPF 200) that produced output energy of up to 35 mJ in an 11 ns pulse (full width at half maximum) was used to expose the various polymer samples. The charging voltage of the laser could be varied in 1 kV steps from 21 to 26 kV, and this allowed the laser energy and hence fluence at the target to be varied. The full-angle beam divergence of the direct output beam was ~3 mrad in its narrow dimension and ~8 mrad in its long dimension.

The polymer samples were held on a motorized stage, which was capable of movements in the *x-y-z* directions in increments from 1 mm to 1 cm under computer control. Due to the high absorption of the 157 nm wavelength in oxygen in air, the laser output had to be delivered either in vacuum or in a rare gas such as argon. In these experiments, the target was placed in a chamber that was capable of being evacuated down to 1×10^{-5} mbar. The chamber was evacuated using a dry pump, and the pressure was measured using Edwards Pirani/Penning 1005 pressure gauges. The chamber could also be purged with He or Ar gas as an alternative for beam delivery but in these experiments this was not used. **Figure 1** shows a schematic diagram of the F_2 laser experimental setup.

http://dx.doi.org/10.5772/66377

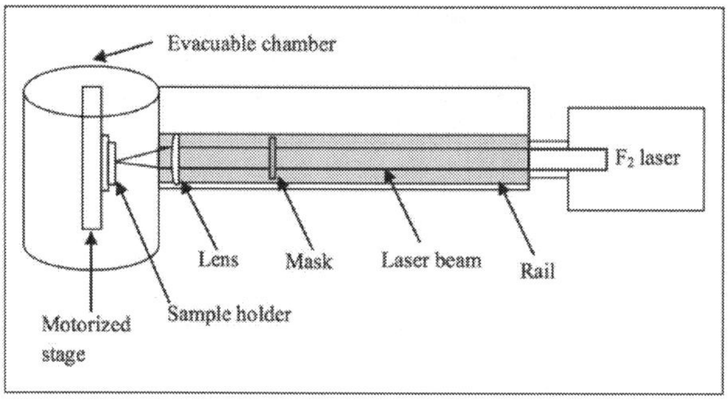

Figure 1. F_2 laser experimental setup.

2.2. 248 nm Pulsed UV laser

KrF excimer laser (Opto Systems Ltd Excimer Laser series CL5100) operating at 248 nm wavelength was the laser tools used in this research. The maximum repetition rate can range up to 100 Hz with a maximum average output power of 5 W and a pulse interval range from 9 to 11 ns. A 50 mm focusing lens is used to focus the light onto the CR39, which has a thickness of 1 mm and is in a sheet form (Solar lens product). A three-axis *x*-*y*-*z* translational platform was used to mount the sample. In the effort to position the polymer at the focus point, the sample location was varied through the micrometer-driven sample holder. The minimum spot size formed on the surface of the CR39 at the focus of the laser beam is noted as $465 \times 255\ \text{nm}^2$. Fluence of laser ablated on the CR39 can be calculated from the ratio of pulse energy (mJ) per laser spot area:

$$\text{Fluence} = \text{Energy}(\text{mJ}) / \text{Area}(\text{cm}^2) \quad (1)$$

For the whole experiments, fluence measurements are within an uncertainty of ±5%.

2.3. Continuous UV laser

For waveguide channel fabrication on CR39, a frequency-doubled argon ion laser emitting at 244 nm was used as a light source. The laser beam is focused by a 25-mm focusing lens onto the CR39. It was placed onto a three-axis stepper motor. By translating the stepper motor along the laser focusing plane, straight waveguides were produced. Fluence of laser irradiation on the CR39 can be calculated using the following equation [9]:

$$Fluence = P_{out}.d / v.A \quad (2)$$

where F is the fluence, P_{out} is the power of fiber output, d is the diameter of the beam, v is the relative traveling speed, and A is the area of the beam.

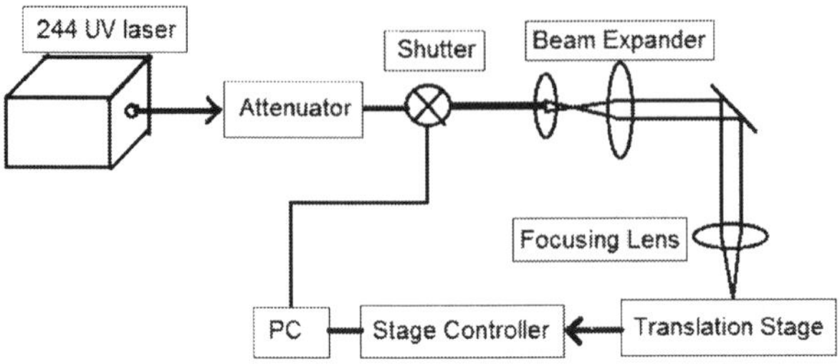

Figure 2. Waveguide channel written experimental arrangement.

The measured numerical aperture (NA) of the written waveguide can be used to compute the refractive index change, ΔRI of the UV irradiated area using the formula below:

$$NA = \left(n_c - n_{cl}\right)^{1/2} \tag{3}$$

where n_c and n_{cl} are the refractive index of UV written area and unwritten area, respectively. ΔRI can be calculated from the difference between n_c and n_{cl}.

A fiber pigtail was used to couple a tunable laser source into the CR39 waveguides to measure NA, whereas an objective lens was used to display the output of CR39 waveguides onto an image capture device. The formula below can be used to measure the NA of a waveguide from the waveguides divergence angle, θ:

$$NA = N_C \sin\theta \tag{4}$$

A straight waveguide with 3 cm length was fabricated using different fluences (between 1 and 5 KJ cm^{-2}). During this process, the laser power was set at a fixed value, and the laser beam was aligned to ensure that the focal plane was positioned on the CR39 sample surface. **Figure 2** shows the schematic diagram of waveguide channel written on the polymer.

3. Results and discussion

3.1. Etch rate analysis at 157 nm F_2 laser

Figure 3 shows a plot of the etch rate per pulse versus fluence for CR39 as derived from the white light interferometer; based on the linear fits for ≥10 pulse exposure, the estimated ablation threshold is ~60 mJ m^{-2}. However, **Figure 3** shows that there is still a small level of etching at a fluence of ~50 mJ cm^{-2}, and thus the estimated ablation threshold for CR39 is taken to lie in the range ~50–60 mJ cm^{-2}. From **Figure 3**, the etch rate per pulse for a single pulse reached ~100 nm at ~120 mJ cm^{-2}, higher than for multiple exposure. The data for 50 pulse exposure gave reasonably consistent values, and the gradient of the corresponding line in **Figure 3** gave an effective absorption coefficient of $\alpha_{eff} \approx 2.9 \times 10^5$ cm^{-1}. This is similar

http://dx.doi.org/10.5772/66377

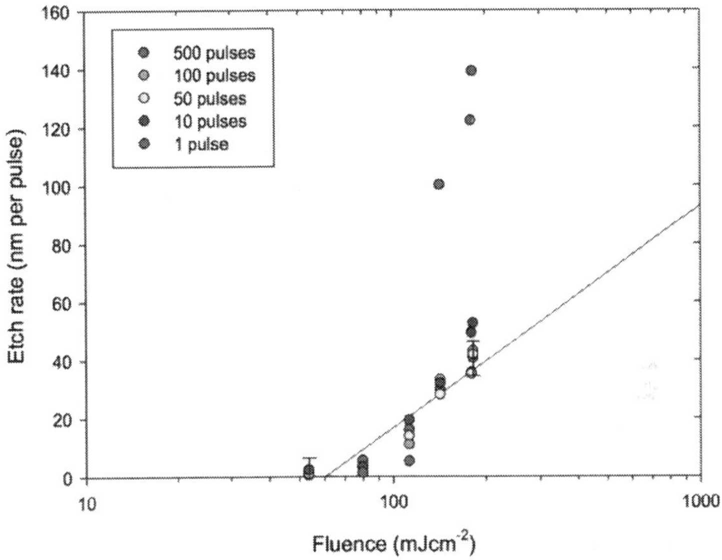

Figure 3. Etch rate as a function of fluence for CR39 polymer using the 157 nm laser. Results for the average etch rate per pulse for various numbers of pulses are shown.

to polycarbonate and indicates CR39 is a strongly absorbing organic polymer at 157 nm. No data relating to the optical constants in the VUV spectral region could be found for this material.

3.1.1. Formation of cones

For this experiment, clean CR39 samples (unseeded) were irradiated using 157 nm laser radiation over a range of pulse number from a single pulse to thousands of pulses and over a range of fluences of ~50 to ~180 mJ cm^{-2}. The dark spots that were seen under optical microscopy in the previous section were confirmed by scanning electron microscopy to be cones on the CR39 surface. These had very well-defined structures and, in general, appeared to have even better definition than those on the irradiated polycarbonate surface. In particular, the cones on CR39 were found to have extremely straight walls and to be exceptionally sharp at their tips as can be seen from the results shown in **Figure 4**. From the SEM images of the ablation sites, small particles appeared to be on the surface though it is difficult to make out if these reside on the top of the cone as "initiating" sites. **Figure 4a** and **b** shows the cones that developed at fluences of 112 and 180 mJ cm^{-2} with 500 pulses. The cones appear to have a similar size and shape at the same fluence. A comparison of **Figure 4a** and **b** shows as expected that the cone apex angle is larger at the lower fluence, that is, the full apex angle is ~70° at 112 mJ cm^{-2} and ~55° at 180 mJ cm^{-2} when corrected for the 60° viewing angle. It also appears that the cone tips get sharper as the fluence is raised. Exposure of the CR39 surface to a higher number of

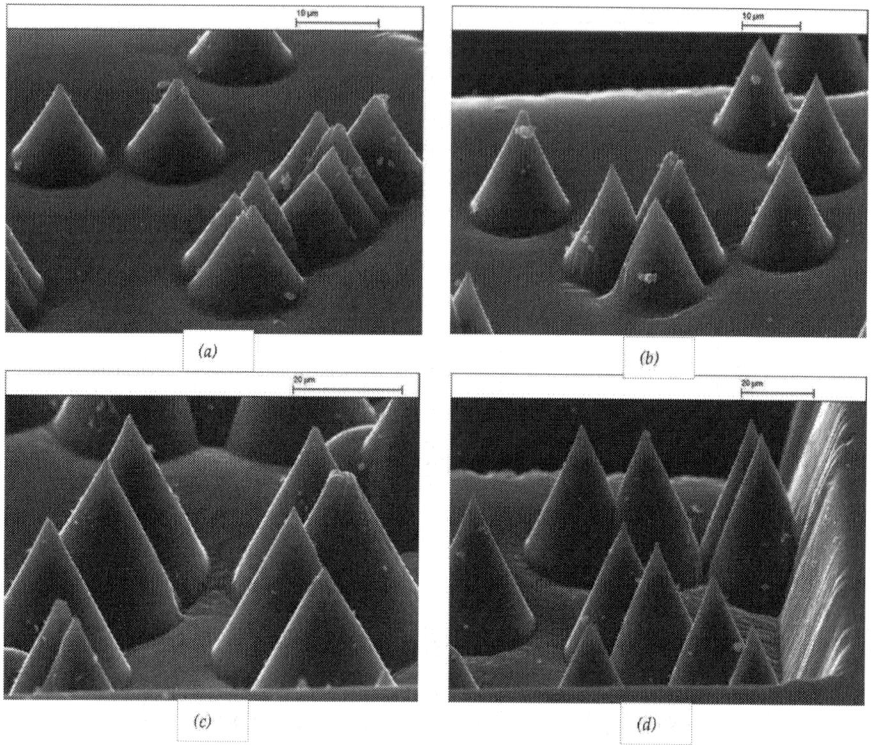

Figure 4. Examples of cone formation on the CR39 surface using the 157 nm laser (a) 500 pulses at 112 mJ cm^{-2} (b) 500 pulses at 180 mJ cm^{-2} (c) 1000 pulses at 142 mJ cm^{-2}, and (d) 1000 pulses at 182 mJ cm^{-2}.

pulses, **Figure 4c** and **d** led to an increase in the areal density of the cones compared to that at lower pulse number, **Figure 4a** and **b**.

Figure 5 shows a group of cones produced with 500 pulses at a fluence of ~80 mJ cm^{-2}. In this case, full apex angle of the cone is 83° corrected for the viewing angle of 60° on the SEM. At this fluence of ~80 mJ cm^{-2}, the cones have not fully developed and are not as well defined as those seen at higher fluences (**Figure 5b** and **c**), where the full apex angle is 66° and 51°, respectively, again illustrating that the angle is reduced at higher fluence. Here, they are fully developed, with sharp tips, and very well-defined structure.

In **Figure 5**, the ablated surface of this polymer well away from the cone bases is seen to be relatively smooth and devoid from significant debris indicating the good surface quality of this material when ablated with the 157 nm laser. The fringes around the bottom of the cones can be clearly seen in **Figure 6a** with 100 pulses at ~60 mJ cm^{-2} and **Figure 6b** with 100 pulses at ~180 mJ cm^{-2}.

http://dx.doi.org/10.5772/66377

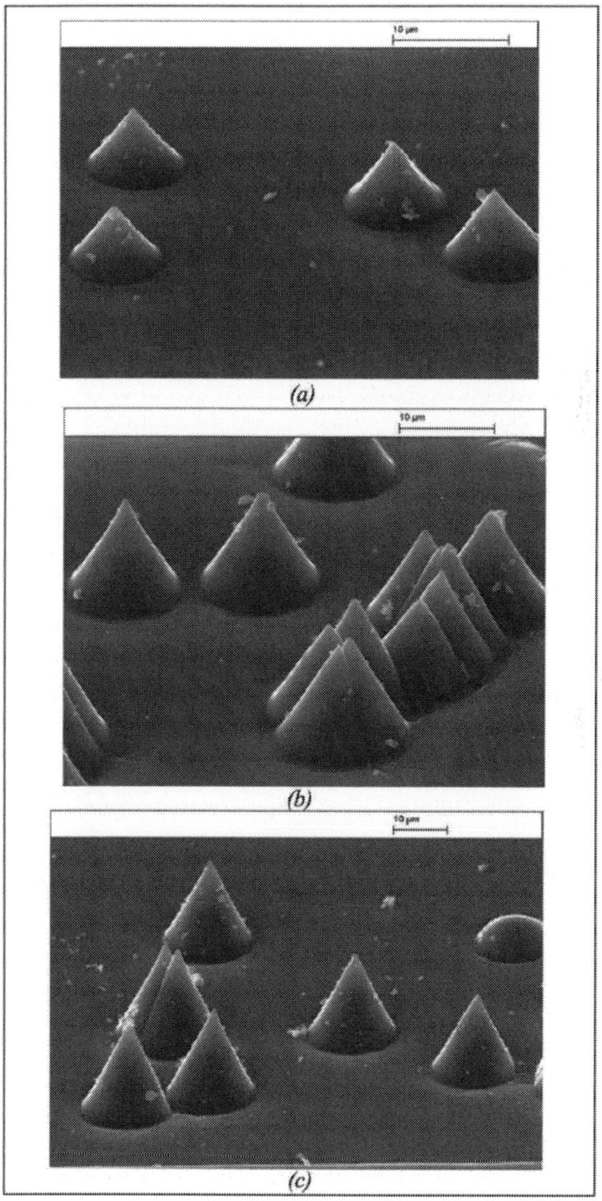

Figure 5. Evolution of conical structures developed on CR39 using the 157 nm laser (a) 500 pulses at ~80 mJ cm^{-2} (b) 500 pulses at ~112 mJ cm^{-2} (c) 500 pulses at ~140 mJ cm^{-2}.

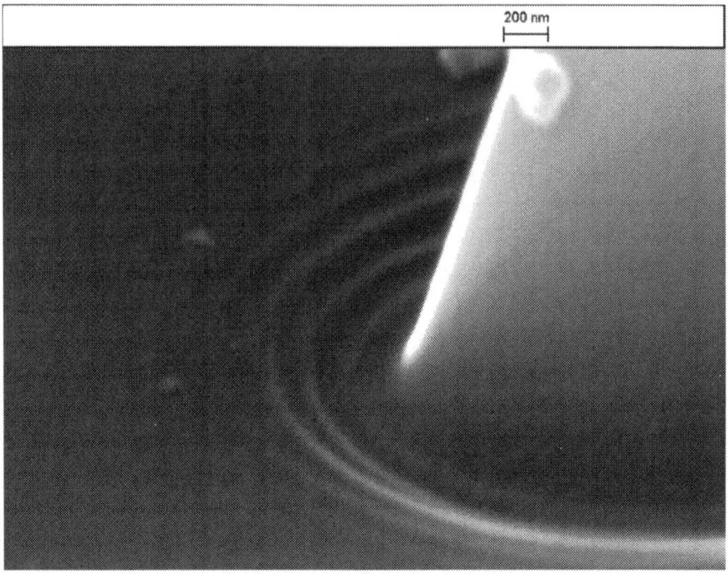

(a)

(b)

Figure 6. Fringes seen at the region of the bottom of the cones on CR39 (a) 100 pulses at ~60 mJ cm^{-2} and (b) 100 pulses at ~182 mJ cm^{-2}.

3.2. Pulsed UV laser at 248 nm

Figure 7 illustrates the microscopic image of the ablated CR39 craters spot size. The elimination of material and creation of the crater due to the ablation process are presented by the dark spot on the surface of the CR39. Partial transparent curves can be seen surrounding the edge of the craters. This could possibly be caused by the heat affected zone (HAZ), which occurs during the process of laser ablation. Heat affected zones (HAZ) are the outcome of materials that are molten due to heat transfer from the crater and then rapidly cooled. The short pulse width of the excimer laser nevertheless reduces the heat transfer from the ablated crater. This phenomenon is thought to be caused by the photothermal effect in which the laser energy absorption by the material is converted into heat energy, which in turn leads to the localized modification of its structure.

Figure 8 alternatively illustrates the average etch depth per pulse as a function of fluence F for CR39 ablated at 248 nm based on 150, 180, and 210 pulses. A linear fit of the form $d = k^{-1} \ln F/F_T$ provides an ablation threshold of 6–7 mJ cm^{-2}. This value is marginally lower than that reported using 157 nm laser which gives F_T as 11 mJ cm^{-2} [10].

3.3. Continuous UV laser at 244 nm

This section describes the use of this particular polymer as a waveguide where here we utilized a CR39 polymer sheet with a thickness of 0.5 mm and a refractive index as 1.486. Tunable laser

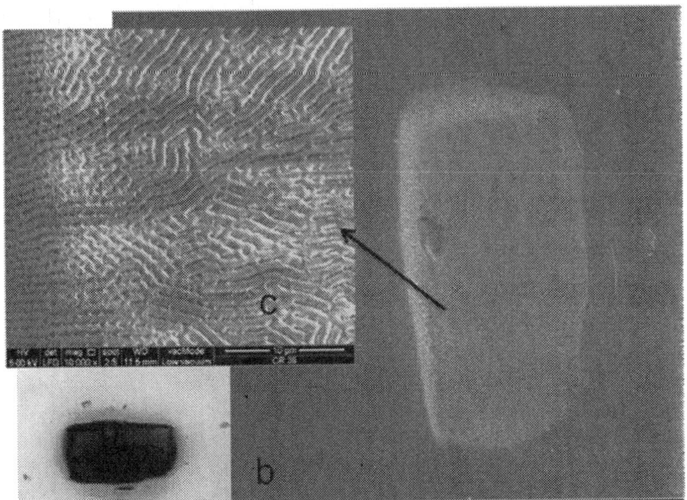

Figure 7. (a) Optical microscope imaging of the surface craters at CR39, (b) FESEM image of the crater, and (c) higher magnification of the ablated surface.

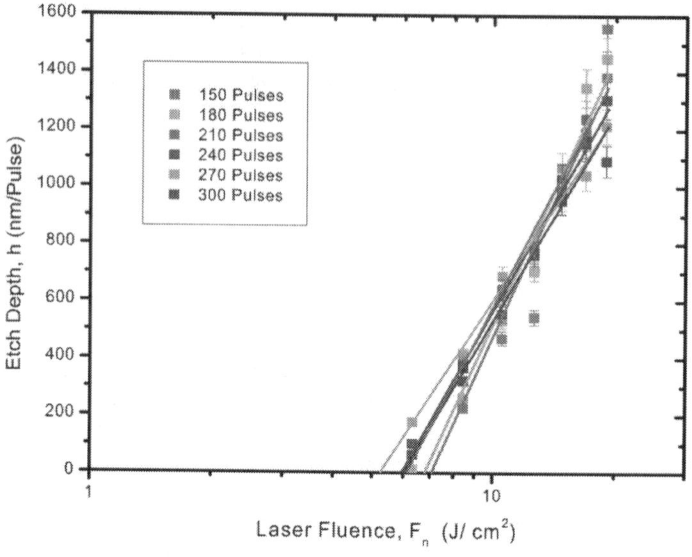

Figure 8. Plot of etch depth/pulse against log fluence at 150, 180, and 210 pulses on CR39.

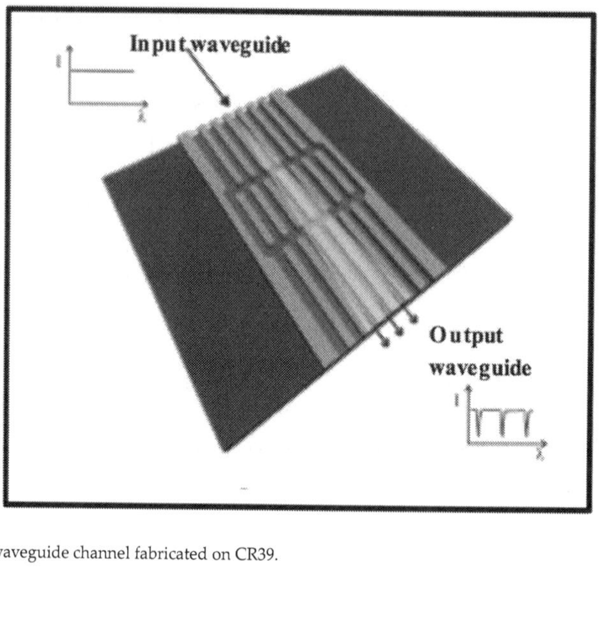

Figure 9. Optical waveguide channel fabricated on CR39.

source at 1550 nm (Sairon Technology SPA-4000 Prism Coupler) was used as a substrate for the polymer waveguide. SU-8 polymer was spin coated on the CR39 sheet and patterned using photolithography technique, formed a core of the channel waveguide structure. The height of the channel waveguide measured using optical microscopy was recorded as 5.0 ± 0.1 μm and width range between 10 and 15 μm. **Figure 9** shows the fabrication of the waveguide using photolithography technique.

Figure 10 shows the mode field diameter (MFD) and refractive index contrast of CR39 waveguide against the laser fluence. The refractive index change is seen to be more significant with the increase in fluence. Higher laser fluence will lead to a higher change of the refractive index, which likens to the reaction of other optical materials including silica glass to laser ablation. Extrapolation of the results in **Figure 10** also forecasts that the refractive index can be increased by the irradiation of CR39 with higher fluence. This increase, however, was observed to be limited as an upper fluence limit (5 KJ/cm^2) exists. When the laser fluence is above this limit, ablation will occur and subsequently UV irradiation of the CR39 will lead to the elimination of materials from its surface, similar to the outcome that occurs in Section 3.

The microscope image of the modification and ablation of CR39 is illustrated in **Figure 11**. As a result of refractive index modification, a partially transparent line can be seen in **Figure 11a**, whereas **Figure 11b** shows a darker line that indicates the removal of material and formation of craters due to ablation. The inset in **Figure 11b** further shows a cross-sectional view of the ablated sample. It is easier to detect the ablation effect from cross-sectional images in which a part of CR39 is removed. Subject to the laser fluence being applied, the removed area depth can be up to several microns.

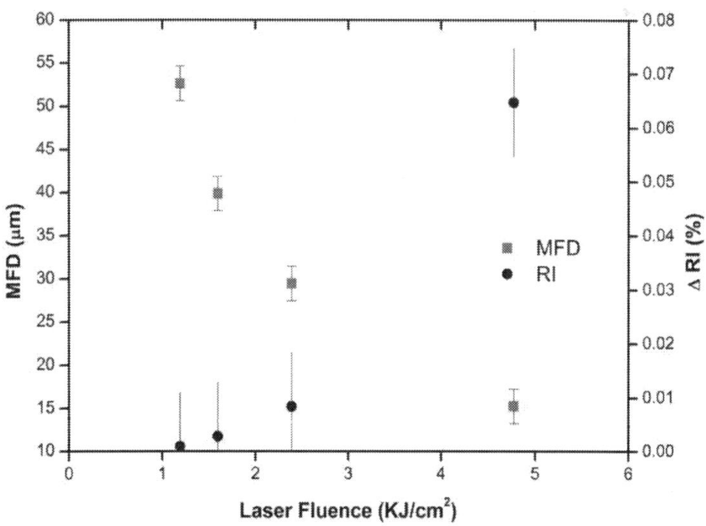

Figure 10. MFD and index contrast of CR39 waveguides versus laser fluence at 30 mW laser power.

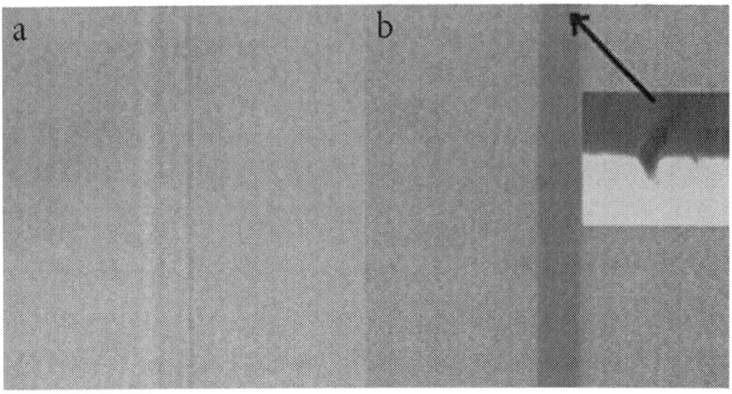

Figure 11. Microscope images of (a) internal modification, (b) ablation of CR39, and (inset) cross-sectional view of ablated CR39.

3.4. Application as optical waveguide

This section describes the use of laser process to fabricate optical waveguide that can be integrated into printed circuit board (PCB). Optical polymer materials offer a good aspect in terms of their relatively low cost and compatibility with traditional manufacturing process of PCB. Laser ablation is one of known approaches in processing PCB. The use of excimer laser has been repeatedly reported as it is capable of producing high-quality micromachining, which is sometimes ascribed as "cold ablation," which is possibly due to the UV absorption and the short pulse duration of the excimer laser.

Figure 12 shows the near-field image of one of the fabricated CR39 waveguides. The fabricated waveguides are observed to be faintly guided due to the light which is not well confined in the core, marked by the crosshair. The guiding characteristic of waveguide in which the largest refractive index variation measured is <0.07% for the range of fluence is verified as in **Figure 12**. It portrays that the fluctuations of refractive index caused by the laser fluence leads to a decrease in mode field diameter (MFD) due to better light confinement at the waveguide core.

The higher the variation of refractive index, the stronger the waveguides confinement in which a larger portion of light is being confined in the core and this in turn leads to the reduction in the MFD of the waveguides.

The successful direct writing of straight waveguides proves that a positive refractive index reaction of CR39 occurs when irradiated by 244 nm UV laser since refractive index for core must be greater than the surrounding for light guiding by the principle total internal reflection. The actual mechanism leading to the variation of the CR39 refractive index is yet to be determined. Nonetheless, we believe that it is caused by the photothermal effect in which the laser energy absorption by the material is converted into heat energy, which in turn leads to the localized modification of its structure and subsequently the refractive index. Nevertheless, further investigation on the origins of refractive index modification in CR39 is still being carried out.

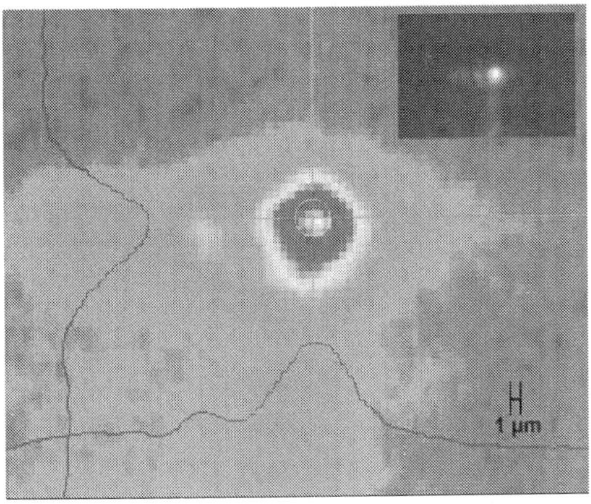

Figure 12. Near-field imaging of the CR39 waveguides.

4. Summary

This study deliberates the CR39 ablation with 248 nm UV laser. The variation in etch depth of craters is measured, and we can determine the fluence from the laser energy and spot size. There will be an increase in etch depth when the laser fluence is altered from 6.3 to 19.0 mJ cm^{-2}, respectively. Fluence threshold for CR39 is determined from the graph is 6 mJ cm^{-2}. Refractive index change from 1×10^{-5} to about 1×10^{-3} (0.0009–0.065%) is achieved on CR39 by varying the laser fluence from 1.2 to 4.8 KJ/cm^2, respectively. We suspect that the mechanism responsible for the refractive index change in CR39 by 244 nm laser irradiation is due to photothermal effect where localized heat generated by absorption of 244 nm laser modifies the structure/densified of CR39.

Author details

Rozalina Zakaria

Address all correspondence to: rozalina@um.edu.my

Photonics Research Centre, Faculty Science, University of Malaya, Kuala Lumpur, Malaysia

References

[1] Dyer, P.E., Excimer laser polymer ablation: twenty years on. Applied Physics A, 2003. 77(2): pp. 167–173.

[2] Jae-Hoon, L., et al., Enhanced extraction efficiency of ingan-based light-emitting diodes using 100-kHz femtosecond-laser-scribing technology. Electron Device Letters, IEEE, 2010. 31(3): pp. 213–215.

[3] Wu, C.-Y., Shu, C.-W., and Yeh, Z.-C., Effects of excimer laser illumination on microdrilling into an oblique polymer surface. Optics and Lasers in Engineering, 2006. 44(8): pp. 842–857.

[4] Lorenz, R.M., et al., Direct laser writing on electrolessly deposited thin metal films for applications in micro- and nanofluidics. Langmuir, 2004. 20(5): pp. 1833–1837.

[5] Cristea, D., et al., Polymer micromachining for micro- and nanophotonics. Materials Science and Engineering: C, 2006. 26(5–7): pp. 1049–1055.

[6] Cheng, Y., et al., Optical gratings embedded in photosensitive glass by photochemical reaction using a femtosecond laser. Optics Express, 2003. 11(15): pp. 1809–1816.

[7] Rafique, M.S., et al., Nonlinear absorption properties correlated with the surface and structural changes of ultra-short pulse laser irradiated CR-39. Applied Physics A, 2010. 100(4): pp. 1183–1189.

[8] Serafetinides, A.A., et al., Ultra-short pulsed laser ablation of polymers. Applied Surface Science, 2001. 180(1–2): pp. 42–56.

[9] Arshad K. Mairaj, et al., Fabrication and characterization of continuous wave direct UV (l= 244 nm) written channel waveguides in chalcogenide (Ga:La:S) glass. Journal of Lightwave Technology, 2002. 20(8).pp 1578–1584.

[10] Dyer, P.E., Walton, C.D., and Zakaria, R., Interference effects in 157 nm laser ablated cones in polycarbonate and application to spatial coherence measurement. Applied Physics A, 2009. 95: pp. 319–323.

Section 4

Other Applications

Chapter 17

Laser Ablation Applications in Ablation-Resistance Characterization of Materials

Yonggang Tong, Xiubing Liang, Shuxin Bai and Qing H. Qin

Additional information is available at the end of the chapter

http://dx.doi.org/10.5772/65108

Abstract

Owing to the rapid heating and large power intensity, the laser beams were successfully used to characterize the ablation-resistant performance of materials, which provided us more knowledge about the usability of materials in the ablation environment and developing protection against laser irradiation. In this chapter, we comparatively introduced some experimental methods for ablation-resistance characterization of materials. The fundamentals of laser-material interactions were discussed from the physical and chemical aspects to help understand the laser ablation mechanism. Finally, we presented some practical applications of laser ablation in ablation-resistance characterization of ultra-high-temperature ceramics (UHTCs) and ceramic matrix composites and discussed the laser ablation behavior and mechanism.

Keywords: Laser ablation, Ablation resistance, Laser-material interaction, Ceramic matrix composite, Ceramic coatings

1. Introduction

A light amplification by stimulated emission of radiation (laser) is a device that emits light through the process of stimulated emission. Owing to the advantages of laser radiation over conventional mechanical and thermal techniques, it was suggested to be used as a manufacturing tool after the development of the first laser. Over the last few decades, great effects have been made to develop various laser devices. According to the lasing medium, the lasers can be classified as gas lasers, liquid lasers, solid-state lasers, semiconductor lasers, and free-electron lasers.

Because of the unique energy sources characterized by spectral purity, spatial and temporal coherence, and high intensity, the laser beams have captured significant attention and widely been used in matrix-assisted laser desorption/ionization, laser surgery, micro-fabrication, pulsed laser deposition, etc. [1]. Recently, they were successfully used to evaluate the ablation-resistant performance of materials, which provided us more knowledge about the usability of materials in the ablation environment and developing protection against laser irradiation. A great amount of work on laser ablation of polymer-based composites, ultra-high-temperature ceramics (UHTCs), ceramic-based composites, and ablation-resistant coatings has been reported. The linear and mass ablation rates of these materials were tested using the laser ablation method, and the laser ablation behavior and mechanism were investigated by experiments and numerical simulation.

In this chapter, we firstly made a simple introduction of the ablation-resistance characterization methods for materials and compared the advantages and disadvantages of these methods with laser ablation method. To help understand the laser ablation mechanism of ultra-high-temperature materials, the fundamentals of laser-material interactions were discussed from the physical aspects and chemical aspects. The physical aspects mainly involved the absorption of laser radiation, heating and propagation, melting, vaporization, and solidification. The chemical aspects mainly involved the decomposition of the phases in materials and the reaction of the materials with the atmosphere. Finally, we presented some practical applications of laser ablation in ablation-resistance characterization of materials based on our research and some relevant literatures. The ablation rates of materials under different laser parameters were tested. The morphologies of the ablated surface were observed by electron microscopes, and the ablation mechanism was discussed.

2. Ablation-resistance characterization methods for materials

Advanced aerospace structures and anti-ablation components, such as nose caps, sharp leading edges, and rocket engines for hypersonic aerospace vehicles, suffer from high heat fluxes and pressure, severe thermal shock, and perhaps high-speed erosion of ceramic particles in their working conditions [2]. The serving temperatures of these components may increase rapidly from the room temperature to over 2000 °C and last from several seconds to several hundreds of seconds. Materials with outstanding mechanical and ablation-resistant properties are required for these structures and components. Refractory metals, carbon-based composites (graphite and C/C composite), ultra-high-temperature ceramics, and composites are potential candidates due to their extremely high melting points, high-temperature mechanical strength, and outstanding ablation resistance. Due to the special serving environments, it is necessary to validate the ablation properties of these materials. Ablation resistance has been one of the most important properties in evaluating the usability of these materials. Great efforts should be devoted to the investigation on the ablation-resistance characterization, microstructure evolution, and ablation mechanism of these materials before their practical applications in the ablation environments.

Normally, ablation resistance of materials can be investigated by ballistic flight experiments and ground-based simulation experiments [3]. Ballistic flight experiments can evaluate the ablation properties of materials in a real serving condition. However, this experiment is seldom used to characterize the ablation resistance of materials because of the considerable cost. By contrast, ground-based simulation experiments are more practical to investigate the ablation resistance of materials. The main ground-based simulation experimental methods are wind tunnel ablation testing, plasma arc-jet ablation testing, and oxyacetylene flame ablation testing [4–6]. Wind tunnel ablation testing can simulate the ablation conditions of high enthalpy and strong gas flows, but it cannot simulate the fully representative flight envelope in terms of Mach and Reynolds number [4]. Plasma arc-jet ablation testing [5] can partially simulate the reentry environment. Nevertheless, the parameters are simple. Additionally, both wind tunnel and plasma arc-jet ablation testing are costly. By contrast, oxyacetylene torch testing is a simple and low-cost method [6, 7]. It is widely used in labs to provide a primary evaluation about the ablation properties of materials. However, the ablation temperature of oxyacetylene torch test is limited by the flame temperature (about 3000°C), and the combustion products of the oxyacetylene flame include O_2, CO_2, O, OH and H_2O [7], which may affect the ablation resistance of materials.

Recently, the laser beams were successfully used to evaluate the ablation-resistant performance of materials. A great amount of work on laser ablation of polymer-based composites, ultra-high-temperature ceramics, ceramic-based composites, and ablation-resistant coating has been reported. The published literatures demonstrate that laser ablation is a simple and cost-effective method to characterize the ablation resistance of materials. Irradiated by the laser beams, materials can be rapidly heated to a very high temperature. Additionally, the laser ablation testing does not introduce any combustion products and mainly provides a rapid thermal impact, which is beneficial to analyze the ablation behavior and mechanism of the materials.

3. Fundamentals of laser-material interactions

Laser-material interactions are very important to understand laser ablation process of materials. Irradiated by the laser beams, the substrate materials absorb the irradiation energy. Absorption of radiation in the materials results in various effects such as heating, melting, vaporization, and plasma formation. The extent of these effects primarily depends on the characteristic of electromagnetic radiation and the thermophysical properties of the substrate materials [8]. The laser parameters include intensity, wavelength, angle of incidence, spatial and temporal coherence, illumination time, and polarization, whereas parameters of the substrate materials include absorption of the laser energy, thermal conductivity, specific heat, and density. They all should be taken into account in order to understand in details the effects of laser ablation processing on the substrate materials tested. In addition to these physical aspect effects, the substrate materials are believed to react with the atmosphere under the thermal impact. The phases in the substrate materials will be changed, and they even decompose and vaporize, which greatly affects the ablation resistance of the substrate material. Thus, the laser-material interactions are very complex, and only in some simple cases, the laser can be merely seen as a heat source. The laser-material interactions should be comprehensively considered from the physical aspects and chemical aspects.

3.1. Physical aspects

3.1.1. Absorption of laser radiation

When laser beam strikes the surface of the substrate material, a portion of laser energy will be reflected from the interface due to the discontinuity in the real index of refraction, and the rest will be transmitted into the material. The reflectivity of a given material depends on the frequency of the light source through the dispersion relation of its index of refraction. **Figure 1** [9] presents the variation of reflectivity with the wavelength of some common metallic materials. As indicated in **Figure 1**, the reflectivity of the material generally increases with increasing wavelength. Thus, the laser energy is strongly absorbed by materials at shorter wavelengths.

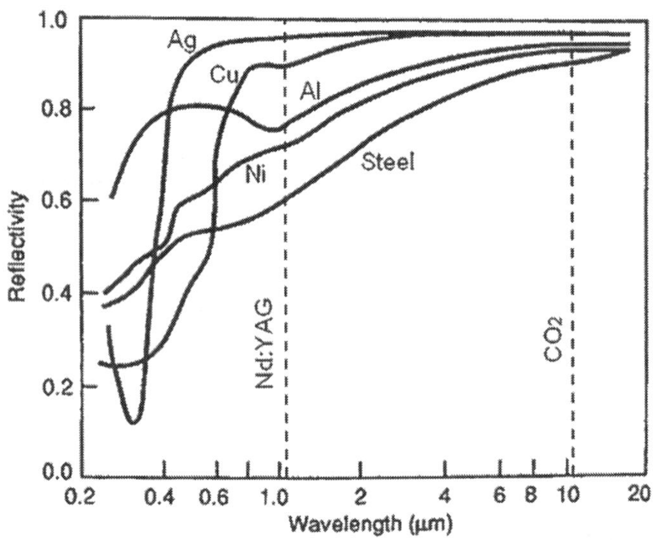

Figure 1. Variation of reflectivity with wavelength for some metallic materials.

In addition to the reflected fraction, the laser irradiation on the surface of substrate material leads to the excitation of free electrons (in metals), vibrations (in insulators), or both (in semiconductors) [8]. The excitation results in an energy increase in the system, and this excitation energy is rapidly converted into heat for a very short time period. That is the fraction of laser irradiation absorbed by the substrate materials.

Once inside the materials, absorption causes the laser intensity to decay with depth at a rate determined by absorption coefficient of the substrate materials α. In general, α is a function of laser's wavelength and temperature. However, for constant α, the laser intensity, I, decays exponentially with depth z according to the Beer-Lambert law [10]:

$$I(z) = I_0 e^{-\alpha z} \tag{1}$$

where I_0 is the intensity just inside the surface after considering reflection loss. Considering the depth at which the intensity of the transmitted light drops to $1/e$ of its initial value at the interface, the optical penetration or absorption depth can be conveniently defined as $\delta = 1/\alpha$. **Figure 2** [11] shows the optical absorption depth as a function of wavelength for some common materials, such as metals and semiconductors. The important thing to note from **Figure 2** is that the absorption depths are quite small relative to bulk material dimensions. Thus, absorption of the laser energy only happens at the surface of materials.

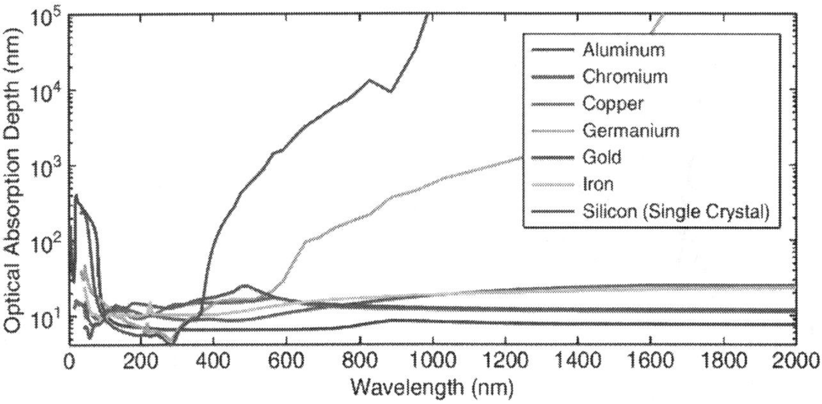

Figure 2. Optical absorption depths of some materials for different wavelengths.

3.1.2. Heating

Once the laser energy is absorbed by the substrate materials, it is followed by various heat transfer processes such as conduction into the materials, convection, and radiation from the surface. The most significant heat transfer process is the heat conduction into the material. The heat conduction from the surface to the inner of materials establishes a temperature distribution in the material depending on the thermophysical properties of the material and laser parameters. Ignoring the convective and radiative energy transport, the temperature distribution T in a material can be given by the heat conduction equation, which, in 2D, can be written as [12],

$$\rho(T)C_p(T)\frac{\partial T(x,y,t)}{\partial t} = \nabla\left[k(T)\nabla T(x,y,t)\right] + Q(y,t) \tag{2}$$

where x and y are the space coordinates and k, C_p and ρ are the thermal conductivity, specific heat at constant pressure, and mass density of the target material, respectively. Here, the source term $Q(y,t)$ is the laser radiation absorbed by the substrate material and is expressed as [13],

$$Q(y,t) = I_s(1-R)\alpha\exp(-\alpha y) \tag{3}$$

where α and R are the absorption and reflectivity coefficient of the substrate material, respectively, y is the spatial coordinate in the direction normal to the sample surface, and I_s is the laser irradiance at the sample surface. According to Bulgakova and Bulgakov [14], I_s is expressed as,

$$I_s(t) = I(t)\exp\left[-\omega(t)\right] \tag{4}$$

Based on the above equations and thermophysical properties of the materials, the temperature distribution caused by the laser energy absorption at the material surface and subsequent conduction into the body material can be calculated. Finite element analysis and some numerical calculation methods can be used to solve these equations. For composite materials, the temperature distribution plays a more important role to analyze the ablation behavior and mechanism. The phases in the composites have different thermophysical properties, which cause a quite different temperature distribution in different phases, and consequently, the phases may present different ablation behavior and mechanism.

3.1.3. Melting

Irradiated by laser beams with relatively large laser power densities, surface temperatures of the materials may reach their melting temperatures (without surface evaporation). The melting of materials significantly depends on their melting points. **Figure 3** shows the melting points of some high-temperature materials [15]. The surface of pure metals melts once the surface temperature reaches their melting temperatures. Nevertheless, the surface melting of composite materials is greatly influenced by the phase composition of the materials. The phases in the composites have different melting points, which can get melting in turn with increasing laser power densities or impulse time. The former melted phases may react with some other phases or the atmosphere. New substances may be produced and affect the surface melting and phase composition of the materials.

http://dx.doi.org/10.5772/65108

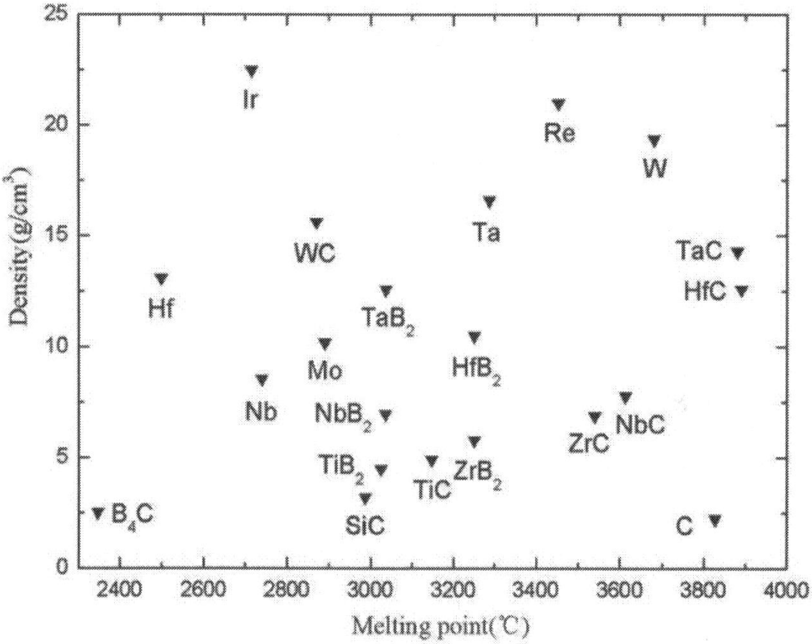

Figure 3. Melting points of some high-temperature materials.

3.1.4. *Vaporization*

If the incident laser intensity is sufficiently high, the surface of materials can be heated to its boiling temperature, and material removal by evaporation is caused by the laser ablation. The melting depth reaches the maximum value once the surface temperature reaches the boiling point. Further increase of laser power density or pulse time causes the more severe evaporative material removal from the surface without any further increase in the melting depth. The maximum melting depth (z_{max}) at which the surface reaches the boiling point can be calculated as follows [16]:

$$ierfc\left(\frac{Hz_{max}}{kT_b\sqrt{\pi}}\right) = \frac{T_m}{T_b\sqrt{\pi}} \quad (5)$$

where Hz_{max} is the maximum melting depth, T_b and T_m are the boiling and melting temperatures of the substrate materials, and k is the Boltzmann constant.

Once the vaporization is initiated at the surface of the material, the continued laser irradiation causes the liquid-vapor interface to move inside the material with the evaporative removal of material from the surface above the liquid-vapor interface. Assuming the thermal ablation as vaporization, the flow of material vaporized from the surface of a body at temperature T can be calculated according to the Hertz-Knudsen equation [13], leading to an ablation rate ϑ as [17],

$$\vartheta(T)=(1-\beta)\sqrt{\frac{m}{2\pi mk_BT}}\frac{p_0}{\rho}\left[\frac{L_V}{k_B}\left(\frac{1}{T_b}-\frac{1}{T}\right)\right] \tag{6}$$

where T_b is the boiling temperature at pressure p_0, k_B is the Boltzmann constant, β is the back flux coefficient, and L_V is the latent heat of vaporization of the material.

3.1.5. Plasma formation

Irradiated by the laser beams with sufficiently large intensity, a great amount of surface evaporation occurs as mentioned in the previous sections. Once the vaporization takes place, the interactions between the as-produced vapor and the incident laser beam become important in determining the overall effect of the laser irradiation on the substrate material. Interaction of the laser with the vapor can lead to the ionizing of the vapor. The highly ionized vapor is termed as plasma. In dynamic equilibrium, the degree of ionization ε in the vapor can be expressed by the Saha equation [18]:

$$\frac{\varepsilon^2}{1-\varepsilon}=\frac{2g_i}{g_aN_g}\left(\frac{2\pi mk_BT}{h^2}\right)^{\frac{3}{2}}\exp\left(-\frac{E_i}{k_BT}\right) \tag{7}$$

with $\varepsilon = N_e/N_g$ and $N_g = N_e + N_a$. Here, N_e and N_a are the number densities of electrons and atoms/molecules, respectively; g_i and g_a are the degeneracy of states for ions and atoms/molecules; and E_i is the ionization energy.

The vapor and the plasma can absorb and scatter the laser radiation, which changes the actual flux received by the substrate surface. Recoil from the vapor and plasma can also generate shock waves in the substrate material, which may cause plastic deformation and work hardening [19]. Expulsion of any remaining molten material as well as initiate shock waves can be further caused by the recoil as well. In this chapter, laser ablation method is used to characterize the ablation-resistance performance of materials. It is not reasonable to choose too large laser intensity. Therefore, we make very little consideration of plasma formation.

3.1.6. Recondensation and resolidification

Irradiated by the laser beams with large power densities, the material surface is heated to a rather high temperature, and significant surface evaporation and sometimes plasma take place, which makes a positive pressure over the ablated surface. Surface temperature increment from 300 to 3500 K can lead to an enormous vapor pressure increase from 10 bar to almost 160 bar

http://dx.doi.org/10.5772/65108

for a time interval comparable with the pulse duration (pulse D 55 ns). The vapor and plasma may eject from the ablated center under the driving of the positive pressure and subsequently recondensate on the surface of the material. Besides, laser ablation is usually carried out in the air atmosphere or some reactive atmospheres. Some new phases may be formed due to the chemical reaction between the vapor and the ambient gas. The new reaction-formed chemical compounds may also condensate on the surface of the materials.

As we mentioned above, recoil of the vapor and plasma can cause further expulsion of the remaining molten material. The molten materials expulsed from the ablated center and remaining in the ablated center resolidify and also form some new phases on the surface of materials.

Resolidification of the molten materials and condensation of the vapor and plasma are very important to analyze the morphologies of the ablated surface of the materials. They may form some thin films and nanoparticles and alter the topography at the rim and surrounding areas of the ablated region.

3.2. Chemical aspects

In addition to the physical aspects, the substrate materials and ablation products are believed to react with the ambient gas, and some new compounds may be formed during the laser ablation process. Thus, the laser ablation should also be discussed from the chemical aspects.

Normally, laser ablation is carried out in the air atmosphere. Irradiated by the laser beams, the substrate material is heated to a high temperature, and the phases in the materials can react with the oxygen in the air atmosphere. For carbon-based materials, such as graphite and C/C composite, the reaction of the materials with oxygen produces gas products, CO and CO_2, which are liable to eject from the ablated surface. The substrate materials are severely ablated with a large decrease of external dimensions. However, for the refractory alloys and refractory-based ceramics, refractory oxides are produced due to the oxidation reaction. The refractory oxides have high melting points and low evaporation rates at high temperatures (see in **Table 1** [20]). Especially, the oxygen permeabilities in these oxides are very small. The substrate materials can be isolated from most of the oxygen, and thus, the substrate material can be prevented from the reaction of oxygen in the air atmosphere. Taking this into consideration, the former reaction-formed oxides provide protection for the substrate materials and reduce the damage of the laser ablation.

Oxides	Melting point (°C)	Density (g/cm³)	Thermal expansion (ppm/°C)	Evaporation rate (10^{-6} m/h) 10^{-5} at 1650 °C	Oxygen permeability (g/cm s) 10^{-10} at 1800 °C
HfO_2	2810	9.68	6.8	6.7	46
ZrO_2	2700	5.6	7.5	670	30
Y_2O_3	2460	5.03	6.8	–	2.0
Al_2O_3	2015	3.98	8.1	–	–
SiO_2	1728	2.32	0.5	–	–

Table 1. Properties of some refractory oxides.

Irradiated by laser beams with very large intensity, the substrate materials may be heated to very high temperatures over the decomposed temperature of some phases. Taking carbon and SiC, for example, they sublimate once heated over 3827 and 2987 K, respectively [21]. The gas products eject from the ablation surface and may also result in a severe damage to the substrate materials. Besides, these gas products may also recondensate and form some thin films and nanoparticles at the rim and surrounding areas of the ablated region. Understanding these chemical reactions during laser ablation plays a great role to analyze the final morphologies of the ablated surface and study the ablation mechanism of the substrate materials. It should be noted that the above reactions are greatly affected by the temperatures during laser ablation, which depends on the laser parameters (pulse duration, energy, and wavelength), the substrate materials' properties, and the surrounding environment condition.

4. Laser ablation applications in ablation-resistance characterization of materials

4.1. Carbon fiber-reinforced ceramic matrix composites

Carbon fiber-reinforced ceramic matrix composites, combined with the excellent properties of carbon fiber and high-temperature ceramics, are one of the most promising candidate materials for high-temperature components due to their unique properties such as relative low density, low coefficient of thermal expansion, high specific strength/modulus, and excellent ablation resistance. In this section, we chose C/SiC composite as an example and comprehensively discussed the laser ablation behavior and mechanism. In addition, some recent work on laser ablation of C/C-ZrC composite in our group and by some other researchers was reviewed.

4.1.1. Production of C/SiC composite and the laser ablation test

C/SiC composite for laser ablation testing was fabricated using a polymer infiltration and pyrolysis (PIP) method [22] by the Key Laboratory of Advanced Ceramic Fibers and Composites (National University of Defense Technology, Changsha, China). Three-dimensional braided carbon fiber preform was used as the reinforcement. Polycarbosilane (PCS) was chosen as the precursor of SiC matrix. Divinylbenzene (DVB) was used as solvent and cross-linking reagent for PCS. The preparation process of the composite contained three steps: (1) the carbon fiber preforms were immersed into the PCS/DVB solution and infiltrated with the PCS/DVB solution in vacuum at room temperature; (2) the preforms filled with PCS/DVB solution were cured at 150 °C; and (3) the cured preforms were pyrolyzed at 1200 °C to form the SiC matrix in an inert atmosphere. In order to densify the composites, the other several infiltration-cure-pyrolysis cycles were repeated. The as-produced C/SiC composite is composed of carbon and SiC (determined by XRD) and generally dense with two kinds of pores. One is small distributing in the intra-fiber bundles, and the other is larger locating in the interfiber bundles (**Figure 4**).

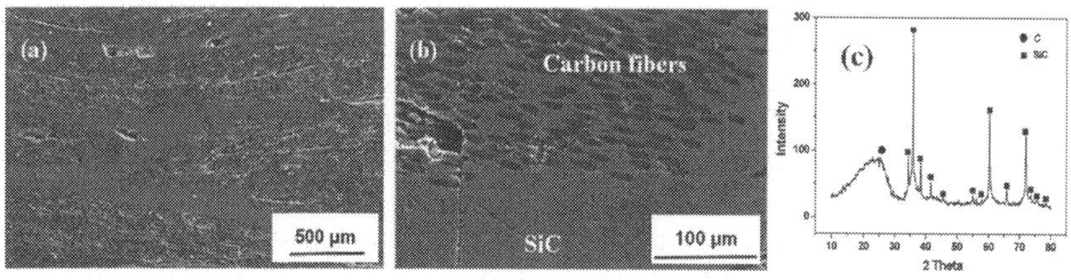

Figure 4. Cross-sectional microstructure (a and b) and XRD diffraction patterns (c) of C/SiC composite.

Ablation properties of the as-produced C/SiC composite were tested using a pulsed laser in the air. The laser ablation equipment is an Nd:YAG pulsed laser (wavelength 1.064 μm) with the following parameters: frequency 20 Hz and pulse width 1 ms. During the ablation testing, the C/SiC composite was located in a test chamber and was then vertically irradiated by the pulsed laser. The ablation depth of C/SiC composite was given by the thickness changes before and after the ablation test, which was measured by a microscope.

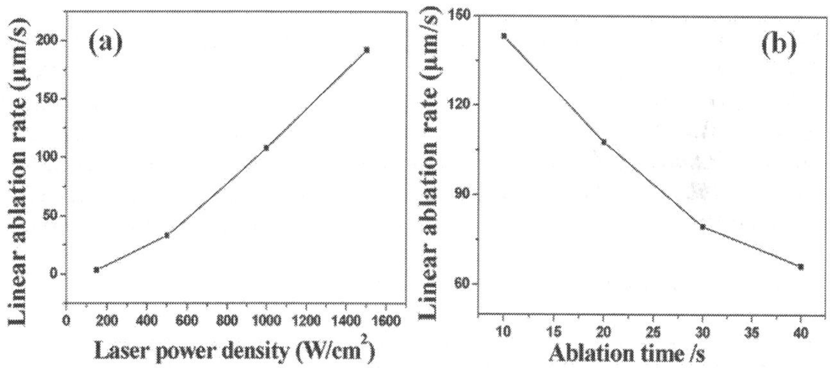

Figure 5. Linear ablation rates of C/SiC composite versus (a) laser power densities and (b) time.

4.1.2. *Laser ablation resistance of C/SiC composite*

Linear ablation rates of C/SiC composite tested with different laser power densities are shown in **Figure 5(a)**. It is indicated that the linear ablation rates of the composites increase with increasing laser power densities. During the ablation process, the laser energy is absorbed by the composite. Along the heat penetration depth and conduction width, the conduction laser energy decreases progressively from their input value, which in turn affects the corresponding temperature distribution. The larger the testing laser power density is, the higher the temperature of the composite is heated to be and the greater the heat penetration depth is. Therefore, the linear ablation rates increase with increasing of the laser power densities.

Ablation resistance of C/SiC composite was also tested for different time periods. **Figure 5(b)** shows the linear ablation rates of C/SiC composites tested for different time periods with laser power density of 1000 W/cm². It is indicated that the linear ablation rates of C/SiC composites decrease with an increase of ablation time.

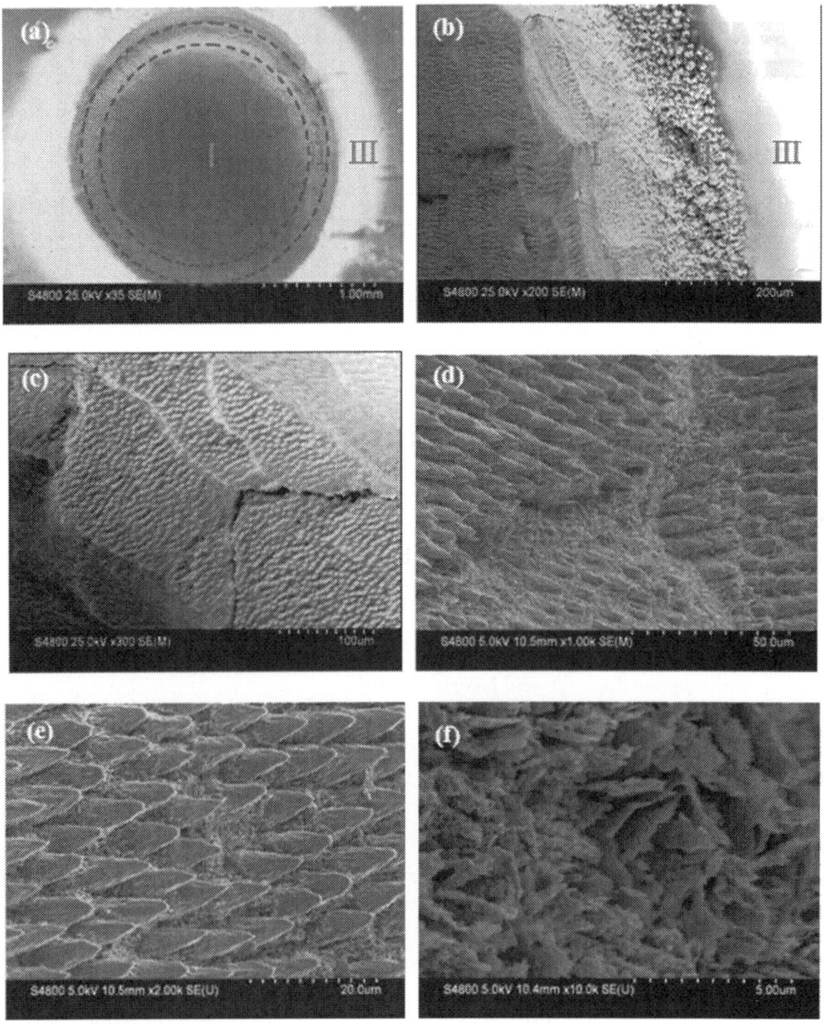

Figure 6. Ablated surface morphologies of C/SiC composite at 1000 W/cm² for 20 s (a and b) three ablated regions on the ablated surface, (c) macro-morphologies of the ablation center, (d and e) large magnification morphologies of ablation center, and (f) large magnification of the marked area in (e).

4.1.3. Microstructure morphologies of the ablated composite [23]

Figure 6 shows the surface morphologies of the C/SiC composite after laser ablation testing at 1000 W/cm^2 for 20 s. Three ablated regions can be identified on the ablated surface of the composite according to the difference of the morphologies: region I ablation center with a deep pit, region II transitional zone with a lot of spheric particles, and region III ablation edge covered by a white glassy layer (**Figure 6(a** and **b**)). Detailed scanning electron microscope (SEM) characterization of region I (**Figure 6(c** and **d)**) shows that the ablation center region exhibits a needle-like structure with taper-ended carbon fibers standing on the ablated surface filled with some nanostructure sheet. In order to determine the composition of the above three ablated regions and the nanostructure sheet, EDS analysis was carried out. The results indicate that the nanostructure sheet is nanocarbon sheet composed of pure carbon. The spheric particles in the transitional region are composed of silicon and carbon owning the atom proportion of 54.47:45.53, which are SiC particles. The white layer covered on the ablated surface of region III ablation edge is SiO_2 composed of silicon and oxygen.

During the laser ablation testing, the center region of the composite was instantly heated to a very high temperature, which was thought to be approximately higher than 3500 °C due to the large amount of ablated carbon fibers observed in the ablated center region. At such a high temperature, SiC matrix in the composite decomposes and sublimates, which forms a hot mixture of gases and vapors. Carbon fibers in the composite also get to its sublimation temperature to form a carbon vapor. The decomposition and sublimation of SiC and carbon fibers can be described as the following reactions:

$$SiC(s) \rightarrow Si(g) + C(g) \tag{8}$$

$$SiC(s) \rightarrow SiC(g) \tag{9}$$

$$C(s) \rightarrow C(g) \tag{10}$$

Because the ablation testing was performed in the air, carbon fibers and SiC matrix are believed to react with the oxygen in the atmosphere and form the products of CO, CO_2, SiO, and SiO_2 due to the reaction with oxygen. The possible reactions that take place in the oxidation of carbon fibers and SiC matrix are as follows:

$$SiC(s) + O_2(g) \rightarrow SiO_2(g) + CO(g) \tag{11}$$

$$SiC(s) + O_2(g) \rightarrow SiO(g) + CO(g) \tag{12}$$

$$C(s) + O_2(g) \rightarrow CO_2(g) \tag{13}$$

$$C(s) + O_2(g) \rightarrow CO(g) \tag{14}$$

Among the products of the above reactions, SiO, CO, and CO_2 are gases. SiO_2 immediately gasifies due to its low boiling temperature, 2230 °C [24]. These gas products result in a positive pressure atmosphere on the ablation surface. Oxygen in the atmosphere on the ablation surface can be instantly exhausted, and the oxygen in the outside atmosphere hardly diffuses into the atmosphere on the ablation surface because of the positive pressure kept by the gasses and vapors produced according to Eqs. 8–14. Therefore, the oxidation of SiC and carbon fibers just takes place at the very beginning and is made no consideration in the subsequent analysis.

Ablation of the composite is greatly influenced by the temperatures functioning on the ablated areas. Irradiated by the laser, the center area of C/SiC composite was heated to the highest temperature, and it was severely ablated. Because the carbon fibers are thermally more stable than the SiC matrix, carbon fibers with taper ends protruded on the ablation surface without the SiC matrix as shown in **Figure 6(c)**. Inside the composite under the ablated center region, the heat absorbed from the laser penetrates into the inner of the body composite along the direction of the laser beam. It is known from Section 3 that the penetrated heat decreases with the increasing of the penetration distance. Thus, the temperature at this region decreases with the increase of the heat penetration depth. The ablation of carbon fibers is gradually alleviated due to the temperature decrease and finally avoided along the heat penetration direction. Therefore, carbon fibers on the ablated surface in the center region present a needled-like structure with taper ends. Though the ablation of carbon fibers is alleviative and avoidable along the heat penetration direction, the temperature at this region is high enough to lead to the SiC decomposition. Therefore, lots of grooves without SiC matrix are formed among the carbon fibers which are lower than the carbon fibers' taper ends. With further increase of the depth, the temperature in the grooves decreases. At a certain depth, the temperature reaches a critical value, and the decomposition of SiC matrix stops. The SiC decomposition according to Eq. 8 and carbon fiber sublimation according to Eq. 10 produce a carbon-rich atmosphere on the ablation surface. In the grooves among carbon fibers, the SiC matrix can decompose while the temperature is not high enough to sublimate the carbon fibers. Some nanocarbon sheet is deposited at the bottom region of the grooves (**Figure 6(c** and **d)**), which is quite similar to the preparation of carbon nanotubes by laser ablation method [25].

Different from the ablated center region, the conducted heat and the corresponding heat penetration depth at region II transitional zone of the ablation surface are much smaller. It is not high enough to lead to the decomposition, vaporization, and sublimation of SiC matrix. The mixed gases of C, Si, and SiC escaped from the ablation center region can be cooled down in this area with relatively lower temperature. SiC grains re-nucleate and grow to spherical particles (see **Figure 6(b)**). Though the temperature at the transitional zone is lower than the ablation center and cannot lead to the decomposition, vaporization, and sublimation of SiC

http://dx.doi.org/10.5772/65108

matrix, it is still believed to be high enough to volatilize the SiO_2 phase. Thus, no SiO_2 was found at the transitional zone.

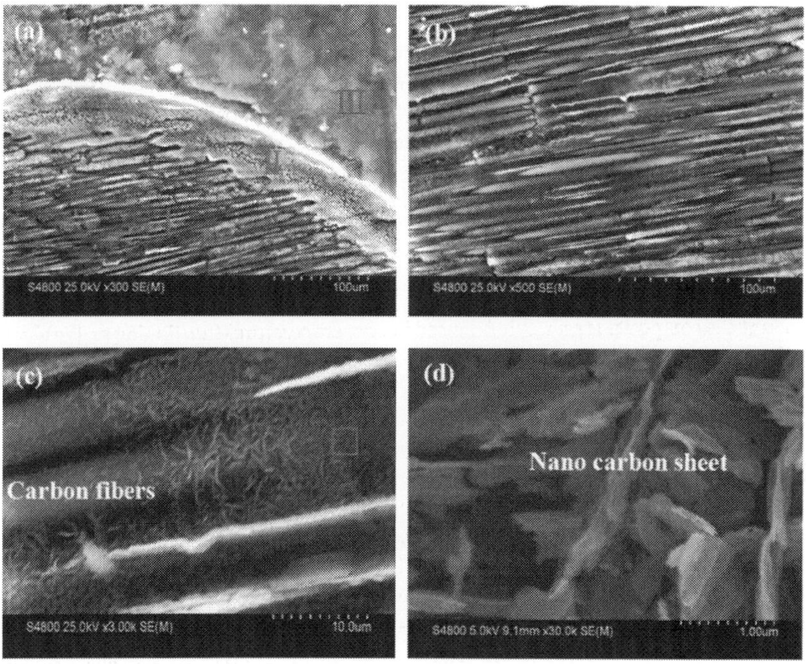

Figure 7. Surface morphologies of C/SiC composite ablated at 150 W/cm² for 20 s.

Compared with the ablated center region and the transitional zone, the heat at region III ablation edge is only conducted form the ablation center, and the temperature is the lowest. The oxygen in the outside atmosphere can continuously diffuse into the atmosphere over the composite surface of this area, and the C/SiC composite is oxidized. Due to the lowest temperature, SiO_2 produced from the oxidation of SiC matrix cannot be volatilized. The composite at this region is covered by a white glassy SiO_2 layer (**Figure 6(a)**). The electrical conductivity of SiO_2 is not very well. Thus, the layer shows a white fuzzy pattern under the scanning electron microscope.

Figure 7 shows the surface morphologies of the C/SiC composite ablated at 150 W/cm² for 20 s. As can be seen, most carbon fibers remain their original shape without obvious ablation damage, which is owing to the low temperature heated by laser beam with low laser power density. Nevertheless, some nanocarbon sheets were found on the ablated surface and the SiC matrix among the carbon fibers decomposed (the same situation on the surface of the composite ablated at 1000 W/cm²), which indicates that the temperature on the composite surface ablated at 150 W/cm² is high enough to induce the decomposition of SiC matrix. However, that temperature is not sufficient to lead to the sublimation of the carbon fibers, and they still remain the original shapes. The ablated surface mainly shows the ablation of SiC matrix.

4.1.4. Laser ablation mechanism

Laser ablation of composite materials is very complicated and influenced with the laser power density and properties of phases in the composite substrate. In order to further understand the laser ablation processes of the C/SiC composite, an ablation model based on the previous characterization results and discussion was proposed (**Figure 8**).

As aforementioned, the temperature at the ablated center during laser ablation is the highest. The composite is heated to a very high temperature over 3500 °C. The SiC matrix reaches its decomposition and sublimation temperatures to form a hot mixture of gasses and vapors, and the carbon fibers get to its sublimation temperature to form a carbon vapor, which results in a positive carbon-rich atmosphere on the ablated surface. The carbon fibers are thermally more stable than the SiC matrix and carbon fibers with taper ends protrude on the ablation surface without the SiC matrix. Nanocarbon sheet is formed in the grooves among the protuberant

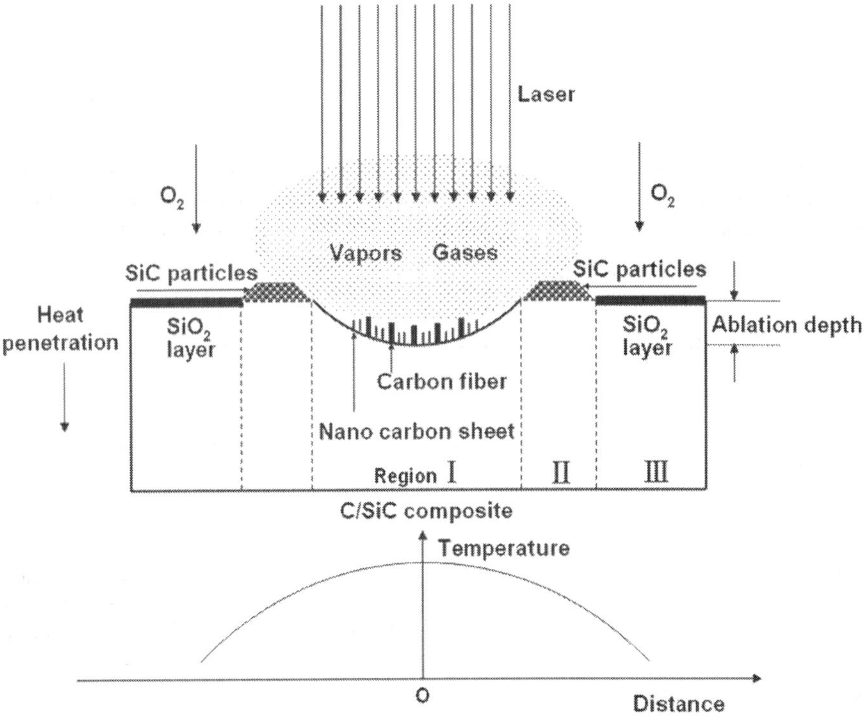

Figure 8. Schematic of the laser ablation processes.

http://dx.doi.org/10.5772/65108

carbon fibers where the temperature is high to decompose the SiC matrix but is not high enough to sublimate carbon fibers. The laser ablation of C/SiC composite in this center area is dominated by the decomposition and simulation process.

With the proceeding of the laser ablation, the positive gases escape from the surface of the ablated center due to the positive atmosphere produced by the decomposition, sublimation, and oxidation in the center region. Some gases deposit at the transitional zone (region II) where the temperature is relatively low and not high enough to lead to the decomposition of SiC phase. SiC grains re-nucleate and grow to spherical particles at the transitional zone, and the region becomes protuberant on the ablated surface.

At the ablation edge (region III), the heat is only conducted form the ablation center, and the temperature is the lowest among the three ablated regions. The temperature in this region cannot lead to the decomposition and sublimation of the phases. The C/SiC composite in this area is only oxidized and is covered by a white glassy SiO_2 layer. The oxidation products of C/SiC composite include CO, CO_2, SiO, and SiO_2. CO, CO_2, and SiO are gaseous and escape from the ablated surface. SiO_2 formed during the oxidation is in the liquid state at such a temperature, which can flow on to the carbon fibers and protects the C/SiC composite from further oxidation damage. The laser ablation of C/SiC composite in this region is dominated by the oxidation reaction of the composite with the atmosphere.

4.1.5. Laser ablation testing of C/C-ZrC composite

Due to the higher melting temperature and high-temperature ablation resistance of ZrC compared with SiC, we produced a C/C-ZrC composite and tested the ablation resistance of the as-produced C/C-ZrC composite using an impulse laser beam. The ablation resistance of C/SiC composite was also tested in the same parameters for comparison.

Figure 9 shows the linear ablation rates of C/C-ZrC composite and C/SiC composite versus the laser power densities. Both the linear ablation rates of the composites increase with increasing laser power densities. The linear ablation rates of C/C-ZrC composite are smaller than that of C/SiC composite. With increasing laser power densities, the difference of linear ablation rates between the two composites increases, which indicates that the ablation resistance of the C/C-ZrC composite is greatly improved at higher temperatures.

The much better ablation resistance of C/C-ZrC composite can be also illustrated by observing the morphologies of the ablated surface. As shown in **Figure 9**, the ablated surface of C/C-ZrC composite is covered by a white melting ZrO_2 layer (determined by EDS) without obvious ablation pit, while the C/SiC composite is severely ablated with a deep ablation pit without any protecting layer on the ablated surface. ZrO_2 owns a high melting point (2700 °C), low oxygen penetration rate, and excellent ablation resistance, which can act as a protecting layer for the C/C-ZrC composite, and thus, C/C-ZrC composite presents much better ablation resistance than the PIP C/SiC composite.

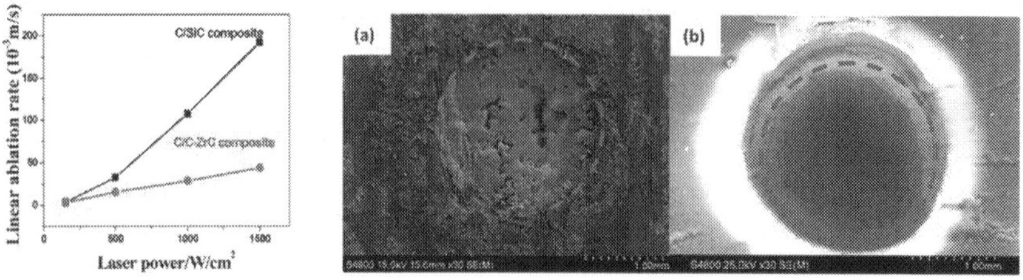

Figure 9. Linear ablation rates and macro-morphologies of (a) C/C-ZrC composite and (b) C/SiC composite after laser testing.

Materials	Test length (s)	Heat flux (W/cm^2)	Blackbody temp (°F)	Adjusted temp (°F)	Weight change/area (g/in^2)
C/ZrC	300	600	4858	5200	−0.029
C/Zr-Si-C	125	500	4556	4875	−0.051

Table 2. Laser-testing parameters and weight changes of the composites before and after testing.

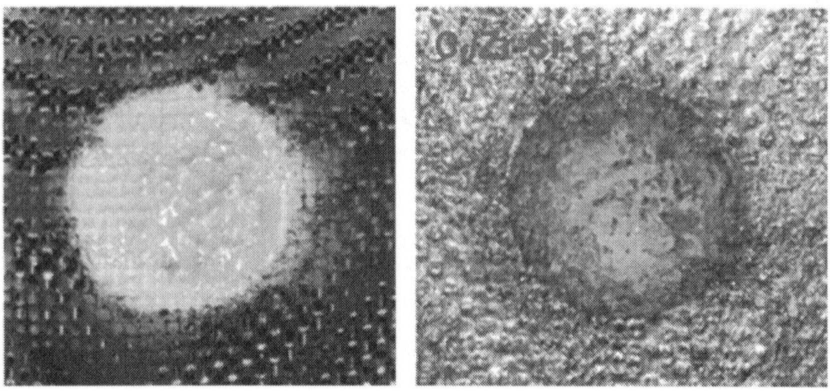

Figure 10. Surface morphologies of C/ZrC and C/Zr-Si-C composites after laser ablation testing.

Because of the excellent properties of carbon fiber-reinforced ZrC-based composites and their potential applications in hypersonic aerospace vehicles, Ultramet [26] developed a fast and economic reactive melt infiltration method to prepared C/ZrC and C/Zr-Si-C composites. Ablation resistance of the composites was tested using a continuous laser beam. Both C/ZrC and C/Zr-Si-C composites survived laser testing at 2871 °C and 2691 °C, respectively, under forced air flow at the Air Force Laser Hardened Materials Evaluation Laboratory. The testing parameters and weight changes of the composites before and after testing are shown in **Table 2**. The composites show small weight change and excellent oxidation stability. Continuous protecting layers were formed on the ablated surface of C/ZrC and C/Zr-Si-C composites (**Figure 10**) and effectively protected the composites from severely ablation damage. Wang et al. [27] also prepared C/C-ZrC composite by reactive melt infiltration of liquid zirconium. They used a continuous wave CO_2 laser (ROFIN DC080W, Germany) to test the ablation resistance of the composite. The ablation depth increased with the increase in laser power densities. The reactive melt-infiltrated C/C-ZrC composite presented much better ablation resistance than C/C composite.

4.2. Ultra-high-temperature ceramics and coatings

Owing to their high melting temperatures, good chemical stability and excellent oxidation, and ablation resistance at elevated temperatures, ultra-high-temperature ceramics have captured great attentions and have potential application in extreme aerothermal and nuclear power-generation environments. Laser heating, known as rapid heating and high power density, is a convenient technique to test the ablation resistance of ultra-high-temperature ceramics.

Yan et al. [28] prepared an ultra-high-temperature ceramic-based composite ZrB_2/20SiC-Cu by spark plasma-sintering method and investigated the ablation behavior of the composite using a 20 MW/m^2 high-intensity continuous laser beam. They characterized the phase and microstructure evolution of the composite during ablation with XRD and SEM. The results revealed that no macroscopic damage but only an ablated layer was observed after being ablated. The composite exhibited good ablation resistance against high-intensity continuous laser beam. Cu phase in the composite evaporated preferentially, and a ZrO_2 layer with different forms such as closely packed nanoparticle and melting layer was generated at the laser spot center. Dendrites and columnar ZrO_2 grains were found on the cross section morphologies of the composite after ablation, which further demonstrated the melting of ZrO_2 layer during ablation. The liquid ZrO_2 layer had capability of thermal insulation, which prevented the inner matrix from further ablation. Both the state transformation of Cu and generated ZrO_2 dissipated a lot of energy, which was another important reason why ZrB_2/20SiC-Cu composite had good ablation resistance in such ablation condition.

Due to the rapid achieving of ultra-high temperatures, Jayaseelan et al. [29] used a laser beam to heat and melt ultra-high-temperature ceramics. The laser beam is a 3 kW Nd:YAG laser with a 10 mm diameter collimated beam capable of delivering a heat flux of up to~20 MW/m^2. Sample temperature was recorded by a pyrometer (Raytek 1MH) with a measurement range of 650 and 3000 °C. They investigated the microstructural evolution of spark plasma-sintered ZrB_2, ZrB_2/20 vol.% SiC (ZS20), and ZrC ultra-high-temperature ceramics (UHTCs) during laser heating. Laser heating at temperatures between 2000 and 3750 °C resulted in extensive bubble and crater formation on the surfaces of 10 mm diameter samples (**Figure 11**). Only crystalline zirconia with a wide range of morphologies including nodules, needles, nanofibers, and lamella was formed on the surface of ZrB_2 and ZS20 samples laser heated in air up to 2700 °C (**Figure 11**). The surface of ZrC samples after laser heated in vacuum up to 3750 °C was characterized by dendritic and eutectic morphologies. Solidification cracks and trapped porosity were also observed on the samples' surface. A complex array of mechanisms involving solid, liquid, and vapor phases led to formation of these morphologies including melting, oxidation, volatilization, and liquid flow.

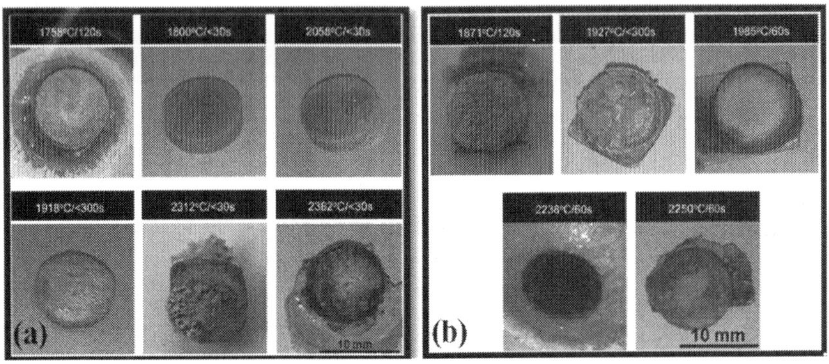

Figure 11. Macro-morphologies of laser-heated (a) ZrB_2/20 vol.% SiC and (b) ZrB_2 samples.

In addition to the laser ablation of ultra-high-temperature ceramics, the laser beams were also used to characterize the ablation resistance of ultra-high-temperature ceramic coatings. Liu et al. [30] prepared a two-layer SiC-ZrC coating on C/C composite by chemical vapor deposition method. They characterized the ablation resistance of the coating using a continuous CO_2 laser beam and investigated the laser ablation behaviors and ablation mechanisms of the SiC-ZrC-coated C/C composites. The results indicated that the ablation depth and width increased with increasing the laser power. Laser ablation resistance of the composites was greatly improved by the SiC-ZrC coating. They found three ablation regions on the ablated coating surface and discussed the formation mechanism. At the ablation center, the ablation was thought to be dominated by the blast and sublimation processes. At the transitional zone, the main ablation mechanism was evaporation. At the ablation edge, the ablation was mainly controlled by the oxidation process. Li et al. [31] also employed the laser beam to characterize ablation resistance of ultra-high-temperature ceramic coatings. A TaC coating was prepared on C/C composite by chemical vapor deposition to improve ablation resistance of the composite at high temperatures. They tested the ablation resistance of TaC coating and investigated ablation mechanism.

5. Conclusions

Laser ablation is a facile, reliable, and economical method to evaluate ablation resistance of materials. The fundamentals of laser-material interactions were discussed from the physical aspects and chemical aspects. The physical aspects mainly involved the absorption of laser radiation, heating and propagation, melting, vaporization, and solidification. The chemical aspects mainly involved the decomposition of materials and the reaction between the materials

http://dx.doi.org/10.5772/65108

and the testing atmosphere. Some practical applications of laser ablation in ablation-resistance characterization of ultra-high-temperature ceramics and composites were presented, and the ablation behavior and mechanism were discussed.

Author details

Yonggang Tong[1,2*], Xiubing Liang[1], Shuxin Bai[2] and Qing H. Qin[3]

*Address all correspondence to: tygiaarh419@163.com

1 National Engineering Research Center for Mechanical Product Remanufacturing, Academy of Armored Forces Engineering, Beijing, China

2 College of Aerospace Science and Engineering, National University of Defense Technology, Changsha, China

3 Research School of Engineering, Australian National University, Canberra, Australia

References

[1] Malek C. Laser processing for bio-microfluidics applications. Anal. Bioanal. Chem. 2006;385:1351–1361.

[2] Kontinos DA, Gee K, Prabhu DK. Temperature constraints at the sharp leading edge of a crew transfer vehicle, AIAA 2001-2886.

[3] Zhao D, Zhang CR, Hu HF, Zhang YD. Ablation behavior and mechanism of 3D C/ZrC composite in oxyacetylene torch environment. Compos. Sci. Technol. 2008;71(11):1392–1396.

[4] Zhang X, Hu P, Han J, Meng S. Ablation behavior of ZrB_2–SiC ultra high temperature ceramics under simulated atmospheric re-entry conditions. Compos. Sci. Technol. 2008;68:1718–1726.

[5] Monteverde F, Savino R, Fumo MDS, Maso AD. Plasma wind tunnel testing of ultra-high temperature ZrB_2–SiC composites under hypersonic reentry conditions. J. Eur. Ceram. Soc. 2010;30:2313–2321.

[6] Tang SF, Deng JY, Wang SJ, Liu WC, Yang K. Ablation behaviors of ultra-high temperature ceramic composites. Mater. Sci. Eng. A 2007;465:1–7.

[7] Shen XT, Li KZ, Li HJ, Fu QG, et al. The effect of zirconium carbide on ablation of carbon/carbon composites under an oxyacetylene flame. Corros. Sci. 2011;53:105–112.

[8] Dahotre NB, Harimkar SP. Laser Fabrication and Machining of Materials. New York: Springer, 2008.

[9] Ready JF. Industrial Applications of Lasers. San Diego: Academic Press, 1997.

[10] Steen W, Mazumde J. Laser Materials Processing. London: Springer, 2010.

[11] Lide DR. CRC Handbook of Chemistry and Physics. Boca Raton: CRC, 2001.

[12] Wu CW, Wu XQ, Huang CG. Ablation behaviors of carbon reinforced polymer composites by laser of different operation modes. Opt. Laser Technol. 2015;73:23–28.

[13] Schaaf P. Laser Processing of Materials: Fundamentals, Applications and Developments. Berlin: Springer, 2010.

[14] Bulgakova NM, Bulgakov AV. Pulsed laser ablation of solids: transition from normal vaporization to phase explosion. Appl. Phys. A 2001;73:199–208.

[15] Zhou RF, Han YF, Li SS. High Temperature Structure Materials. Beijing: National Defend Industry Press, 2006.

[16] Wilson J, Hawkes JF. Lasers: Principles and Applications. New York: Prentice-Hall, 1987.

[17] Oliveira V, Colaco R, Vilar R. Simulation of KrF laser ablation of Al_2O_3–TiC. Appl. Surf. Sci. 2007;253:7585–7590.

[18] Bäuerle D. Laser Processing and Chemistry. Berlin: Springer, 2000.

[19] Clauer AH, Fairand BP, Wilcox BA. Laser shock hardening of weld zones in aluminum alloys. Metall. Trans. A 1977;8(12):1871–1876.

[20] Ultramet. Ceramic Matrix Composites [Internet]. Available from: http://www.ultramet.com/ceramic_matrix_composites.html

[21] Pan YS, Xu YD, Chen ZF, Chen LF, Zhang LT, Xiong DS. Ablation properties analysis of 2D C/SiC composites. Ordnance Mater. Sci. Eng. 2006;29:17–21.

[22] Jian K, Chen ZH, Ma QS, Hu HF, Zheng WW. Effects of polycarbosilane infiltration processes on the microstructure and mechanical properties of 3D-Cf/SiC composites. Ceram. Int. 2007;33:905–909.

[23] Tong YG, Bai SX, Zhang H, Ye YC. Laser ablation behavior and mechanism of C/SiC composite. Ceram. Int. 2013;39:6813–6820.

[24] Fang D, Chen ZF, Song YD, Sun ZG. Morphology and microstructure of 2.5 dimension C/SiC composites ablated by oxyacetylene torch. Ceram. Int. 2009;35:1249–1253.

[25] Yudasaka M, Komatsu T, Ichihashi T, Iijima S. Single wall carbon nanotube formation by laser ablation using double-targets of carbon and metal. Chem. Phys. Lett. 1997;278:102–106.

[26] Ultramet. Ceramic Matrix Composites [Internet]. Available from: http://www.ultramet.com/ceramic_matrix_composites_performance.html

[27] Wang YG, Zhu XJ, Zhang LT, Cheng LF. Reaction kinetics and ablation properties of C/C–ZrC composites fabricated by reactive melt infiltration. Ceram. Int. 2011;37(4):1277–1283.

[28] Yan ZY, Ma Z, Liu L, Zhu SZ, Gao LH. The ablation behavior of ZrB_2/Cu composite irradiated by high-intensity continuous laser. J. Eur. Ceram. Soc. 2014;34:2203–2209.

[29] Jayaseelan DD, Jackson H, Eakins E, Brown P, Lee WE. Laser modified microstructures in ZrB2, ZrB2/SiC and ZrC. J. Eur. Ceram. Soc. 2010;30:2279–2288.

[30] Liu QM, Zhang LT, Jiang FR, Liu J, Cheng LF, Li H et al. Laser ablation behaviors of SiC-ZrC coated carbon/carbon composites. Surf. Coat. Technol. 2011;205(17–18):4299–4303.

[31] Li GD, Xiong X, Huang KL. Ablation mechanism of TaC coating fabricated by chemical vapor deposition on carbon-carbon composites. Trans. Nonferrous Met. Soc. China 2009;19:689–695.

Chapter 18

Standardization and Normalization of Data from Laser Ablation Inductively Coupled Plasma Mass Spectrometry

Ricarda Uerlings, Andreas Matusch and Ralf Weiskirchen

Additional information is available at the end of the chapter

http://dx.doi.org/10.5772/64769

Abstract

Laser ablation inductively coupled plasma mass spectrometry is a useful technique for the precise determination of major, minor and trace element distributions or isotope ratios in solid samples and biological tissue sections. However, measured ion intensities of selected mass-to-charge ratios, may vary considerably from run to run and might also underlie non-linear drift within a run. Therefore, beside the calibration of the measurement, normalization of ion intensities to a reference such as an internal standard is necessary. Other strategies use an endogenous reference element of which a homogenous distribution in the sample is assumed, or derive a more complex reference parameter from a given dataset. Generally, normalization methods depend on the experimental setup and sample material and are usually based on one or few isotopes or the total ion current. This chapter reports different normalization methods that either used a separate reference value for each data point – constituting a pixel in the isotope image – or used a constant normalization factor per measurement run. In conclusion, normalization is essential to minimize deviations of element concentrations due to measurement-related fluctuations. Normalization and definition of an area of interest are powerful tools to obtain high-contrast isotope images with absolute element concentrations.

Keywords: standardization, normalization, spiking, visualization, quantification, ELAI

1. Introduction

Laser ablation inductively coupled plasma mass spectrometry (LA-ICP-MS) is today a well-established technique for direct analysis of solid samples [1, 2]. It allows the uncomplicated elemental analysis of many different kinds of sample materials with inexpensive preparation methods. It can be used for the precise determination of major, minor, and trace element distributions or isotope ratios and is therefore already applied in many research fields [3, 4]. The basic principle of LA-ICP-MS is the multielement analysis of a mostly solid sample. The sample is placed in an ablation chamber purged with a constant argon gas flow and ablated line-by-line with a laser beam focused on the sample surface. The amount of ablated material and thus the composition of the gas phase are highly depending on the sample and laser properties [5, 6]. The aerosol consisting of the ablated material and argon is transported to the plasma torch for evaporation, atomization, and ionization [7]. After refocusing the gas stream in the sampler/skimmer interface, the generated ions are separated in the quadrupole mass spectrometer according to their mass-to-charge (m/z) ratio. The acquired signal-ion intensities are then measured and recorded as a continuous list of data points without any trigger signal indicating the line shift. This is necessary because the laser ablation unit and the mass spectrometer work independently from each other [8].

In LA-ICP-MS measurements, the ion intensities depend on many parameters. Since it is not possible to maintain perfectly constant conditions per sample and throughout a measurement run, the obtained data have to be normalized allowing the comparison of absolute element concentrations [9, 10]. There are several ways for normalization that depend on the sample material and the experimental setup. Common normalization methods [10] are either based on one isotope, the total ion current (TIC), or on selected isotopes that should represent the ion current attributable to the sample matrix termed extracted ion current (EIC). The use of internal standards as reference is desirable, but challenging with respect to the choice of a matching material and to its application or introduction [10]. Normalization can either be performed using a separate reference for each data point or a normalization factor constant throughout a measurement run or sample. In addition, for the calibration of the laser ablation process, matrix-matched standards are essential to compensate changes in the amount of ablated material even if the concentration remains constant [5].

In this chapter, different normalization methods were tested and compared:

- sum of all measured isotopes (more or less reflecting "total ion current", TIC),
- sum of different selected isotopes ("extracted ion current", EIC),
- different homogeneously distributed isotopes as reference,
- pixel per pixel normalization versus overall factor normalization,
- masking of background data.

The final generation of element distribution maps from the different normalization methods was done with the aid of Excel Laser Ablation Imaging (ELAI) representing a software tool that is based on Microsoft Excel 2007 Visual Basic for Applications. This software is freely

http://dx.doi.org/10.5772/64769

customizable and allows fast and easy semi-manual image generation and normalization without any further additional commercial software [8].

2. Material and methods

2.1. Sample preparation

Mice organs (brain and liver) were cryo-cut into 30 μm slices with a standard research cryostat (CM3050 S, Leica Biosystems, Wetzlar, Germany) at −18°C box temperature and −16°C object temperature. In brief, the frozen organ was fixed on a sample holder with Tissue-Tek® O.C.T. compound (Sakura Finetek Europe B.V., The Netherlands). Slices were thaw-mounted onto StarFrost® adhesive microscope slides (Waldemar Knittel Glasbearbeitungs GmbH, Braunschweig, Germany), dried, and stored at room temperature until used for LA-ICP-MS measurement.

2.2. Measurement setup

The LA-ICP-MS measurements were performed using a quadrupole-based ICP-MS (XSeries 2, Thermo Fisher Scientific, Bremen, Germany) without operating the collision cell. The Nd:YAG laser ablation unit (New Wave UP 213, New Wave, Fremont, CA, USA) was coupled directly to the ICP-MS. Measurement parameters such as RF power and gas flow were optimized for each sample with a daily performance test. Typical LA-ICP-MS measurement parameters that we used in our experiments for simultaneous detection of 21–29 different isotopes were: RF power, 1400–1600 W; carrier gas flow, 0.85–1.05 l/min; dwell time per isotope, 0.02–0.3 s; maximal energy at sample surface, > 3 mJ; fluence at sample surface, >30 J/cm^2; scan speed, 60–90 μm/s; spot size, 30 μm; and y-distance between the center of the lines, 60 μm, respectively. To keep the integration time of the mass spectrometer as low as possible and for receiving a high spatial resolution in the future image, only isotopes of interest were selected. The dwell time was adjusted to the sensitivity of the detector for each selected isotope.

For the measurement, the glass slide bearing the sample tissue section was positioned on the xyz-stage in the laser ablation chamber and moved into the focus plain (z-direction) of the laser beam. Within the focus correction option, the z-position was adjusted to the starting and ending point of every line to compensate minor variations in inclination and tilt of the sample. The ablation was performed line by line with a continuous motion in the x-direction and a discontinuous motion in the y-direction. At the end of each line, the laser was interrupted and moved back to the zero x-position with y-offset of 30 μm to the next line, while sample material was continuously transported to the mass spectrometer. The primary output data were a continuous list of ion intensities with no trigger signal indicating line shifts, because the mass spectrometer operated independently from the laser unit.

Before each measurement, the LA-ICP-MS was equilibrated to obtain constant conditions. Therefore, the ablation chamber was washed by constant carrier gas flow for about one hour to remove air entered during loading. As a function of equilibration and washing, y-drifts occurred in some data sets, but also during prolonged measurement runs, i.e., over 12 h. For calibration, matrix-matched standards were measured at the end of the same run. These were produced from homogenized tissue of a homologous organ spiked with different concentrations of a standard solution [11]. Sample and standard were present at separate slides positioned side by side into the chamber.

For image generation, the LA-ICP-MS raw data of individual measurements were copied into a standardized Excel summary file, whereupon the user-defined functions (UDF) of ELAI could be accessed. Once the image dimensions (line length) have been found, the absolute element concentrations of each data point were determined from its ion intensity and the slope of a six-point calibration curve. The software tool "ELAI" [8] that we have recently developed is open software offering free customizable semimanual image generation. The complete program, including detailed documentation for installation and disclosure of all program codes is freely available and can be downloaded from the supplementary material of [8].

3. Results

Normalization requires a reference element or another parameter which is relatively homogeneous distributed throughout the sample and standard tissue sections. The concentration of a given pixel (C_{px}) is then calculated according to Eq. (1), whereby $\bar{r}_{Std}$ is the reference parameter averaged across the calibration standard measurement and r_{px} the reference parameter of the respective pixel, I_{px} the ion intensity of the isotope (m/z) of interest of the respective pixel, and m the slope of the calibration line. All these parameters are given as net values after subtraction of the respective glass background.

In the ideal case, when equal conditions apply for the considered pixel and the standard, the normalization factor $\bar{r}_{Std}/r_{px}$ becomes one when the normalization reference is one single isotope $\bar{r}_{Std}$ and r_{px} correspond to the net ion intensity thereof. In case of TIC as normalization reference, r is the sum of all ion intensities, in case of EIC the sum of the selected isotopes. When the reference parameter does not enter per pixel, but is averaged throughout the sample, r_{px} becomes constant ($\bar{r}_{px}$).

$$C_{px} = \frac{\bar{r}_{Std}}{r_{px}} \frac{I_{px}}{m} \quad (1)$$

TIC as a reference should also comply with the requirements to homogeneity and constancy in concentration throughout samples of the same type. **Figure 1** shows that after applying a TIC-based normalization strategy, the absolute element concentrations and the background noise are reinforced inversely proportional to the TIC and thus become distorted. The TIC shows a high dependence to the structure of the sample tissue because it is dominated by the intensities of potassium (^{39}K, **Figure 2**) in measurements of liver and brain tissue. Additionally, for potassium despite low dwell times selected, saturation effects of the detector may occur in the ICP-MS. For these reasons, TIC-based normalization does not confer a significant gain of biological information.

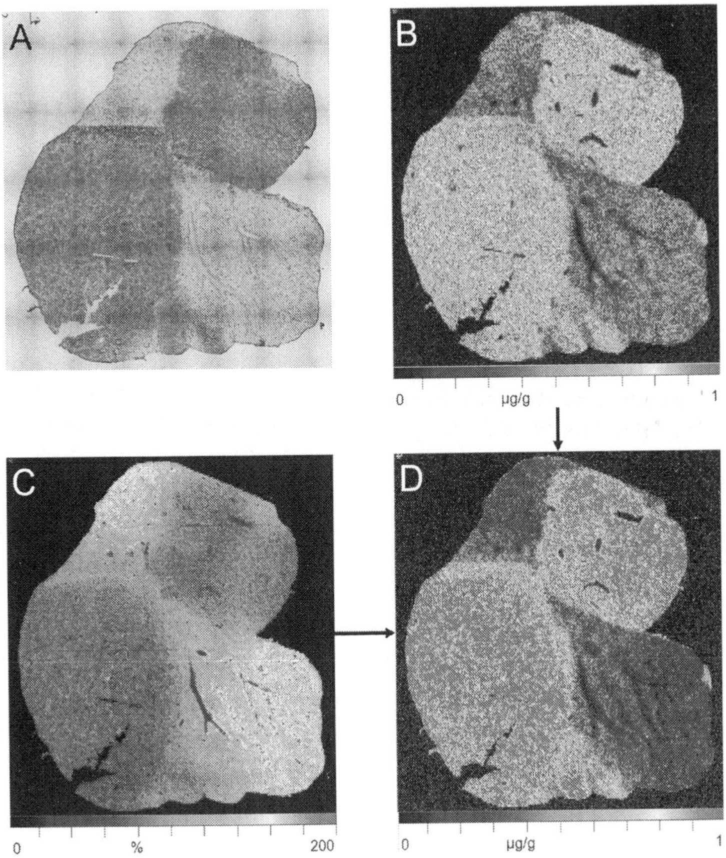

Figure 1. Normalization of ^{55}Mn in tumorous mouse liver tissue. (A) Depicted is the original liver tissue with normal hepatocytes (*dark tissue*), blood vessels, and tumors (*bright tissue*). The not normalized image of ^{55}Mn (B) shows higher manganese concentrations in regions containing normal tissue architecture than in the tumors and very low manganese concentrations in the blood vessels. (C) The pixelwise TIC divided by the average TIC of the calibration standard that is given in %. The higher signal in the tumor regions is due to the lower water content and therefore a higher amount of dry material in a given area of the slide. Pixelwise division of image (B) by image (C) results in the TIC-normalized image of ^{55}Mn that is depicted in (D).

For an EIC-based normalization strategy, a set of the most homogeneously distributed isotopes was chosen. These comprised carbon, magnesium, silicon, and sulfur (**Figure 3**).

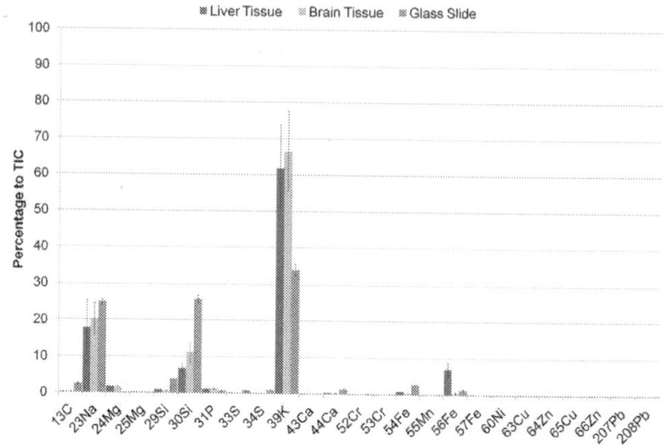

Figure 2. Percentage contribution of isotope intensities to the TIC in liver and brain tissue and to the background signal of the glass slide. The error bars indicate the standard deviation (SD) throughout all data points of the sample.

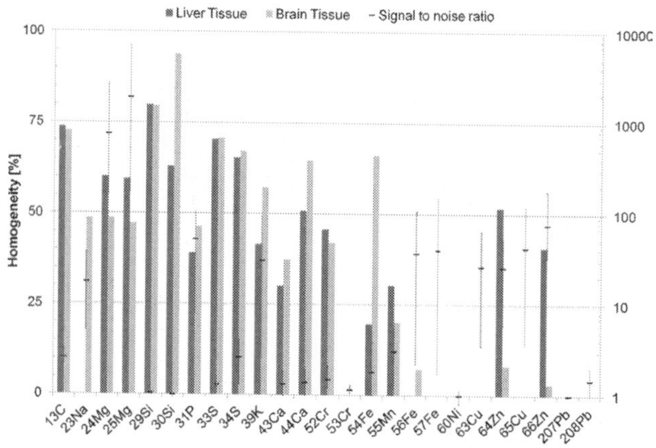

Figure 3. Homogeneity of the isotope distribution in the liver and brain tissues and the glass slide. The signal-to-noise ratio is logarithmically scaled on the secondary *y*-axis with error bars representing the range of the average signal-to-noise ratio in different measurements.

Silicon originates from the partially ablated glass slide surface subjacent to the sample. In blank background areas, the signal-to-noise ratio was about one. Silicon could therefore be a good candidate as an internal standard. However, the Si signal nonlinearly depends on the geometry of laser spot, focus, and tissue laying over the glass surface. The extreme cases are subtotal ablation of the tissue and no ablation of glass and deep ablation of glass and sparkling of tissue fragments to the side. Furthermore, the most abundant Si isotopes ^{28}Si and ^{29}Si interfere with bi-atomic $^{14}N_2$ and $^{14}N^{15}N$. In liver and brain tissue, carbon and sulfur also show a good homogeneity with moderate signal-to-noise ratios for ^{13}C and ^{34}S. ^{33}S is homogeneously distributed in both sample types, but its signal-to-noise ratio is only slightly above one, making the image of ^{33}S in general very noisy. Although magnesium seems relatively homogeneously distributed in the liver and brain tissue and its signal-to-noise ratio is very high, it is not suitable for normalization because it cannot be assumed to be similar concentrated in samples of the same type. For this reason, all other metals were also omitted, independent from their homogeneities and signal-to-noise ratios. Former measurements also showed biologically and structurally meaningful and heterogeneous distributions of that magnesium, other metals and phosphorus. Therefore, these were considered unsuitable for normalization. In conclusion, taking into account homogeneity, signal-to-noise ratio and constancy of concentration, only silicon, carbon, and sulfur qualified for an extracted ion current.

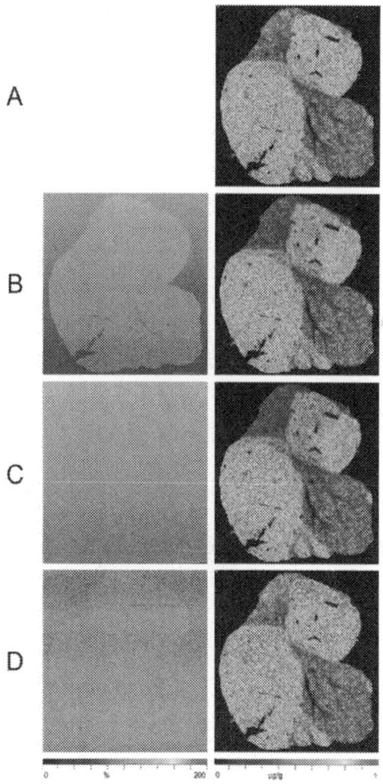

Figure 4. Normalization of ^{55}Mn in tumorous mouse liver. (A) Depicted is the non-normalized image of absolute manganese concentrations. Left images present the normalization reference images, right images the normalized manganese concentrations based on the EIC (B), ^{30}Si (C), and ^{29}Si (D).

Figure 4 shows different normalization methods for ^{55}Mn, again in comparison to the non-normalized image (**Figure 4A**) generated from absolute concentrations. The EIC image contains the summarized intensities of ^{13}C, ^{29}Si, ^{30}Si, ^{33}S, and ^{34}S relative to the respective sum extracted from the standard measurement. Since the EIC image (**Figure 4B**) shows slightly decreasing intensities (a signal drift) from the top to the bottom, the EIC-normalized image of ^{55}Mn shows too low absolute concentrations in the upper half of the image. However, the lower half of the image looks similar to the not normalized image. The same effect is seen in the image of ^{30}Si and the corresponding normalized image of ^{55}Mn (**Figure 4C**). Since the EIC is dominated by the intensities of ^{30}Si (see below), both methods produce similar results. Additionally, since the normalization factor ($\bar{r}_{Std}/r_{px}$) for ^{30}Si was lower than for the EIC, the overall concentrations in the ^{30}Si-normalized image were lower compared to the EIC and to the non-normalized image. The opposite effect can be seen in ^{29}Si-based normalization (**Figure 4D**), where overcorrection of the vertical drift and overestimation of overall concentrations occurred. Additionally, the image of $m/z = 29$ shows slightly increased intensities over sample covered areas pointing to interferences of ^{29}Si and $^{14}N^{15}N$ [12].

A higher contribution of $^{13}C/^{16}O$ can be excluded as in this case also at $m/z = 30$, a contribution of $^{12}C/^{18}O$ should be observed. The contribution of five different ion intensities (^{13}C, ^{29}Si, ^{30}Si, ^{33}S, and ^{34}S) to the EIC is resumed in **Figure 5**.

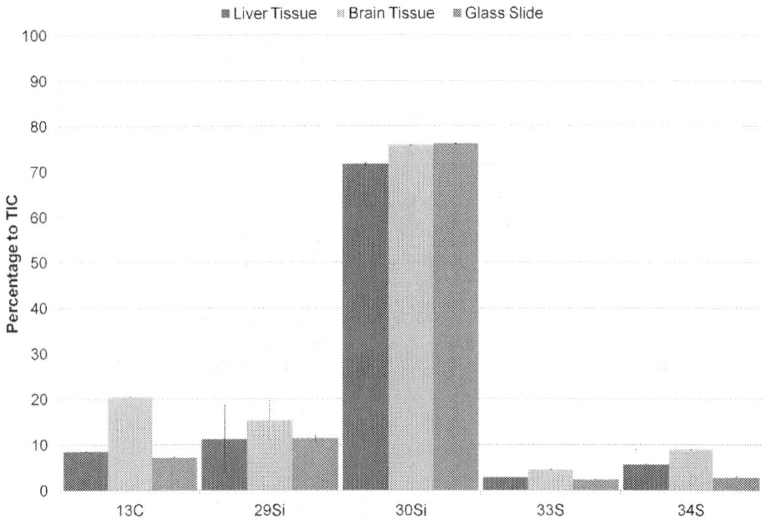

Figure 5. Contribution of isotope intensities to the EIC for liver and brain tissue and the glass background. Error bars indicate the standard deviation of all pixels across the area considered.

Figure 6 presents the results for normalization to other reference isotopes, which are also contained in the set for EIC. In comparison to the not normalized image of ^{55}Mn (**Figure 6A**), the concentrations of the normalized images that are based on ^{13}C (**Figure 6B**), ^{33}S (**Figure 6C**), and ^{34}S (**Figure 6D**) show no distortion and look much more similar than the not normalized image. This is due to the fact that the respective elements (^{13}C, ^{33}S, and ^{34}S) are more homogeneously distributed and only differ in the background height and image noise.

http://dx.doi.org/10.5772/64769

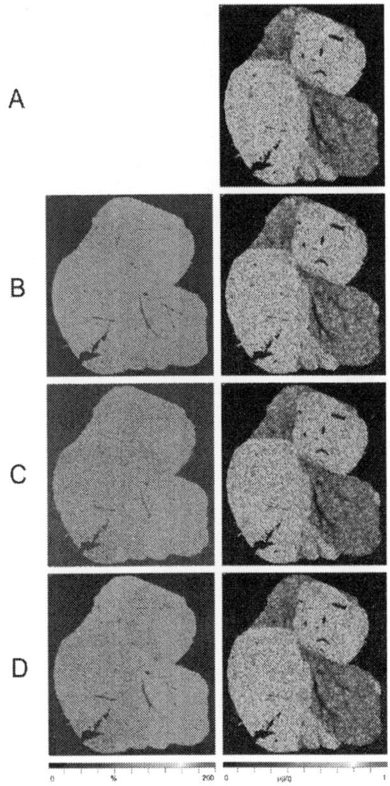

Figure 6. Normalization of ^{55}Mn in tumorous mouse liver tissue. (A) Depicted is the not normalized image of absolute manganese concentrations. Images on the left present the normalization reference images, the images on the right the normalized manganese concentrations based on ^{13}C (B), ^{33}S (C), and ^{34}S (D) resulting from pixelwise division of image (A) by the respective left hand images.

Consequently, normalization that is based EIC of carbon and sulfur can be a useful method to obtain comparable absolute concentrations for all measurements of one specific sample type. Since ^{13}C and ^{34}S normally provide higher intensities than ^{33}S, with an additionally good signal-to-noise ratio (cf. **Figure 3**), normalization with only ^{13}C or ^{34}S is a good and easy to handle option for organic tissue. The final election for ^{13}C or ^{34}S depends on which element is more homogeneous in the analyzed tissue type. However, due to the lower water and therefore higher C and S content in tumor regions, Mn concentrations were underestimated there. This effect is especially obvious in the example depicted in **Figure 6D**.

The principal advantage of pixel per pixel normalization is the attenuation of signal drift and occasionally occurring line artifacts that are caused by sparkling of insufficiently adhering material fragments (**Figure 7A–C**) and therefore appearing at all sample related m/z in parallel or intervention by the user. Of course, drifts and artifacts cannot be overcome by a constant normalization factor (**Figure 7D**). On the other hand, pixelwise normalization introduces and multiplies noise, especially if the divisor image is not graphically smoothed (**Figure 7E**).

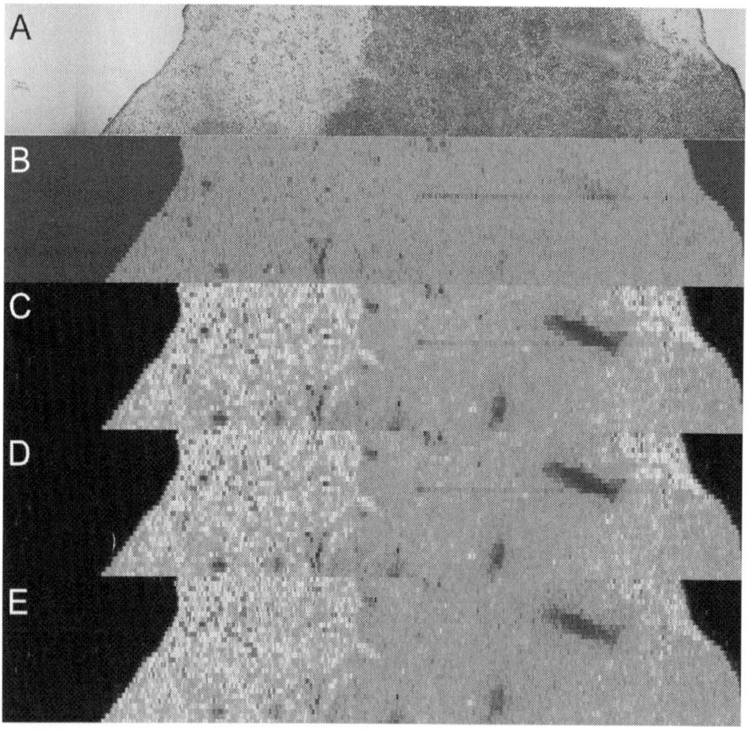

Figure 7. Image detail containing a line artifact (*dark stripe*) that occurred for all measured isotopes. (A) Depicted is the light microscopic image of the analyzed tissue sample. Exemplarily shown are the images generated for (B) ^{13}C, (C) ^{24}Mg (not normalized), (D) ^{24}Mg (normalized with a constant factor based on ^{13}C), and (E) ^{24}Mg normalized pixel per pixel based on ^{13}C.

In measurements of isotopes with low signal-to-noise ratios (i.e., the averaged ion intensity divided by the standard deviation in areas covered by sample/in glass background areas) or elements that are concentrated just above the detection limit, the normalization could increase the background and therefore make the evaluation of the isotope image more difficult. Here, a correct determination and subtraction of the glass background has special impact. A series of works used image analysis software for hand drawing regions of interest (ROIs) including glass background and read out of average values thereof [10].

In our studies, we used a histogram function within Excel Visual BASIC to identify background pixels and for segmentation of a background and a foreground mask. In typical measurements, there is a discontinuous histogram showing a bimodal intensity distribution with a clear cut between the sample and background pixel populations. In the "de-spiked data," the scaling was weighted to the mid range. Most data points assigned to the sample were in the range of 50–60% of the maximum percentage ("Max%" in **Figure 8**). Background data points were in the range of 20–30% of the maximum percentage in typical images. The output of the integral of the histogram "cumulative %" allows additional control of plausibility, whether this corresponds to the visual impression of the fraction of background (**Figure 8**). In a second step, the pixels assigned to sample (foreground) by cutting the histogram are further evaluated individually. Therefore, at least one neighbored data point (i.e., left, right, above, and below) must also be greater than the threshold value, otherwise the considered data point is assigned as spike and not added to the final mask.

The mask should edge the sample tissue clearly and ideally exclude cracks and holes in the tissue. If a mask generation is performed based on an image with inhomogeneous distribution, many data points in the sample tissue could be falsely excluded (**Figure 9A**). In contrast to this, the resulting mask may contain many data points that do not belong to the sample tissue, but to the background noise, if the reference image for the mask generation has a low signal-to-noise ratio (**Figure 9B**). Thus, analog to the determination of the normalization factor, best results for the mask can be achieved with a homogeneous reference image and a good signal-to-noise ratio (**Figure 9C**). Once the optimum area of interest is found, it is converted into a binary image, in which all background data points are colored black and assigned a value of "0.01" and all sample tissue data points are colored gray and set to "1."

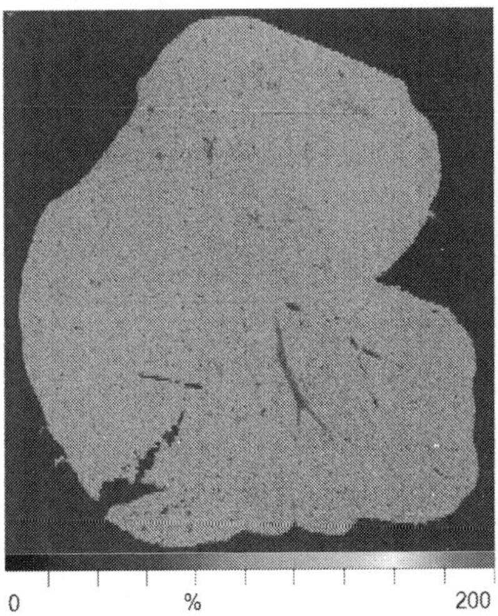

Max%	class	abundance	cumulative %
0.1	0	3	0%
1	2	0	0%
5	10	0	0%
10	20	3	0%
20	40	29858	25%
30	60	8602	32%
40	80	1618	33%
50	100	38117	65%
60	120	42027	100%
70	140	522	100%
80	160	8	100%
90	180	3	100%
95	190	0	100%
100	200	0	100%

Figure 8. Image and histogram of ^{13}C ion intensities in mouse liver weighted to the mid range. Please note the clear separation of pixel populations from background (*blue*, 30% and below) and sample (*green*, 50% and above).

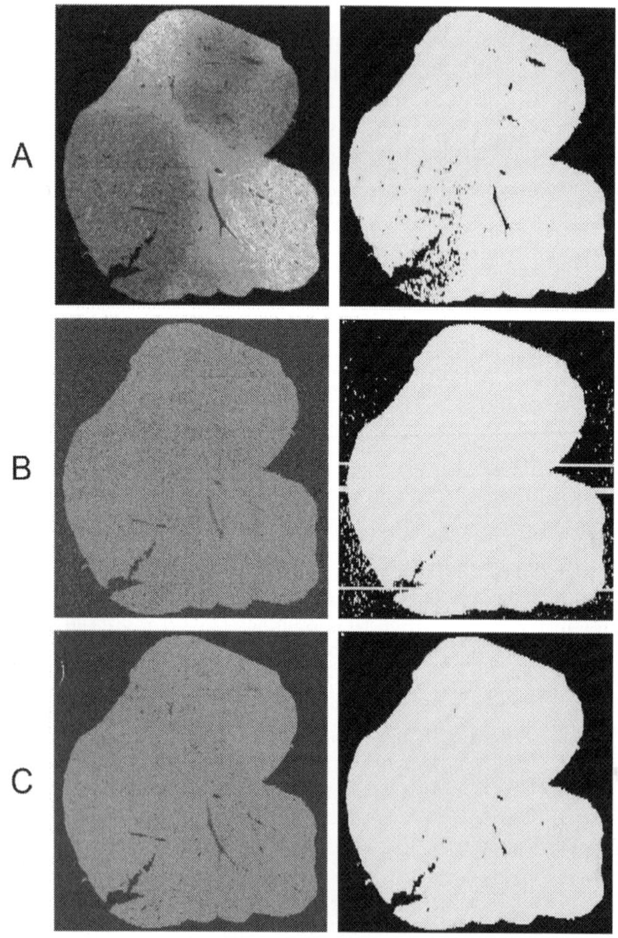

Figure 9. Mask generation based on different reference images. Please note the differences in the masks that were prepared with images that are based on (A) the total ion current (TIC), (B) ^{33}S, and (C) ^{13}C.

Thus, all background data points outside of the area of interest are divided by hundred and therefore significantly reduced. On the other hand, all sample tissue data points retain their concentration level, but the achievable contrast between sample tissue and background is much higher.

Of course, there are also some limitations in the generation of images from the measured data sets. Since the length of a formula in Excel 2007 is limited to 1000 characters [13] and a typical data set contains more than 32,000 data points, it is not useful to process every single cell position of all data points that meet the requirements. Therefore, for data compression, matching data points are combined into blocks whose cell coordinates are stored in a two-dimensional matrix. For example, in the third line of an image, data points "cells(3;5)" to "cells(3;12)" fit the conditions, then the cell coordinates are saved as "E3" and "L3" and combined to "E3:L3." One line can include several blocks if the sample tissue contains any gaps. Finally, all blocks are combined in one overall area of interest and stored as cell name reference "total_mask." The mask is transferred to all previously applied images and stored for each specific image sheet. These masks can be selected in the "name bar" and used for further calculations, for example, the average element concentration in the area of interest. To visualize the mask for verification, sample data points (**Figure 10A**) are colored gray and background is colored black (**Figure 10B**). On this basis, the reference image is recalculated with reduced background (**Figure 10C**). Moreover, background masking is especially useful for reducing background noise in images of low contrast, such as of some trace elements (**Figure 10D–F**).

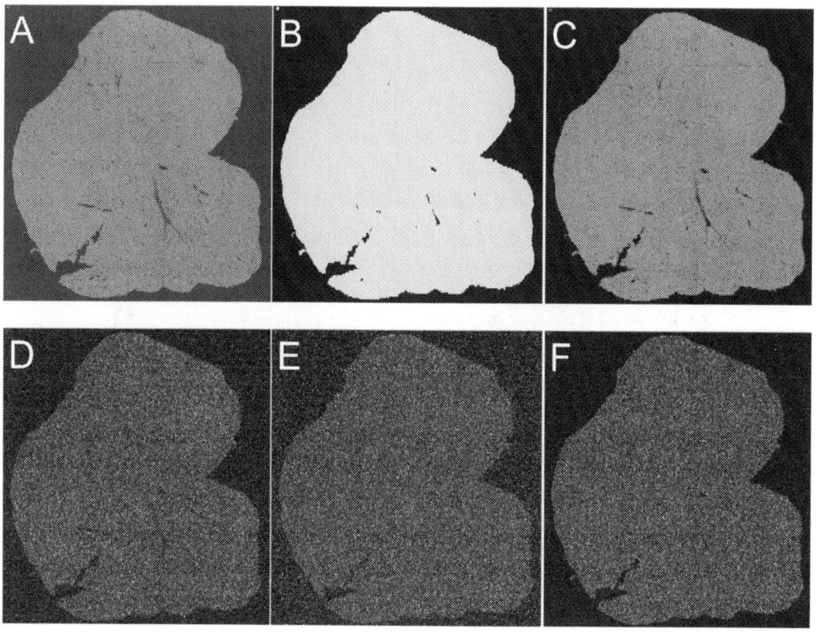

Figure 10. Creation of a foreground sample mask from the reference image. The ^{13}C measurement shown in (A) was used to prepare the mask depicted in (B). The reference image was recalculated with reduced background (C). Low concentrated ^{52}Cr (D) shows a higher background (E) after normalization based on (A), but significantly lower background (F) when normalized with (C) as normalization reference.

Pixelwise normalization can be used as a powerful tool to minimize deviations of element concentrations due to measurement-related fluctuations. Horizontal brain sections as shown in **Figure 11** are expected to be approximately anatomically symmetrical and thus, signal drifts in the non-normalized images most likely result from insufficient washing and equilibration and also prolonged measurement times (that were here 14 h). This is of course not overcome by background subtraction and normalization with a constant factor. In contrast, pixelwise normalization reduces excessive values and increases too small intensities in relation to the reference isotope which significantly improves the image quality (**Figure 11**).

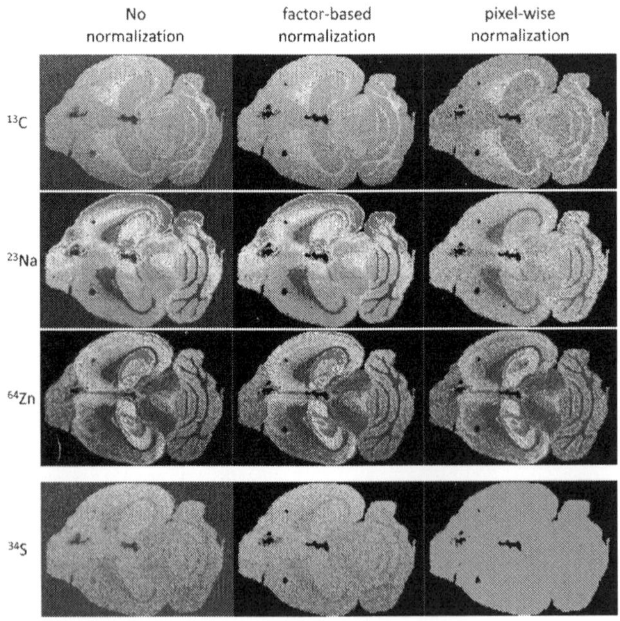

Figure 11. Comparison between images that were normalized using different strategies. Shown are non-normalized images (*left*), congruent to constant factor normalized images (*middle*), and pixelwise normalized images (*right*) of ^{13}C, ^{23}Na, and ^{64}Zn in section derived from mouse brain. The normalization was performed on the basis of the image of ^{34}S (*left*) and the corresponding area of interest (*middle*). The pixelwise normalization of ^{34}S (*right*) shows the theoretical homogeneity obtained by decreasing and increasing signal-drift distorted values.

4. Conclusion

In order to take into account that absolute element concentrations vary due to slightly different measurement conditions, normalization is essential to make data sets comparable. It is necessary to minimize deviations originated by measurement errors or measurement-related fluctuations, especially in the form of signal drifts.

Generally, normalization is highly dependent on the sample type and the selected reference, but in each case, the normalization reference should be distributed as homogeneous as

http://dx.doi.org/10.5772/64769

possible. Ideally, an internal standard is used for normalization, but this is difficult to introduce into sectioned, nonhomogenized sample material. Within narrow limits and high requirements at the constancy of ablation conditions ^{30}Si but not ^{29}Si may be used as an internal standard in the broadest sense.

TIC-based normalization proved to be not recommended for biomedical sample tissue of this study because it is dominated by inhomogeneously distributed potassium. Inhomogeneity leads to distorted absolute element concentrations due to structural and matrix effects. Mostly, one single isotope with a good homogeneity and signal-to-noise ratio is sufficient as normalization reference. EIC-based normalization underlay the same restrictions, but may be helpful if inverse variations of isotope signals such as ^{13}C, ^{33}S, and ^{34}S form a relatively homogeneous reference which cannot be achieved by one single isotope.

Moreover, factor-based normalization is completely sufficient, but device-dependent errors in the data may be attenuated with pixelwise normalization. However, pixelwise normalization may increase the image noise, especially for isotopes with low signal-to-noise ratios. In this case, the definition of an area of interest is very useful to improve the image contrast significantly.

ELAI that we use for image generation, proved as a powerful software tool, capable of quickly responding to user demands and solving variable and individual analytical questions in the area of LA-ICP-MS of tissue sections. Both, development and support of ELAI are ongoing in our laboratory.

Abbreviations

EIC	extracted ion current
ELAI	Excel Laser Ablation Imaging
LA-ICP-MS	laser ablation inductively coupled plasma mass spectrometry
m/z	mass-to-charge
TIC	total ion current

Acknowledgements

Ralf Weiskirchen is supported by the German Research Foundation (SFB/TRR 57, projects P13 and Q3) and the Interdisciplinary Centre for Clinical Research within the Faculty of Medicine at the RWTH Aachen University (IZKF Aachen, project E7-6). None of the funding sources were involved in the writing of the report or in the decision to submit this book chapter for publication.

Author details

Ricarda Uerlings[1], Andreas Matusch[2] and Ralf Weiskirchen[1*]

*Address all correspondence to: rweiskirchen@ukaachen.de

1 Institute of Molecular Pathobiochemistry, Experimental Gene Therapy and Clinical Chemistry, RWTH University Hospital Aachen, Aachen, Germany

2 Institute of Neurosciences and Medicine (INM-2), Forschungszentrum Jülich, Germany

References

[1] Sussulini A, Becker JS, Becker JS. Laser ablation ICP-MS: application in biomedical research. *Mass Spectrom Rev.* 2015 Sep 23. DOI: 10.1002/mas.21481.

[2] Limbeck A, Galler P, Bonta M, Bauer G, Nischkauer W, Vanhaecke F. Recent advances in quantitative LA-ICP-MS analysis: challenges and solutions in the life sciences and environmental chemistry. *Anal Bioanal Chem.* 2015;407:6593–6617. DOI: 10.1007/s00216-015-8858-0.

[3] Konz I, Fernández B, Fernández ML, Pereiro R, Sanz-Medel A. Laser ablation ICP-MS for quantitative biomedical applications. *Anal Bioanal Chem.* 2012;403:2113–2125. DOI: 10.1007/s00216-012-6023-6.

[4] Weiskirchen R, Uerlings R Laser ablation inductively coupled plasma mass spectrometry in biomedicine and clinical diagnosis. *Cell Mol Med: Open Access.* 2015;1:3.

[5] Russo RE, Mao X, Borisov OV. Laser ablation sampling. *TrAC Trends Anal Chem.* 1998;17:461–469. DOI: 0.1016/S0165-9936(98)00047-8.

[6] Limbeck A, Galler P, Bonta M, Bauer G, Nischkauer W, Vanhaecke F. Recent advances in quantitative LA-ICP-MS analysis: challenges and solutions in the life sciences and environmental chemistry. *Anal Bioanal Chem.* 2015;407:6593–6617. DOI: 10.1007/s00216-015-8858-0.

[7] Becker JS. Applications of inductively coupled plasma mass spectrometry and laser ablation inductively coupled plasma mass spectrometry in materials science. *Spectrochim Acta B.* 2002;57:1805–1820. DOI: 10.1016/S0584-8547(02)00213-6.

[8] Uerlings R, Matusch A, Weiskirchen R. Reconstruction of laser ablation inductively coupled plasma mass spectrometry (LA-ICP-MS) spatial distribution images in Microsoft Excel 2007. *Int J Mass Spectrom.* 2016;395:27–35. DOI: 10.1016/j.ijms.2015.11.010.

[9] Liu Y, Hu Z, Gao S, Günther D, Xu J, Gao C, Chen H. *In situ* analysis of major and trace elements of anhydrous minerals by LA-ICP-MS without applying an internal standard. *Chem Geol.* 2008;257:34–43. DOI: 10.1016/j.chemgeo.2008.08.004.

[10] Fonville JM, Carter C, Cloarec O, Nicholson JK, Lindon JC, Bunch J, Holmes E. Robust data processing and normalization strategy for MALDI mass spectrometric imaging. *Anal Chem.* 2012;84:1310–1319. DOI: 10.1021/ac201767g.

[11] Hare DJ, Lear J, Bishop D, Beavis A, Doble PA. Protocol for production of matrix-matched brain tissue standards for imaging by laser ablation-inductively coupled plasma-mass spectrometry. *Anal Methods.* 2013;5:1915–1921. DOI: 10.1039/c3ay26248k.

[12] May TW, Wiedmeyer RH. A table of polyatomic interferences in ICP-MS. *Atom Spectrosc.* 1998;19:150–155.

[13] Microsoft. Excel 2010 Performance: Performance and limit improvements. Available from: https://msdn.microsoft.com/en-gb/library/office/ff700514%28v=office.14%29.aspx [Accessed: 2016-06-30].